Operational Amplifiers
and
Linear Integrated Circuits

K. Lal Kishore
Professor in Electronics and Communication Engineering
and
Registrar
JNTU, Hyderabad

PEARSON
Education

ISBN 81-7758-566-5

First Impression, 2008

Published by Dorling Kindersley (India) Pvt. Ltd., licensees of Pearson Education in South Asia.

Head Office: 482, F.I.E., Patparganj, Delhi 110 092, India.
Registered Office: 14 Local Shopping Centre, Panchsheel Park, New Delhi 110 017, India.

Laser typeset by Anvi Composers, Delhi.

Printed in India by Baba Barkha Nath Printers.

Shree vakratunda mahakaya koti soorya samaprabha
Nirvighnam kurumedeva shubha karyeshu sarvada

(O lord with the twisted trunk, with the effulgence of a billion suns,
always remove the obstacles when I am on an auspicious undertaking.)

CONTENTS

FOREWORD

I consider it a privilege to write the foreword for this book by Prof. K. Lal Kishore. Since the author has taught this subject for many years, it is apt that such a book be authored by him.

The Jawaharlal Nehru Technological University has often revised their curriculum to maintain currency with the requirements of the industry. Also, the practical aspects of the subject have been increasingly incorporated into the curriculum. This book is an essential text as it fulfils both, theoretical and practical, requirements equally.

K. RAJAGOPAL
VICE-CHANCELLOR
JNTU
HYDERABAD

PREFACE

The fag end of the 20th century saw a technology revolution, after which integrated circuits have come to play an important role in various industries. Although digital circuits are used widely, the role of analog circuits and linear integrated circuits (ICs) is no less important. These topics, therefore, need to be addressed from a student's point of view and their applications explained in more detail. In the light of these developments, I felt the need for a textbook on operational amplifiers and linear integrated circuits. Through this book, I have tried to fill the gap in this area and familiarize the students with their design and application.

This book covers operational amplifiers, their applications, other linear integrated circuits like 555 timer, voltage regulator ICs, phase-locked loop ICs, and waveform generator ICs. Importance is given to the explanation of concepts and the working of circuits, so that teachers find it easy to explain the finer points and students are able to understand the same easily. Wherever possible, due importance is given to practical aspects connected with ICs. This book also contains a good number of solved problems. To enable students to prepare for competitive examinations, objective-type questions are also given with answers. This textbook meets the requirements of the curriculum of the Indian universities on the subject and can also be useful for students of M.Sc. (Electronics), B.Sc. (Electronics), AMIETE, AMIE (Electronics), diploma courses in electronics and many such courses.

While writing this book, I have referred to a number of textbooks written on this subject by both Indian and foreign authors. I am grateful to all the authors and publishers of those books. Though every care has been taken to minimize mistakes, the book may have some errors and omissions. Suggestions on this aspect are welcome.

I am thankful to my family members, my wife, Gayatri, and daughters, Kalpana and Pratyusha, for the patience they have shown when I was busy with my book.

K. Lal Kishore

ACKNOWLEDGEMENTS

I am thankful to Sri. G. N. Ganesh, Associate Professor, Department of E.C.E., BSA Crescent Engineering College, Chennai, for helping me in preparing objective-type questions and solved problems. I am also thankful to Sri. E.V.L N. Rangacharyulu, Associate Professor, Department of ECE, and Mrs. N. Mangala Gowri, Associate Professor, Department of ECE, JNTU College of Engineering, Hyderabad for the support they have given in the preparation of this book.

I am thankful to my innumerable students and colleagues whose doubts and questions on the subject helped me in compiling this book.

Finally, I dedicate this book to my wife, Gayatri, and daughters, Kalpana and Pratyusha, for their understanding and encouragement.

K. Lal Kishore

Operational Amplifiers—Basics

Objectives:

In this chapter…

- *Basic and practical considerations of operational amplifiers as well as definitions and measurement of operational amplifiers are explained.*
- *Definitions of various parameters, their typical values and units are discussed. The students will also get familiar with the internal block schematic and derivations of expressions for different applications of operational amplifiers.*

1.1 INTRODUCTION

The concept of an **integrated circuit** (IC) was laid in 1958 by Jack St. Clair Kilby of Texas Instruments. The circuit combined basic elements in a single piece of germanium. The device was held together by glue, with gold wires providing interconnections. The first planar (flat) transistor was made by Swiss Physicist Jean Hoemi of Fairchild Semiconductors. This technology became the basis for the present ICs where silicon is the semiconductor material used.

In 1959, Robert Noyce of Fairchild constructed an IC on a thin slice of silicon. This model made mass production of ICs easier.

Integrated circuits are so called because various components of an electronic circuit such as bipolar junction transistors (BJTs), MOSFET, resistors, capacitors, and so on are integrated or combined together on a single silicon wafer. ICs are also referred to as chips since their actual size is very small.

Largely, based on the application and output response, ICs are classified as (1) linear integrated circuits (linear ICs) and (2) digital integrated circuits (digital ICs).

Linear ICs accept analog inputs and deliver analog outputs. Examples of linear ICs are given below.

1. Operational amplifiers (op-amps) 741
2. Timer IC 555
3. Phase locked loop IC 565
4. Voltage regulator IC 723
5. Waveform generator ICs (8038)

Digital ICs accept input in two discrete voltage levels, viz., logic 0 (zero) or logic 1 (+5 V). The output is also discrete in two specific voltage levels only. Due to this, noise immunity is better in digital circuits compared to analog circuits. The noise signal is analog in nature and it also gets processed along with input in analog ICs and linear ICs, unlike in digital ICs. Examples of digital ICs are given below.

1. Logic gates: 7400, 7404
2. Multiplexers
3. Microprocessors: 8085, 8086, 80486

Classification of ICs is as shown in Fig. 1.1.

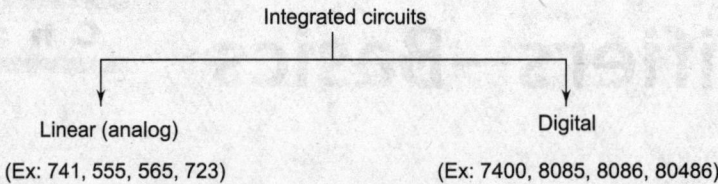

Fig. 1.1 *General Classification of ICs*

1.2 OPERATIONAL AMPLIFIERS

The credit of desiging internal circuit using vacuum tubes goes to George Philbrick. This was in 1948. Originally, **operational amplifiers** were meant for analog computers to perform **mathematical operations** like integration, summation, inversion, and so on. The internal circuit is a differential amplifier and cascaded stages and hence the name, operational amplifier. These are also abbreviated as **op-amp** and popularly called by that name.

The op-amp internal circuit design was further improved by replacing some BJTs with junction field effect transistors (JFETs). JFETs at the input stage of the op-amp draw very little current. The input can be varied between supply limits because the supply voltage will not fall as very less current is drawn. MOS transistors in the output circuit will also allow voltage to swing closer to supply limits. When both BJTs and JFETs are used, these are called as BiFET op-amps—LF 356, CA 3130 are of this type. With the development of technology, two op-amps sharing the same V_{cc} have also been fabricated. These are known as **dual op-amp ICs**. Similarly four op-amps in the same package of pins have also been fabricated. These are called **quad op-amps:** LM 358 is a dual op-amp, and LM 324 is a quad op-amp.

1.3 CLASSIFICATION OF OP-AMPS

Based on their application and features, op-amps are classified as shown in Fig. 1.2.

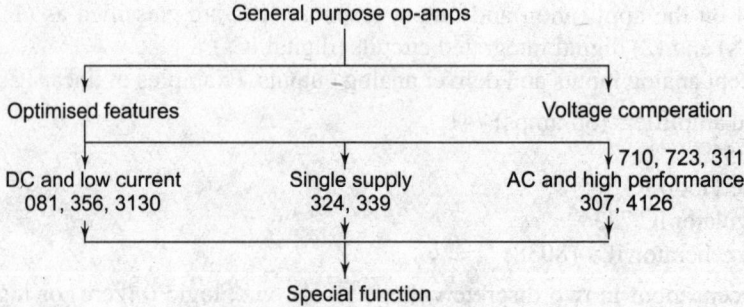

Fig. 1.2 *Classification of ICs Based on Features*

Some of the special function op-amps are:
1. Video/audio ICs
2. Instrumentation applications
3. Sonar send/receive modules
4. Communication ICs

1.4 SYMBOL

The symbol for operational amplifiers is shown in Fig. 1.3.

The internal circuit of an op-amp consists of a differential amplifier to get high common mode rejection ratio (CMRR) and reduce the effect of noise. Output stage is designed to deliver output power and provide impedance matching.

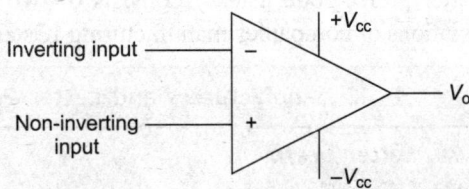

Fig. 1.3 *Symbol for Operational Amplifier*

1.4.1 Package

Op-amps are available in three commonly used packages—T0-5 (Metal can), dual-in-line package (DIP), and ceramic flat package. These are shown in Fig. 1.4.

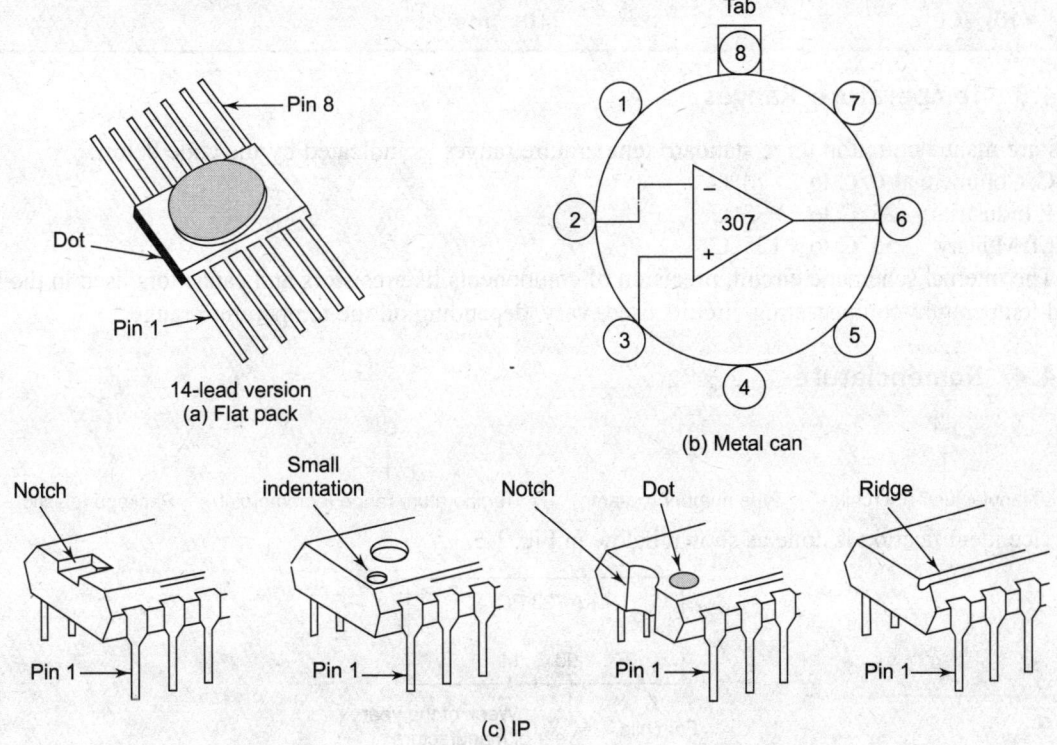

Fig. 1.4 *IC Packages Used for Operational Amplifiers*

The package of the IC is denoted by a code as follows.

D: Plastic dual-in-line package

J: Ceramic

N, P: Plastic

Dual-in-line for insertion into sockets.

1.4.2 Identification Code

The letter prefix code usually consists of two or three letters that identify the manufacturer. The abbreviations of companies manufacturing ICs are indicated in Table 1.1.

Table 1.1   IC Manufacturers and Letter Prefixes

Letter prefix	Manufacturer
(1) CA	(1) Radio Corporation of America (RCA)
(2) LN	(2) Natural Semiconductors
(3) MC	(3) Motorola
(4) NE/SE	(4) Signatures
(5) OP	(5) Precision Monolithic
(6) AD	(6) Analog Devices
(7) μA	(7) Fairchild
(8) TI	(8) Texas Instruments
(9) BB	(9) Burr Brown
(10) ICL	(10) Intersil

1.4.3 Temperature Ranges

ICs are manufactured in three standard temperature ranges as indicated by the code below.

C: Commercial 0°C to   70°C

I: Industrial −25°C to +85°C

M: Military −55°C to +125°C

The internal schematic circuit, precision of components like resistors and capacitors used in the IC, and temperature-compensating circuits used, vary, depending on the temperature range.

1.4.4 Nomenclature

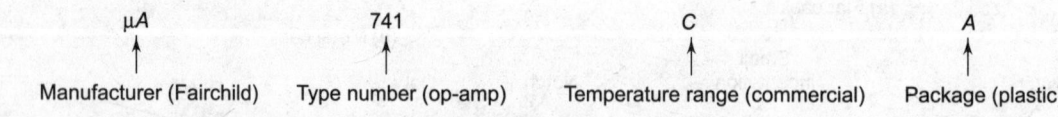

μA	741	C	A
↑	↑	↑	↑
Manufacturer (Fairchild)	Type number (op-amp)	Temperature range (commercial)	Package (plastic)

Device identification is done as shown below in Fig. 1.5.

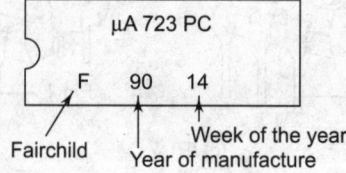

Fig. 1.5 *Nomenclature of ICs*

A very commonly used, general-purpose op-amp is 741. Its pin configuration is shown in Fig. 1.6.

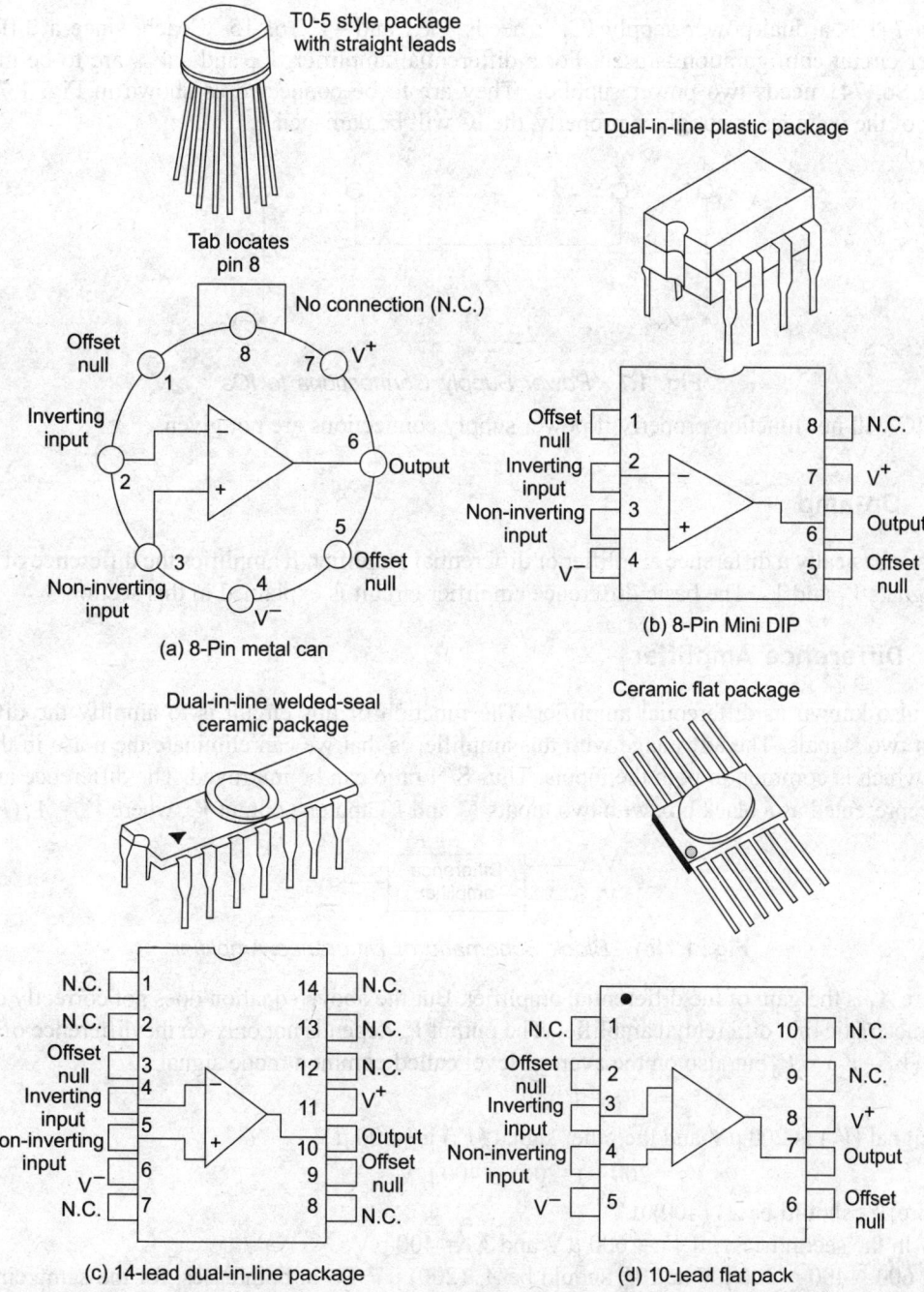

Fig. 1.6 *Various IC Packages of μA 741 Op-amp Along with Connection Diagrams (Top view)*

A ridge or small identification is also provided to identify pin 1 as shown in Fig. 1.6(b) and (c).

1.4.5 Power Supply Connections

Op-amp 741 is a dual power supply IC. It needs $+V_{cc}$ and $-V_{cc}$ of 15 V each, since a differential amplifier circuit configuration is used. For a differential amplifier, V_{cc} and $-V_{EE}$ are to be given for biasing. So, 741 needs two power supplies. They are to be connected as shown in Fig. 1.7. If the polarity of the voltages is not given properly, the IC will be damaged.

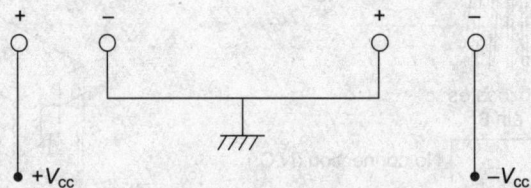

Fig. 1.7   *Power Supply Connections to ICs*

The IC will not function properly if power supply connections are not given.

1.4.6 Op-amp

Op-amp is basically a difference amplifier or differential amplifier. It amplifies the difference of the two input signals V_1 and V_2. The basic difference amplifier circuit is explained in this section.

1.4.7 Difference Amplifier

This is also known as differential amplifier. The function of this circuit is to amplify the difference between two signals. The advantage with this amplifier is that we can eliminate the noise in the input signals which is common to both the inputs. Thus S/N ratio can be improved. The difference amplifier can be represented as a black box with two inputs V_1 and V_2 and one output V_o, where $V_o = A_d (V_1 - V_2)$.

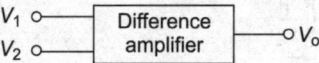

Fig. 1.7(a)   *Block Schematic of Difference Amplifier*

Where A_d is the gain of the differential amplifier. But the above equation does not correctly describe the characteristic of a differential amplifier. The output V_o depends not only on the difference of the two signals $(V_1 - V_2) = V_d$ but also on the average level called common mode signal.

$$V_c = (V_1 + V_2)/2$$

If one signal (V_1) is 200 μV and the other signal (V_2) is −200 μV

$$V_d = 200 - (-200) = 400 \ \mu V$$

Therefore, V_o should be A_d (400) μV

Now in the second case, if $V_1 = 600$ μV and $V_2 = 400$ μV

$V_d = 600 - 400 = 200$ μV and V_o should be A_d (200) μV. So in both cases, for the same circuit, V_o should be the same. But in practice it will not be so because the average of these two signals V_1 and V_2 is not the same in both the cases.

$$V_d = (V_1 - V_2) = \text{differential voltage}$$

$$V_c = \frac{1}{2}(V_1 + V_2) = \text{common mode voltage}$$

From the two equations above, we can write that

$$V_1 = V_c + \frac{1}{2}V_d \quad \text{[If we substitute the values of } V_c \text{ and } V_d \text{ we get the same.]}$$

$$V_1 = V_c + \frac{1}{2}V_d = \frac{V_1}{2} + \frac{\cancel{V_2}}{\cancel{2}} + \frac{V_1}{2} - \frac{\cancel{V_2}}{\cancel{2}} = V_1$$

$$V_2 = V_c - \frac{1}{2}V_d$$

V_o can be represented in the most general case as

$$V_o = A_1 V_1 + A_2 V_2$$

Substituting the values of V_1 and V_2

$$V_o = A_1 \left[V_c + \frac{1}{2}V_d \right] + A_2 \left[V_c - \frac{1}{2}V_d \right]$$

$$= A_1 V_c + \frac{A_1}{2}V_d + A_2 V_c - \frac{A_2}{2}V_d$$

$$V_o = V_c (A_1 + A_2) + V_d \left[\frac{A_1 - A_2}{2} \right]$$

Therefore,

$$V_o = V_c A_c + V_d A_d$$

where

$$A_d = \frac{A_1 - A_2}{2} \quad \text{and} \quad A_c = A_1 + A_2$$

for op-amps input is always given to the inverting node to get $\frac{A_1 - (-A_2)}{2}$ (so that A_d is very large and A_c is very small).

A_1 and A_2 are the voltage gains of the two amplifiers each separately.

The voltage gain for the difference signal is A_d.

The voltage gain for the common mode signal is A_c.

$$V_o = A_d V_d + A_c V_c$$

To measure A_d directly set $V_1 = -V_2 = 0.5$ V so that

$$V_d = 0.5 - (-0.5) = 1 \text{ V}$$

$$V_c = \frac{(0.5 - 0.5)}{2} = 0$$

So,

$$V_o = A_d \times 1 = A_d \text{ itself}$$

Therefore, if we set $V_1 = -V_2 = 0.5$ V,
output voltage directly gives the value of A_d. Similarly, if we set $V_1 = V_2 = 1$ V then $V_d = 0$

$$V_c = \frac{V_1 + V_2}{2} = \frac{2}{2} = 1 \text{ V}$$

So,

$$V_o = 0 + A_c \times 1 = A_c$$

Therefore, the measured output voltage directly gives A_c. We want A_d to be very large and A_c to be very small because only the difference of the two signals should be amplified and the average of the signals should not be amplified.

Hence, the ratio of these two gives $P = \left| \dfrac{A_d}{A_c} \right|$ which is called the CMRR. This should be large for a good difference amplifier.

$$V_o = A_d V_d + A_c V_c$$

$$P = \frac{A_d}{A_c} \quad \text{So,} \quad A_c = \frac{A_d}{P}$$

So,

$$V_o = A_d V_d + \frac{A_d}{P} V_c$$

$$V_o = A_d V_d \left(1 + \frac{1}{P} \frac{V_c}{V_d} \right)$$

Circuit diagram: In the previous DC amplifiers, namely, CB, CC, and CE, the output is measured with respect to ground. But in difference amplifier, the output is the difference of the inputs. So, V_o is not measured w.r.t. ground but w.r.t. to the output of one $I_v Q_1$ or output of the other $I_v Q_2$.

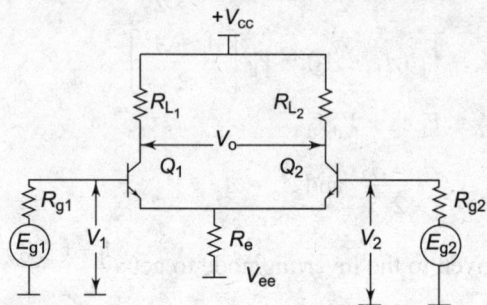

Fig. 1.7 (b)   *Circuit Diagram of Difference Amplifier*

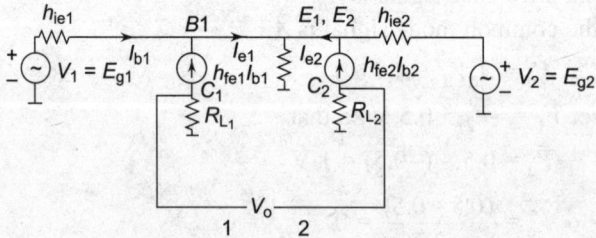

Fig. 1.7 (c)   *Equivalent Circuit*

Drift voltage means, even when there is no input voltage V_i, there can be some output voltage V_o which is due to the internal thermal noise voltage of the circuit components getting amplified and appearing at the output terminals.

Due to temperature and resulting thermal energy, carriers (electrons or holes) are generated in semiconductor devices. Due to the movement of these carriers, current flows and the voltage is called as thermal noise voltage. Though there is no external input signal V_i, certain output voltage results, which is called as noise voltage.

Any electrical signal which is not desirable is called as noise signal. The output voltage changes with temperature. This is referred to as *drift*. The term drift is used to refer to variation of a parameter with temperature.

Drift is reduced in differential amplifier circuits because in these circuits the two transistors must be identical. So, h_{fe} of two transistors will also be the same. If I_{C_1} rises due to increase in temperature, and V_{cc} is fixed, voltage drop across R_L ($I_{C_1} R_{L_1}$) increases. So the voltage at collector of Q_1 decreases. If Q_2 is also identical to Q_1, its collector voltage also drops by the same amount. Hence, V_o, which is the difference of these two voltages remains the same. Thus, the drift of these two transistors gets cancelled.

The advantage with this type of amplifier is that the drift problem is eliminated. Drift voltage means even when there is no input voltage V_i, there can be some output voltage V_o which is due to the internal thermal noise voltage of the circuit components getting amplified and appearing at the output. Drift is reduced in this type of circuit, because the two transistors should be exactly identical. Hence, also, h_{fe} will be the same for the two transistors. Now, if I_{C_1} rises due to increase in temperature, V_{cc} is fixed.

Drop across R_L ($I_{C_1} R_{L_1}$) increases with inverse in I_{C_1}. So the voltage at collector of Q_1 decreases. If Q_2 is also identical to Q_1, its collector voltage also drops by the same amount. Hence, V_o which is the difference of these voltages remains the same. Thus, the drift increase of Q_1 these transistors gets cancelled.

The inputs given to a differential amplifier are of two types.

1. Differential mode   2. Common mode

If V_1 and V_2 are the inputs, the differential mode input $= V_2 - V_1$

Here two different AC signals V_1 and V_2 are being applied. So, there will be interference of these signals and so both the signals will be present simultaneously at both input points, that is, if V_1 is applied at point 1, it also picks up the signal V_2 and so the net input is common mode input $= \dfrac{V_1 + V_2}{2}$

An ideal difference amplifier must provide large gain to the differential mode inputs and zero gain to common mode inputs.

So,
$$V_o = A_2 V_2 - A_1 V_1 \tag{1}$$

$A_2 =$ Voltage gain of the transistor Q_2

$A_1 =$ Voltage gain of the transistor Q_1

We can also express the output in terms of the common mode gain A_c and differential gain A_d.

Therefore,
$$V_o = A_d (V_2 - V_1) + A_c \left(\frac{V_1 + V_2}{2} \right) \tag{2}$$

$$= A_d V_2 - A_d V_1 + A_c \left(\frac{V_1}{2} \right) + A_c \left(\frac{V_2}{2} \right) \tag{3}$$

$$V_o = V_2 \left(A_d + \frac{A_c}{2} \right) - V_1 \left(A_d - \frac{A_c}{2} \right) \tag{4}$$

Comparing Equations (4) and (1)

$$A_2 = A_d + \frac{A_c}{2}$$

$$A_1 = A_d - \frac{A_c}{2}$$

Solving these two equations

$$A_d = \frac{A_1 + A_2}{2}$$

$$A_c = A_2 - \frac{A_1}{2}$$

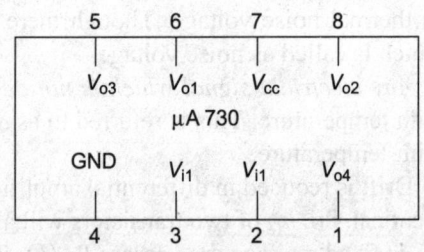

Fig. 1.7 (d)   *μA 730 Differential Amplifier*

μA 730 is an IC differential amplifier with eight pins.

Input is given to pins 2 and 3. $+V_{cc}$ to pin 7.4 is ground. Output is taken at pins 6 and 8 or 5 and 1 is high Z output for input 1. Pin 1 is high Z output 2.6 and 8 is low Z output for input 1 and 2, respectively. In the difference amplifier, the difference of the input voltages V_1 and V_2 is amplified. The collectors of the transistors Q_1 and Q_2 are floating. They are not at ground potential. So, the output voltage is measured w.r.t. one of the collectors voltages, which is not at ground potential. Hence, the output voltage is the difference of the collectors voltages (AC) of transistors Q_1 and Q_2. Difference amplifiers are used in measuring instants and instrumentation systems. The difference of V_{i1} and V_{i2} may be 1 μV which is difficult to measure. So, if this is amplified to 1 μV or 1 V, the measurement will be accurate. So, difference amplifiers are used to measure small increased voltages.

While computing A_1 and A_2 of individual transistors, the other input should be made zero, i.e., while computing A_1, $V_2 = 0$. Because there should be no common mode signal, while computing A_1. A_1 is the actual gain, not differential gain. Therefore, the other input is made zero.

In the case of op-amps, for single-ended operation, the positive end is always grounded (non-inverting input) and input is applied to the inverting input (negative). It is because at this part the feedback current and input current get added algebraically. So this is known as the summary junction. When sufficient negative feedback is used, the closed-loop performance becomes virtually independent of the characteristics of the op-amp and depends on the external passive elements, which is desired.

1.5   OP-AMP PARAMETERS

An op-amp is a differential amplifier that provides noise immunity. It also gives large voltage gain through a multi-stage amplifier circuit configuration. In addition, in the internal schematic, the input stage circuit provides high-input impedance. The output stage circuit provides low-output impedance and the required current drive to deliver output current to supply output power P_o, ($P_o = V_o I_o$) to the load (R_n or Z_n). For internal transistor circuits, to provide biasing, $+V_{cc}$ and $-V_{cc}$ are to be applied. The pin configuration of 741 IC is shown in Fig. 1.8.

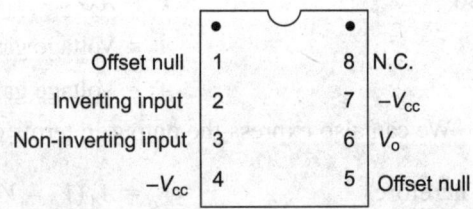

Fig. 1.8   *Pin Configuration of μA 741*

The op-amp has high-input impedance and low-output impedance. Therefore, it will not draw much current from the connected external input voltage signal source. As a result, the loading effect is avoided. Due to low output resistance it delivers the maximum output. Based on these characteristics, due to biasing voltages applied to the IC and the current drawn by the IC from the DC-supplying (bias) voltages, various parameters are defined. These are explained in the following section.

1.5.1 Input Offset Voltage (V_{io})

If no external **input signal** is applied to the op-amp at the inverting and non-inverting input terminals the output must be zero. That is, if $V_i = 0$, $V_o = 0$. But as a result of the given biasing supply voltages, $+V_{cc}$ and $-V_{cc}$, a finite bias current is drawn by the op-amp, and as a result of unsymmetry on the differential amplifier configuration, the output will not be zero. This is known as **offset**. Since V_o must be zero when $V_i = 0$, the input signal must be applied such that the output offset is cancelled and V_o is made zero. This is known as **input offset voltage**. **It is the voltage that must be applied between the two input terminals of an op-amp to nullify output**. This is shown in Fig. 1.9. The value for ideal op-amp is $V_{io} = 0$ V. Practical value $\cong 100\ \mu V$ (typical).

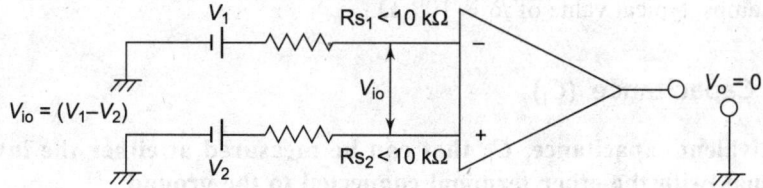

Fig. 1.9 *Input Offset Voltage*

1.5.2 Input Offset Current (I_{io})

Though for an ideal op-amp the input impedance is ∞, it is not so practically. So the IC draws current from the source of the voltage, however small it may be.

The algebraic difference between the currents into the inverting and non-inverting terminals is referred to as input offset current I_{io}.

$$I_{io} = |\,I_{B_1} - I_{B_2}\,|$$

This is shown in Fig. 1.10.

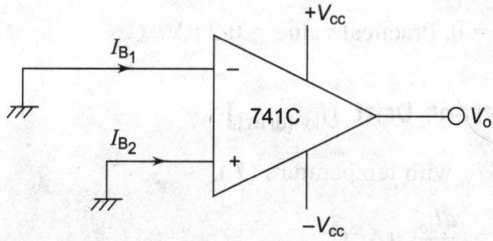

Fig. 1.10 *Input Offset Current*

For an ideal op-amp, $I_{io} = 0$. The typical value for a practical op-amp = 100 nA
For precision op-amp 741C, the value typically is 6 nA.

1.5.3 Output Bias Current (I_B)

This is the average of the currents that flow into the inverting and non-inverting input terminals of the op-amp.

$$I_B = \frac{I_{B_1} + I_{B_2}}{2}$$

For an ideal op-amp $I_B = 0$ A

For a practical op-amp $I_B = 500$ nA

For a precision op-amp, I_B is typically 10 nA.

1.5.4 Input Resistance (R_i)

This is **the equivalent resistance of the IC measured at either the inverting or non-inverting input terminal, with the other terminal connected to the ground**.

$$\text{Ideal value} = \infty \ \Omega, \text{Practical value} = 2 \ M\Omega$$

For JFET op-amps, typical value of R_i is $10^{12} \ \Omega$.

1.5.5 Input Capacitance (C_i)

This is **the equivalent capacitance, C_i, that can be measured at either the inverting or non-inverting terminal, with the other terminal connected to the ground.**

$$\text{Ideal value of } C_i = 0 \text{ pF, Practical value} = 1.5 \text{ pF}$$

Due to change in the Q point of the biasing circuits in the internal circuits of an op-amp, the input offset voltage V_{io} and input offset current I_{io} parameters change with temperature. The term **drift** is used to indicate the same.

1.5.6 Input Offset Voltage Drift [$V_{io(drift)}$]

This is the rate of change of V_{io} with temperature (T).

$$V_{io(drift)} = \frac{dV_{io}}{dT} \ \mu V/^\circ C$$

$$\text{Ideal value} = 0, \text{Practical value} = 0.2 \ \mu V/^\circ C$$

1.5.7 Input Offset Current Drift [$I_{io \ (drift)}$]

This is the rate of change of I_{io} with temperature (T).

$$I_{io \ (drift)} = \frac{dI_{io}}{dT} \ \mu A/^\circ C$$

$$\text{Ideal value} = 0, \text{Practical value} = 0.1 \ nA/^\circ C$$

1.5.8 Common Mode Rejection Ratio [CMRR (ρ)]

This parameter indicates the capability of the op-amp to reject noise. **The higher the value of CMRR, the better it is.**

CMRR is defined as the ratio of the differential voltage gain, A_d, to the common mode voltage gain, A_{cm} (cm = common mode)

$$\text{CMRR} = \frac{A_d}{A_{cm}}$$

The op-amp has a differential voltage amplifier configuration. A_d is usually large. A_{cm} is small. Hence, the value of CMRR is large. CMRR is expressed in decibels.

$$A_{cm} = \frac{V_{o\,cm}}{V_{i\,cm}}$$

V_{ocm} = Output common mode voltage

V_{icm} = Input common mode voltage

A_{cm} = Common mode voltage gain

In the differential amplifier configuration, the noise signals which are common to the two inputs of the differential amplifier will also get amplified. If the common mode gain of the amplifier is less, then the noise will not get amplified, but will be altered. So, only the differential input will be amplified and the output V_o will have less noise. This is the advantage of the differential amplifier which is used in op-amps. Thus, the CMRR value is high.

Ideal value = ∞, Practical value = 90 db

The CMRR value is determined under test conditions with R_s = 10 kΩ.

1.5.9 Power Supply Rejection Ratio (PSRR)

The input offset voltage V_{io} of the op-amp changes if the bias power supply of the op-amp changes. The change of V_{io} with $+V_{cc}$ or $-V_{cc}$ is called the **PSRR**. The term is also called the **supply voltage rejection ratio (SVRR)** or **power supply sensitivity (PSS)**.

PSRR is defined as the ratio of change in the input offset voltage V_{io} with a change in one of the bias power supplying V_{cc}, when the other power supply is held constant.

$$PSRR = \frac{\Delta V_{io}}{\Delta V} \; \mu V/V$$

Ideal value = 0, Practical value = 150 μV/V

1.5.10 Slew Rate (SR)

This is defined as **the maximum rate of change of output voltage per unit time.**

The input capacitance of the op-amp circuit prevents it from responding instantaneously to high frequency signals. If a square wave input or pulse of high frequency is applied as input, at the output the slope of the leading edge and trailing edge is measured and the larger value is computed. This gives SR.

$$SR = \left.\frac{\Delta V_o}{\Delta t}\right|_{max} \; V/\mu sec$$

Ideal value = 0, Practical value = 0.01 V/μsec

1.5.11 Gain BW Product (A_v BW)

This is the bandwidth (BW) of the op-amp when the voltage gain is 1 (unity).

As the frequency increases, the gain decreases because as f increases, $X_c = \left(\dfrac{1}{2\pi fc}\right)$ decreases. So Z_L also decreases. Hence, $V_o = I_L \times Z_L$ also decreases. So the gain A_v decreases. The gain BW product indicates the capability of the op-amp to amplify the input signals of larger frequency.

The other terms used to describe A_v BW are **closed-loop bandwidth and unity gain bandwidth**.

Ideal value $= \infty$, Typical value $= 1$ MHz

$$V_0 = I_L \times Z_L$$

1.5.12 Offset Voltage Adjustment

For op-amp 741, pin numbers 1 and 5 are provided for making offset adjustment. A 10 kΩ potentiometer is connected as shown in Fig. 1.11 and the wiper of the potentiometer is connected to the negative supply. There is no external input signal applied, only bias voltages $+V_{cc}$ and $-V_{cc}$. The wiper of the potentiometer is adjusted till V_0 becomes zero.

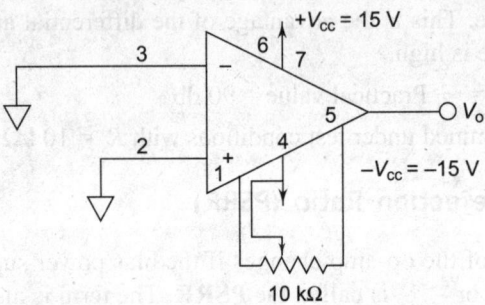

Fig. 1.11 *Offset Voltage Adjustment*

Table 1.2 Op-amp Parameters: Ideal and Typical Values

S. No.	Parameter	Symbol	Ideal value	Typical value	Units
1.	Input offset voltage	V_{io}	0	100	μV
2.	Input offset current	I_{io}	0	100	nA
3.	Input bias current	I_o	0	500	nA
4.	Input resistance	R_i	∞	2	MΩ
5.	Input capacitance	C_i	0	1.5	pF
6.	Input offset voltage drift	$V_{io\ (drift)}$	0	2	μV/$^\circ$C
7.	Input offset current drift	I_{io}	0	0.1	nA/$^\circ$C
8.	Common Mode Rejection Ratio	CMRR	∞	90	dls
9.	Power Supply Rejection Ratio	PSRR	0	150	μV/V
10.	Slew Rate	SR	0	0.01	V/μsec

1.6 FREQUENCY ROLL OFF

The decrease in voltage gain of the op-amp with frequency is called **frequency roll off**. The rate of decrease of gain with frequency is expressed in **db/octave** or **db/decade**.

1.6.1 Octave

When the frequency changes from f_1 to f_2 such that $f_2 = 2f_1$, it is referred to as *octave number*. That is, when frequency doubles, if the gain decreases by 20 db, it is referred to as db/octave.

1.6.2 Decade

When the frequency changes from f_1 to f_2 such that $f_2 = 10 f_1$, it is referred to as **decade**.

If the gain decreases by 100 db when the frequency changes from 1 MHz to 10 MHz, then the fall in gain is referred to as 100 db/decade.

1.7 OP-AMP IN OPEN-LOOP CONFIGURATION

The circuit shown in Fig. 1.12 shows an op-amp in open-loop configuration, that is, the output terminal is not connected to the input terminal. So the feedback loop is open. The voltage gain in this configuration is the overall voltage gain of the different stages inside the op-amp internal circuit of A_{VOL}. Typical value is 10^5. Ideal value is ∞.

$$A_{VOL} = \text{Voltage gain in open-loop configuration}$$

Fig. 1.12 *Operational Amplifier in Open-loop Configuration*

Fig. 1.13 shows an op-amp in closed-loop configuration. Here the feedback is negative, since the input is given to the inverting input terminal. A phase shift of 180° is produced in the output. The closed-loop voltage gain, A_{VCL}, or simply A_v, is limited by the ratio of R_f and R_1.

R_f = Feedback resistance because it is connected in the feedback path of the circuit (connecting input and output points).

R_1 = Resistance connected in the input side of the circuit.

$$A_{VCL} = A_v = \frac{V_o}{V_i} = \left(\frac{R_f}{R_1} \right)$$

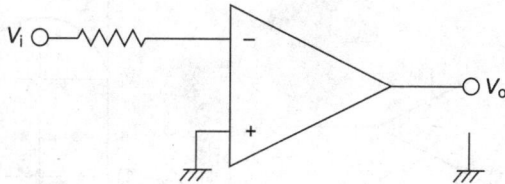

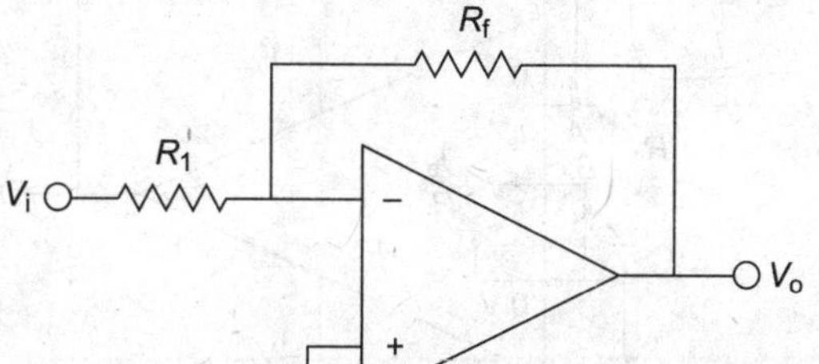

Fig. 1.13 *Operational Amplifier in Closed-loop*

1.8 OP-AMP GOING TO SATURATION

If an op-amp is in open-loop configuration and if an input of 1 V is given, what is the output? The open-loop voltage gain is say $10^5 \times V_i = 1$ V. So, $V_o = V_i \times A_{VOL} = 1 \times 10^5 = 10^5$ V. It is impossible to get or imagine to get such output voltage from a tiny op-amp. The output voltage swing will be limited by $+V_{cc}$ and $-V_{cc}$.

Usually a voltage drop across various components inside the op-amp is assumed to be 1 V. This voltage drop in op-amp circuit is assumed due to the biasing current flowing through different components and the subsequent voltage drop across these **elements**. So the output will be limited to 14 V only if V_{cc} = +15 V. This is known as op-amp *going to saturation*. If a sine wave of 1 V rms is applied to the inverting terminal in the open-loop configuration, the output wave form will be as shown in Fig. 1.14.

If V_o is very large, or A_v is very large, the expression $V_o = A_v \times V_i$ will not hold good. V_o will be limited to the maximum permissible value of $(V_{cc} - 1)$ volts.

Even in closed-loop configuration, if the gain is too large or input voltage is too large, so that

$$V_o = (A_{VCL})(V_i)$$

product goes above $+V_{cc}$ or $-V_{cc}$, the op-amp goes to saturation. If a sine wave input is given in such conditions, the output appears like square wave.

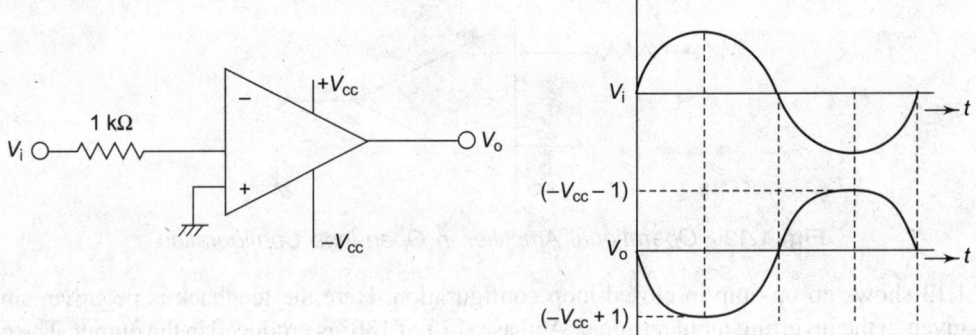

Fig. 1.14 *Operational Amplifier Going to Saturation in Open-Loop Configuration*

1.9 VIRTUAL GROUND

To understand the op-amp virtual ground concept, consider Fig. 1.15.

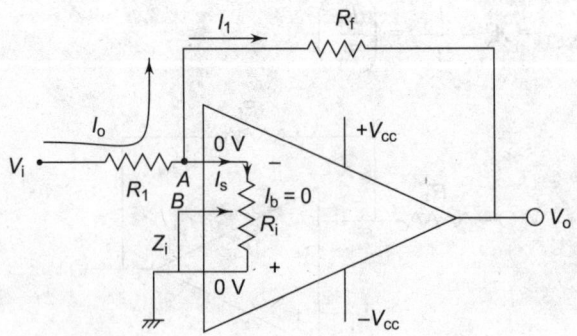

Fig. 1.15 *Operational Amplifier Virtual Ground*

V_i is the input applied at the inverting terminal through external resistance R_1. The op-amp is in closed-loop configuration.

$$A_{VCL} = \frac{R_f}{R_i} \ (V_i).$$

Z_i is the input impedance of the op-amp. It is represented as R_i seen on the input side of the op-amp between inverting and non-inverting terminals. If we consider the ideal op-amp. R_i or Z_i is ∞. So the current drawn by the op-amp is 0 A. If the non-inverting terminal is grounded, that is at 0 V, since there is no current flowing through R_i, the potential at the inverting terminal must also be 0 V. The potential difference must be 0 V as no current is flowing. Hence, the inverting terminal is also at 0 V. Thus, though the inverting input terminal is not really grounded, it behaves as if it is at ground potential when the non-inverting terminal is grounded because of the high input impedance of the op-amp. This is referred to as **virtual ground**. Since the input impedance of op-amp is ∞, it draws no current from the external signal source V_i and hence, the current from V_i flows through R_1 and R_f. The current $I_b = 0$. If $V_B = 0$ V, V_A is also 0 V. This is called virtual ground. If $V_B = 1$ V, then V_A also will be at 1 V, since the potential difference V_{AB} must be 0 V, as op-amp does not draw current and $I_b = 0$.

This holds good even in the case of a practical op-amp circuit, this holds because R_i of op-amp is usually very high—of the order of 1 MΩ. So the current branching at node A,

$$I = I_1 + I_b$$
$$I_b \simeq 0$$

So,
$$I \simeq I_1$$

Even in the case of a practical op-amp circuits, if $V_B = 0$, V_A also can be considered to be 0 V.

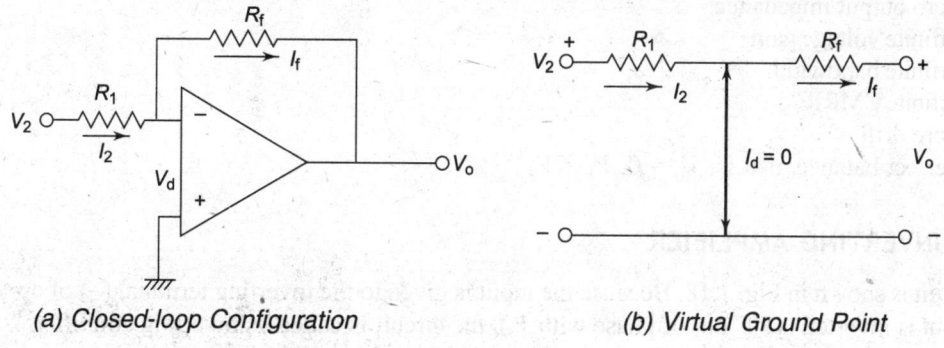

| (a) Closed-loop Configuration | (b) Virtual Ground Point |

Fig. 1.16

If R_f is connected between output and inverting terminal, the feedback is negative. If it is connected to the non-inverting terminal, the feedback is positive.

By definition $V_d = \dfrac{V_o}{A_o} \approx 0$ since $A_o \to \infty$

that is, the inverting terminal is grounded or the two terminals are virtually short-circuited or grounded (Fig. 1.16).

It does not mean that they are physically short-circuited but whatever potential exists at the terminal (1) the same will also appear at terminal (2).

The assumptions for an ideal op-amp are: (1) the current to each input terminal (inverting and non-inverting) is zero and (2) the voltage between the two input terminals is zero.

1.10   OP-AMP: A DIRECT COUPLED HIGH GAIN AMPLIFIER

Op-amps were used initially in the area of analog computation. These were constructed with discrete components. In mid-1960s, the IC op-amp was produced. This circuit (μA 709) was made up of a large number of transistors and resistors on a single chip.

The popularity for op-amp is due to its versatility.

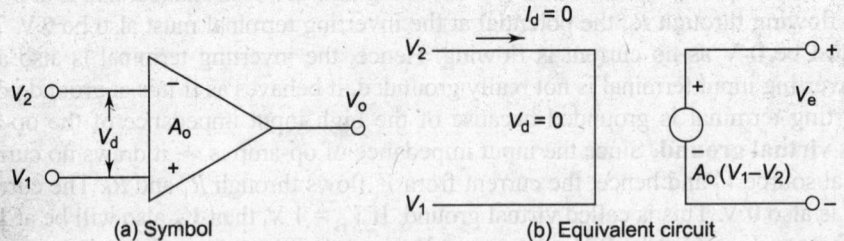

(a) Symbol   (b) Equivalent circuit

Fig. 1.17   *Op-amp Symbol and Equivalent Circuit*

If the output is in phase with the input, the terminal is non-inverting ('+') and if it is out of phase it is inverting ('−').

Its characteristics closely approach the assumed ideal; its circuits are easy to design; and the circuits work at levels close to their predicted theoretical performance.

Ideal characteristics of op-amp.

1. Infinite input impedance
2. Zero output impedance
3. Infinite voltage gain
4. Infinite bandwidth
5. Infinite CMRR
6. Zero drift
7. Perfect balance, that is, $V_o = 0$, $V_1 = V_2$

1.11   INVERTING AMPLIFIER

The circuit is shown in Fig. 1.18. Because the input is given to the inverting terminal (−) of op-amp and the output is inverted (180° out of phase with V_i), the circuit is called as inverting amplifier.

$$V_o = -\left(\frac{R_f}{R_1}\right) V_2$$

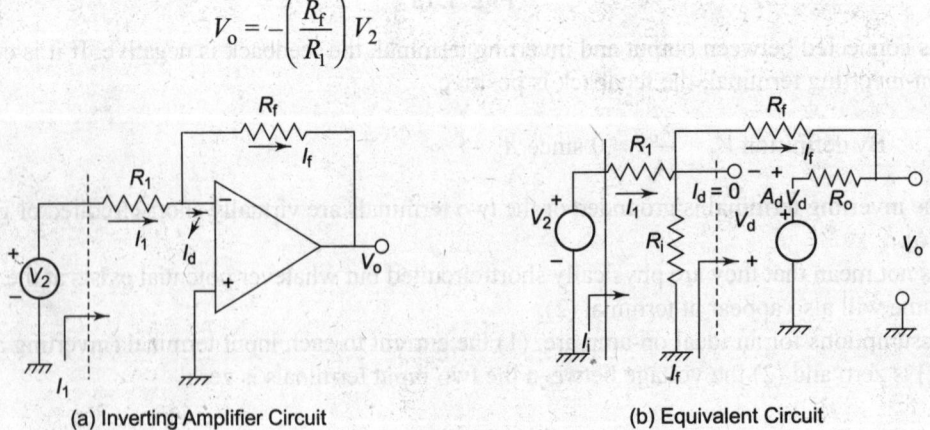

(a) Inverting Amplifier Circuit   (b) Equivalent Circuit

Fig. 1.18

1.11.1 Ideal Case

Since for ideal amp $I_d = 0$

$$I_1 = I_f, \quad I_1 = \frac{V_2 + V_d}{R_1} = I_f = \frac{-V_d - V_o}{R_f}$$

Since $V_d \to 0$

$$\frac{V_2}{R_1} = -\frac{V_o}{R_f}; \quad A_v = \frac{V_o}{V_2} = \frac{-R_f}{R_1}$$

$$A_v = \frac{V_o}{V_2}; \quad \therefore \quad \frac{V_o}{V_2} = -\frac{R_f}{R_1} \quad \boxed{A_v = -\frac{R_f}{R_1}}$$

1.11.2 Input Impedance

$$r_i = \frac{V_2}{I_1} = \frac{I_1 R_1 - V_d}{I_1} \text{ since } V_d = 0 \quad r_i \approx R_i$$

1.11.3 Non-ideal Case

From the equivalent circuit $r_i = R_1 + (r_f \| R_i)$

$$r_f = -V_d / I_f$$

$$-V_d = R_f I_f + I_f R_o + A_d V_d$$

$$-V_d(1 + A_d) = I_f(R_o + R_f); \quad \frac{-V_d}{I_f} = r_f = \frac{R_o + R_f}{(I + A_d)}$$

since $r_f \ll R_L \ll R_i$

$$r_i \approx R_1$$

1.11.4 Output Impedance

$$r_o = V_o / I_o$$

$$I_o = \frac{V_o - A_d V_d}{R_o} + \frac{V_o}{R_1 + R_f} \tag{1}$$

since $R_i \gg R_1$ (Fig. 1.19)

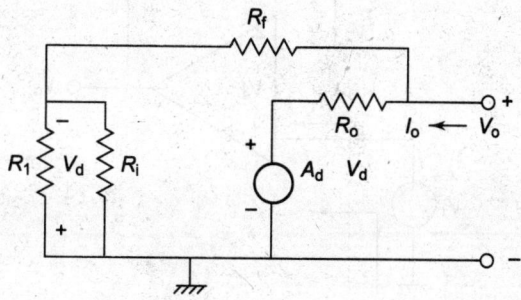

Fig. 1.19 *Equivalent Circuit to Show Output Impedance*

$$-V_\mathrm{d} = \left(\frac{R_1}{R_1 + R_f}\right)V_\mathrm{o} \tag{2}$$

$$I_\mathrm{o} = \left[\frac{V_\mathrm{o} + \left(\dfrac{A_\mathrm{d}\,R_1}{R_1 + R_f}\right)V_\mathrm{o}}{R_\mathrm{o}} + \frac{V_\mathrm{o}}{R_1 + R_f}\right]\frac{I}{r_\mathrm{o}} = \frac{I_\mathrm{o}}{V_\mathrm{o}} = \frac{1 + \dfrac{A_\mathrm{d}\,R_1}{R_1 + R_f}}{R_\mathrm{o}} + \frac{1}{R_1 + R_f}$$

equivalent circuit will be

$$r_\mathrm{o} \approx \frac{R_\mathrm{o}}{1 + \dfrac{A_\mathrm{d}R_1}{R_1 + R_f}} \quad \text{since } R_1 + R_f \text{ is large (Fig. 1.20)}$$

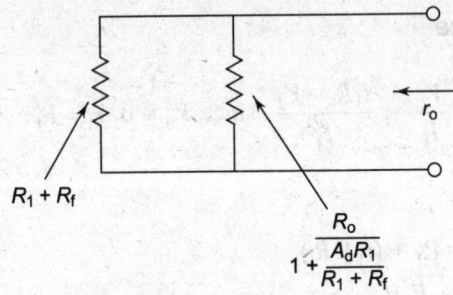

Fig. 1.20 *Simplified Circuit*

1.11.5 Ideal Case: Non-Inverting Amplifier

$$V_1 = V_2$$

$$V_2 = \left(\frac{R_1}{R_1 + R_f}\right)V_\mathrm{o}$$

$$\frac{V_\mathrm{o}}{V_1} = \frac{R_1 + R_f}{R_1} = \left(1 + \frac{R_f}{R_1}\right)$$

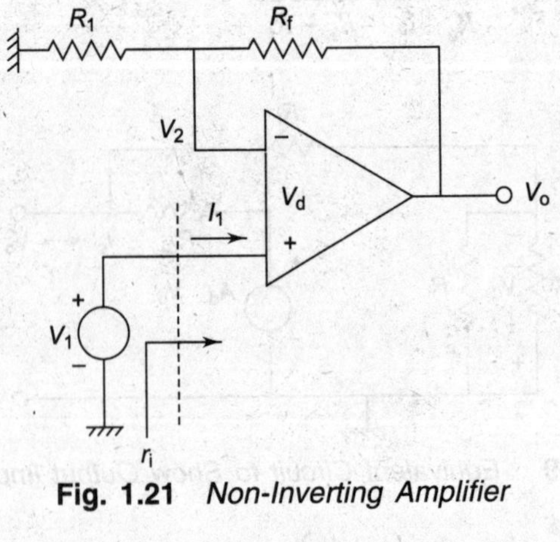

Fig. 1.21 *Non-Inverting Amplifier*

1.11.6 Input Impedance

Ideal op-amp takes no current from the source and hence presents infinite input impedance. Even if A_{VOL} is not infinity, it is a sufficient condition to make the input impedance ∞.

1.11.7 Non-ideal Case: Input Impedance

$$I_1 = \frac{V_d}{R_i} \Rightarrow I_1 = \frac{V_o}{A_d R_i} \left[\because V_d = \frac{V_o}{A_d} \right]$$

But

$$V_o = \left(1 + \frac{R_f}{R_1}\right) V_1$$

$$I_1 = \left(1 + \frac{R_f}{R_1}\right) \frac{V_1}{A_d R_i}$$

$$r_i = \frac{A_d R_i}{(1 + R_f/R_1)}$$

1.11.8 Output Impedance

From the equivalent circuit of inverting amplifier shown in Fig. (1.20) expression for 'r_o', the expression for output impedance can be written as,

$$r_o \approx \frac{R_o}{1 + \dfrac{A_d R_i}{R_1 + R_f}}$$

1.12 BASIC LINEAR CIRCUITS USING OP-AMPS

The block schematic of op-amp is shown in Fig. 1.22.

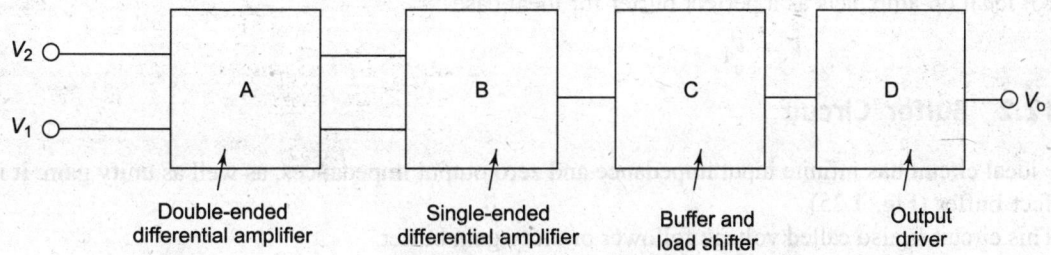

Fig. 1.22 *Block Diagram*

The input stage is differential amplifier having two inputs and its CMRR should be very high.

One more stage is added to provide additional amplification. The level shifter is necessary since for zero input, the output should be zero.

The final stage supplies the desired signal current or voltage to drive the load.

An ideal op-amp acts as a current to voltage converter since $I_{in} = I_f$ (ideal case)

$$V_o = -I_{in} R_f$$

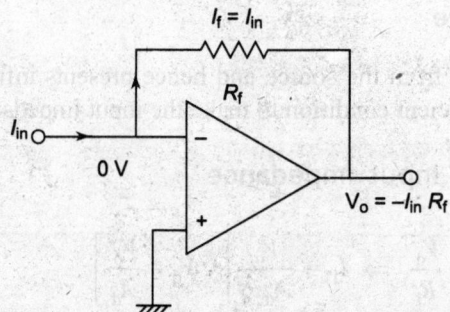

Fig. 1.23 *Op-amp in Closed-loop Configuration*

1.12.1 Adder Circuit

An ideal op-amp adds voltages or currents independently (Fig. 1.24).

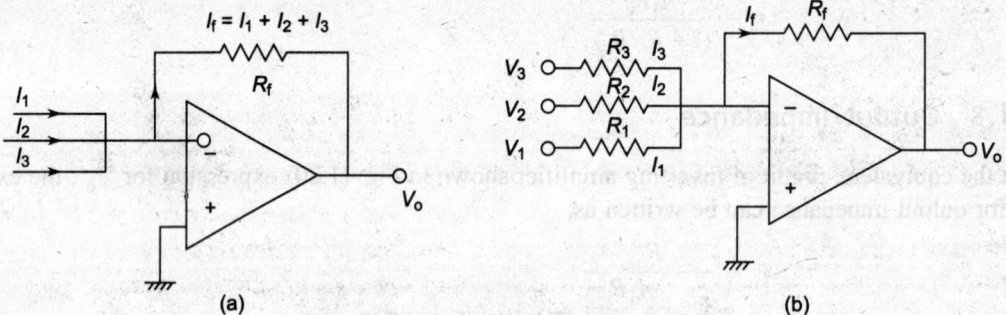

(a)   (b)

Fig. 1.24 *Op-amp Adder Circuits*

$$V_{\mathrm{o}} = -(I_1 + I_2 + I_3)R_{\mathrm{f}}, \quad V_{\mathrm{o}} = -\left(\frac{V_1}{R_1} + \frac{V_2}{R_2} + \frac{V_3}{R_3}\right)R_{\mathrm{f}}$$

An ideal op-amp. acts as a perfect buffer for ideal case

$$V_{\mathrm{o}} = V_1$$

1.12.2 Buffer Circuit

The ideal circuit has infinite input impedance and zero output impedances, as well as unity gain. It is a perfect buffer (Fig. 1.25).

This circuit is also called voltage follower or unity gain buffer.

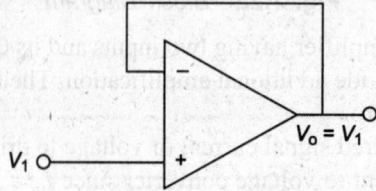

Fig. 1.25 *Buffer Amplifier*

1.12.3 General Analysis of Op-Amp Circuits

Consider the general configuration of op-amp circuit shown below, in Fig. 1.26.

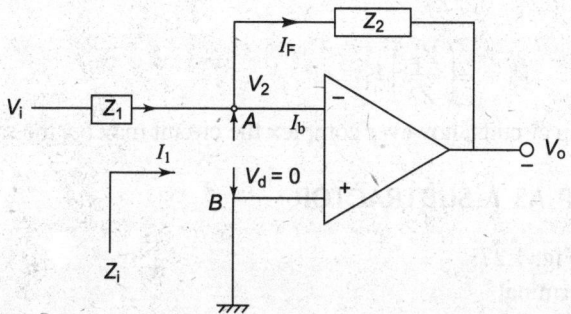

Fig. 1.26 *General Form of op-amp Circuit*

Assume ideal op-amp, so $Z_i = \infty = Z_o = 0$. So, the op-amp does not draw any current from the input-voltage source V_i, that is $I_b = 0$. So, the input current due to voltage source I_1 flows entirely into the feedback circuit.

That is, $$|I_1| = |I_F|$$

I_1 is entering node A.

I_F is leaving node A.

(These directions can be assumed arbitrarily.)

So, following the convention, $I_1 = -I_F$.

The non-inverting terminal is grounded, that is, $V_B = 0$.

Op-amp circuit responds such that $V_B = V_A$ or $(V_A - V_B) = V_d$, the differential potential is zero. Even if non-inverting terminal is not at ground potential and $V_B = 0$, the voltage at 'A' will also be the same

or $V_A - V_B = 0$ V.

In analysis of op-amp circuits, however complex the circuit may be, follow the guideline given below, when ideal op-amp is assumed.

1. Choose the directions for I_1 and I_F arbitrarily (entering the node or leaving the node).
2. Potential at 'A' will be the same as potential at 'B'. If 'B' is grounded, 'A' will be at virtual ground point.
3. Now write the expression for I_1, keeping in mind the direction of current chosen. In Fig. 1.26 as I_1 is entering node A,

$$I_1 = \frac{V_i - V_A}{Z_1} \tag{1}$$

I_2 is leaving node A.

Therefore, $$I_2 = \frac{V_A - (-V_o)}{Z_F} \tag{2}$$

V_o is negative because the input is given to inverting terminal

$$V_A = V_B = 0.$$

So, $$I_1 = \frac{V_i}{Z_1} \tag{3}$$

$$I_2 = +\frac{V_o}{Z_F} \tag{4}$$

$$I_1 = -I_F$$

$$\frac{V_i}{Z_i} = -\frac{V_o}{Z_F}$$

So,

$$V_o = -\left(\frac{Z_F}{Z_i}\right) V_i$$

For the analysis of op-amp circuits, however complex the circuit may be, the same approach can be used.

1.13 IDEAL OP-AMP AS A SUBTRACTOR

The circuit is shown in Fig. 1.27.
Voltage at the inverting terminal

$$V^- = \left(\frac{R_2}{R_1 + R_2}\right) V_2 + \left(\frac{R_1}{R_1 + R_2}\right) V_o$$

at the non-inverting terminal

$$V^+ = \left(\frac{R_2}{R_1 + R_2}\right) V_1$$

since

$$V^- = V^+$$

$$V_o \left(\frac{R_1}{R_1 + R_2}\right) + \left(\frac{R_2}{R_1 + R_2}\right) V_2 = \left(\frac{R_2}{R_1 + R_2}\right) V_1$$

$$V_o = \frac{R_2}{R_1} (V_1 - V_2)$$

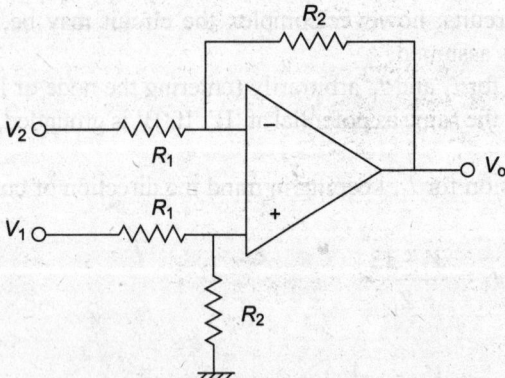

Fig. 1.27 *Op-amp Subtractor Circuit*

1.14 IDEAL OP-AMP AS AN INTEGRATOR

The circuit is shown in Fig. 1.28.

  I_{in} is input current
  V_{in} is input voltage
  V_o is output voltage

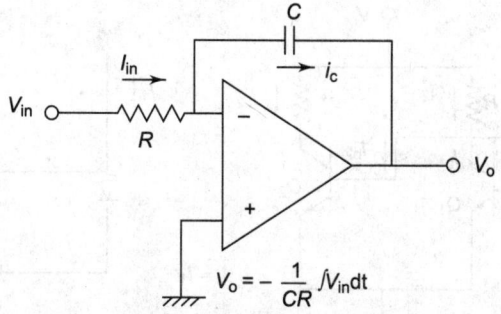

Fig. 1.28 *Op-amp Integrator Circuit*

$$I_{in} = \frac{V_{in}}{R} = C \frac{dvc}{dt}$$

$$\frac{V_{in}}{R} = -C \frac{dV_o}{dt}$$

$$V_o = -\frac{1}{CR} \int V_{in}\, dt$$

1.15 IDEAL OP-AMP AS A DIFFERENTIATOR

The circuit is shown in Fig. 1.29.

$$i_R = i_c$$

$$\frac{-V_o}{R} = C \frac{dV_{in}}{dt}$$

$$V_o = -CR \frac{dV_{in}}{dt}$$

L_R = Current through resistor R in the feedback path
Op-amp differentiator circuits are not used as S/N reduces or noise increases at the output.

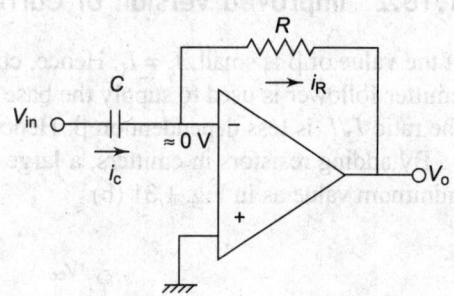

Fig. 1.29 *Op-amp Differentiator Circuit*

1.16 OP-AMP DESIGN TECHNIQUES

This unit consists of some of the basic concepts of design techniques that are employed in the operational amplifiers available today.

We shall first consider the input section.

In order to increase the input impedance of a differential amplifier, we have to increase the gain of the transistors and also reduce the value of I_c as seen from the equation $R_{id} = 2h_{ie} = 2h_{fe} \times V_T/|I_c|$

By using supergain transistors at I_c value of 1 μA, the input impedance can be increased. A matched discrete FET differential stage at the input can be fabricated on the same chip. Hence, a very high value of input resistance can be obtained.

1.16.1 Biasing Method

For self bias, a transistor is used as a diode instead of large values of resistances and capacitances which cannot be fabricated.

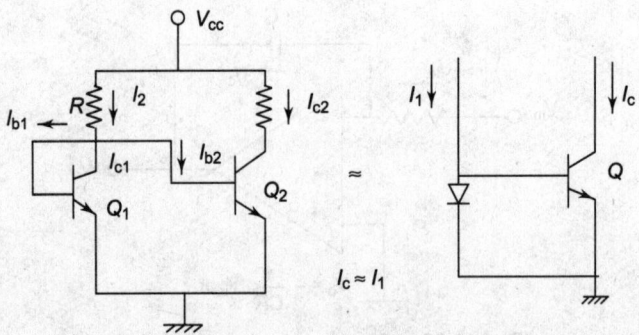

Fig. 1.30   *Biasing Method*

If the transistors Q_1 and Q_2 have the same value of V_{BE} (Fig. 1.30),

$$I_{c1} = I_{c2} = I_c$$

If $\beta \gg 2$ $$I_c = I_1 = (V_{cc} - V_{BE})/R$$

Since the variation of V_{cc} with temperature is small compared to V_{cc}, I_c is independent of temperature. This configuration is called a current mirror circuit.

1.16.2 Improved Version of Current Mirror Circuit

If the value of β is small, $I_c \neq I_1$. Hence, considerable error will be present and in order to reduce it, an emitter follower is used to supply the base currents as shown in Fig. 1.31 (a). If the circuit is analyzed, the ratio I_2/I_1 is less dependent on β. Hence, the error involved in assuming $I_c = I_1$ is small.

By adding resistors in emitters, a large ratio of I_2/I_1 can be obtained and the error can be kept to a minimum value as in Fig. 1.31 (b).

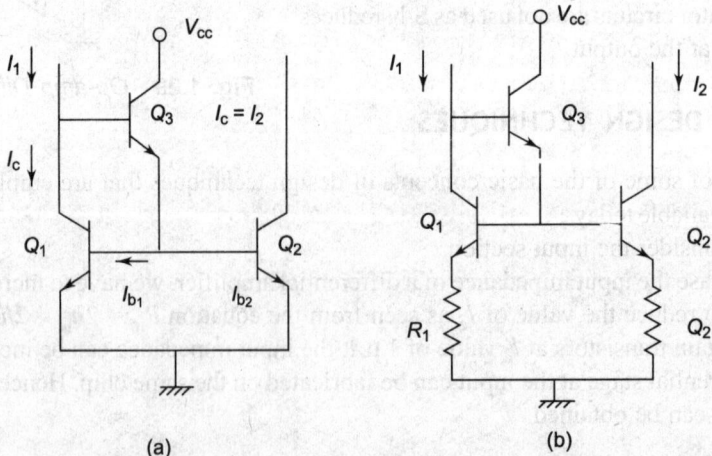

(a)

(b)

Fig. 1.31   *Current Mirror Circuit*

1.16.3 Active Loads

The constant current I_o may be obtained by using current mirror circuit with Q_3 and Q_4 transistors. This increases the voltage gain. (Fig. 1.32)

If
$$V_1 = V_2 = 0$$
$$I_1 = I_2 = I_0/2$$

Since Q_3 and Q_4 are acting as current mirror circuit, $I = I_1$.

Hence
$$I = I_2$$
$$I_d = I - I_2 = 0$$

V_1 and I_1 increase, while V_2 and I_2 decrease, to keep
$$I_1 + I_2 = I_0$$

Because of current mirror effect,
$$I_d = I - I_2 = I_1 - I_2$$
$$= g_m V_1 - g_m V_2 = g_m V_d$$

Where g_m = mutual conductance $= \dfrac{1}{R}$

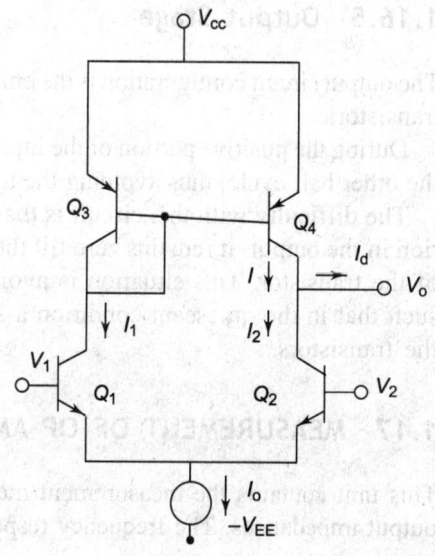

Fig. 1.32 *Active Load Circuit*

1.16.4 Level Shifting

Capacitors are less frequently used in IC fabrication. Hence, it is necessary to shift the quiescent voltage before applying input to the next stage. Further, load shifting is necessary to keep the output close to zero when there is no input. Therefore, an emitter follower circuit is used (Fig. 1.33).

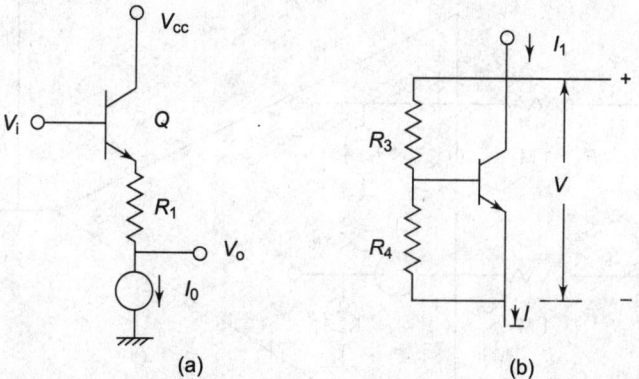

Fig. 1.33 *Level Shifting Circuits*

The shift in level obtained in Fig. 1.33 (a) will be
$$V_o = - (V_{BE} + I_0 R_1)$$
From Fig. 1.33 (b), if the base current is neglected

$$V = \frac{V_{BE}}{R_4} (R_3 + R_4) = V_{BE} \left(1 + \frac{R_3}{R_4} \right)$$

This voltage source may be used in place of R_1 to obtain any value of voltage shift.

1.16.5 Output Stage

The output circuit configuration is the emitter follower with complementary transistors.

During the positive portion of the input, Q_1 conducts and Q_2 conducts in the other half cycle, thus avoiding the use of an output transformer.

The difficulty with this circuit is that because of the crossover distortion in the output, it remains zero till the input exceeds the cut-in voltage of the transistor. This situation is avoided by keeping the bias voltage such that in the quiescent condition a small amount of current flows in the transistors.

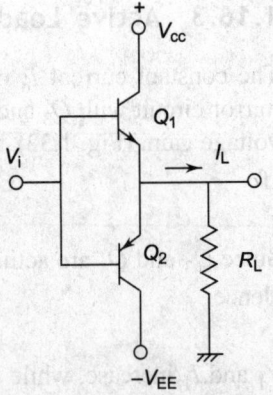

Fig. 1.34 *Output Stage*

1.17 MEASUREMENT OF OP-AMP PARAMETERS

This unit contains the measurement methods of offset input voltage and current, CMRR input, and output impedances. The frequency response is also discussed.

1.17.1 Measurement of Offset Voltage and Current

When one of the switches is open, bias current flows through the resistance R_3. Hence, the input voltage is applied. When both the switches are closed, the input voltage is V_{io}, as the resistance R_1 is small. $\pm I_b\, R_1$ can be neglected (Fig. 1.35).

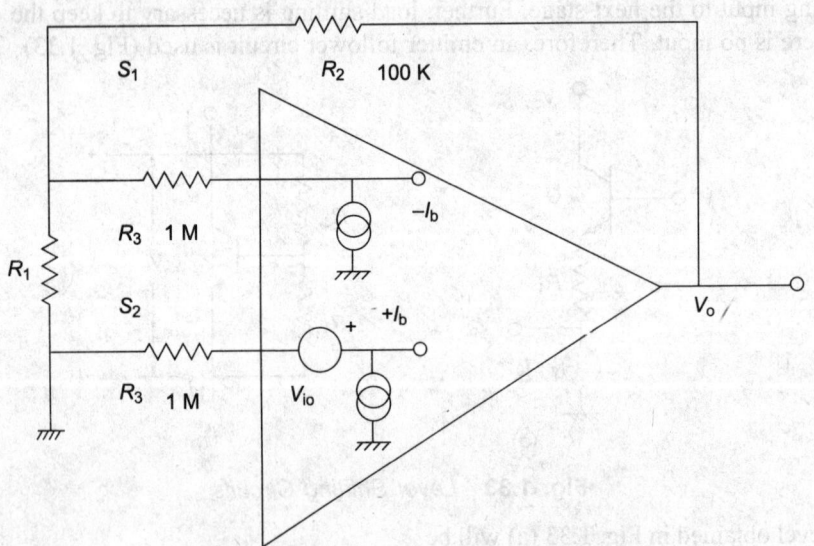

Fig. 1.35 *Measurement of Offset Voltage*

1. When S_1 and S_2 are closed

$$V_{o1} = \left(1 + \frac{R_2}{R_1}\right) V_{io} \tag{1}$$

2. When S_1 and S_2 are open

$$V_{o2} = \left(1 + \frac{R_2}{R_1}\right)[V_{io} + (I_b^- - I_b^+)\,R_3]$$

$$= \left(1 + \frac{R_2}{R_1}\right)(V_{io} + I_{io}R_3) \tag{2}$$

Solving the equations in 1 and 2 above give the offset voltage and current.

3. When S_1 is closed and S_2 open

$$V_{o3} = \left(1 + \frac{R_2}{R_1}\right)(V_{io} + I_b^+\,R_3)$$

4. When S_1 is open and S_2 is closed

$$V_{o4} = \left(1 + \frac{R_2}{R_1}\right)(V_{io} - I_b^-\,R_3)$$

The bias currents can be calculated from these two equations.

1.17.2 Measurement of CMRR (ρ)

Assuming $R_2 \gg R_1$ and $R_2' \gg R_1'$ (Fig. 1.36)

$$V_c \approx V_s$$
$$V_1 = V_c = 0, \text{ provided } A_c = 0$$

Since $A_c \neq 0$, $V_i = \dfrac{V_o}{A_d}$; $A_c = \dfrac{V_o}{V_c}$

$$\rho = \frac{A_d}{A_c} = \frac{V_o/V_i}{V_o/V_c} = \frac{V_c}{V_i}$$

But $\quad V_i = \left(\dfrac{R_1}{R_1 + R_2}\right)V_o \text{ and } V_c \approx V_s$

$$\text{CMRR} = \left(\frac{R_1 + R_2}{R_1}\right)\frac{V_r}{V_o}$$

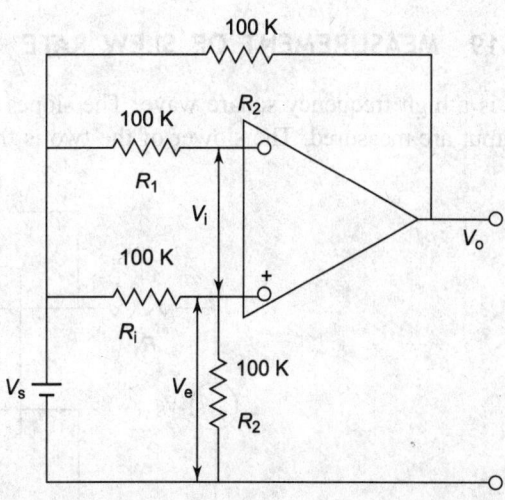

Fig. 1.36 *Measurement of CMRR*

1.17.3 Measurement of Open-loop Input and Output Impedances

Input Impedance. With the switch open, $R_s = 0$, some amount of input is given to obtain a convenient value of V_o. R_s is adjusted till V_o falls to $\dfrac{V_o}{2}$. Then $R_s = R_{in}$ (Fig. 1.37).

Output Impedance. Keeping $R_s = 0$, the same procedure is repeated varying R_L after closing the switch.

Then $\quad R_L = R_o$

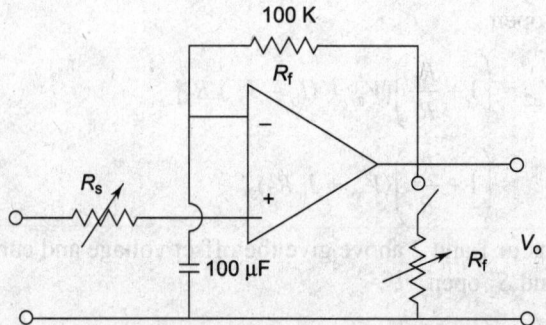

Fig. 1.37   *Measurement of Impedances*

1.18   MEASUREMENT OF PSRR

Keeping one power supply constant, two values of V_{io} are taken by changing the other power supply.

$$PSRR = \frac{\Delta V_{io}}{\Delta V_{cc}}\bigg|_{\text{at the specified power supply}}$$

1.19   MEASUREMENT OF SLEW RATE

V_s is a high frequency square wave. The slopes with respect to time of the leading, trailing edges of output are measured. The slower of the two is the slew rate (Fig. 1.38).

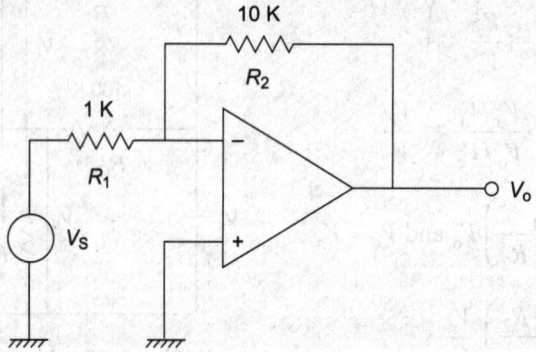

Fig. 1.38   *Measurement of Slew Rate*

1.20   MEASUREMENT OF OPEN-LOOP GAIN

Neglecting the bias current of BUF (Buffer), $V_o = -V'$ (Fig. 1.39)

Since $V_B = 0$

Select $V' = -10$ V

So, $V_o = 10$ V

Since the output impedance is small compared to 100 k

$$A_o V_i = V_o$$

Input to amplifier under test (AUT) is $V_{io} + V_i$

$$V = \frac{R_1 + R_2}{R_1} (V_{io} + V_i) \approx 10^3 \left(V_{io} + \frac{V_o}{A_v} \right) = V_2$$

$$V_2 = \left(V_1 + \frac{10^4}{A_v} \right), \text{ where } V_1 = 10^3 \, V_{io}$$

$$A_v = \frac{10^4}{(V_2 - V_1)}$$

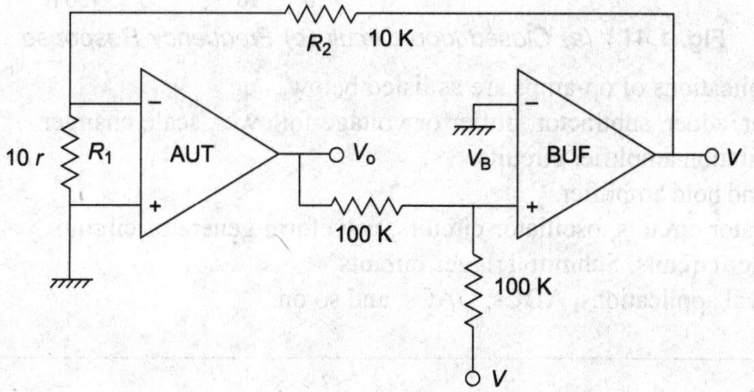

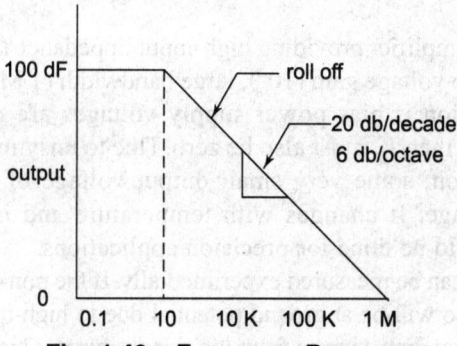

Fig. 1.39 *Measurement of Open-loop Gain*

1.21 FREQUENCY RESPONSE

The frequency response is given by a Bode plot. The gain of the op-amp is 10^5 (100 db) and rolls off at a uniform rate of 20 db/decade till 1 MHz where the gain becomes unity. The gain attenuation with increase in frequency is referred to as the roll off in the frequency response (Fig. 1.40).

Fig. 1.40 *Frequency Response*

If an extension of frequency response is required, the open-loop gain of the amplifier should be sacrificed.

The bandwidth can also be increased by changing the rate of roll off (Fig. 1.41).

Present day op-amps are frequency compensated and for a majority of applications, there is no need to extend the frequency response beyond that normally available value.

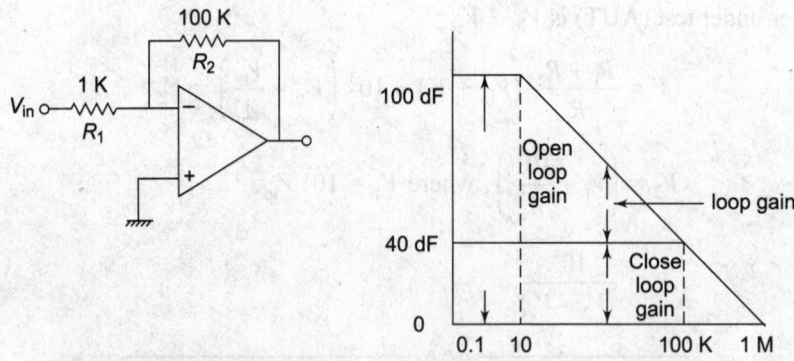

Fig. 1.41 *(a) Closed-loop Circuit (b) Frequency Response*

The various applications of op-amps are as listed below.
1. As integrator, adder, subtractor, buffer or voltage follower, scale changer.
2. In instrumentation amplifier circuits.
3. As sample and hold amplifier.
4. In multivibrator circuits, oscillator circuits, waveform generator circuits.
5. In active filter circuits, Schmitt Trigger circuits
6. In bio-medical applications, ADCs, DACs, and so on.

SUMMARY

Specifications of μA 741 and other op-amps from data sheets must be given here.

- An operational amplifier, abbreviated as op-amp, is a multi-stage, differential amplifier, with an output driver circuit packaged in IC form. All the circuit components are fabricated on a very small area of silicon wafer and packaged. A popular eight-pin op-amp IC is 741C.
- Op-amp is so named because various mathematical operations like addition, subtraction, scale changing, integration, and so on can be performed. This IC was initially developed for use in analog computers.
- This IC is a differential amplifier providing high-input impedance (1 MΩ), low-output impedance (500 Ω), large open-loop voltage gain (10^5), large bandwidth (1 MHz), and high CMRR (90 db).
- The IC does not function if bias power supply voltages are not given. When there is no external input signal V_i, then V_o must also be zero. Due to unsymmetry in differential amplifier configurations and so on, some very small output voltage of the order of μV may result, inferred as offset voltage. It changes with temperature and is referred to as *drift*. Offset voltage adjustment has to be done for precision applications.
- The op-amp parameters can be measured experimentally. If the non-inverting terminal is grounded, the inverting terminal also will be at ground potential due to high-input impedance of the op-amp. Ideally, an op-amp does not draw current from the signal source. This is referred to as *virtual ground*.
- For an op-amp IC to function, biasing supply voltages must be given with proper polarity. If the gain is very large or input signal is very large, the op-amp will go to saturation. The impression $V_o = A_{VCL} \, V_i$ will not hold good because the output swing of V_o will be limited to a maximum of $+ V_{cc}$ and $- V_{cc}$ less the drop across the internal circuit of the IC, typically 1 V.
- Op-amps can be used as integrator, differentiator, scale changer, buffer or voltage followers, adder or summer, and subtractor. As differentiator, the S/N ratio gets deteriorated. Therefore, this

configuration is not used. If voltage follows configuration, it provides higher input impedance and low-output impedance. So it is used as a buffer.

- Op-amps are used in inverting configuration because in that mode the negative feedback circuit configuration will hold good. So the advantages of a negative feedback are made use of. Input is given to a non-inverting terminal for oscillator circuits where positive feedback is required.
- Due to the input capacitance associated with the op-amp circuit, it will not be able to respond instantaneously. The capacitor slows down the response due to finite charging time and associated time constant. So, there is a limit to the highest frequency of the input signal that can be applied to the op-amp. This is indicated by the slew rate. It denotes the capability of the op-amp to respond to high-frequency signals.

SOLVED EXAMPLES

Example 1.1 *For an op-amp integrator with R = 100 MΩ and C = 1 μF, an input of 2 sin 1000 t is applied. Determine the value of V_o.*

Given

$V_i = 2 \sin 1000\ t$

$R = 100\ \text{M}\Omega$

$C = 1\ \mu\text{F}$

Required; V_o.

$V_o = 20\ \mu V$

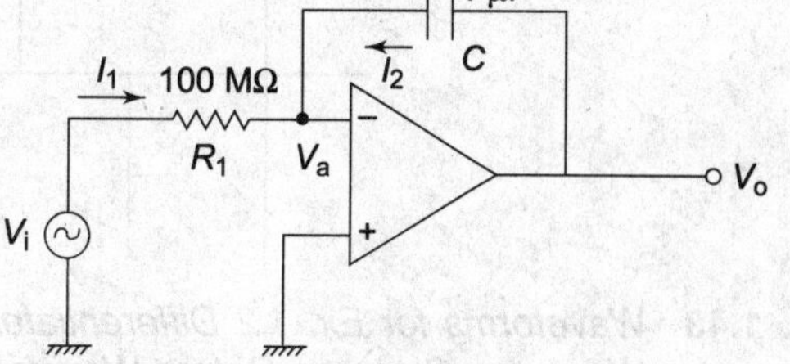

Fig. 1.42 *Circuit for Ex. 1.1*

$$I_1 + I_2 = 0$$

$$I_1 = -I_2$$

$$I_2 = \frac{V_o - V_a}{Z_f}$$

$$I_2 = \frac{c\,dV_o}{dt}$$

$$I_1 = -I_2$$

$$\frac{V_i}{R} = -\frac{c\,dV_o}{dt}$$

$$I_1 = \frac{V_i - V_a}{R},\ V_o = 0$$

$$I_1 = \frac{V_i}{R}$$

$$\frac{V_i}{R} = -c\frac{dV_o}{dt}$$

V_i = Input voltage

V_o = Output voltage

V_a = Voltage at the inverting terminal

= 0 V since non-inverting terminal is grounded.

$$\Rightarrow \qquad -\frac{V_i}{RC} = \frac{dV_o}{dt} = \int \frac{dV_o}{dt} - \int \frac{V_i}{RC}\,dt \qquad V_o = -\frac{1}{RC}\int_0^1 V_i dt$$

$$\Rightarrow \qquad V_o = -\frac{1}{RC}\int_0^1 V_t dt$$

$$= \frac{-1}{(100 \times 10^6)(10^{-6})}\int_0^1 2 \sin 1000\ t\ dt$$

$$= \frac{-1}{100}\int 2 \sin 1000\ t\ dt$$

$$= \frac{-1}{50} \int \sin 1000 \, t \, dt$$

$$= \frac{-1}{(50)(1000)} \cos 1000 \, t$$

$$|V_o| = \frac{1}{50000} = 20 \, \mu V$$

Example 1.2 *For an op-amp differentiator with R = 1 kΩ and C = 0.01 µF, a square wave input of 200 Hz is applied. Sketch the output waveform giving reasons.*

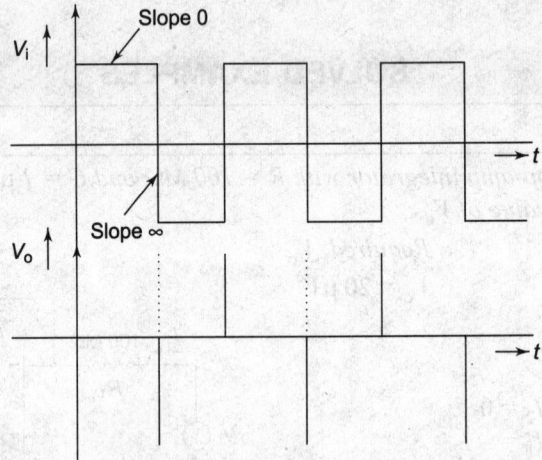

Fig. 1.43 *Waveforms for Ex. 1.2 Differentiator Circuit, the Positive and Negative Slopes of Square Wave are Given as Output Waveform*

The reason is $T = RC \ll T$. So near ideal differentiation is done. The slope of the square wave (output of differentiation) is ∞ for vertical line and zero for horizontal zone.

Differentiation means the slope of the line. For a perfect vertical line, the slope is ∞. For a perfect horizontal line, the slope is 0. Therefore, if a square wave input is given to ideal differentitor circuit, the output waveform is as shown in figure for $V_o \, V_{st}$, a series of spikes.

Example 1.3 *(a) Give the specifications and typical values of op-amp 741.*

	Minimum	Typical	Maximum
Input bias current		200 nA	
Input offset current		30 nA	
Input offset voltage		1 mV	
Input requistance		1 MV	
Open-loop voltage gain		100,000	
Common-mode rejection ratio		100 db	
Slew rate		0.7 V/µsec	
Unity gain bandwidth		1 MHz	
Occupy voltage rejection ratio		90 db	
PSRR 15 µV/V		15 µV/V	

(b) Give the pin configuration of 741 IC. How do you identify the pins of the IC?

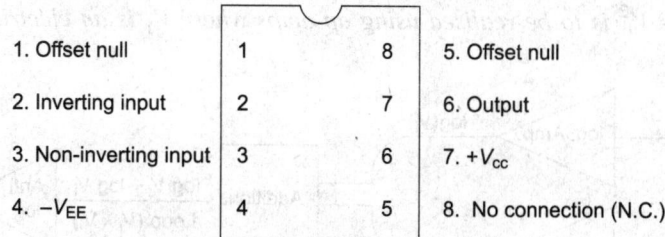

1. Offset null	1	8	5. Offset null
2. Inverting input	2	7	6. Output
3. Non-inverting input	3	6	7. $+V_{cc}$
4. $-V_{EE}$	4	5	8. No connection (N.C.)

Fig. 1.44 *Pin Configuration of 741 IC*

Example 1.4 *Explain different IC packages.*

The IC packages are
1. Ceramic flint packages
2. T0-5 package (with straight leads)
3. T0-5 package (with dual-on-line format)
4. DIP (dual-in-line package)

The tab in the IC package indicates pin 8. The other pins, starting from 1, are counted in an anticlockwise direction for the T0-5 package. The dot on the notch side indicates pin 1 in a DIP. The first pin on the ridge side is pin 1 in some packages.

The package types are indicated by the code
 I: Mini DIP
 P: Plastic DIP
 F: Flat pack

Example 1.5 *(a) A triangular waveform of (5x + 3) with a peak-to-peak value of 2 V is applied to the inverting terminal of an op-amp in open-loop configuration. Sketch the input and output wave format to the same time scale.*

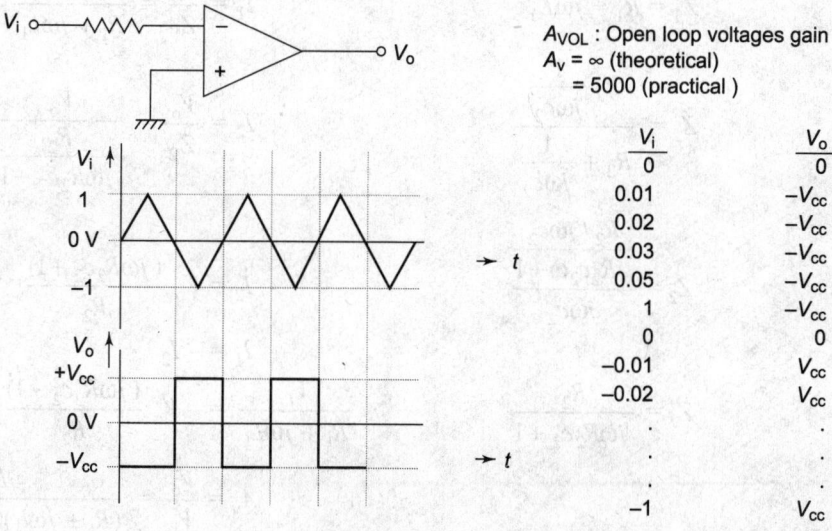

A_{VOL} : Open loop voltages gain
$A_v = \infty$ (theoretical)
   = 5000 (practical)

V_i	V_o
0	0
0.01	$-V_{cc}$
0.02	$-V_{cc}$
0.03	$-V_{cc}$
0.05	$-V_{cc}$
1	$-V_{cc}$
0	0
−0.01	V_{cc}
−0.02	V_{cc}
.	.
.	.
.	.
−1	V_{cc}

Fig. 1.45 *Waveforms for Ex. 1.6*

(b) An output $V_o = V_i^2$ is to be realised using op-amps where V_1 is an electrical signal. Give the schematic.

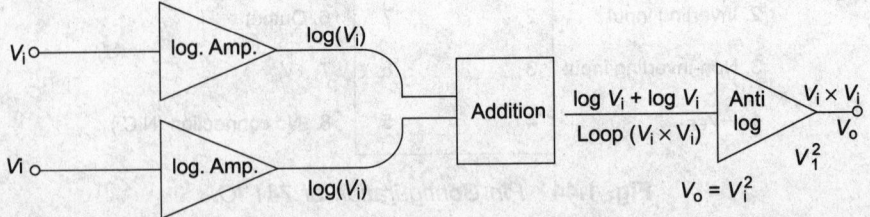

Fig. 1.46   *Realization of Expression $V_o = V_i^2$*

Example 1.6   *An op-amp has R_1 and L_1 in series on the input side, connected to the inverting terminal and R_2, C_2 connected in parallel in the feedback path. The non-inverting terminal is grounded. Derive the transfer function.*

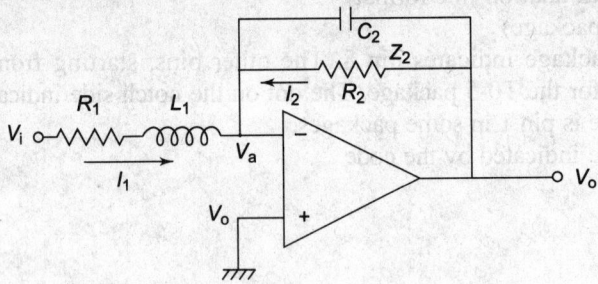

Fig. 1.47   *Circuit for Ex. 1.6*

$$I_1 + I_2 = 0 \Rightarrow I_1 = -I_2$$

$$Z_1 = R_1 + j\omega L_1 \qquad\qquad I_1 = \frac{V_i}{Z_1} = \frac{V_i}{R_1 + j\omega L_1}$$

$$Z_2 = \frac{R_2 \dfrac{1}{j\omega c_2}}{R_2 + \dfrac{1}{j\omega c_2}} \qquad\qquad I_2 = \frac{V_o}{Z_2} = \frac{V_o}{\dfrac{R_2}{j\omega R_2 c_2 + 1}}$$

$$Z_2 = \frac{\dfrac{R_2 / j\omega c_2}{j R_2 c_2 \omega + 1}}{j\omega c_2} \qquad\qquad I_2 = V_o \frac{(j\omega R_2 c_2 + 1)}{R_2}$$

$$I_1 = -I_2$$

$$Z_2 = \frac{R_2}{j\omega R_2 c_2 + 1} \qquad\qquad \frac{V_i}{R_1 + j\omega L_1} = -V_o \frac{(j\omega R_2 c_2 + 1)}{R_2}$$

$$A_v = \frac{V_o}{V_i} = \frac{R_2}{(R_1 + j\omega l_1)(j\omega R_2 c_2 + 1)}$$

Example 1.7 *For the circuit shown in Fig. 1.48, obtain the expression for e_0.*

The input impedance of an op-amp is very high. Therefore, the currents into the op-amp are negligible. Hence,

$$I_1 = I_2$$

and

$$V_1 = V_2$$

Applying Kirchhoff's law of currents at points 1 and 2,

at 1 $\Rightarrow \dfrac{e_2 - V_1 + e_1 - V_1}{R} = \dfrac{V_1}{R}$

$\Rightarrow \qquad\qquad \dfrac{e_1 + e_2}{3} = V_1 = V_2$

at 2 $\qquad \dfrac{e_4 + e_3 - 2V_2}{R} = \dfrac{V_2 - e_0}{R}$

$\Rightarrow \qquad\qquad e_0 = 3V_2 - (e_4 + e_3)$

$$= 3 \times \dfrac{e_1 + e_2}{3} - (e_4 + e_3)$$

$\Rightarrow \qquad\boxed{e_0 = (e_1 + e_2) - (e_3 + e_4)}$

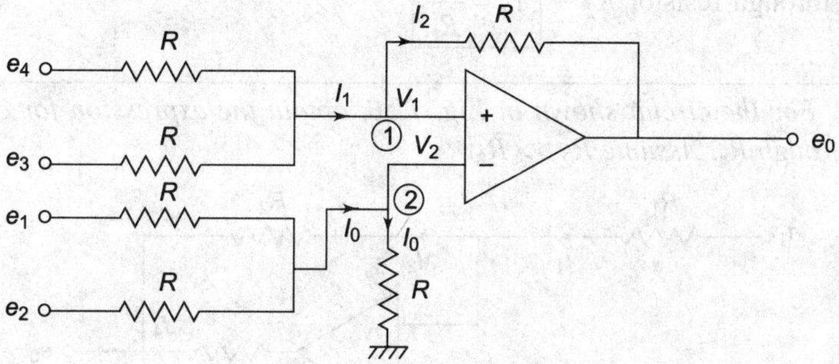

Fig. 1.48 *Circuit for Ex. 1.7*

Example 1.8 *For the circuit shown in Fig. 1.49, show that current through R_2 is independent of R_2 and is equal to* $\left| \dfrac{e_1}{R_1} \right|$.

$$V_B = \dfrac{R_1 R_2}{R_1 + R_2}(e_0/R_1) + \dfrac{R_1 R_2}{R_1 + R_2} \Rightarrow \dfrac{\dfrac{e_0 R_1 R_2}{R_1 + R_2}}{R_1\left(1 + \dfrac{R_2}{R_1 + R_2}\right)}$$

$$V_B = e_0 \dfrac{R_2}{R_1 + 2R_2}$$

Applying Kirchhoff's law of currents at point 1

$\Rightarrow \qquad\qquad \dfrac{e_1 - V_B}{R_1} = \dfrac{V_B - e_0}{R_1}$

$$e_1 = 2V_B - e_0$$

$$= 2e_0 \frac{R_2}{R_1 + 2R_2} - e_0$$

$$= e_0 \left(\frac{2R_2}{R_1 + 2R_2} \right) - 1 = -\frac{R_1}{R_1 + 2R_2} e_0$$

$$e_0 = -\frac{e_1}{R_1}(R_1 + 2R_2)$$

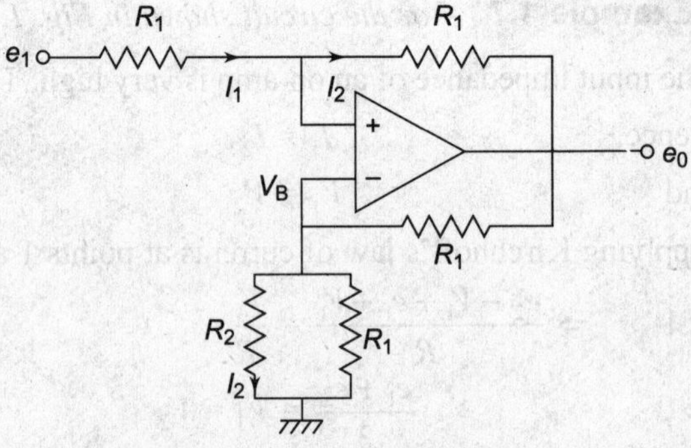

So, $\quad V_B = e_0 \dfrac{R_2}{R_1 + 2R_2} = -\dfrac{e_1}{R_1}\left(\dfrac{R_2}{R_1 + 2R_2} \right)$

$$V_B = -\frac{e_1}{R_1} R_2$$

The current through the resistor R_2

$$= \frac{V_B}{R_2} = -\frac{e_1}{R_1}\frac{R_2}{R_2} = -\frac{e_1}{R_1}$$

The negative sign indicates that the direction of current is opposite to what was assumed.

Therefore, current through resistor $R_2 = \boxed{I_2 = \dfrac{e_1}{R_1}}$

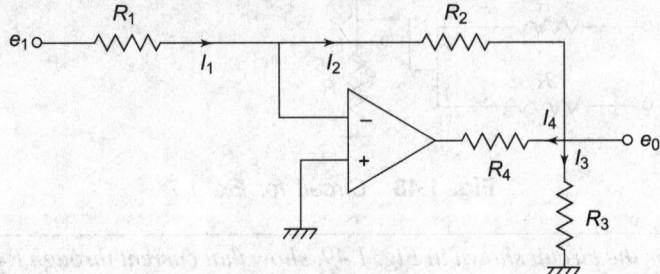

Fig. 1.49   *Circuit for Ex. 1.8*

Example 1.9   *For the circuit shown in Fig. 1.50, obtain the expression for current through R_4. Find the current through R_4. Assume $R_2 \gg R_3$.*

Fig. 1.50   *Circuit for Ex. 1.9*

We know that $\qquad I_1 = I_2 \qquad$ since op-amp does not draw any current.

$$\frac{e_1}{R_1} = -\frac{e_0}{R_2}$$

$$I_2 = I_3 + I_4$$

$$I_4 = I_2 - I_3$$

$$e_0 = I_2 R_2 = I_3 R_3 \Rightarrow I_3 = -I_2 \frac{R_2}{R_3}$$

e_0 is the drop across R_2. Non-inverting terminal is at virtual ground point.

So, $\qquad\qquad I_4 = I_2 + I_2 \dfrac{R_2}{R_3} \; I_2\left(1 + \dfrac{R_2}{R_3}\right) = \dfrac{e_1}{R_1}\left(\dfrac{R_1 + R_3}{R_3} \right)$

Assuming $\qquad R_2 \gg R_3$

$$I_4 = \frac{e_1}{R_1}\frac{R_2}{R_3}$$

So, the current through $\quad R_4 = \boxed{I_4 = \dfrac{e_1 R_2}{R_1 R_3}}$

Example 1.10 *Show that $V_o = 2(V_1 - V_2)$.*

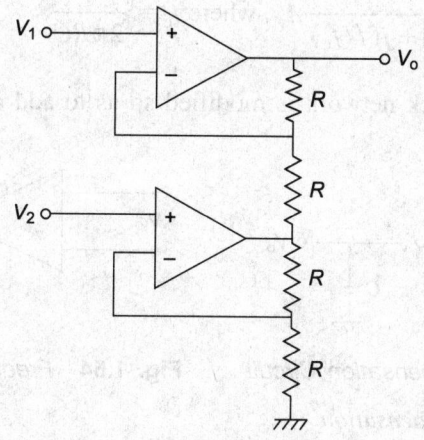

Fig. 1.51 *Circuit for Ex. 1.10*

The circuit can be drawn as given below.

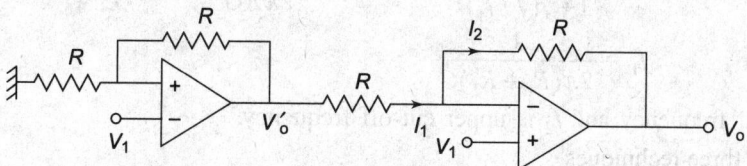

Fig. 1.52 *Redrawn Circuit for Ex. 1.10*

$$\frac{V'_o}{V_2} = \text{Gain} = \frac{R_1 + R_2}{R_1} = \frac{2R}{R} = 2$$

$$V'_o = 2V_2$$

$$\frac{V'_o - V_1}{R} = \frac{V_1 - V_o}{R} = 2V_1 = -V'_o - V_o$$

$\Rightarrow \qquad\qquad 2V_1 = (V'_o + V_o)$

$\qquad\qquad\qquad 2V_1 = 2V_2 + V_o$

$\Rightarrow \qquad\qquad V_o = 2(V_1 - V_2)$

Example 1.11 *Write a short note on frequency compensation techniques used in op-amps.*

The essential idea of compensation is to reshape the magnitude and phase plots of βA so that $|\beta A| < 1$ when the angle of βA is $180°$. There are three general methods of accomplishing this goal.

(i) *Dominant pole or lag compensation*

This method exerts an extra pole into the transfer function at a later frequency than the existing poles. Such a circuit introduced a phone log into the amp. The loop gain drops 0 db with the slope of 6 db/octave at a frequency where the poles of A_v contribute negligible phase shift. The only disadvantage is it wastes the workable bandwidth.

$$A_v = \frac{1}{1 + j(f/f_c)} A_v, \text{ where } f_c = \frac{1}{2\pi RC}$$

(ii) *Lead compensation*

The amplifier or the feedback network is modified so as to add a zero to the transfer function, thereby increasing the phase.

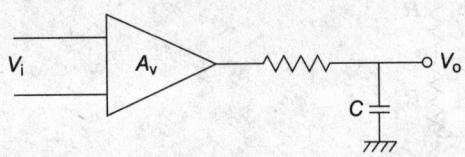

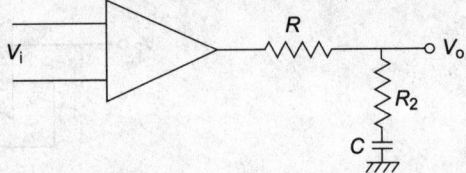

Fig. 1.53 *Frequency Compensation Circuit*   **Fig. 1.54** *Frequency Compensation Circuit*

(iii) *Pole–zero or lead–lead compensation*

This technique adds both a pole (a lag) and zero (lead) to the transfer gain. The zero is to cancel the lowest pole. The transfer function of the phase network is found to be

$$= \frac{1 + j(f/f_2)}{1 + j(f/f_1)}, \text{ where } f_2 = \frac{1}{2\pi RC}$$

$$f_1 = \frac{1}{2\pi(R_1 + R_2)C}$$

f_1 is lower cuf-off frequency and f_2 is upper cut-off frequency.

Compensation of three techniques.

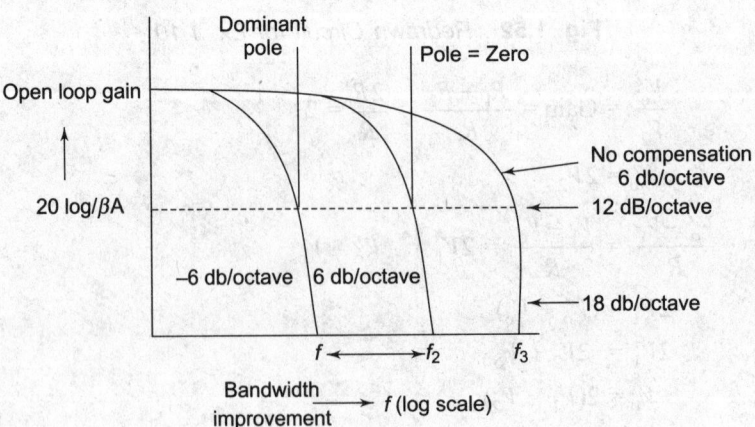

Fig. 1.55 *Frequency Response*

Example 1.12 *Explain the advantages and disadvantages of ICs. List typical specifications of an IC op-amp.*

All the components in each integrated circuit are fabricated on the same chip. ICs have become a vital part of modern electronics circuit design. They are used in the computer industry, automobile industry, home appliances, communications, and central systems where they permit minimisation and superior performance not possible with discrete components. They provide long, trouble-free service and are economical. Digital ICs are used to form circuits as gates, counters, multipliers, shift registers, and so on. Linear ICs are equivalent to discrete transistor networks. They are used for amplifier filters, modulations, integrators, timers, and other special purposes.

ICs (a) are small in size, (b) are low in cost, (c) have a low offset voltage, (d) have low offset current, (e) have high reliability, and (f) have good temperature tracking.

Disadvantages

(a) Fabrication of inductor with large value of quality factor Q in ICs has not been successful. Building of inductor with reasonable value and by IC has not been successful.
(b) Integrated resistor and capacitor have limited moderate values and are available with wide tolerance.
(c) Circuit adjustments are difficult.

Typical specifications of an IC amplifier are as follows.

$$\text{Open-loop gain } A_d = 50,000$$
$$\text{Input offset voltage } V_{io} = 1 \text{ mV}$$
$$\text{Input offset current } I_{io} = 10 \text{ nA}$$
$$\text{Input bias current } I_B = 100 \text{ nA}$$
$$\text{CMRR } (\rho) = 100 \text{ db}$$
$$\text{PSRR} = 20 \,\mu V/V$$
$$I_{io} \text{ drift} = 0.1 \text{ nA/}°C$$
$$V_{io} \text{ drift} = 1.0 \,\mu V/°C$$
$$\text{Slew rate} = 1 \text{ V/}\mu sec$$

ESSAY-TYPE QUESTIONS

1. Define and explain various op-amp parameters.
2. How do you experimentally determine the values of various op-amp parameters? Explain.
3. Derive the impervious to show that an op-amp can be used as (i) adder, (ii) subtractor, (iii) scale changer, (iv) integrator, (v) differentiator, and (vi) voltage follower.
4. Give the ideal values and practical values of various op-amp parameters.
5. Explain frequency compensator techniques used in op-amps.
6. Explain the advantages and disadvantages of op-amps.

OBJECTIVE-TYPE QUESTIONS

1. An op-amp is so named because _____.
2. The values of A_{VOL} and BW for an ideal op-amp are _____.
3. If $V_i = 1$ V, $A_{VCL} = 100$, V_o of op-amp with $+V_{cc} = 10$ V and $-V_{ce} = 10$ V is _____.
4. The value of PSRR and CMRR for ideal op-amp are respectively _____.

5. The op-amp voltage follower circuit is also known as _____.

6. The temperature range in which 741C op-amp is used is _____.

7. Offset adjustment in an op-amp is done with the pin numbers _____.

8. The maximum values of $+V_{cc}$ and $-V_{cc}$ that can be given to op-amp are _____.

9. The expression for input offset voltage drift is _____.

10. If $V_o = V_{in} \sin \omega t$, the expression for slew rate (SR) is _____.

ANSWERS

1. Various mathematical operations like addition, subtraction, scale changing, integration, and so on can be done.

2. ∞ and ∞

3. $V_o = 9$ V

4. 0 and ∞

5. Buffer or unity gain follower

6. 0°C to 70°C

7. 1 and 5

8. +15 V and −15 V

9. $\Delta V_{io}/\Delta T$

10. $V_o = V_{in} \sin \omega t$; $SR = \left.\dfrac{dV_o}{dt}\right|_{max} = 2\pi f V_m \dfrac{V}{\mu \sec}$

PROBLEMS

Prob. 1. Show that

$$e_0 = (e_1 + e_2) - (e_3 + e_4)$$

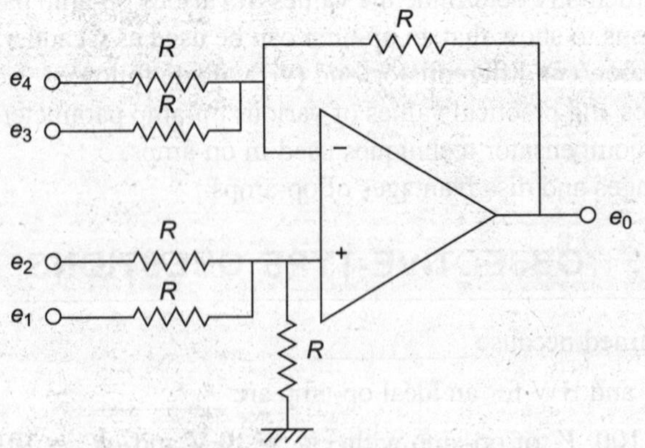

Fig. 1.56   *For Prob. 1*

Prob. 2. Show that a current flows through R_2 which is independent of R_2 and is equal to $-e_1/R_1$.

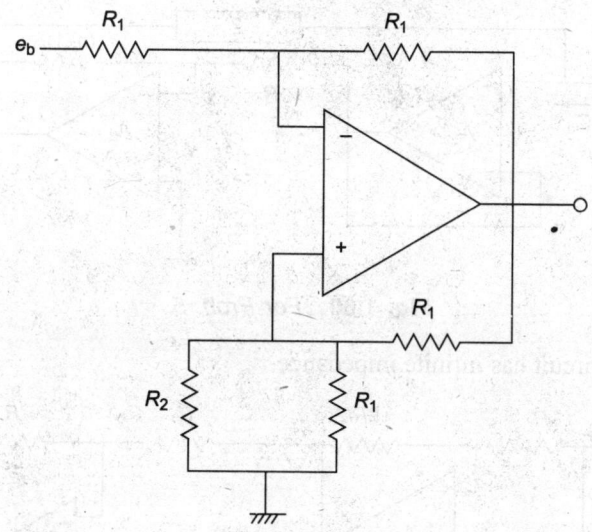

Fig. 1.57 *For Prob. 2*

Prob. 3. Show that the circuit generates through the floating load R_4, a current that is independent of R_4

• and equal to $\dfrac{e_1 R_2}{R_1 R_3}$. Assume $R_2 \gg R_3$

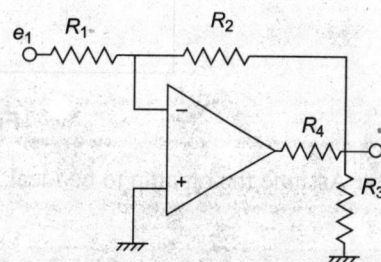

Fig. 1.58 *For Prob. 3*

Prob. 4. Show that

$$Z_{in} = \frac{A_0 R_i}{\left(1 + \dfrac{R_2}{R_1}\right)}$$

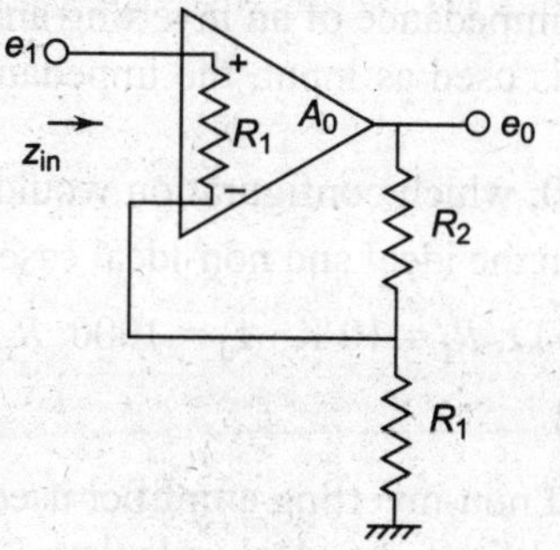

Fig. 1.59 *For Prob. 4*

Prob. 5. Show that Z_{in} can be varied by varying the resistor $(n - 1)R$ and $Z_{in} = Z/n$.

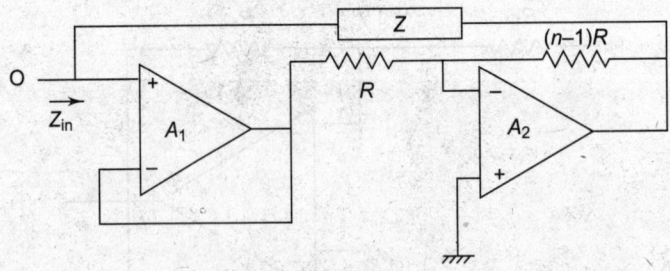

Fig. 1.60   *For Prob. 5*

Prob. 6. Show that the circuit has infinite impedance.

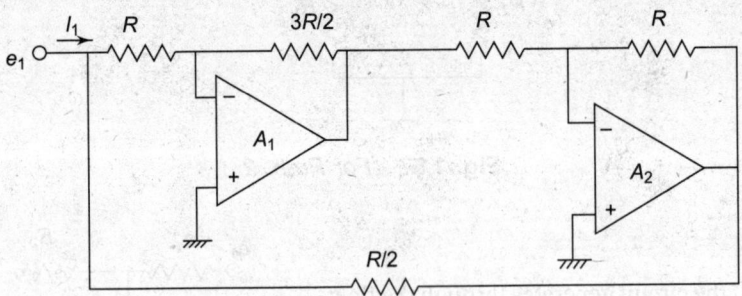

Fig. 1.61   *For Prob. 6*

Note: Assume the op-amp to be ideal.

SELF-ASSESSMENT QUESTIONS

1. What is the open-loop gain of op-amp (741)?
2. What is value of CMRR of 741?
3. What is PSRR? What is the unit in which it is expressed?
4. What is slew rate? What is the value for 741?
5. What are the voltage and current offset? What are the typical values for μA 741 op-amp?
6. On what value does the input impedance of an inverting amplifier depend upon?
7. If the non-inverting terminal is used as input, the impedance looking into the terminals will be higher or lower?
8. If the closed-loop gain is 1000, which configuration would you prefer?
9. What is the input impedance in the ideal and non-ideal cases if an inverting amplifier is used?

$$R_1 = 1000 \ \Omega, \ R_f = 10 \ K, \ A_d = 1000, \ R_i = 500 \ K$$
$$R_0 = 10 \ \Omega$$

10. What is the input impedance if non-inverting amplifier used with the above values?
11. Assuming the operational amplifier to be ideal, calculate V_o for circuits a to g.

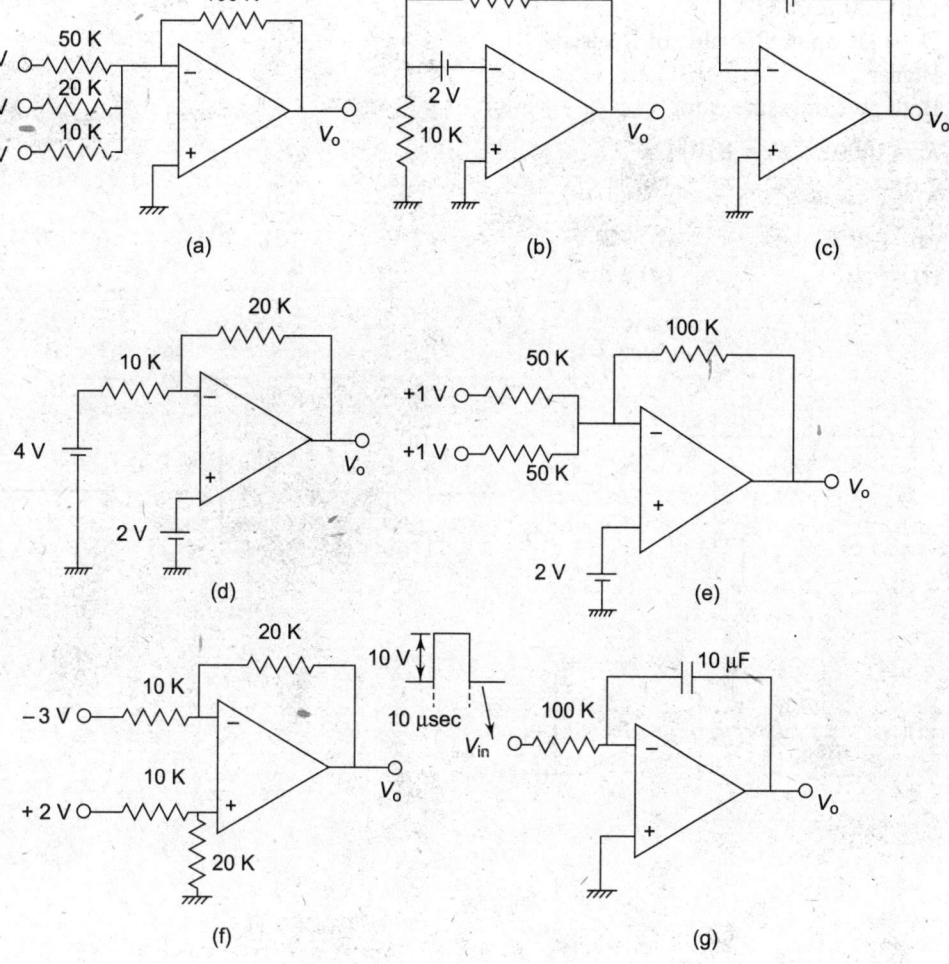

Fig. 1.62 *Op-amp Circuits*

12. Assuming the operational amplifier to be ideal, find the values of circuit for (i), (ii) and (iii).
    (i) Amp voltage gain = – 5 and input resistance 100 K
    (ii) Amp voltage gain = – 20 and input resistance 2 K
    (iii) Voltage gain = + 100 and high gain input resistance

ANSWERS

1. 2,00,000
2. 90 db

3. $\left.\dfrac{\Delta V_{io}}{\Delta V_{oc}}\right|_{\text{other } V_{cc} \text{ kept contant}}$ μV/V

4. The maximum rate of change of the output voltage when supplying the rates O/p. 0.5 V/μsec

5. Voltage and current required to make output zero for no signal condition.
   2 mV, 20 A.
6. Depends upon the value of R_1 used.
7. Higher
8. Both give the same gain.
9. $R_i = 1000\ \Omega$, $R_i = 1010\ \Omega$
10. $R_i \approx 50\ M\Omega$
11. (a) -6 V   (b) $+6$ V   (c) -2 V   (d) -2 V   (e) $+6$ V
    (f) $+10$ V   (g) 1 mV

12. (i)

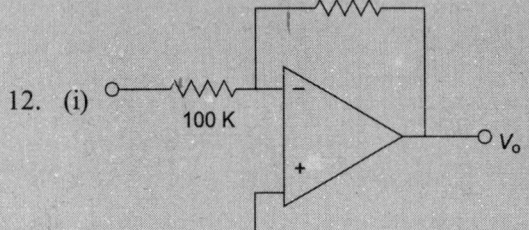

(ii)

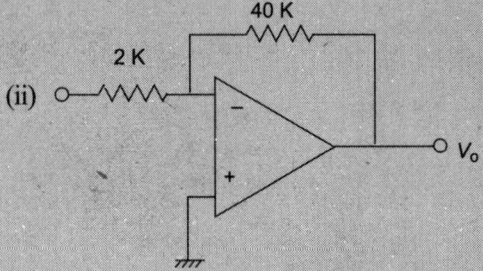

(iii)

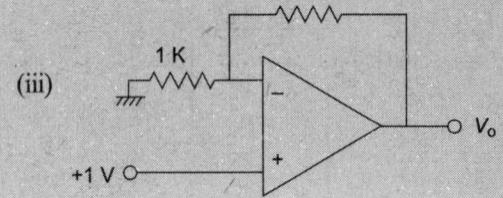

ASSIGNMENT

For the given 741 circuit, find the various stages and discuss its working briefly.

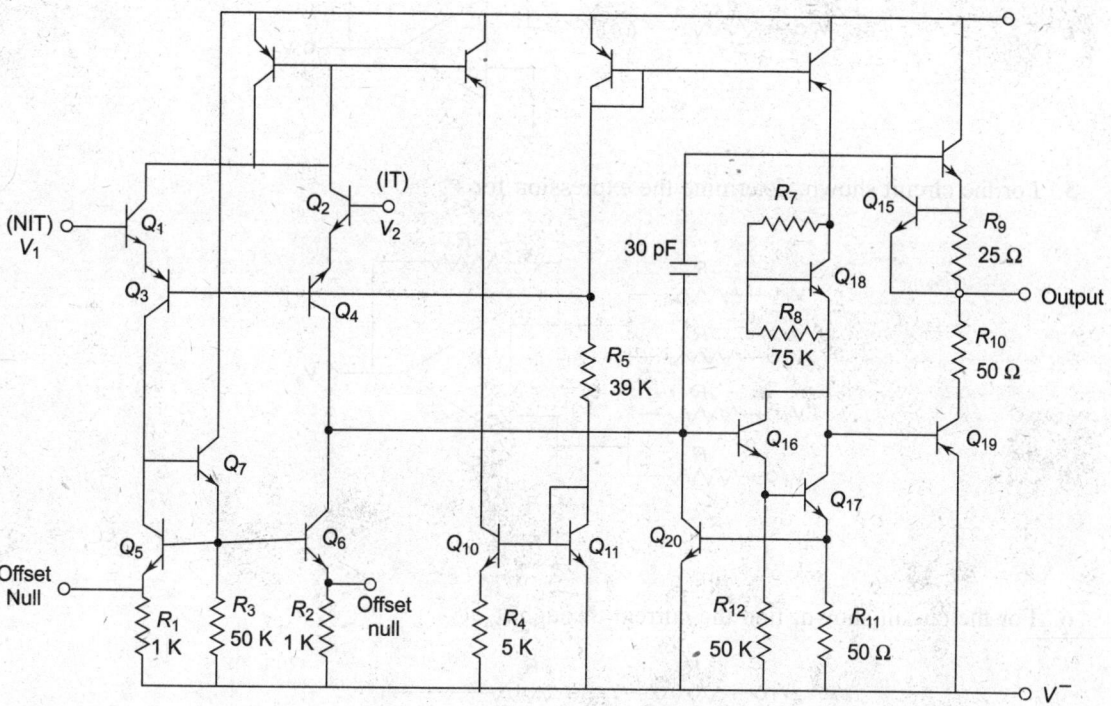

Fig. 1.63 *Op-amp 741 Internal Circuit*

UNSOLVED PROBLEMS

1. For an op-amp integrator circuit, sine wave input of 4 Sin 1000 t is given. In the circuit, $R = 200\ M\Omega$ and $C = 0.1\ \mu F$. Determine the value of output voltage.

2. Sketch the output waveforms, if a square wave input is applied to op-amp differentiator circuit. Given $R = 0.1\ K\Omega$, $C = 0.01\ \mu F$. Frequency of the square wave input is 100 Hz.

3. If a triangular waveform is applied to an op-amp in open-loop configuration, with non-inverting terminal grounded, sketch the output waveform.

4. Derive the transfer function for the circuit shown.

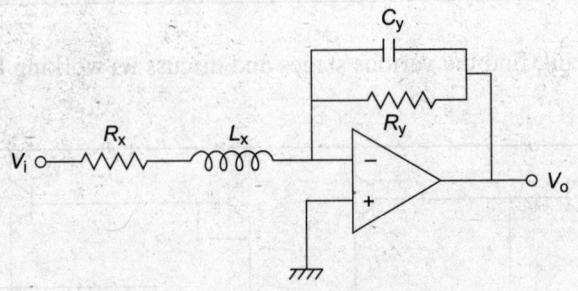

5. For the circuit shown, determine the expression for V_o.

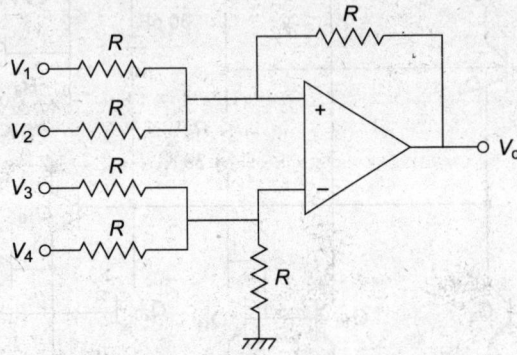

6. For the circuit shown, find the current through R_4.

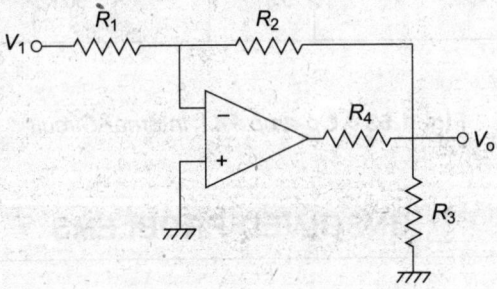

Given: $R_1 = 10$ kΩ; $R_2 = 20$ kΩ; $R_3 = 1$ kΩ, $R_4 = 4.8$ kΩ; $V_1 = 1$ V

7. Determine the value of V_o for the circuit shown, if $V_1 = 2$ V and $V_2 = 1$ V.

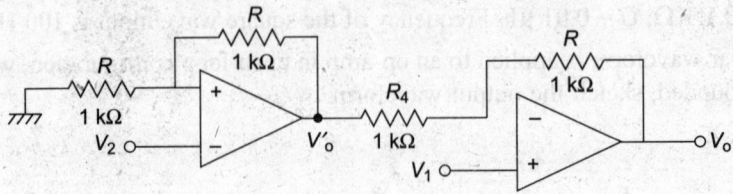

8. For the circuit shown considering ideal op-amp, determine the value of output voltage for the given input voltages $+ V_{cc} = 15$ V, $- V_{cc} = - 15$ V.

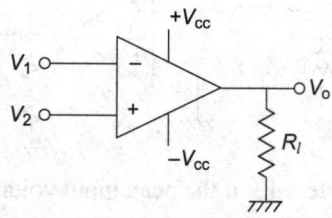

$$A_{OL} = 100,000$$

   (i) $V_1 = 10$ μV, $V_2 = 15$ μV
   (ii) $V_1 = 20$ μV, $V_2 = 10$ μV
   (iii) $V_1 = - 5$ μV, $V_2 = + 5$ μV
   (iv) $V_1 = + 10$ m, $V_2 = - 5$ μV

9. For the circuit shown, calculate I, V_o, and A_{VCL}.

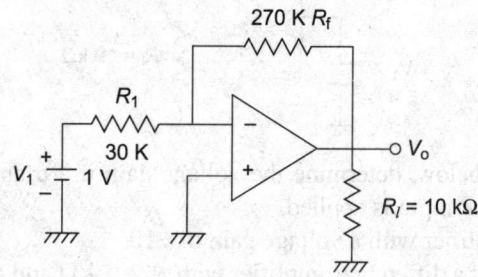

10. For the above circuit shown in Problem 9, if $R_1 = 47$ kΩ, determine I_1 and total current into the output pin of op-amp.

11. For the circuit shown, $R_f = 48$ kΩ, $R_1 = 10$ kΩ. Calculate the closed-loop voltage gain A_{VCL}.

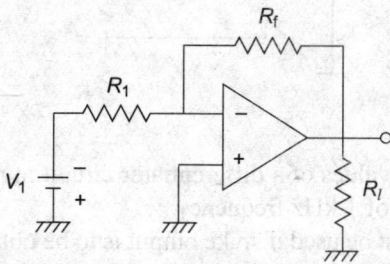

12. For the above circuit shown in Problem 11, determine R_1 for a load current of 3 mA, I_0, and circuit input resistance.

13. For the circuit shown, $R_f = 470$ kΩ. $R_1 = 10$ kΩ, $V_1 = -0.5$ V. Calculate closed-loop voltage gain A_{VCL}.

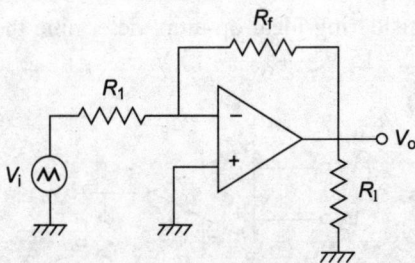

14. For the circuit shown for Problem 13, if the peak input voltage is –5 V, determine the peak output voltage.
15. Design an amplifier with a gain of –30. The input resistance R_{in} must be equal to or greater than 20 kΩ.
16. Design a two-input (three-channel) inverting amplifier.
17. For the circuit of three-input inverting amplifier, if $R_1 = R_2 = R_3 = 68$ kΩ, and $R_f = 136$ kΩ, find V_o, if $V_1 = +1$ V, $V_2 = 2$ V, and $V_3 = -1$ V.
18. For the circuit shown, determine V_o, I_L, and I_o.

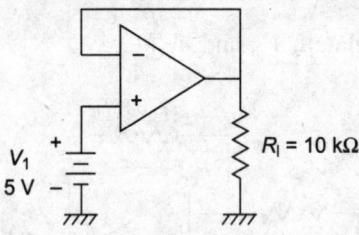

19. For the circuit shown below, determine the voltage gain if a triangular wave input of 200 Hzs frequency with 1 V peak input is applied.
20. Design an op-amp amplifier with a voltage gain of +10.
21. Determine the output of a difference amplifier with $R_1 = 1$ kΩ and $R_f = 10$ kΩ. $V_1 = V_2 = 5$ V DC
22. For the circuit shown, determine V_o, given $V_{in} = 1$ V DC, $R_f = 12$ kΩ, $R_1 = 2.2$ kΩ.

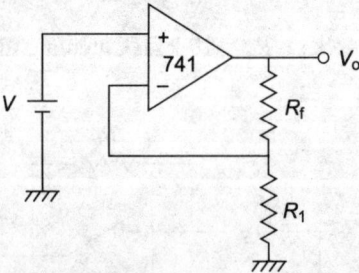

23. Determine the component values of a differentiator circuit to perform true differentiation, when the input is a square wave of 1 kHz frequency.
24. What value of R and C must be used if spike output is to be obtained from a square wave input of 5 kHzs frequency? Draw the circuit.
25. Determine the component values of R and C to perform true integration, when the input given is a square wave of 1 kHzs frequency.
26. It is desired to get triangular waveform output from square wave input of 5 kHzs. Draw the circuit and give values of R and C.

Op-amp Applications

Chapter

2

Objectives:

In this chapter...

- ■ *Op-amp application for DC and AC voltage and current measurements is explained.*
- ■ *Op-amp circuits as voltage reference sources and V/I converters are explained.*
- ■ *Half-wave rectifier (HWR) and full-wave rectifier circuits using op-amp are explained. By using op-amp for rectification, the reader is expected to realize the importance of op-amp in rectifying voltage signals below the cut-in voltage V_γ of the diode. Hence, the name super diode circuit.*
- ■ *That the $V_o - V_i$ characteristic of HWR circuit with op-amp is linear is explained. Hence, the name linear rectifier circuit until normal diode circuit which is non-linear.*
- ■ *Realization of $V_o = log\ (V_i)$ and $V_o = antilog\ (V_i)$ relationships through op-amp logarithmic and antilogarithmic amplifiers, respectively. The expressions for the relationships are derived.*
- ■ *Using two op-amps, one as Schmitt trigger and the other as integrator, the generation of triangular and rectangular waveforms is explained. The expression for frequency of waveforms and design of such circuits is explained.*

2.1   OP-AMP CIRCUITS

When negative feedback is given to op-amp, it can be used as
  (a) Current source
  (b) Voltage source
  (c) Summer or adder
  (d) Subtractor
  (e) Integrator
  (f) Differentiator
  (g) Logarithmic amplifier
  (h) Antilogarithmic amplifier
  (i) Active filters

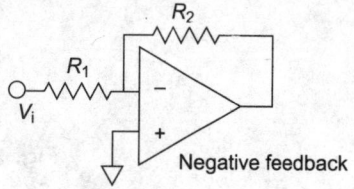

Fig. 2.1   *Op-amp Closed-Loop Circuit*

Positive feedback is used to develop circuits for
  (a) Multivibrators
  (b) Sinewave oscillators
  (c) Triangular waveform generators

2.2 OP-AMP FOR VOLTAGE AND CURRENT MEASUREMENTS

An ideal voltmeter should have ∞ resistance/impedance because it should not draw any current from the voltage source whose V is to be measured and thus reduce the voltage being measured.

$$I = \frac{V}{R_m} \qquad R_m \approx \infty$$

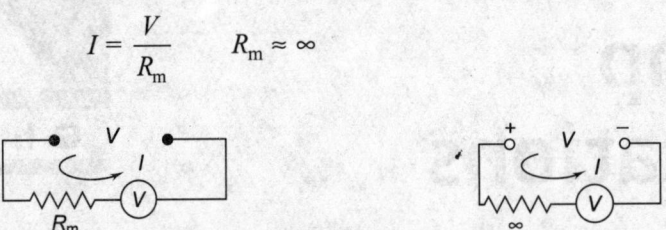

Fig. 2.2 *Voltage Measurement*   **Fig. 2.3** *Current Measurement*

An ideal ammeter should have zero resistance because there should not be any voltage drop across the ammeter, thus changing the current passing through the circuit (Figs. 2.2 and 2.3).

By using an op-amp circuit along with a moving coil meter with finite Z, the effect of the meter on the measurement can be reduced. The resolution of this circuit for measurement will be the same as the resolution of the meter. But the negative effects of the meter on the measurement will be eliminated. A moving coil meter is a current-sensitive device. The current flowing through it causes deflection, which is calibrated in terms of voltage. So R_m will be less.

2.3 MEASUREMENT OF DC VOLTAGE

A moving coil meter is used to measure voltage. Current flowing through it causes a deflection. The deflection is proportional to voltage. The meter resistance R_m of the moving coil meter is small.

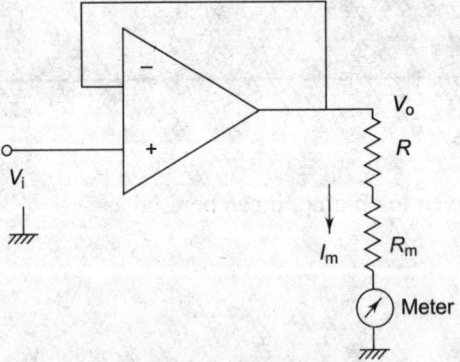

Fig. 2.4 *Measuring DC Voltage Directly*

Here op-amp is used as a voltage follower. It has high input Z. Since the meter is not directly connected to the output V_i, there will be no loading effect. The current flowing through the meter I_m is proportional voltage (V_i) (Fig. 2.4).

$$I_m = V_o/R + R_m, \qquad V_o = V_i$$

R is selected for voltage scaling.

$$R \gg R_m$$

If the op-amp is used with the same gain, very small V_i of the order of μV can be measured using the same meter.

For the source V_i, op-amp offers very high input $|z|$ because of voltage followers configuration. So, the loading effect is reduced while resolution remains the same.

Another method to measure DC voltage is shown in Fig. 2.5.

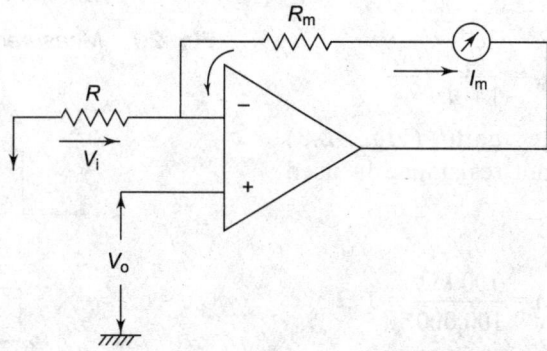

Fig. 2.5 *Measurement of Current*

Here, the meter is connected in a feedback loop. As a result of the negative feedback, the voltage at the inverting terminal will be the same as the voltage at the non-inverting terminal because $V_d \simeq 0$.
So, voltage at the inverting terminal is V_i
Therefore, I_m, the current in the circuit (Fig. 2.6), is

$$I_m = \frac{V_i}{R}; \ R \gg R_m$$

$$R_m \simeq 0$$

$$I = \frac{V_i - (-V_o)}{R_f}$$

$$I = \frac{V_i + V_o}{R_f}$$

$$V_o = AV_i$$

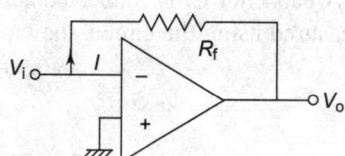

Fig. 2.6 *Measurement of Current*

$$R_i = V_i/I; \ \text{Therefore,} \ R_i = \frac{R_f}{1 + A}$$

2.4 MEASUREMENT OF DC CURRENT

Current to voltage converter or current amplifier.

For the measurement of currents, the circuit should have low input Z. A voltage follower has high input Z. The circuit shown below in Fig 2.7 has low input Z (inverting voltage feedback).

$$r_{in} = \frac{R_f}{A_v}$$

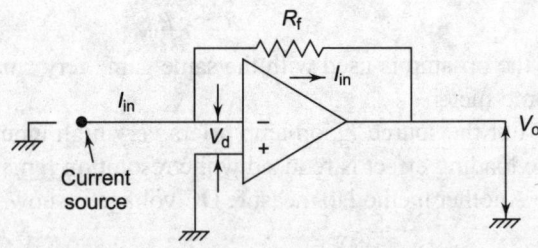

Fig. 2.7 *Measurement of DC Current*

So, $\quad\quad\quad r_{in} = \frac{V_d}{r_{in}}; V_d \simeq 0$

V_d is differential input voltage

Therefore, $\quad\quad\quad r_{in} \simeq 0$

and $\quad\quad\quad r_{in} = \frac{R_f}{1+A} \simeq \frac{R_f}{A}$.

If open-loop voltage gain (Fig. 2.8) $A_{VOL} = 100,000$, then input resistance in used configuration,

$$r_{in}(CL) = \frac{100\ k\Omega}{100,000} = 1\ \Omega$$

$$V_o = 100\ k\Omega \times i_{in}$$

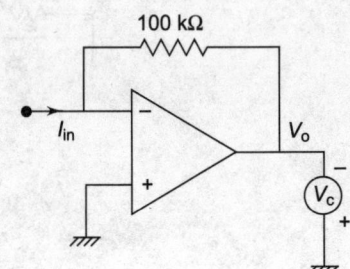

Fig. 2.8 *Measurement of Current*

V_o is a measure of input current i_{in}.

$$V_o = -I_{in}R_f$$

In the ideal case, $Z_{in} = 0;$ $\quad\quad Z_{in} = R_f \parallel R_{in}$

Input current I_{ON} is proportional to output voltage.

Output Z is very low because of 100 per cent feedback.

Therefore, it is also called as a current amplifier.

A capacitor C_f is connected across R_f to reduce high-frequency current noise. So the current can be measured using the circuit shown in Fig. 2.9.

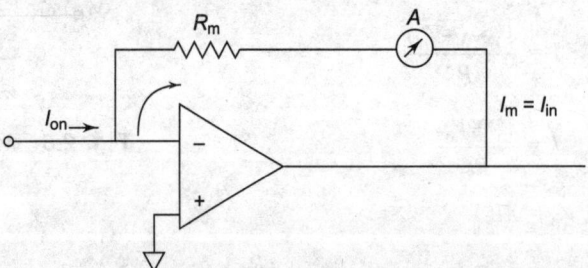

Fig. 2.9 *Measurement of Current*

The input resistance of the circuit as a whole will be less. It is $\frac{R_m}{A_v}$. Therefore, the circuit acts as an ideal ammeter.

The op-amp will not draw any current. So, all the input current flows into the feedback path through the meter. The circuit can be modified as shown in Fig. 2.10.

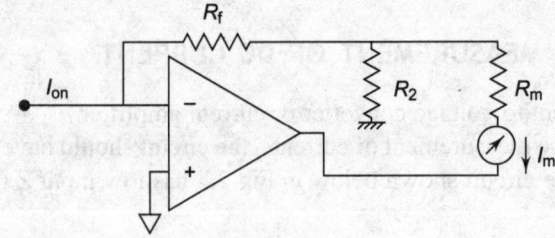

Fig. 2.10 *Circuit for Measurement of Current*

R_2 is connected between one terminal of the meter and the non-inverting terminal. The non-inverting terminal is at ground. So the other terminal of R_2 is also at ground. This circuit increases the range of the current that can be measured. Input current gets divided between R_2 and the meter. The meter can be calibrated to read I_{in}.

The range of measurement is increased by a factor $\left(1 + \dfrac{R_f}{R_2}\right)$.

2.5 AC MEASUREMENT

The DC measuring circuits for voltage and current can also be used for AC. A bridge rectifier circuit is incorporated in the feedback path. So, the meter deflection is proportional to the average value of the AC current. RMS value can also be measured by calibrating the scale to the RMS value (Fig. 2.11).

$$\frac{\text{RMS value}}{\text{Average value}} = \text{Form factor}$$

Square Waveform factor = 1.11

Therefore, RMS value is proportional to average value.

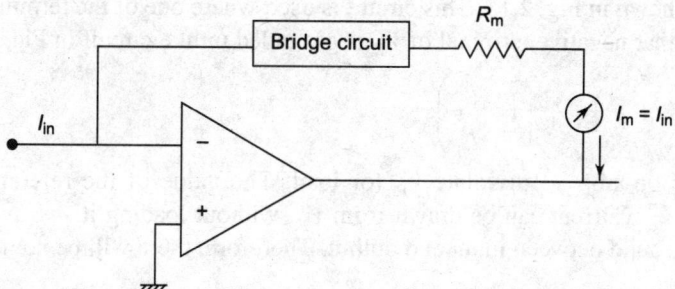

Fig. 2.11 *Circuit for AC Current Measurement*

2.6 OP-AMP REFERENCE VOLTAGE SOURCE

Op-amps can be used as a reference voltage source with very low output resistance and large output current capability. Op-amps have high input Z, and their gain can be adjusted. One such circuit is shown in Fig. 2.12. The reference source is isolated from the output terminal.

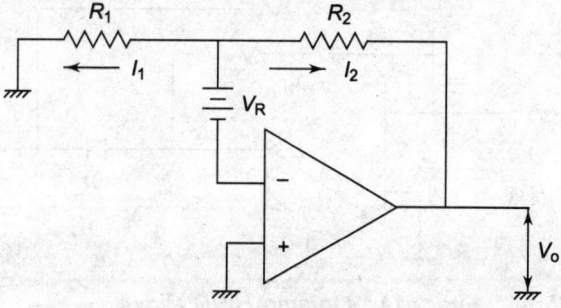

Fig. 2.12 *Op-amp Reference Voltage Source*

Current is not being drawn from the voltage source directly. So regulation will improve (Fig. 2.12).

$$I_1 = -I_2$$

$$\frac{V_o + V_R}{R_2} = \frac{-V_R}{R_1}$$

$$+\frac{V_o}{R_2} = V_R\left(\frac{1}{R_1} - \frac{1}{R_2}\right)$$

$$+ V_o = V_R R_2 \left(\frac{1}{R_1} + \frac{1}{R_2}\right)$$

$$+ V_o = V_R \left(1 + \frac{R_2}{R_1}\right)$$

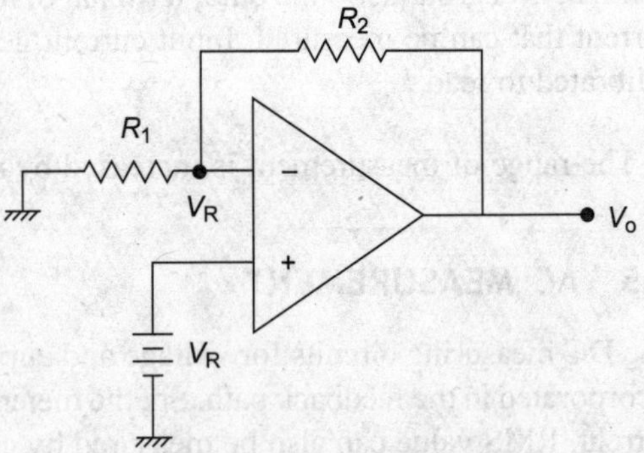

Fig. 2.13 *Reference Voltage Circuit*

The reference value can be changed by the ratio $\dfrac{R_2}{R_1}$. Output is isolated from V_R.

Another circuit is shown in Fig. 2.13. This circuit is used where one of the terminals of the supply is to be grounded. Note that negative terminal of V_R is grounded unlike circuit in Fig. 2.12.

$$+ V_o = V_R \left(1 + \frac{R_2}{R_1}\right)$$

The purpose of an op-amp is to isolate V_R for load. The value of the reference voltage is also increased from V_R to V_o. Current can be drawn from V_R, without loading it.

There is common ground between input and output. Therefore, there will be no noise grounding and shielding problems.

2.7 OP-AMP HALF-WAVE RECTIFIER (HWR) CIRCUIT

If a non-linear element such as a *p–n* junction diode is incorporated in the feedback path of the op-amp, the circuit becomes a non-linear one, half-wave rectifier (HWR) and full-wave rectifier (FWR) circuits are examples of such circuits (Fig. 2.14).

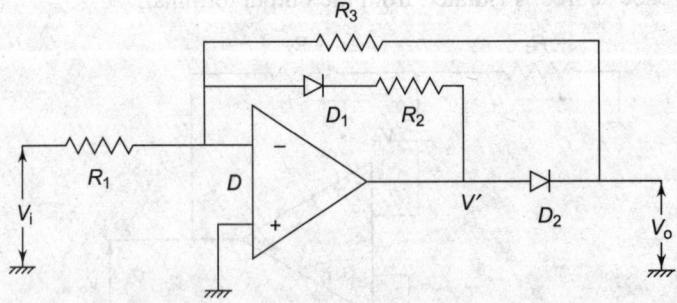

Fig. 2.14 *Op-amp HWR Circuit*

The threshold voltage for rectification is V_r open-loop gain of op-amp. This circuit combination is also called a *super diode*.

If V_d is the differential potential, $V' = A_v V_d$. So D_2 gets forward biased.

When V_i is negative, the output of the op-amp is positive because V_i is given to the inverting terminal. D_1 is off because the voltage V' at the cathode of D_1 is positive.

D_2 is forward biased because the voltage V' at the anode of D_2 is positive. So, the circuit operates like a normal inverting amplifier (Fig. 2.15).

$$\text{Gain} = R_3/R_1$$

D_1R_2 path is open.

D_2 gets forward biased because of the open-loop gain and finite V'_d. The current path is through R_3.

$V_i - V_e$. V_o is positive.

D_2 is forward biased, D_1 is reverse biased.

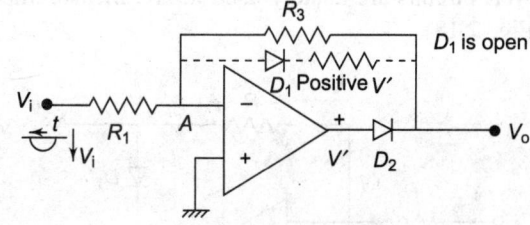

Fig. 2.15 *HWR Circuit*

V' is positive. So D_2 is forward biased.

V_i is positive, $V_o = 0$. D_2 is reverse biased, D_1 is forward biased.

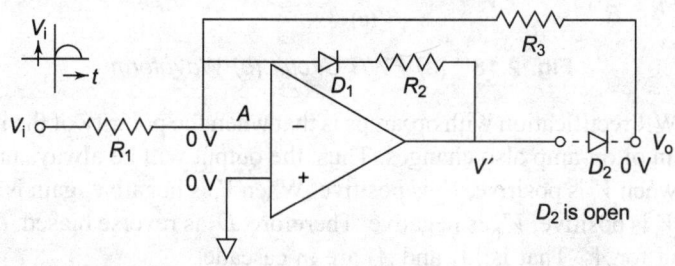

Fig. 2.16 *HWR Circuit*

When V_i is positive (Fig. 2.16), D_2 is reverse biased because V' is negative and the voltage at the anode of D_2 is negative. Therefore, current flows through R_2 and D_1 is forward biased. Output voltage V_o is derived via resistor R_3 from the inverting terminal point A. Voltage before R_1 is V_i. The voltage at point A of inverting terminal is 0 V because non-inverting terminal is at 0 V. Since D_2 is reverse biased, the op-amp is not connected to the output terminal V_o. Non-inverting terminal, which is at 0 V, is connected to V_o.

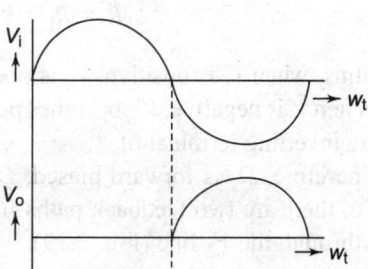

Fig. 2.17 *Waveforms*

So, $$V_o = 0 \text{ V}$$

Therefore, V_o when the input is positive is 0 V.

Thus, the output V_o is HWR output.

Without R_2, the gain will be the open-loop's and the output of the op-amp may go to saturation. So R_2 provides some finite gain.

If the gain $\dfrac{R_3}{R_1}$ is large, even small voltage of the order of mV and μV can be rectified.

Disadvantages:
1. Two diodes are used for HWR.
2. If V_o is more, op-amp will go to saturation.

2.8 OP-AMP FULL-WAVE RECTIFIER

FWR circuits are usually used in AC measurements. It gives the FWR output a higher average value (Fig. 2.18).

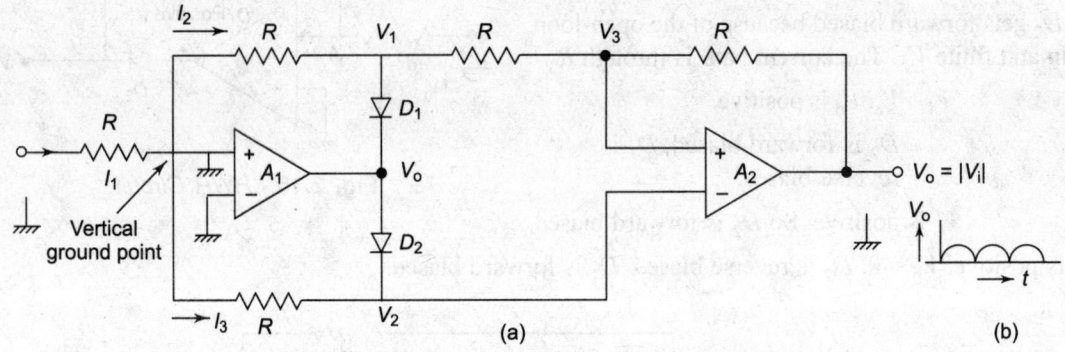

Fig. 2.18 *(a) FWR Circuit (b) Waveform*

The principle in FWR rectification with op-amps is that when the polarity of the input signal changes, the polarity of the gain of op-amp also changes. Thus, the output will be always unidirectional.

If gain is positive when V_i is positive, V_o is positive. When V_i is negative, gain is negative. Therefore, V_o is positive. When V_i is positive, V'_o is negative. Therefore, D_2 is reverse biased. D_1 is forward biased. Then V'_o is the output for A_2. That is, A_1 and A_2 are in cascade.

$$\therefore \qquad R_f = R_i = R, \text{ gain} = \frac{R}{R} = 1. \text{ For } A_1, \text{ gain} = 1, \text{ and for } A_2, \text{ gain} = 1.$$

Thus, when V_i is positive, V_o is positive. $V_o = V_i$.

When V_i is negative, V'_o becomes positive because it is given to the inverting terminal of A_1.

Therefore, D_2 is forward biased; D_1 is reverse baised.

So, there are two feedback paths for I_1, through the V_1 line and through the V_2 line (Fig. 2.19).

For an ideal circuit, $I_1 = I_2 + I_3$

For a positive V_i, $\dfrac{V_i}{R} = I_1$; $\qquad I_2 = +\dfrac{V_o}{R+R+R} = \dfrac{V_o}{3R}$

For a positive V_i, V'_o is negative.

Therefore, I_2 and I_3 are negative, $\quad I_3 = +\dfrac{V_2}{R}$

$$V_3 = \frac{2V_o}{3} \text{ potential divider networks.}$$

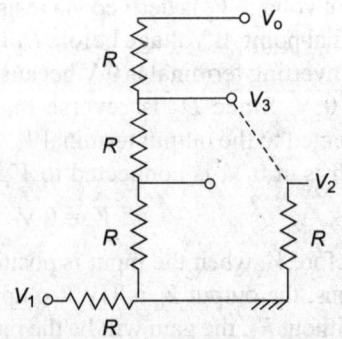

Fig. 2.19 *Equivalent Circuit, When V_i is positive, for FWR Circuit*

$V_2 = V_3$ (Since for A_2, negative is at V_3, positive is at V_2.)

So, $\qquad I_1 = I_2 + I_3$

That is, $\qquad \dfrac{V_i}{R} = +\dfrac{V_o}{3R} + \dfrac{2V_o}{3R} = +\dfrac{V_o}{R}$

So, $V_o = + V_o$ for positive V_i; $V_o = - V_i$ for negative V_i.

When V_i is negative, V'_o is positive and V_o is positive. Therefore, output =

When V_i is positive, V'_o is negative and V_o is positive.

$$V_o = V_i$$

Thus, full-wave rectification is obtained.

2.9 LINEAR RECTIFIER

If in the Fig. 2.20, $R_1 = R_2 = R_3$, when e_1 is negative, e_2 is positive.

Therefore, D_2 is forward biased, D_1 is reverse biased. Thus the path D_1R_2 is open.

Current will flow through R_3, D_2, e_2 path. Therefore, $e_0 = - e_1$. Even if e_1 is very small, less than the cut in voltage of the diodes, the gain of the op-amp will make D_2 forward biased because D_2 is connected in the forward biased path. Therefore, the input voltage of the order of μV can be rectified.

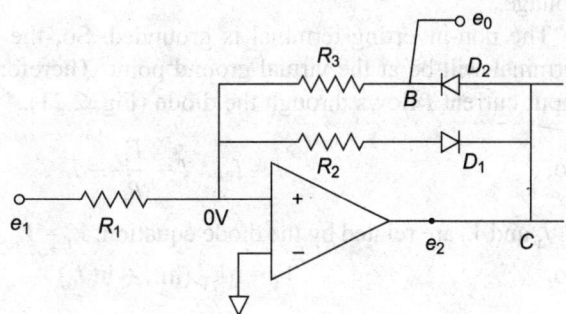

Fig. 2.20 *Linear Rectifier Circuit*

When e_1 is positive and e_2 is negative, D_2 is reverse biased and D_1 is forward biased. Since R_3 is connected to the virtual ground 0 V point, e_0 is 0 V.

Because e_0 is being taken at point B, which is now not connected to point C, the HWR output is obtained. The output is being taken from e_0 and not from e_2. But without an op-amp, μV cannot be rectified. In normal diode rectifier circuits, the diode does not conduct until $V_i > V_r$. As I increases, drop across the diode changes since its r changes. So, e_0 changes. It is a non-linear circuit. But the above circuit is a linear circuit. R_2D_1 path keeps the circuit symmetrical and prevents the op-amp from saturating in positive input half cycle.

2.10 LOGARITHMIC AMPLIFIER

In the case of a diode, $I = I_0\, e^{\frac{V}{\eta V_T}} - 1$. It is the diode current equation.

I_0 = reverse saturation current

η = constant, 1 for Ge and 2 for *Si*

$$\boxed{V_T = \text{Volt eq. of temperature } \dfrac{KT}{e}}$$

V = applied forward bias voltage (Fig. 2.21)

If V is large, $e^{\frac{V}{\eta V_T}} \gg 1.$

So, $I = I_0 \, e^{\frac{V}{\eta V_T}}$

$$\ln I = \ln I_0 + \frac{V}{\eta V_T}$$

$$\frac{V}{\eta V_T} = \ln I - \ln I_0$$

Therefore, $V = \eta V_T \, (\ln I - \ln I_0)$

If temperature T is constant, η, V_T, and I_0 will be constants. If the diode is connected in the feedback path of the op-amp, the output voltage will be a logarithmic function of the input voltage.

The non-inverting terminal is grounded. So, the inverting terminal will be at the virtual ground point. Therefore, all the input current I flows through the diode (Fig. 2.21).

So, $I = I_f, \quad I = \dfrac{V_i}{R_1} = I_f$

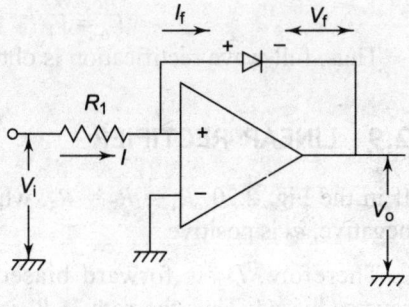

Fig. 2.21   *Rectifier Circuit*

I_f and V_f are related by the diode equation, $V_f = V_o$

So, $V_f = \eta V_T \, (\ln I - \ln I_0)$

$$V_f = \eta V_T \left(\ln \frac{V_i}{R_1} - \ln I_0 \right)$$

$$V_o = -V_f$$

Hence, $V_o \propto \ln (V_i)$

because of feedback current $I = I_f$. It changes exponentially with V.

If T changes, then I_0 will change and V_T will change. So, the logarithmic variation of V_o with V_i is valid at constant T. These are used in logarithmic voltmeters to measure voltage from 0 V to 1000 V on a log scale (with a single scale).

However, in the case of a diode, the log relation between *voltage and current is not valid over a very wide range due to the finite resistance in the diode*. So, in practice, instead of a diode a transistor is used.

If a transistor is operated with V_C, I_C, and V_{BE}, they can be related in the form (Fig. 2.22),

$$V_{BE} = K \ln (I_C),$$

where K is a constant, $K \propto T$

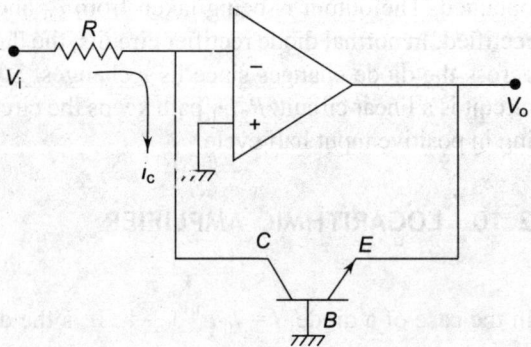

Fig. 2.22   *Using BJT for a Diode*

Collector is at virtual ground point. (Inverting terminal is at 0 V.) Base is grounded. Therefore, $V_{CB} \simeq 0$.

$$V_i = I_C R$$

$$V_o = V_{BE}$$

So,

$$V_o = K \ln \left(\frac{V_i}{R} \right)$$

Applications of logarithmic amplifiers.

1. In logarithmic voltmeters
2. In multipliers
3. In square root function generation, square function generation
4. In analog computers

2.11 ANTILOGARITHMIC AMPLIFIER

The circuit is shown in Fig. 2.23.

Voltage V_1 at the non-inverting terminal of op-amp (1) is

$$V_1 = \frac{V_i R_2}{R_1 + R_2} \tag{1}$$

Op-amp (1) is being used as differential amplifier $V_A \neq V_1$ because the I_f source is connected to the inverting terminal (Fig. 2.23). If R is connected, $V_A = V_1$. Voltage V_2 at the output of op-amp (1) is the difference of the voltages at the inverting and non-inverting terminals. When the diode D_1 is forward biased, feedback path is a short-circuit terminals. Therefore, gain = 1. $V_2 = V_1 - V_A$. V_A is not at virtual ground point.

$$V_2 = V_1 - V_A$$

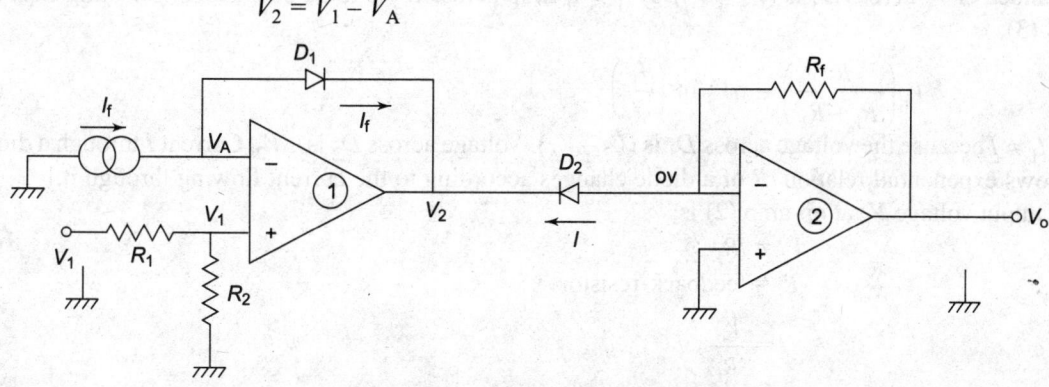

Fig. 2.23 *Antilogarithmic Amplifier*

So,

$$I_f = I_0 \; e^{\frac{V_o}{\eta V_T}}$$

Diode is forward biased, so it is in closed-loop configuration.

Since diode is short, $V_2 = V_1 - V_A$.

The voltage across the diode D_1 is the output of op-amp 1. It is V_2. Expression for V_2 is $\eta V_T [\ln (I_f) - \ln (I_0)]$. The cathode of D_1 need not be grounded. I_f is the current that flows due to potential difference $(V_A - V_2)$.

So,

$$V_2 = \frac{V_i R_2}{R_1 + R_2} - \eta V_T [\ln I_f - \ln I_0] \tag{2}$$

Therefore,

$$V_2 = V_1 - V_A$$

I_f is the current through diode D_1. Therefore, in terms of this current, the expression for $V_A = \eta V_T [\ln I_f - \ln I_0]$. $V_2 = (V_1 - V_A)$.

I is the current through diode D_1.

The voltage across it is V_2.

The symbol for voltage source or current source is ─⦶─, and the symbol for a current meter or voltage meter ─Ⓜ─.

If two inputs V_1 and V_2 are given to the negative and positive input terminals, it is differential amplifier configuration. If any one input is given, it acts as a virtual ground point.

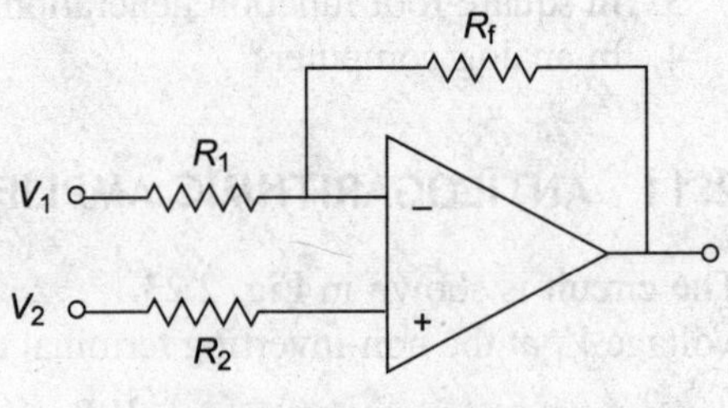

$$I = \boxed{I_0 \left(e^{-\frac{V_2}{\eta V_T}} - 1 \right)}$$

$$V_2 = -\eta V_T [\ln I - \ln I_0]$$

Eq. (3) is in terms of I and I_0.

Fig. 2.24 *Differential Amplifier*

Voltage across diode D_2 is

$$V_2 = -\eta V_T [\ln I - \ln I_0] \tag{3}$$

D_2 is reverse biased.

Voltage across diode D_2 is negative because the anode is at 0 V and cathode is at negative potential. Therefore, Eq. (2) is in term of I_f and I. Since the cathode of D_2 is at negative potential, it is as good as grounded or V_f across D_1 is $(V_A - V_2)$. If $I_f \neq I$, drop across D_2 is to be considered. Equating Eqs. (2) and (3),

$$V_i \left(\frac{R_2}{R_1 + R_2} \right) = \eta V_T \ln \left(\frac{I_f}{I} \right) \tag{4}$$

[$I_f \neq I$ because the voltage across D_1 is $(V_2 - V_A)$. Voltage across D_2 is $-V_2$. Current I through a diode follows exponential relation. R of a diode changes according to the current flowing through it.]

Output voltage V_o at op-amp (2) is,

$$V_o = I R_f \tag{5}$$

$$R_f = \text{feedback resistor}$$

So,

$$I = \frac{V_o}{R_f} \tag{6}$$

Substituting Eq. (6) in (4),

$$V_i \frac{R_2}{R_1 + R_2} = \eta V_T \ln \frac{I_f R_f}{V_o} \tag{7}$$

or

$$-V_i \frac{R_2}{R_1 + R_2} = \eta V_T \ln \frac{V_o}{I_f R_f} \tag{8}$$

Therefore,

$$V_o = R_f I_f \ln^{-1} \left[-V_i \frac{R_2}{(R_1 + R_2) \eta V_T} \right]$$

D_2 is forward biased. The anode of D_2 is at 0 V.

Therefore, the cathode is at negative potential.

Two types of differential amplifier circuit configurations and the output V_o from the circuits are indicated in Figs. 2.25 and 2.26.

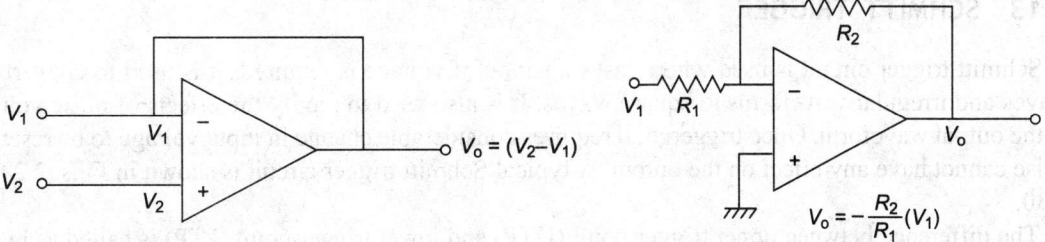

Fig. 2.25 *Differential Amplifier*   **Fig. 2.26** *Differential Amplifier*

If the diodes D_1 and D_2 are identical, the temperature-sensitive offset currents ηV_T and $\ln (I_s)$ are cancelled.

Log and antilog amplifiers are used in analog computers and instrumentation systems.

2.12 LOG MULTIPLIERS

The output e_0 is the product of the inputs e_1 and e_2. Such multipliers are used in analog computers. e_1 and e_2 can be DC voltages or AC signals (Figs. 2.27 and 2.28).

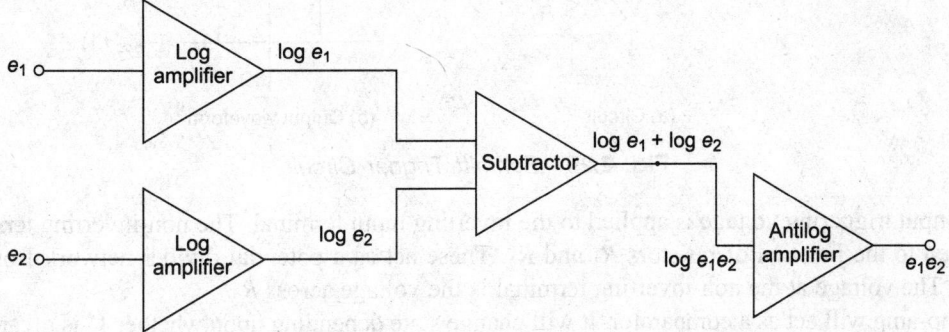

Fig. 2.27 *Log Multiplier Circuit*

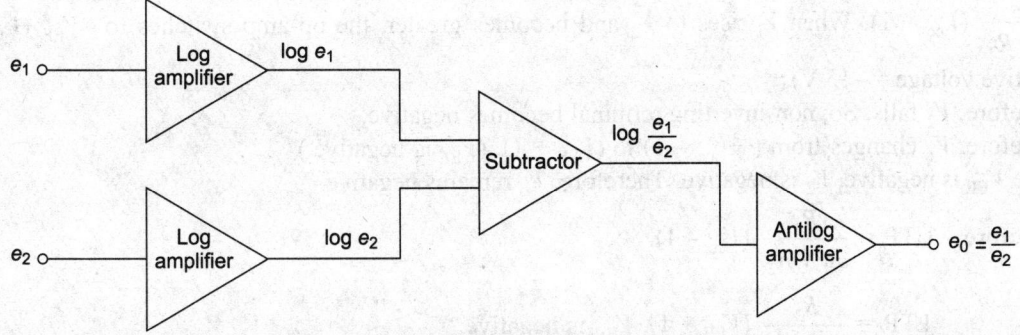

Fig. 2.28 *Log Divider Circuit*

Two input voltages are given.

2.13 SCHMITT TRIGGER

A Schmitt trigger circuit is used where fast transition of voltage is required. It is used to convert sine waves and irregular waveforms to square waves. It is also used to rectify the effects of noise voltages in the output waveform. Once triggered, it requires considerable change in input voltage to be reset. So, noise cannot have any effect on the output. A typical Schmitt trigger circuit is shown in Figs. 2.29 and 2.30.

The difference between upper trigger point (UTP) and lower trigger point (LTP) is called as hysteresis.

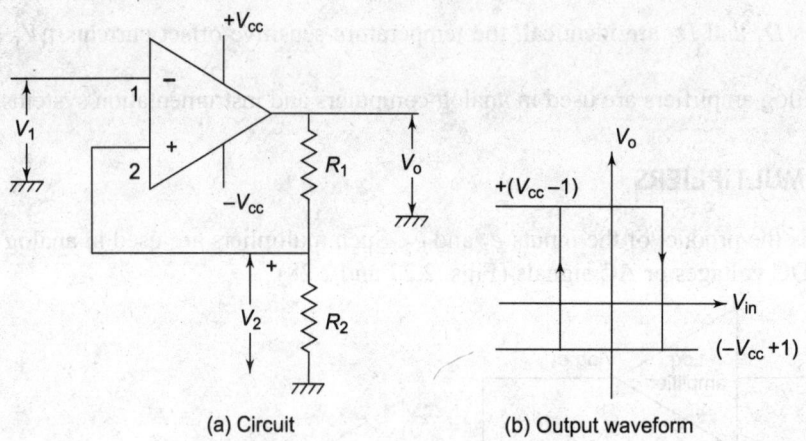

(a) Circuit   (b) Output waveform

Fig. 2.29   *Schmitt Trigger Circuit*

The input triggering voltage is applied to the inverting input terminal. The non-inverting terminal is connected to the junction of resistors R_1 and R_2. These act as a potential divider network from V_0 to ground. The voltage at the non-inverting terminal is the voltage across R_2.

The op-amp will act as a comparator. It will change state depending upon whether V_i is greater or V_2 is greater. When V_i is greater than V_2, the op-amp will saturate to the positive voltage value $\simeq (V_{cc} - 1)$ volts. One volt is the drop across the components of the op-amp. So the voltage at point 2 is $V_2 = \dfrac{R_2}{R_1 + R_2} (V_{cc} - 1)$. When V_i rises to V_2 and becomes greater, the op-amp switches to $-V_{EE}$ (V_{EE} is negative voltage $= -15$ V).

Therefore, V_2 falls. So, non-inverting terminal becomes negative.

Therefore, V_0 changes from $(+V_{cc} - 1)$ to $(V_{EE} + 1)$. (V_{EE} is negative.)

Since V_{EE} is negative, V_2 is negative. Therefore, V_0 remains negative.

Therefore, $\text{UTP} = \dfrac{R_2}{R_1 + R_2} (V_{cc} - 1)$.

$$\text{LTP} = \dfrac{R_2}{R_1 + R_2} (V_{EE} + 1), \; V_{EE} \text{ is negative.}$$

In the circuit, other terminal of R_2 is grounded. Instead, if it is connected to a reference voltage V_R, UTP and LTP can be varied by varying V_R. We can change UTP and LTP by changing R_1 and R_2 also (Fig. 2.30).

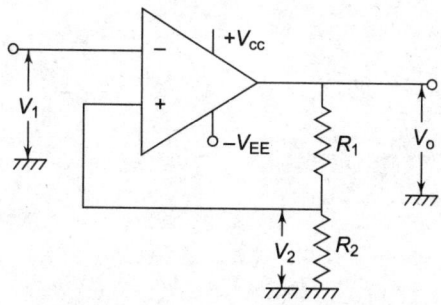

Fig. 2.30 *Schmitt Trigger Circuit*

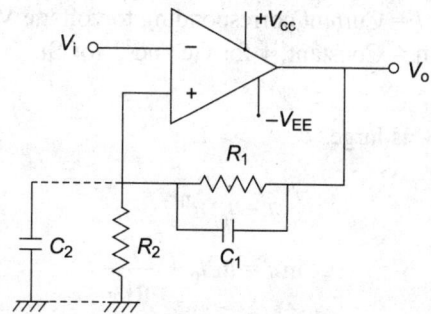

Fig. 2.31 *Schmitt Trigger Circuit*

2.13.1 Speed-Up Capacitor

C_1 in Fig. 2.31 is called a speed-up capacitor. C_1 will enable the lag network that is formed due to strong capacitance C_2 across R_2 to be cancelled. C_1 ranges from 10 pF to 100 pF.

To neutralise the stray capacitance, $R_1C_1 \geq R_2C_2$. So, $C_1 > \dfrac{R_2C_2}{R_1}$

2.13.2 Negative Feedback in Schmitt Trigger

Only positive feedback is used in Schmitt trigger circuit because it forces the feedback voltage to have the same polarity as the output. So we get UTP and LTP. Without feedback, hysteresis = 0. Hysteresis is needed to prevent noise causing false triggering.

2.14 LOGARITHMIC VOLTMETERS

Logarithmic voltmeter employs logarithmic amplifier. Logarithmic scales are employed where large voltages, such as those ranging up to 1,000 V, have to be measured on a single scale, or where measurement of small voltages is also important. The scale divisions will be non-uniform and will be divided according to the logarithmic scale.

Block diagram of logarithmic voltmeter is shown in Fig. 2.32.

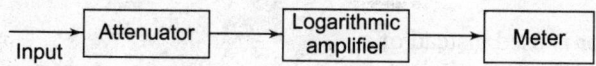

Fig. 2.32 *Logarithmic Voltmeter—Block Schematic*

2.14.1 Principle

In the case of a p–n junction diode, $I = I_0 \left(e^{\frac{V}{\eta V_T}} - 1\right)$

    I_0 = Reverse saturation current

    V = Applied voltage

I = Current corresponding to voltage V.

η = Constant, 1 for Ge and 2 for Si.

If V is large, $e^{\frac{V}{\eta V_T}} \gg 1$

So, $I = I_0\, e^{\frac{V}{\eta V_T}}$

$\ln I = \ln I_0 + \dfrac{V}{\eta V_T}$

Therefore, $\dfrac{V}{\eta V_T} = \ln I - \ln I_0$

So, $V = \eta V_T\,(\ln I - \ln I_0)$

If T is constant, η, V_T, I_0 will be constants. If the diode is connected in the feedback path of the op-amp, the output voltage will be a logarithmic function of the input voltage (Fig. 2.33).

$$I = I_f;\ I = \frac{V_i}{R_1} = I_f$$

I_f and V_f are related by the diode equation.

$V_f = V_o$

So, $V_f = \eta V_T\,(\ln I - \ln I_0)$

$$= \eta V_T \left(\ln \frac{V_C}{R_1} - \ln I_0 \right)$$

$V_o = -V_f$

Therefore, $V_o \propto \ln (V_i)$

If temperature changes, $\ln(I_0)$ will change and V_T will change.

So, the logarithmic variation of V_o with V_i is valid at constant T.

In practice, a transistor is used instead of a diode (Fig. 2.34).

$V_{BE} = K \ln(I_C)$ [where K is a constant.]

Collector is at virtual ground.

Therefore, $V_{CB} \simeq 0$

$$V_i = I_C R$$

$$V_o = V_{BE},$$

Therefore, $\boxed{V_o = K \ln\left(\dfrac{V_o}{R} \right)}$

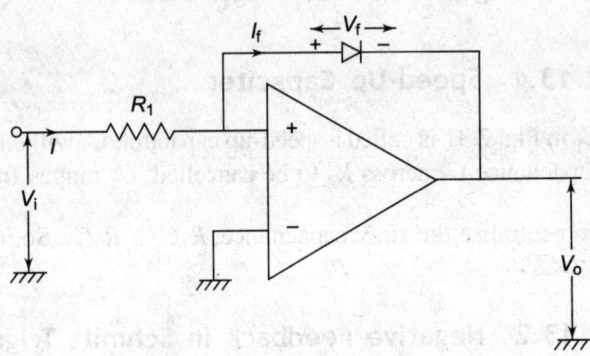

Fig. 2.33 *Logarithmic Amplifier*

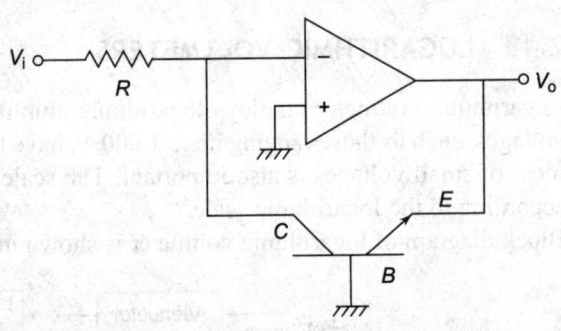

Fig. 2.34 *BJT Being Used as Diode*

2.15 TEXAS INSTRUMENTS TL441C IC

TL441C of Texas Instruments is logarithmic amplifier. Its specifications are as follows:

Input: 0.01 V to 1.0 V

If the input is in excess of 1 V, output will be distorted.

Output: <0.6 V

The output swing and the slope of the output response can be adjusted by verifying the gain by means of slope control.

$$+V_{cc} = +8 \text{ V}; \qquad -V_{cc} = -8 \text{ V}$$

Output sink current: 30 mA

Continuous total disruption: 500 mW (below 70°C)

TL440: Zero-voltage switch. It is the combination of a threshold detector and zero-crossing detector. It is used for AC power control circuits. It allows a triac or SCR to be fired when the AC signal crosses zero volt.

Using log and antilog amplifier, explain how the square root function can be realized.

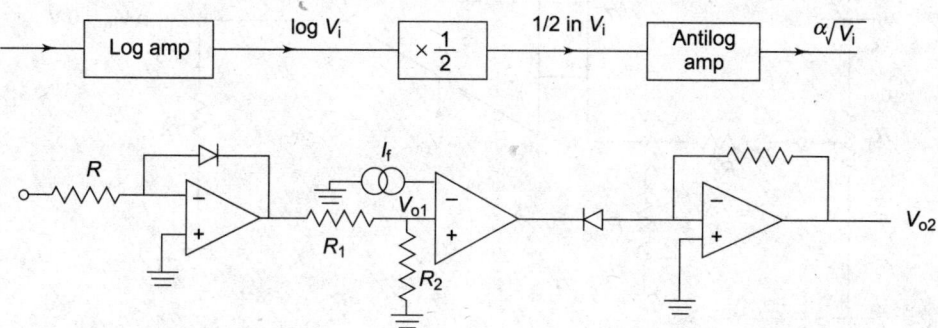

Fig. 2.35 *Realising Square and Square Root Functions*

The output of the log amp $V_{o1} = -\eta V_1 \ln \left[\dfrac{V_i}{RI_0} \right]$

The output of the antilog amp is $V_{o2} = \dfrac{R_F}{I_f} e^{-V_{o1}} \left(\dfrac{R_2}{R_1 + R_2} \right) \dfrac{1}{\eta V_1}$

So, $$V_{o1} = -\eta V_T \ln \left[\frac{V_i}{RI_0} \right]$$

and to make $V_{o1}' = \dfrac{1}{2} V_{o1}$, take $R_1 = R_2 = R$

Therefore, $-V_{o1} = \dfrac{R_2}{R_1 + R_2} \left(\dfrac{1}{\eta V_T} \right) = \left[-\eta V_T \ln \left(\dfrac{V_i}{RI_0} \right) \times \left(-\dfrac{1}{2} \dfrac{1}{\eta V_T} \right) \right]$

$$= \frac{1}{2} \ln \left[\frac{V_i}{RI_0} \right] = \ln \sqrt{\frac{V_i}{RI_0}}$$

$$V_{o2} = \frac{R_F}{I_f} \ln \sqrt{\frac{V_i}{RI_0}} = \frac{R_F}{I_f} \sqrt{\frac{V_i}{RI_0}}$$

$$V_{o2} = k\sqrt{V_i}$$

$$\boxed{V_{o2} \; \alpha \; \sqrt{V_i}}$$

2.16 OFFSET VOLTAGE-COMPENSATING DESIGN

Offset is due to the mismatch of the two input transistors in the differential amplifier configuration. The compensating network is shown in Fig. 2.36.

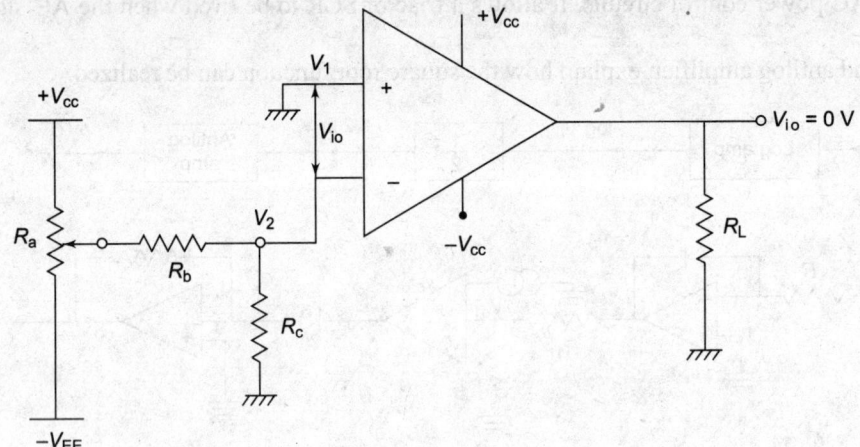

Fig. 2.36 *Offset Voltage Compensation*

R_a = potentiometer

R_b, R_c = resistors

R_{max} = maximum resistance of the potentiometer.

$$R_{max} = \frac{R_a}{2} \| \frac{R_a}{2} = \frac{R_a}{4}$$

$$V_2 = \frac{V_{max} R_c}{R_{max} + R_b + R_c}$$

$V_1 - V_2 = V_{io}$ $V_1 = 0$ because it is grounded.

So,

$$V_{io} = \frac{R_c}{R_{max} + R_b + R_c} V_{max}$$

$V_{max} = V_{ce} = V_{EE}$ maximum values.

$R_{max} + R_b + R_c \simeq R_{es}$

The values must be such that R_b is very large compared to R_{max} and R_c. R_c must be very small.

So,

$$V_{io} = \frac{R_c V_{max}}{R_b}$$

$$V_{max} = V = |V_{cc}| = |-V_{EE}|$$

$$V_{io} = \frac{R_c V}{R_b}$$

Therefore, V_{io} depends on the magnitude of V_{cc} and V_{EE}.

$$V_{io} \propto V$$

Design Criteria: $R_b \gg R_c$, $R_c \simeq 10\ \Omega$

$$R_b = 10\ R_{max}$$

Example 2.1   *Design a compensatory network for an op-amp using ±15 V supply voltages with* $V_{io} = 10\ mV$, *maximum.*

$$V_{io} = \frac{R_c V_{max}}{R_b}$$

$$V_{max} = 15\ V$$

$$10\ mV = \frac{R_c 15\ V}{R_b}$$

Choose R_c as $10\ \Omega$

So, $$R_b = \frac{10 \times 15}{10 \times 10^{-3}} = 15\ k\Omega$$

Let $R_b = 10\ R_{max}$

Therefore, $$R_{max} = \frac{R_b}{10} = \frac{15\ k\Omega}{10} = 1.5\ k\Omega$$

2.17 OP-AMP CLIPPER CIRCUITS

For positive half cycle of input, op-amps can also be used in clipping or wave-shaping circuits. These circuits are shown in Fig. 2.37, and Fig. 2.38 is inverted. V_o is negative. Diode is reverse biased (Fig. 2.37).

So, V_o is in a negative half cycle side (Fig. 2.38).

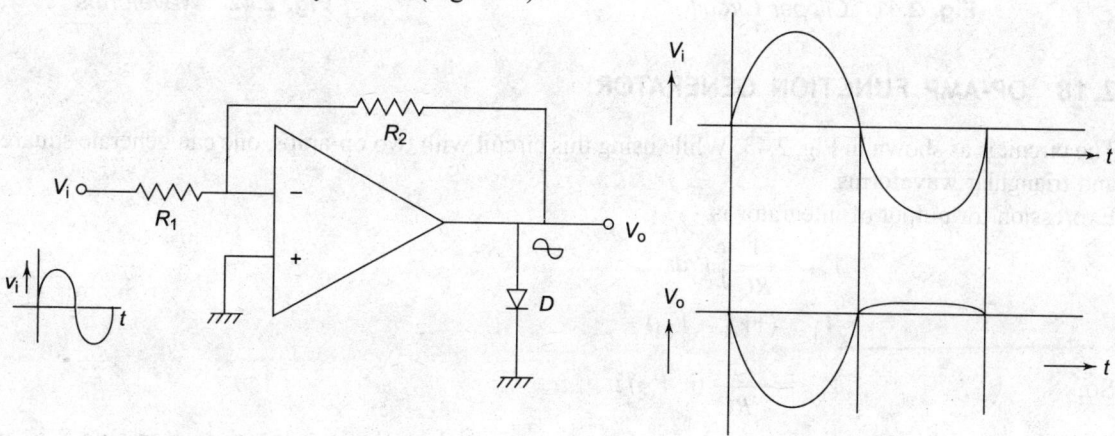

Fig. 2.37 *Clipper Circuit*   **Fig. 2.38** *Waveforms*

For a negative half cycle of input, V_o is positive. Diode D is forward biased.
The diode is shorted, $V_o = V_f = 0$. So, the output waveform for the negative half cycle of input is almost zero (shown in Fig. 2.39).

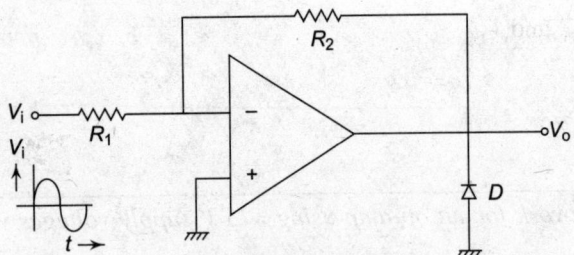

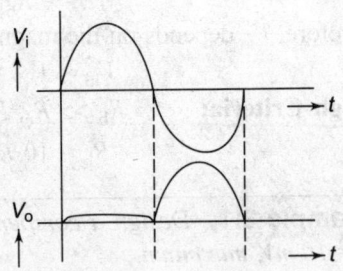

Fig. 2.39   *Clipper Circuit*   **Fig. 2.40**   *Waveforms*

If the diode D is connected as shown in Fig. 2.39, the output waveforms are as shown in Fig. 2.40.

If the op-amp clipper circuit is as shown in Fig. 2.41, with bias supply of 2 V, the diode will get forward biased only when the output V_o is more than +2 V. Till that time the diode is reverse biased. Once the diode is forward biased, output is constant at +2 V as shown (neglecting the drop across the diode).

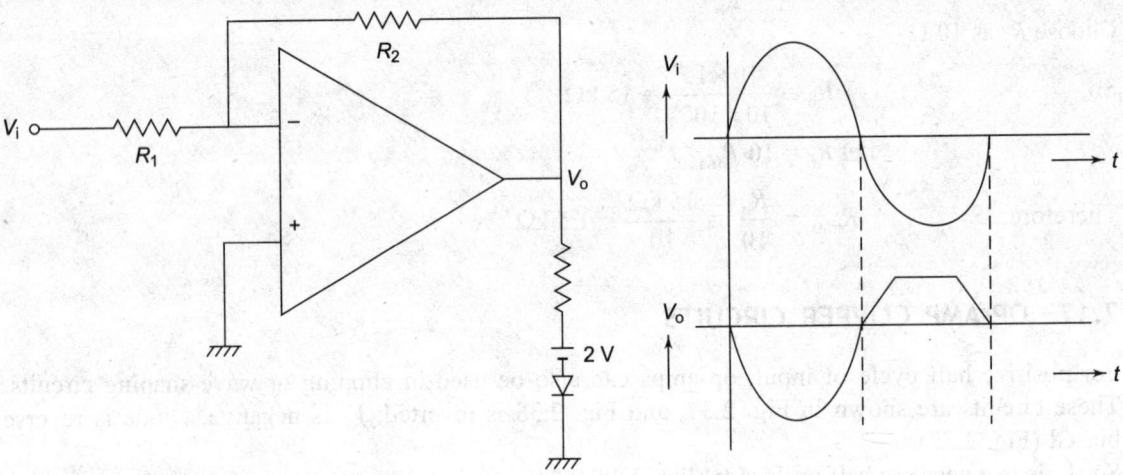

Fig. 2.41   *Clipper Circuit*   **Fig. 2.42**   *Waveforms*

2.18 OP-AMP FUNCTION GENERATOR

The circuit is as shown in Fig. 2.43. While using this circuit with two op-amps, one can generate square and triangular waveforms.

Expression for output of integrator is,

$$V_{o1} = \frac{1}{RC} \int V_i \, dt$$

$$V_i \simeq (+V_{cc} - V_{o2})$$

So, $$V_{o1} = -\frac{1}{RC} (+V_{cc}) \, t$$

Therefore, output of integrator is going towards negative when V_{o2} is $+ V_{cc}$, that is op-amp 2 is at positive saturation level.

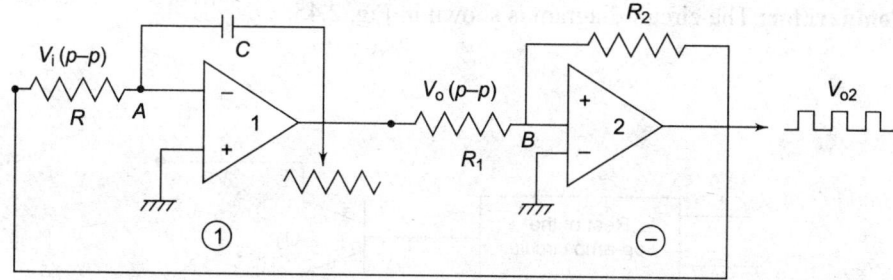

Fig. 2.43 *Function Generator Circuit*

The circuit generates triangular ($\wedge\wedge\wedge$) waveform and ($\sqcap\sqcup\sqcap$) square waveform. It consists of a corporator (op-amp 2) and an integrator (op-amp 1). These are connected in a positive feedback loop. The square wave is a feedback from the output of the comparator to the input of the integrator.

When the output of the comparator is at its positive saturation level, the integrator output decreases at the rate of $V_S^-/RC = -V_{o2(sat)}/RC$ (volts/sec) until it reaches the lower trip point of the comparator =

$V_s^+ \left(\dfrac{R_1}{R_2} \right) = V_{o1(sat)} \dfrac{R_1}{R_2}$. When the input at point B of op-amp 2 is less, the op-amp will change state and the output goes negative.

Though $\left(\dfrac{R_2}{R_1} \right)$ of an op-amp comparator is not very large, the input is large. So, the comparator is driven to saturation. The output of integrator is large, which is the input to the comparator. So, the integrator output level increases at the rate of V_s^+/RC (volts/sec).

When the integrator output reaches the upper trip point of the comparator which is $\left(-V_s \dfrac{R_1}{R_2} \right)$, the comparator again switches state. This repeats and so a square or rectangular waveform is obtained at the output of op-amp 2, as shown in Fig. 2.44.

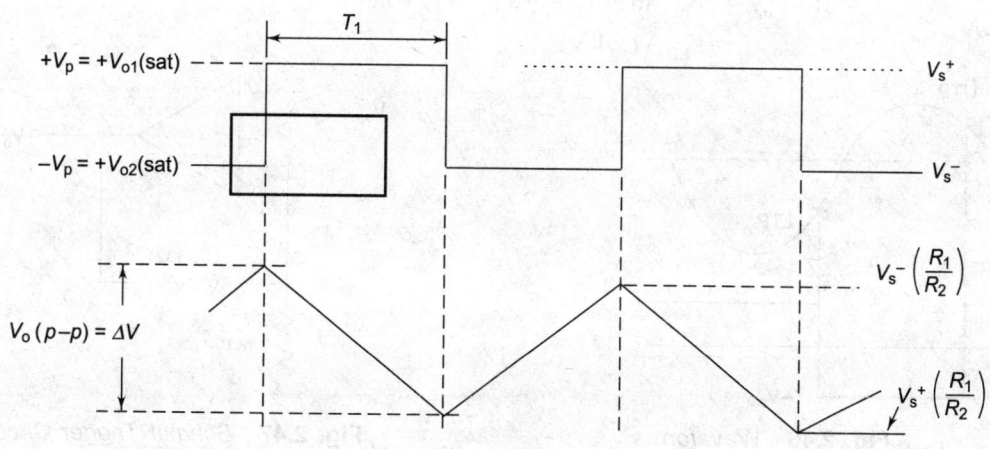

Fig. 2.44 *Waveforms*

Op-amp Comparator: The circuit diagram is shown in Fig. 2.45.

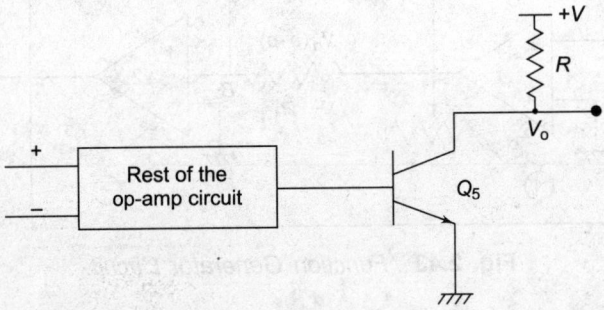

Fig. 2.45 *Output-Stage Comparator Circuit*

When the non-inverting input is more positive than the inverting input, the base voltage for Q_5 decreases. Therefore, Q_5 is cut off. $V_o = +V$. When the non-inverting input is less positive than the inverting input, the base voltage of Q_5 increases. Therefore, Q_5 goes to saturation. Thus, $V_o = V_{ce\ (sat)} =$ low $= 0.1$ V or so. The expression for frequency of oscillator f is

$$f = \frac{R_2}{4RR_1C}$$

2.18.1 Schmitt Trigger

The waveforms are shown in Fig. 2.46 and the circuit is shown in Fig. 2.47.

Differential input V_d of a fraction of an mV is sufficient to saturate the output of an op-amp, which is in open-loop configuration, and A_{VOL} is very high.

The Schmitt trigger produces a rectangular wave irrespective of the shape of the input signal. f will be the same as that of the input. The voltage drop across the op-amp components will decrease the output.

If $+V_{sat} = +V_{cc}$, then $V_o = (V_{cc} - 1)$ volt.

If $-V_{sat} = -V_{EE}$, then $V_o = (-V_{EE} + 1)$ volt.

$$= -V_{EE} - (-1\ V)$$

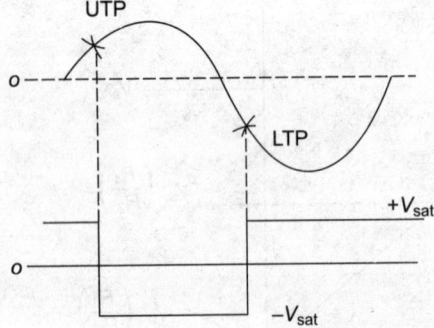

Fig. 2.46 *Waveforms*

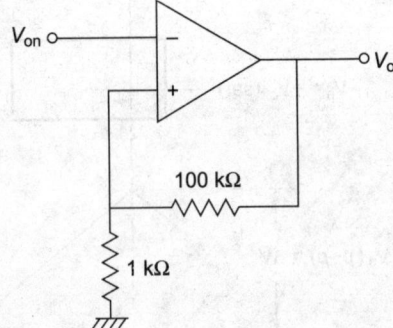

Fig. 2.47 *Schmitt Trigger Circuit*

The drop in voltage is -1 V because V_{EE} is negative voltage; $\Rightarrow (-V_{EE} + 1)$ volt. The 1 V drop is across the various internal components of op-amp.

2.18.2 Rectangular to Triangular Waveform Conversion

The circuit diagram is shown in Fig. 2.48 and output waveforms are shown in Fig. 2.49.

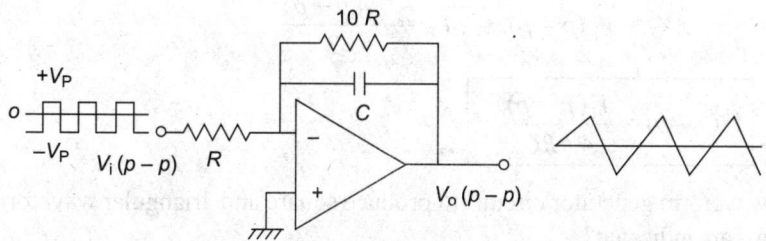

Fig. 2.48 *Circuit for Waveform Conversion*

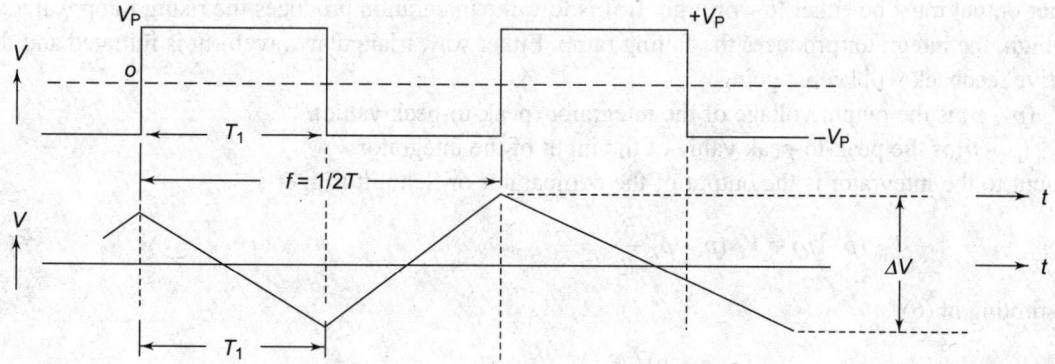

Fig. 2.49 *Waveforms*

Since the rectangular input has a DC average of 0 V, the output triangular waveform ⋀⋁⋀⋁⋀ should also have an average value of 0 V. The slope of ramp waveform will be negative during the positive half cycle of the input and positive during the negative half cycle. So, the output is a periodic wave, with the same f as the input waveforms.

Change in the capacitor voltage $V = \dfrac{IT_1}{C}$ \hfill (1)

Since $$C = \frac{Q}{V} \text{ or } V = \frac{Q}{C}; \; Q = IT$$

In the positive half cycle of the input voltages, the capacitor charging current is $I = \dfrac{V_p}{R}$ \hfill (2)

Since it is a square wave, V_p is the same throughout T.

T_1 = Rundown time of the output ramp
   = Half of output period

If f = Frequency of the input square wave

Then $$T_1 = \frac{1}{2f}$$ \hfill (3)

Therefore,   Eq. (1) is $V = \dfrac{V_P}{2f\,RC}$ \hfill (4)

The input voltage has a peak-to-peak value of $2V_P$.

The output voltage has a peak-to-peak value of $V = V_o \, (p - p)$

$$2V_P = V_i \, (p - p). \text{ so, } V_P = \frac{V_i(p-p)}{2} \tag{5}$$

$$\boxed{V_o(p - p) = \frac{V_i(p - p)}{4fRC}}$$

How does the waveform generator circuit (to produce square and triangular waveforms) get started? How the waveforms are indicated?

When the circuit is first switched on, the power supply to the op-amp, $+V_{cc}$ and $-V_{cc}$, and the Schmitt trigger output must be either low or high. If it is low, the integration produces the rising ramp wafer. If it is high, the integrator produces the falling ramp. Either way, triangular waveform is initiated and the positive feedback will keep it going.

$V_o \, (p - p)$ is the output voltage of the integrator (peak-to-peak value).

$V_{in} \, (p - p)$ is the peak-to-peak value of the input of the integrator.

Input to the integrator is the output of the comparator or Schmitt trigger.

$$V_{in} \, (p - p) = V_o \, (p - p) \, \frac{R_2}{R_1} \tag{7}$$

Substituting in (6),

$$V_o \, (p - p) = \frac{V_o(p - p)}{4fRC} \left(\frac{R_2}{R_1} \right)$$

Therefore,

$$\boxed{f = \frac{R_2}{4RR_1C}}$$

Figure 2.50 shows the silicon wafer, wire bonding connections, and IC package in general for all ICs.

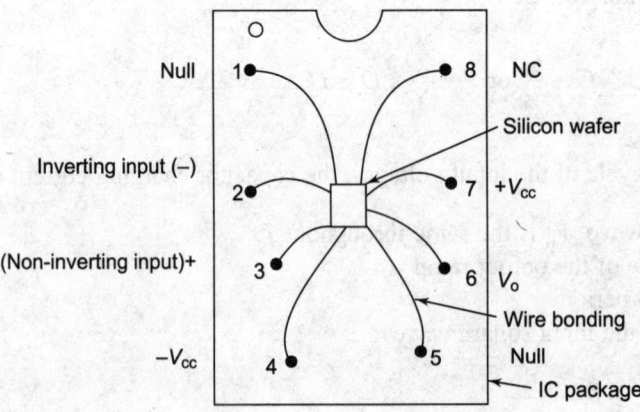

Fig. 2.50   *IC Chip and Leads*

2.19 DIFFERENTIAL VOLTAGE-TO-CURRENT CONVERTER OR CONSTANT CURRENT SOURCE WITH GROUNDED LEAD

Here two input voltages E_1 and E_2 are given in differential configuration. The output current remains constant over a range, with variation in V_o. The output current is proportional to the differential voltage (E_2-E_1). So this circuit is called as *differential voltage-to-current converter*. The circuit is shown in Fig. 2.51. The expressions governing the circuit are as follows:

$$I_1 = \frac{V_L - E_2}{R} \qquad\qquad I_1 = -I_2$$

$$I_2 = \frac{V_L - V_o}{R} \qquad \text{So,} \qquad \frac{V_L - E_2}{\not R} = \left(\frac{V_L - V_o}{\not R}\right)$$

Therefore, $\qquad\qquad V_o = 2V_L - E_2$

$$I_L = \frac{E_1 - E_2}{R} \qquad\qquad I_L = \frac{V_L}{R_L}$$

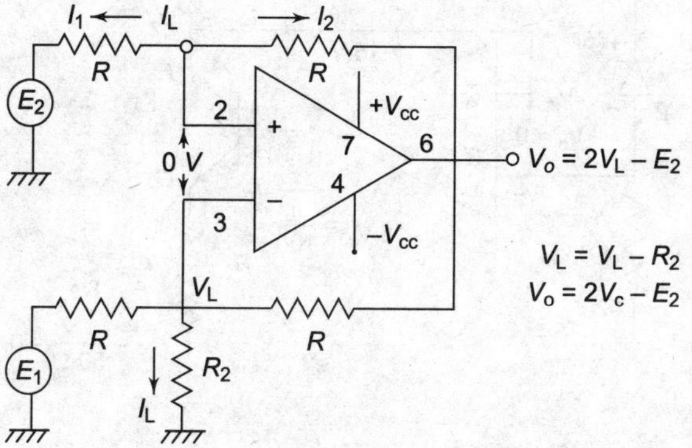

Fig. 2.51 *Differential Voltage-to-Current Converter*

If $\qquad\qquad E_2 = 0, R_L = 5\ \text{k}\Omega, E_1 = 5\ \text{V}, R = 10\ \text{k}\Omega,$

$$I_L = \frac{5\,\text{V} - 0\,\text{V}}{10\ \text{K}\Omega} = \frac{5\,\text{V}}{10\ \text{K}\Omega} = 0.5\ \text{mA}$$

2.20 CONSTANT HIGH-CURRENT SOURCE, GROUNDED LEAD CIRCUIT

The circuit diagram is shown in Fig. 2.52.

If the transistor Q_1 can dissipate 5 W of power, I_L can be made 500 mA.

The zener diode, Z, is reverse biased. If it is 5 V zener, the voltage across it is 5 V.

For the op-amp, the differential voltage $V_d = 0$.

Therefore, the voltage drop across R_s will also be V_Z.

I_E of the transistor $Q_1 = \dfrac{V_Z}{R_s}$

$$I_E \simeq I_C, \text{ so, } I_C = I_L = \frac{V_Z}{R_s}$$

If $\qquad V_Z = 5\,\text{V}, R_s = 50\,\Omega, \text{ therefore, } I_L = \dfrac{5\,\text{V}}{50\,\Omega} = 0.1\,\text{A} = 100\,\text{mA}.$

If β of the transistor is 100, $I_b = \dfrac{I_c}{\beta} = \dfrac{I_c}{100}.$

The output current delivered by the circuit is large and it is maintained constant over a particular range with variation in R_L. The output current value is high compared to previous circuit. So, it is called as constant *high-current source*.

This circuit acts as a constant high-current source.

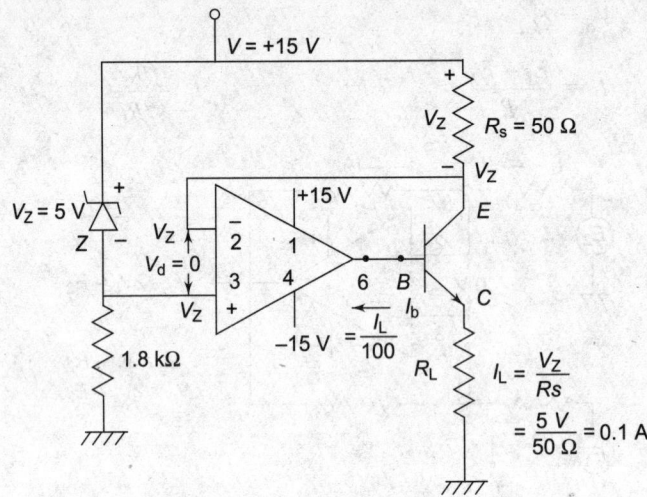

Fig. 2.52   *Constant High-Current Source*

2.21 INSTRUMENTATION AMPLIFIERS: OP-AMP APPLICATIONS

A general purpose instrumentation amplifier should have:
1. A differential input
2. Single-ended output
3. High input Z
4. Simple means of adjusting the gain
5. High CMRR

A simple subtractor circuit will provide the first two requirements (Fig. 2.53)

$$e_o = -\frac{R_2}{R_1}\,(e_1 - e_2)$$

One of the inputs can be a reference voltage.

This circuit has differential input. It has single-ended output. But it doesn't have the characteristics of 3, 4, 5, started of W.

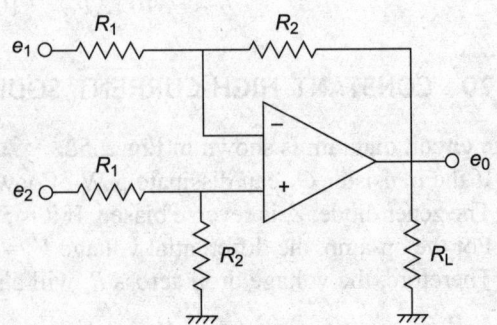

Fig. 2.53   *Subtractor Circuit*

Input i is not high because the higher input Z may be op-amp shunted by R_1 and R_2. So, it cannot serve the purpose of instrumentation amplifiers.

Consider the circuit shown in Fig. 2.54.

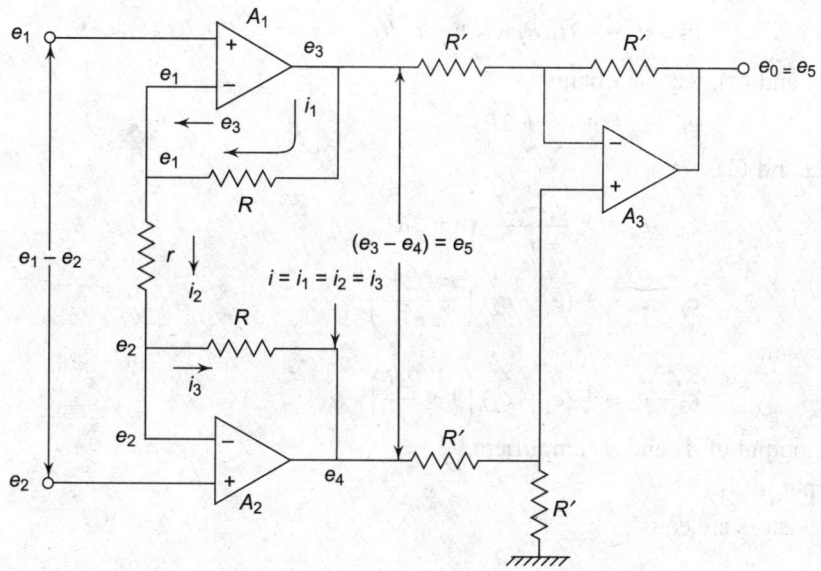

Fig. 2.54 *Instrumentation Amplifier*

The op-amp draws no current from the input. So, current flowing in the loop through R', r, and R must be the same.

This circuit has all the requirements of an instrumentation amplifier.

Gain can be adjusted easily by charging 'r'. It has double-ended input and single-ended output. There are no resistors connected to the input terminals at e_1 and e_2. Therefore, high Z of op-amp is not shunted. Non-inverting terminal has higher input Z. So, input is being given to these points.

High CMRR

Because the input is being amplified by A_1 and A_2, which is further amplified by A_3, the differential gain A_d will be made larger than common mode gain A_c.

$$\text{CMRR} = \frac{A_d}{A_c} \text{ will be high.}$$

If A_d is made large, $\dfrac{A_d}{A_c}$. CMRR will be high.

2.22 ANALYSIS

In the ideal case, op-amp does not draw any current from input signal source.

$$i_1 = i_2 = i_3 = i \qquad (1)$$

Current through resistor 'r' is $i_2 = \left(\dfrac{-e_2 + e_1}{r}\right) = i$ (2)

The voltage drop across R', r, and R equals $e_3 - e_4$.

$$e_3 - e_4 = -(i_1 R + i_2 r + i_3 R) \tag{3}$$

From Eqs. (1) and (3), we can obtain

$$e_3 - e_4 = + i(r + 2R) \tag{4}$$

From Eqs. (2) and (4),

$$e_3 - e_4 = + \frac{e_1 - e_2}{r}(r + 2R) \tag{5}$$

or $$e_3 - e_4 = + (e_1 - e_2)\left(\frac{r + 2R}{r}\right)$$

Rewriting the same,

$$e_3 - e_4 = + (e_1 - e_2)\left(1 + \frac{2R}{r}\right)$$

This is the output of A_1 and A_2 amplifiers.

It is the input to A_3.

If all the resistors are equal,

$$e_5 = -\left(1 + \frac{2R}{r}\right)(e_1 - e_2)$$

$$e_o = e_5$$

Thus, $$\text{gain} = 1 + \frac{2R}{r}$$

By changing 'r', gain can be changed.

Op-amp the function of op-amp. A_3 is to give single-ended out put.

For the circuit shown, even input through R_3 is connected to the non-inverting terminal, the current drawn by op-amp is zero. So, drop across $R_3 = 0$. Hence, non-inverting terminal is at ground potential. R_3 will make the Z at the non-inverting and inverting terminals same. So, $i_b = 0$. Input bias current (ib) $R_3 = R_1$ in parallel with R_2.

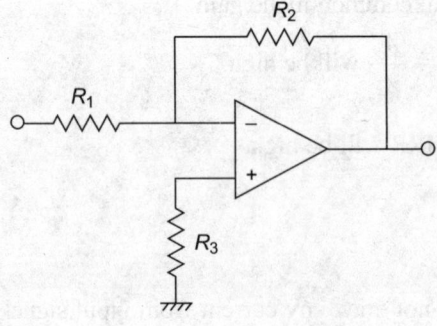

Fig. 2.55 *Op-amp Circuit*

Input is given to the inverting terminal to get negative feedback. Positive feedback will cause oscillations in the circuit.

2.23 SAMPLE-AND-HOLD CIRCUITS

Sample-and-hold (S/H) circuits are used to sample an analog higher at a particular constant of time and hold it as long as required. At what constant sampling is to be done and at how long the sampled value is to be held depends on the application and are determined by a logic circuit.

A capacitor is used to hold the charge. An electronically controlled switch provides a means for rapidly charging the capacitor and then removing the only charge the capacitor can retain the desired voltage. V_A is the analog voltage source. E_g its normal impedance (Fig. 2.56).

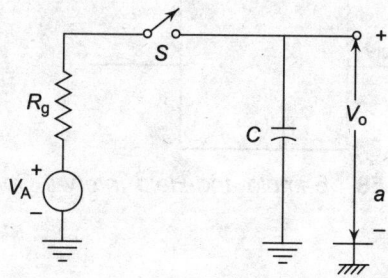

Fig. 2.56 *S/H Circuit Principle*

Applications

1. Instrumentation system
2. Measurement system
3. Communication system

Sampling of analog signal is shown in Fig. 2.57.

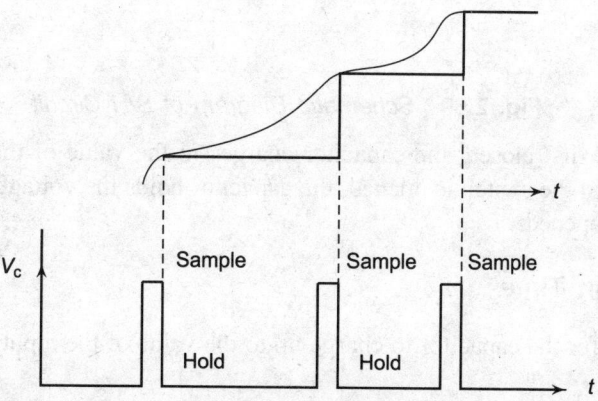

Fig. 2.57 *Sampling Waveform*

Aperture Time: It is the maximum delay between the time logic that the control command is given to device the switching to open and it takes to actually open.

Acquisition time: It is the shortest time after a sample command has been given, that a HOLD command can be given, and gives an output voltage which approximates the input voltage with necessary accuracy.

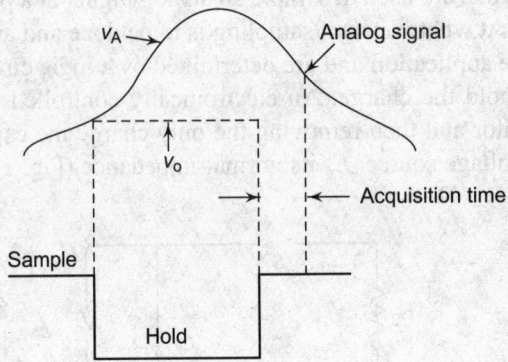

Fig. 2.58 *Sample-and-Hold Interval Conversion*

2.24 TERMINOLOGY

A S/H circuit is used in A/D converter, when it is necessary to convert a high-frequency signal, which is varying too rapidly, to allow an accurate conversion. S/H circuit is basically an op-amp which charges a capacitor during the sample mode and retains the charge during the hold mode. The S/H circuit can be represented by a switch and a capacitor.

The schematic diagram to explain the principle is shown in Fig. 2.59.

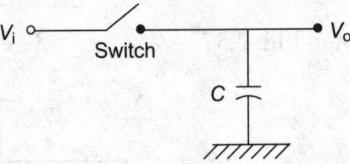

Fig. 2.59 *Schematic Diagram of S/H Circuit*

When the switch is first closed, the capacitor charges to the value of the input voltage and then follows the input. When the switch is opened, the capacitor holds the voltage which it had at the time when the switch was opened.

2.24.1 Acquisition Time

It is the time required for the capacitor to charge up to the value of the input signal after the switch is first shorted.

2.24.2 Aperture Time

It is the time required for the switch to change state and the uncertainty in the time that this change of state occurs.

2.24.3 Holding Time

The length of time the circuit can hold the voltage without dropping by more than a specified per cent of its initial value is called holding time.

Practical circuits use transistor switches and op-amp to increase the available driving current into the capacitor or to isolate the capacitor from an external load on the output.

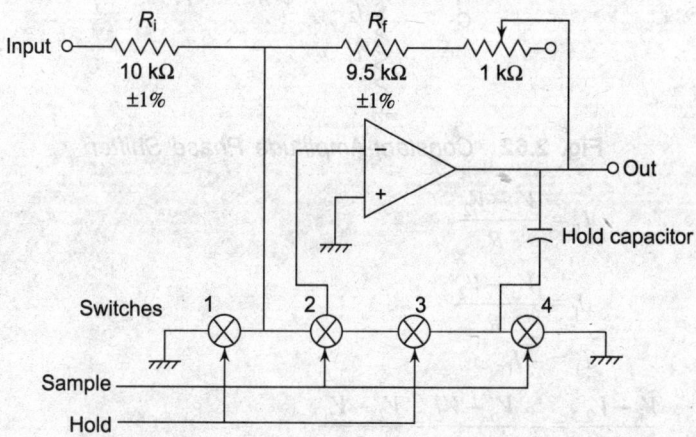

Fig. 2.60 *S/H Circuit Schematic Diagram*

Consider the circut shown in Fig. 2.60.

Sample pulse operates switches 2 and 4. Switches 2 and 4 are closed when sampling is being done. Hold pulse operates switches 1 and 3.

Sample-and-hold pulses are complimentary.

In sample mode, hold capacitor is charged up by the op-amp.

In hold mode, the capacitor is switched into the feedback loop. Input resistor R_o and feedback resistor R_f are switched to ground.

Since the input to the amplifier remains within a few μV to ground (except during switching), the input impedance $Z_i = 10 \text{ k}\Omega$ in both sample-and-hold modes.

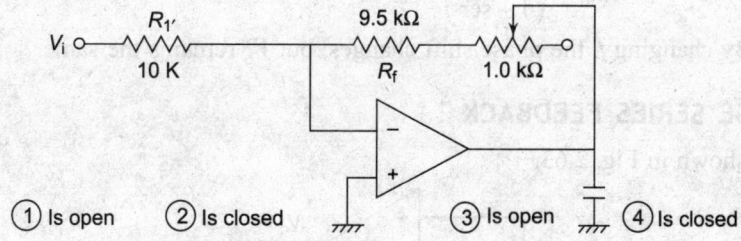

Fig. 2.61 *Sample-and-Hold Circuit*

2.25 CONSTANT-AMPLITUDE PHASE SHIFTER

$V_o = V_i$. But there is phase shift between V_o and V_i, which can be changed by changing C and r. (The circuit is shown in Fig. 2.62.)

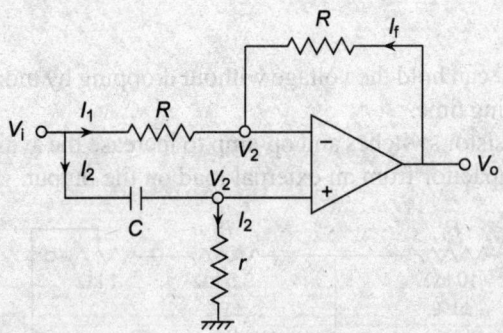

Fig. 2.62   *Constant-Amplitude Phase Shifter*

$$I_1 = \frac{V_i - V_2}{R}$$

$$I_f = \frac{V_o - V_2}{R}$$

$$I_1 = -I_f$$

So,
$$\frac{V_i - V_2}{R} = -\frac{V_o - V_2}{R} = \frac{V_2 - V_o}{R}$$

Therefore,
$$V_2 = \frac{V_o + V_i}{2} \qquad\qquad (1)$$

Op-amp draws no current. Therefore, I_2 flows through 'C' and 'r'.

So,
$$\frac{V_i - V_2}{1/sc} = \frac{V_2}{r} \qquad (S = j\omega) \qquad (2)$$

Substituting (1) for V_2 in (2),

$$V_i = \left(\frac{V_o + V_i}{2} \right) = \frac{V_o + V_i}{2r} \qquad (3)$$

Solving,
$$V_o = -\left(\frac{1 - scr}{1 + scr} \right)$$

Thus $|V_o| = |V_i|$. By changing f, the phase shift changes, but V_o remains the same.

2.26   VOLTAGE SERIES FEEDBACK

The circuit is as shown in Fig. 2.63.

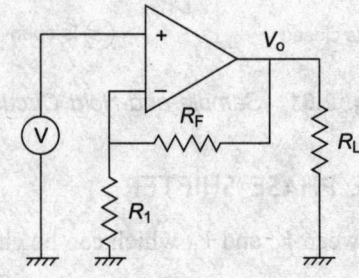

Fig. 2.63   *Voltage Series Feedback*

2.27 VOLTAGE-SHUNT FEEDBACK

The circuit is as shown in Fig. 2.64

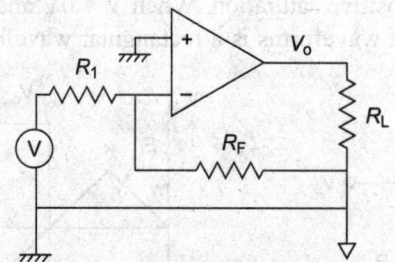

Fig. 2.64 *Voltage-Shunt Feedback*

2.28 ZERO-CROSSING DETECTORS (COMPARATORS)

2.28.1 Non-Inverting Zero-Crossing Detector

Consider the circuit shown in Fig. 2.65 (a). It represents non-inverting zero-crossing detector. The waveforms are shown in Fig. 2.65 (b). The transfer characteristic is shown in Fig. 2.66.

The op-amp is in open-loop configuration. So, op-amp goes to positive saturation level if $(E_i - V_{ref})$ is > 0 V, and it goes to negative saturation level $(-V_{sat})$ if $(V_{ref} - E_i) > 0$ V.

In practice, when $(E_i - V_{ref})$ is $> V_T$, the threshold voltage, op-amp changes state. The value of V_T is about 1 mV to 10 mV.

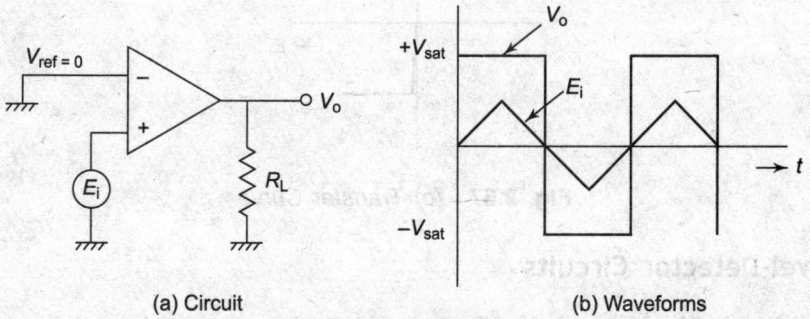

(a) Circuit	(b) Waveforms

Fig. 2.65 *Non-Inverting Zero-Crossing Detector*

Initially, when op-amp is switched on, V_o is at positive saturation level. Internal circuit is set that way. When E_i is positive, $V_o = +V_{sat}$. When E_i is negative, $V_o = -V_{sat}$. Op-amp is in open-loop configuration.

If the input E_i is given to the non-inverting terminal, it is non-inverting ZCD.

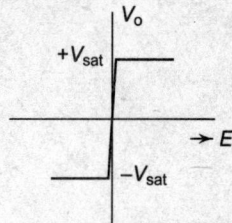

Fig. 2.66 *Transfer Characteristic*

2.28.2 Inverting Zero-Crossing Detector

The circuit is as shown in Fig. 2.67(a). The waveforms are shown in Fig. 2.67 (b). The transfer characteristic is shown in Fig. 2.67 (c).

The op-amp is in open-loop configuration.

As the input 'V' is applied to the inverting terminal and the reference voltage is 0 V, when $V > 0$ V and positive going, op-amp goes to positive saturation. When $V < 0$V and negative going, op-amp goes to positive saturation. So, the output waveforms is a rectangular waveform.

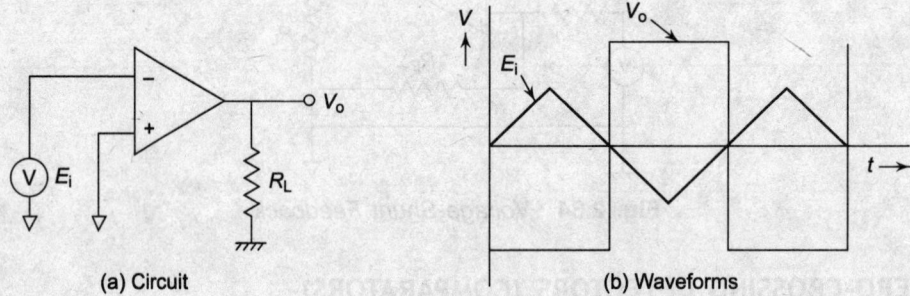

| (a) Circuit | (b) Waveforms |

Fig. 2.67   *Inverting Zero-crossing Detection*

If voltage at the non-inverting terminal is high, V_o must be positive. If the voltage at the inverting terminal is high, V_o must be negative.

If E_i is given to the inverting terminal, it is inverting *ZCD*.

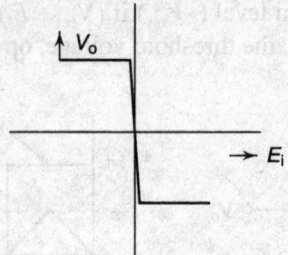

Fig. 2.67   *(c) Transfer Curve*

2.28.3   Level-Detector Circuits

The circuit is shown in Fig. 2.68 and waveforms in Fig. 2.69 for positive-level detector. Non-inverting type. When E_i is above V_{ref}, $V_o = +V_{sat}$.

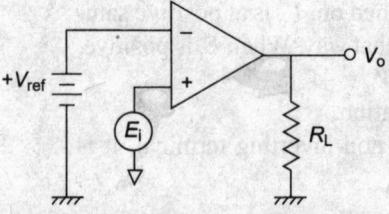

Fig. 2.68   *Positive-Level Detector*

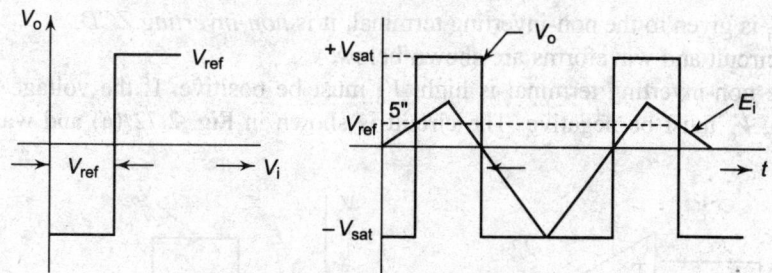

Fig. 2.69 *Output Waveforms*

When $\qquad E_i < V_{ref}, \qquad V_o = -V_{sat}$

When $\qquad E_i > V_{ref}, \qquad V_o = +V_{sat}$

The op-amp detects when the E_i has crossed $+V_{ref}$ value.

This circuit can be used in industries to know when a particular pr. temperature etc are being unloaded.

$$V_T \text{ (threshold voltage)} = 10 \text{ mV}$$

Zero-crossing Detector: The circuit is shown in Fig. 2.70 (a) and waveforms in Fig. 2.70 (b) for non-inverting circuits.

Initially, when op-amp is switched on, V_o is at positive saturation. Internal circuit is set that way when E_i is positive $V_o \equiv +V_{sat}$. When E_i is negative $V_o = -V_{sat}$.
Op-amp is in open-loop configuration.

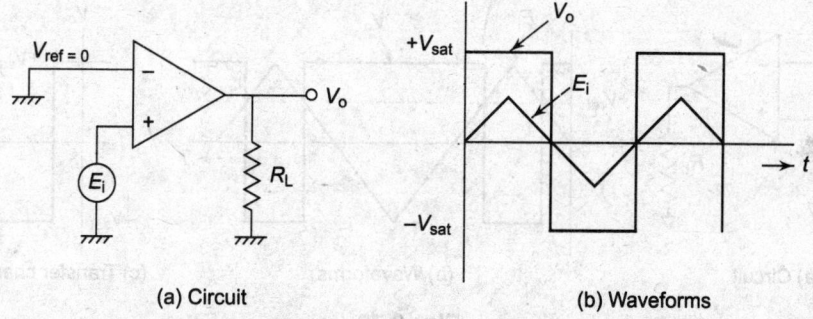

(a) Circuit (b) Waveforms

Fig. 2.70 *Non-inverting Zero-crossing Detector*

When $E_i > 0$ V, $V_o = +V_{sat}$. When $E_i < 0$ V, $V_o = -V_{sat}$. So, for the given triangular input waveform E_i, the output waveform V_o is a rectangular waveform. Whenever the input waveform crosses 0 V, the output waveform changes state from positive to negative and vice versa.

The transfer characteristic is shown in Fig. 2.71.

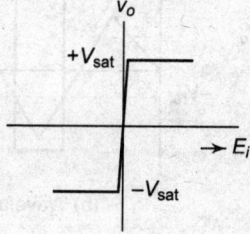

Fig. 2.71 *Transfer Curve*

If the input E_i is given to the non-inverting terminal, it is *non-inverting ZCD*.
Inverting *ZCD* circuit and waveforms are shown below.

If voltage at the non-inverting terminal is high, V_0 must be positive. If the voltage at the inverting terminal is high, V_0 must be negative. The circuit is shown in Fig. 2.72 (a) and waveforms in Fig. 2.72(b).

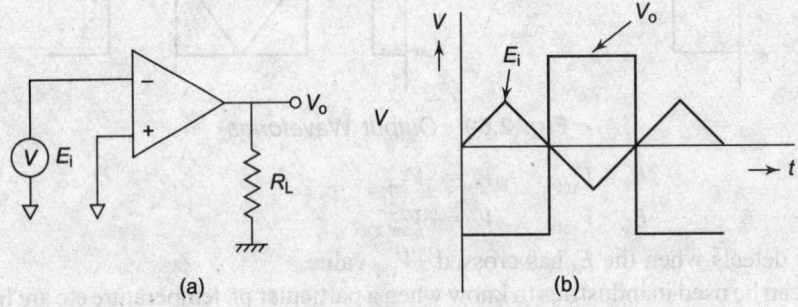

Fig. 2.72 *Inverting Zero-crossing Detector*

If E_i is given to the inverting terminal, it is inverting *ZCD*.

2.28.4 Inverting Positive-Level Detector

The circuit is shown in Fig. 2.73 (a), waveforms in 2.73 (b), and transfer curve in 2.73 (c).

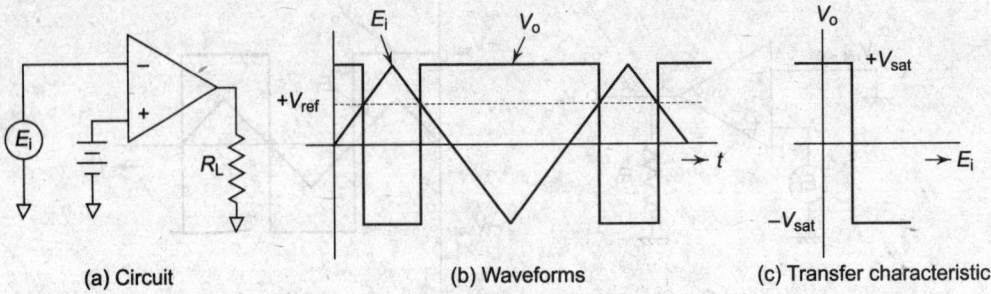

(a) Circuit   (b) Waveforms   (c) Transfer characteristic

Fig. 2.73

2.28.5 Negative-Level Detector (Non-inverting)

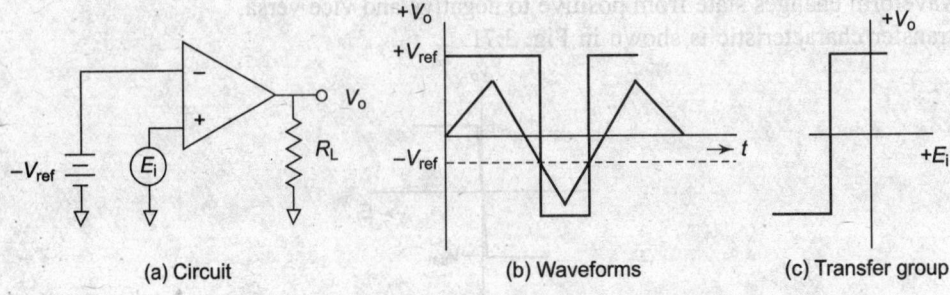

(a) Circuit   (b) Waveforms   (c) Transfer group

Fig. 2.74 *Negative-level Detector (Non-Inverting)*

2.28.6 Negative-Level Detector (Inverting)

When $E > V_{ref}$, $V_o = -V_{sat}$. The circuit is shown in Fig. 2.75(a), waveforms in 2.75(b), and transfer curve in 2.75 (c).

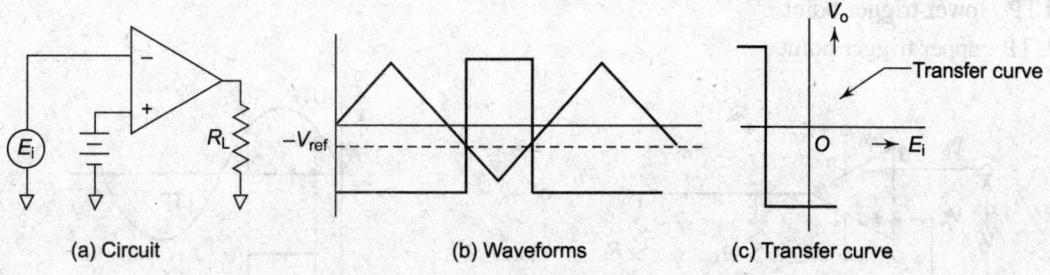

| (a) Circuit | (b) Waveforms | (c) Transfer curve |

Fig. 2.75

2.29 OP-AMP SCHMITT TRIGGER CIRCUIT

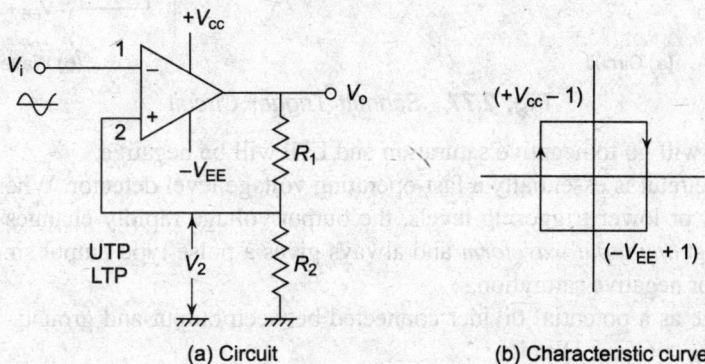

(a) Circuit   (b) Characteristic curve

Fig. 2.76 *Op-amp Schmitt Trigger Circuit*

If $V_i < V_2$: positive saturation

If $V_i > V_2$: negative saturation

$$V_2 = \frac{R_2}{R_1 + R_2} (V_{cc} - 1)$$

There will be six Schmitt trigger circuits in the IC 7414 hex Schmitt trigger: When the closed-loop gain of op-amp is very large, op-amp then also will go to saturation. Because op-amp output $V_o = A_{CL} V_i$ cannot be greater than $+V_{cc}$ or $-V_{cc}$. This is the principle of working of Schmitt trigger circuit.

If $V_i < V_2$, op-amp goes to positive saturation.
If $V_i > V_2$, op-amp goes to negative saturation.

When the difference between V_i and V_2 is equal to V_T, the threshold voltage, then only op-amp goes to saturation in open-loop configuration. V_T solves typically ranges from 1 mV to 10 mV.

The op-amp Schdmitt trigger circuit diagram is shown in Fig. 2.77 (a) and waveforms are shown in Fig. 2.77 (b).

LTP : lower trigger point

UTP : upper trigger point

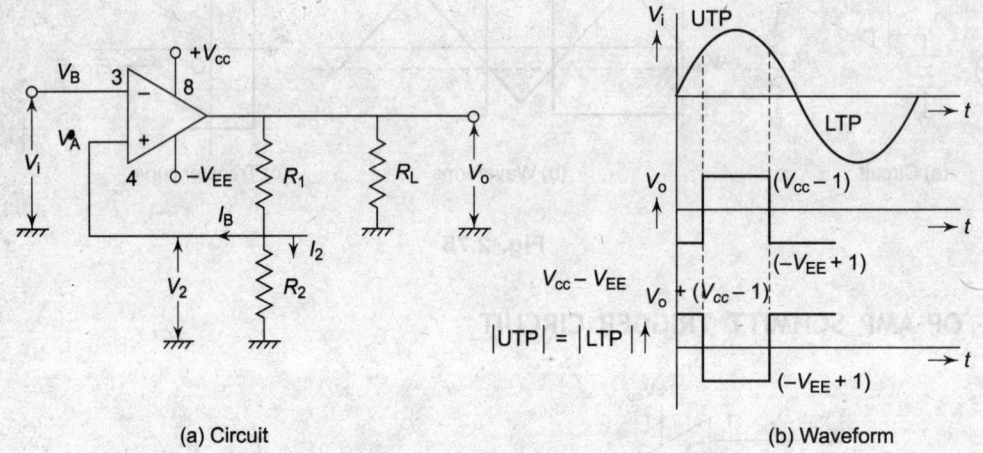

(a) Circuit   (b) Waveform

Fig. 2.77 *Schmitt Trigger Circuit*

If $V_A < V_B$, op-amp will go to negative saturation and LTP will be negative.

Operation: This circuit is essentially a fast-operating voltage level detector. When the input voltage arrives at the upper or lower triggering levels, the output voltage rapidly changes level. This circuit operates from almost *any input waveform* and always gives a pulse-type output since op-amp goes to positive saturation or negative saturation.

R_1 and R_2 operate as a potential divider connected between output and ground. When $V_i < V_2$, the output is positive. $V_o \simeq (V_{cc} - 1)$ volts.

So,
$$V_2 = \frac{R_2}{R_1 + R_2} (V_{cc} - 1)$$

When $V_i > V_2$, output goes negative. So, V_2 decreases. Thus, V_2 becomes negative with respect to V_i. (There will be the voltage drop across the circuit in the op-amp.)

So, output changes very rapidly from $(V_{cc} - 1)$ to $(V_{EE} + 1)$ V. Therefore, *UTP* is $-V_2$ when V_o is positive. When output is $(V_{EE} + 1)$, there is 1 V drop.

$$V_2 = \frac{R_2}{R_1 + R_2} (V_{EE} + 1)$$

Since, V_{EE} is negative, V_2 is negative and V_o remains negative until the voltage at the inverting input terminal is reduced to the new level of V_2.

$$UTP \simeq \frac{R_2}{R_1 + R_2} (V_{cc} - 1)$$

$$LTP \simeq \frac{R_2}{R_1 + R_2} (V_{EE} + 1), V_{EE} \text{ is negative.}$$

So,   LTP is negative.

Example 2.2 *Design a Schmitt trigger circuit with UTP = 3 V.*
$R_L = 10$ kΩ. Supply voltage = ± 15 V. $I_{b(max)} = 500$ μA.
Choose $I_2 = 100$ $I_{b(max)}$. Calculate actual values of UTP and LTP.

$$I_2 = 100 \times 500 \text{ nA} = 50 \text{ }\mu A$$

$$V_{e_1} = (15 - 1) - 3V = 11 \text{ V} = V_o - V_{e_2}$$

$$R_2 = \frac{VR_2}{I_2} = \frac{3 \text{ V}}{50 \text{ }\mu A} = 60 \text{ k}\Omega$$

$$R_1 = \frac{V_{R1}}{I_2} = 205 \text{ k}\Omega$$

$$V_{R1} = (V_o - V_2), \quad V_o = V_{cc} - 1$$

In μA 741, if V_{cc} is 15 V, V_o will be $(V_{cc} - 1)$ V. There will be 1 V drop across the circuit. Thus, the output is less by 1 V for V_{cc} and V_{EE}.

7414 is hex Schdmitt Trigger IC. Six circuits are available in the IC bias. I_B is the current drawn by op-amp.

Recovery time:

Op-amp Schdmitt trigger exhibits maximum hysteresis, since if UTP is +5 V, LTP = –5 V. But magnitudes of UTP and LTP will be the same. If $V_{cc} = V_{EE}$. R_L will not affect UTP or LTP. Output Z of op-amp is low. So, R_L value will not effect the UTP or LTP value.

2.30 ASTABLE MULTIVIBRATOR USING OP-AMP

Capacitor 'C' gets charged through 'R' towards $+V_{o1(sat)}$, when $V_{o1} = V_{o1(sat)}$. It gets charged towards $-V_{o2(sat)}$, when $V_o = -V_{o2(sat)}$. The circuit diagram is shown in Fig. 2.78 (a). The equivalent RC charging circuit when C gets charged towards $-V_{o2(sat)}$ is shown in Fig. 2.78 (b).

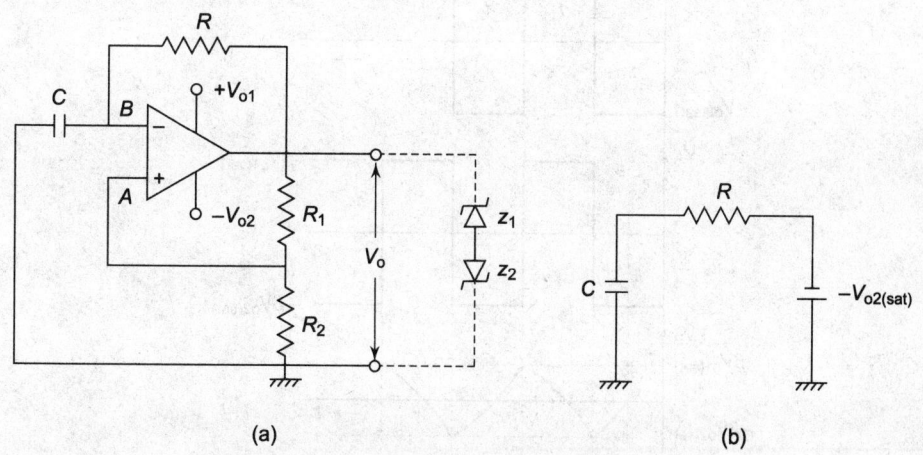

(a)   (b)

Fig. 2.78 *Op-amp Astable Multivibrator Circuit*

When
$$V_A > V_B$$
$$V_o = +V_{o1(sat)}$$

When $\qquad V_B > V_A$

$$V_o = -V_{o2(sat)}$$

This is differential input op-amp acting as free symmetrical multivibrator. The output V_o may be at positive saturation level $+V_o$ or negative saturation level $-V_{o2}$. [normally $V_{o1} = V_{o2}$]. R_1 and R_2 provide a fixed positive feedback since the drop across R_1 of V_o is applied to the non-inverting terminal A. Negative feedback is provided from output to the input in an R and C, since V_o is applied to the inverting terminal. But at high frequencies of the output square waveform, negative feedback is negligible compared to positive feedback since X_c will be small.

Suppose at $t = t_0$, the output of op-amp is in negative saturation and is equal to $-V_{o2}$ (saturation). This is fed to the terminal A.

So, voltage at terminal $A = -V_{o2(sat)}\dfrac{R_1}{R_1 + R_2}$

$\qquad$ ($R_1 - R_2$ is a potential divider network.)

Let $\qquad\qquad\qquad \beta = \dfrac{R_1}{R_1 + R_2}$

So, $\qquad\qquad\qquad V_A = -\beta\, V_{o2(sat)}$

Since the voltage at terminal A is negative terminal B is positive. So capacitor C charges down through R (to negative values) since V_{o2} is negative (Fig. 2.78). Hence, with voltage at B also reduces and becomes negative. When the potential difference between points A and B is zero, the positive feedback at 'A' causes regenerative *switching* and so the output of op-amp switches from negative saturation to positive saturation. Then, the capacitor starts getting charged, and voltage at point B rises exponentially. So during this period, the output of op-amp stays high at $+V_{o1(sat)}$ Again when the voltage at terminal B increases because of the charging of the capacitor, switching action takes places and output becomes $-V_{o2(sat)}$.

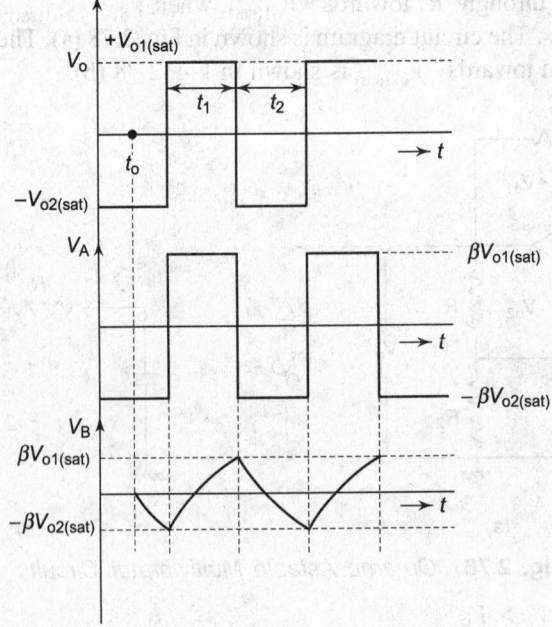

Fig. 2.79 *Waveforms of Astable Multivibrator*

→ This circuit is used effectively in the frequency range from 10 Hz to 100 kHz. Above this frequency, the delay of op-amp in changing states will affect waveform.

→ Two zener diodes can be connected in opposition to clamp the V_o to a different value of V_Z, less than $V_{o1(sat)}$ or $V_{o2(sat)}$. In such a case, the output V_o will be equal to the zener breakdown voltage V_Z (Fig. 2.79).

V_B is the voltage across the capacitor also, as the other terminal is connected to the ground.

Expression for T: (for op-amp astable multivibrator)

See the RC charging circuit in Fig. 2.80.

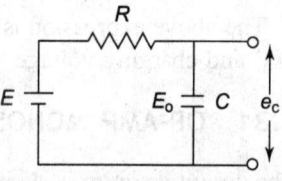

$$E = \text{charging voltage}$$

$$e_c = \text{voltage towards } C \text{ gets charged}$$

$$t = RC \ln\left[\frac{E - E_0}{E - e_c}\right], \quad E_0 = \text{initial voltage.}$$

Fig. 2.80

t = time taken by the capacitor 'C' to get charged through 'R' from the voltage source 'E'. The initial voltage across the capcitor is E_0. e_c is the voltage to which the capacitor gets charged.

During time period t_1, $\quad E = V_{o1(sat)}$

$$E_0 = -\beta V_{o2(sat)} \quad \text{for the astable multivibrator circuit}$$

$$\beta = \left(\frac{R_1}{R_1 + R_2}\right)$$

Charging voltage is the output $e_c = \beta V_{o1(sat)}$

But C gets charged to only $\beta V_{o1(sat)}$

$$\therefore \qquad t_1 = RC \ln\left[\frac{V_{o1(sat)} + \beta V_{o2(sat)}}{V_{o1(sat)} - \beta V_{o1(sat)}}\right]$$

If $V_{o1(sat)} = V_{o2(sat)}$ [negative sign need not be used because $e_c = E - (E - E_0)\, e^{-t/RC}$ expression is used. If $e_c = E + (E - E_0)e^{-t/RC}$, $V_{o1(sat)}$ must be $= -V_{o2(sat)}$].

$$t_1 = RC \ln\left\{\frac{1 + \beta}{1 - \beta}\right\}$$

$$t_2 = RC \ln\left[\frac{-V_{o2(sat)} - \beta V_{o1(sat)}}{-V_{o2(sat)} + \beta V_{o2(sat)}}\right]$$

$$= RC \ln\left\{\frac{1 + \beta}{1 - \beta}\right\} = t_1$$

$$E_2 - V_{o2(sat)}$$

$$e_c = -\beta V_{o1(sat)}$$

$$e_c = -\beta V_{o2(sat)}$$

$$T = t_1 + t_2$$

$$= 2RC \ln\left\{\frac{1 + \beta}{1 - \beta}\right\}$$

$$\beta = \frac{R_1}{R_1 + R_2}$$

So,

$$T = 2RC \ln\left(1 + \frac{2R_1}{R_2}\right)$$

$$e_c = E - (E - E_0)e^{-t/RC}$$

The above expression is for the voltage across the capacitor 'e_c' at any time 't' with initial voltage 'E_0' and charging voltage 'E'.

2.31 OP-AMP MONOSTABLE MULTIVIBRATOR

The circuit diagram is shown in Fig. 2.81.

Diode D_1 across C_1 will prevent inverting terminal of op-amp going positive. So monostable operation is obtained, otherwise it will be astable operation.

$R_3 > R_1$ so that loading effect is not there.

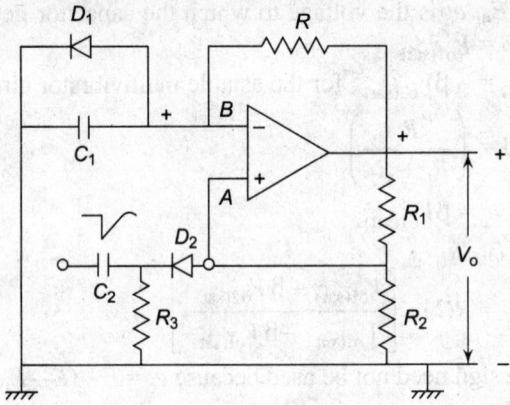

Fig. 2.81   *Circuit Diagram of an Op-amp Monostable Multivibrator*

The op-amp 741 is set such that op-amp goes to positive saturation initially.

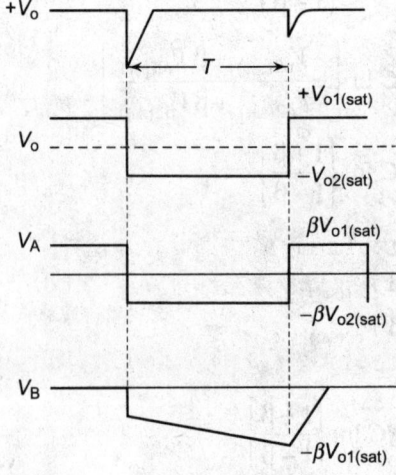

Fig. 2.82

The differential input op-amp is being used in monostable multivibrator configuration. In the normal condition, the output $V_o = + V_{o1(sat)}$. (R_2 and R_1 values are determined that way.) Cathode of diode D_1 is at negative potential. So D_1 is forward biased. Therefore, terminal B of op-amp is at ground potential. (Since D_1 is short and B is connected through D_1 and C to ground.) Terminal A is positive since the

$$\text{voltage at } A = \frac{V_{o1(sat)} R_2}{R_1 + R_2}$$

$$= \{\beta V_{o1(sat)}\}, \text{ where } \beta = \frac{R_2}{R_1 + R_2}$$

(Since R_1, R_2 is a potential divider network and drop across R_2 is feedback to A.)

$R_3 \gg R_1$. So, its loading effect will not be there. Now, if a negative pulse is through capacitor C_2 (to block AC) and diode D_2 which passes only the negative pulses and not positive pulses.

Since it gets reverse biased, the potential at A becomes positive. The magnitude of the pulse should be greater than the net positive potential at A. When the potential at A becomes negative the output of op-amp will change from $+V_{o1(sat)}$ to $-V_{o2(sat)}$. (Since potentials at A and B are same.) Since $V_o = - V_{o2(sat)}$. Now, capacitor C_1 gets charged through R (Fig. 2.83).

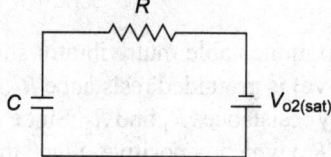

Fig. 2.83   *Capacitor Charging to* $-V_{o2(sat)}$

So, the anode of D_1 will be at negative potential. So, D_1 is reverse biased. The capacitor C_1 gets charged to negative voltage since $V_o = - V_{o2(sat)}$. So, the potential becomes negative when potential at $B = - \beta V_{o2(sat)}$ equal to the voltage at A, now the op-amp changes state and $V_o = +V_{o1(sat)}$ again. Thus, the output of amp changes when trigger pulse is given and comes back to the normal state. So, it is monostable circuit. It has one stable state after sometime.

$$e_c = E - (E - E_0)e^{-t/RC} \qquad E = \text{Charging voltage}$$
$$E_0 = \text{Initial voltage}$$

So,
$$t = RC \ln \left\{ \frac{E - E_0}{E - e_c} \right\}$$

$$\text{at } t = \text{T}, \quad e_c = - \beta V_{o2(sat)}$$
$$E_0 = 0$$
$$E = -V_{o2(sat)}$$

So,
$$T = RC \ln \frac{\left[-V_{o2(sat)} - 0 \right]}{\left[-V_{o2(sat)} + \beta V_{o2(sat)} \right]}$$

Therefore,
$$T = RC \ln \left(\frac{1}{1 - \beta} \right)$$

$$= RC \ln\left(1 + \frac{R_2}{R_1}\right)$$

Where $\qquad \beta = \dfrac{R_2}{R_1 + R_2}$

and diode D_2 which passes only negative pulses and not positive pulses since it gets reverse biased, the potential R_2 and R_1 values are determined that way.

1. D_2 in the monostable circuit will prevent positive triggering pulse being applied by mistake to the non-inverting terminal. If positive pulse is given, the value of input current to the op-amp will be very large, damaging the op-amp.
2. The function of C, near D_2, is when short duration pulse is given, op-amp may not change state. Therefore, C will get charged to the required voltage and will trigger the op-amp.
3. Before the monostable circuit returns to its stable state, if another pulse of large duration and magnitude is given, the time duration T will increase. If it is a short duration pulse, it will not have any effect.

2.31.1 Op-amp Monostable Multivibrator Circuit

Consider another circuit for op-amp monostable multivibrator shown in Fig. 2.84.

The inverting input terminal (negative) is grounded resistance R_3. The non-inverting terminal (positive) of op-amp is biased above ground by resistances R_1 and R_2. Since V_{cc} is positive voltage at the (positive) terminal of IC is (V_{ce} – drop across R_1) which is positive. Since the non-inverting terminal (positive) of op-amp has positive input, the output V_0 will be saturated to $\simeq (V_{ce} - 1)$ V. 1 V is the voltage drop even though the input is positive and of higher value, the output will be only $+V_{cc}$ across op-amp.

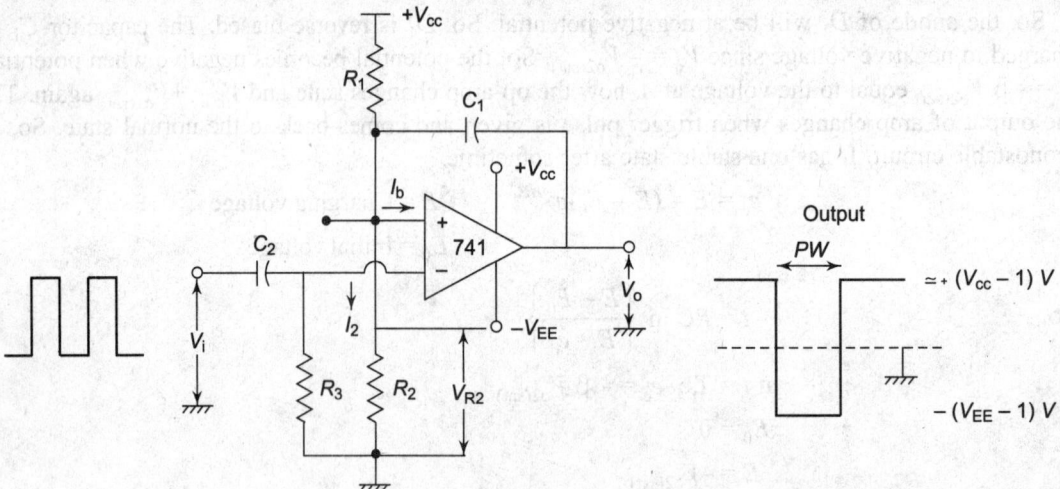

Fig. 2.84 *Op-amp Monostable Multivibrator Circuit*

Since V_0 is positive and equal to $(V_{ce} - 1)$ V, the voltage on the RHS of C_1 is positive. Therefore, E_0 polarity is as shown in Fig. 2.93. E_0 is negative on LHS. Under these conditions, the voltage at the inverting terminal is less as compared to the voltage at the non-inverting terminal.

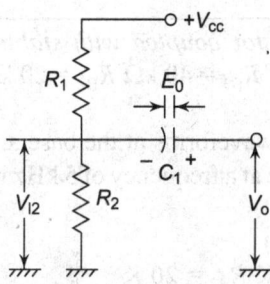

Fig. 2.85

V_o is positive and is $+(V_{cc} - 1)$ V.

Now if a positive going pulse is applied to the inverting terminal in a C_2, the inverting terminal voltage is raised above the level of the non-inverting terminal. So, the output rapidly switches to $-(V_{ce} - 1)$ V. 1 V is the drop across op-amp. Since V_{EE} is negative, V_o will become negative. So, the voltage at the non-inverting terminal of I_c is, $\{-(V_{EE} - 1), -E_0\}$ because V_o is negative and previous voltage across C_1 to E_0. So, V_o will remain negative till C_1 discharges. When the output is negative, C_1 will discharge through R_1 and R_2. So, the voltage level on the LHS will become positive, and negative on RHS. Because of this, when the voltage level on the LHS of C_1 is above the voltage level at the inverting terminal, the non-inverting terminal (positive) of the IC becomes positive. Therefore, V_o rapidly returns to $\simeq (V_{cc} - 1)$ V and the *ckt* has returned to original state. Thus, the output state is changed when a positive pulse appears at the non-inverting terminal of the IC, but quenching returns to the original state as soon as C_1 discharges. So it is called monostable circuit.

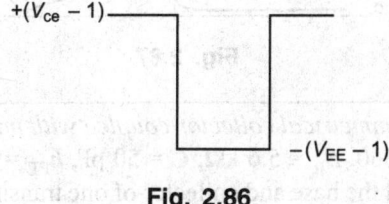

Fig. 2.86

The output is $+(V_{cc} - 1)$ V originally and changes to $(V_{EE} - 1)$ when triggered and returned to $+(V_{ce} - 1)$ V level afterwards. Thus, a negative pulse is generated. The output pulse width depends on C_1 or R_1 and R_2.

Since R_1 and R_2 and C_1 determine the time taken by the capacitor to discharge, and hence, the time for which the output remains at the negative level and thus the pulse width. If the bias level at the inverting terminal of the op-amp is charged by means of a potentiometer, the pulse width can be controlled. This is achieved by intending a potentiometer between R_2 and the supply or ground.

2.32 IC MONOSTABLE MULTIVIBRATOR

Instead of having two transistors and so many resistors and capacitors externally, monostable multivibrator is available in a single chip IC form. It is Motorola **MC951/MC851**.

V_{cc} is +5 V. The output will have an amplitude of 2 V. The pulse width PW is 100 nsec. This can be extended by externally connecting capacitors and resistors. It provides two output pulses which are complimentary.

Example 2.3 *Consider a collector coupled with stable multivibrator. The circuit and device parameters are $V_{cc} = 15$ V, $R_L = 3$ kΩ, $R_{B1} = 40$ kΩ $R_{B2} = 20$ kΩ, $h_{FE} = 30$, $r_{bb} = 0$, $I_{CB} = 0$. Neglect all forward bias voltages.*

(a) Calculate and plot to scale the waveforms at the base and collector of one transistor.

(b) If the multivibrator is to oscillate at a frequency of 5 kHz what is the value of ongoing capacitance? ($C_1 = C_2$).

(b) $T = 0.69\,(R_{B1}C_1 + R_{B2}C_2)$
          $f = 5$ kHz, $R_{B1} = 40$ K $R_{B2} = 20$ K

So, $C = 4{,}830$ pF
          $T' = R_C C = (3 \times 10^3)\,C = 14.5$ μsec

(a) $T_1 = 0.69\,R_{B1}C_1$
          $T_2 = 0.69\,R_{B2}C_2$.

But $R_{B2} = 20$ kΩ and $R_{B1} = 40$ kΩ

Therefore, $T_1 = 2T_2$

This is the purpose. So the waveforms are as shown.

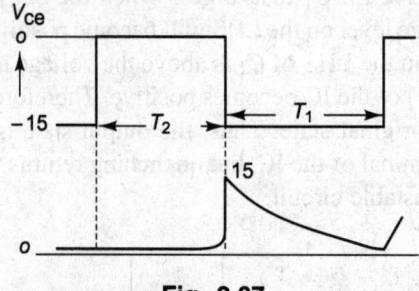

Fig. 2.87

Example 2.4 *Consider a symmetrical collector coupled with multivibrator. The circuit and device parameters are $V_{ce} = 6$ V, $R_L = 560$, $B_R = 5.6$ kΩ, $C = 50$ pF, $h_{FE} = 40$, $r_{bb'} = 2000$.*

Calculate (a) the waveforms at the base and collector of one transistor and plot to scale (b) frequency of oscillation and (c) recovery time.

$\delta =$ Overshoot at the base

$\delta' =$ Overshoot at the collector

$\delta = I_{B'}\,r_{bb'} + V_\alpha - V_r$

(b) Frequency of oscillation $f = 1.38\,RC$
          $= 2.52$ MHz

(c) $t_r = 2.2\,RC$ $R = (RC + r_{bb'})$; $t_r =$ Recovery time.
          $= 2.2\,(760)\,(5 \times 10^{-11}) = 83.6$ nsec

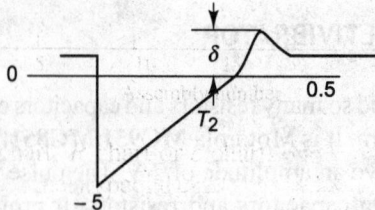

Fig. 2.88 *Overshoot in Monostable Multioutput Waveforms*

If we want to give both positive and negative triggering pulses, we must connect a diode in series. The formation of C_3 is to block DC only.

2.33 BISTABLE MULTIVIBRATOR WITH OP-AMPS

The circuit is shown in Fig. 2.98 (a) and waveforms in Fig. 2.89.

When $V_A > V_B$, V_o is $+ V_{o1(sat)}$

When $V_B > V_A$, V_o is $- V_{o2(sat)}$

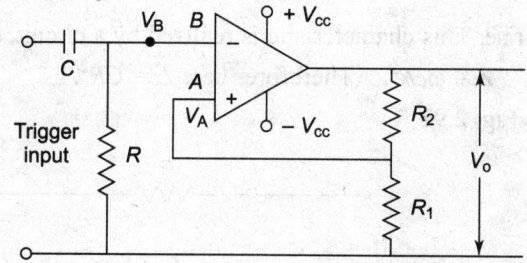

Fig. 2.89 *Op-amp Bistable Multivibrator Circuit*

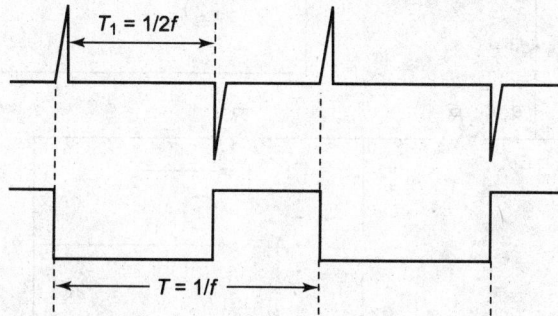

Fig. 2.90 *Bistable Multi-waveforms*

Trigger pulse

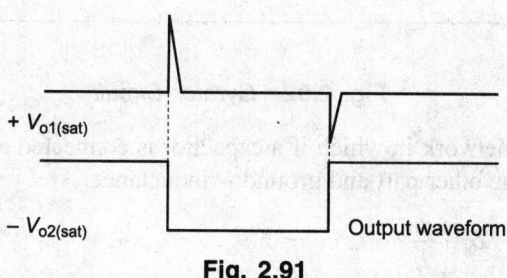

Fig. 2.91

Initially, the input to op-amp at A is positive with respect to B, because positive feedback is there through R_2 and R_1.

$\therefore$ $V_{sat} = + V_{o1(sat)}$. When a positive trigger is applied through C, the potential at B is greater than that at A ($V_B > V_A$). Since $V_{out} = -V_{o2(sat)}$ when a negative trigger is applied at C, $V_B < V_A$. So, $V_o = +V_{o1}$. The frequency of the output is the same as the frequency of the pulse (trigger) waveform.

Frequency of square wave is half the frequency of trigger pulses or T (square wave) = $2T_1$ (pulse).

2.34 GYRATOR

In integrated circuits, incorporating an inductor into the circuits is difficult from the fabrication point of view. But there are many circuits which employ inductors like filter circuits, tuned circuits, and so on.

Gyrator is a circuit using op-amps, which can simulate an inductance property with the help of resistors and capacitors. The property of inductance is $Z = \omega L = 2\pi fL$. As f increases, α_L or Z must increase.

With f rising, Z should rise. This characteristic is realized by a circuit, called gyrator, in the form

$$Z = \omega cR^2. \quad \text{Therefore, here } L = CR^2.$$

The circuit is as shown in Fig. 2.92

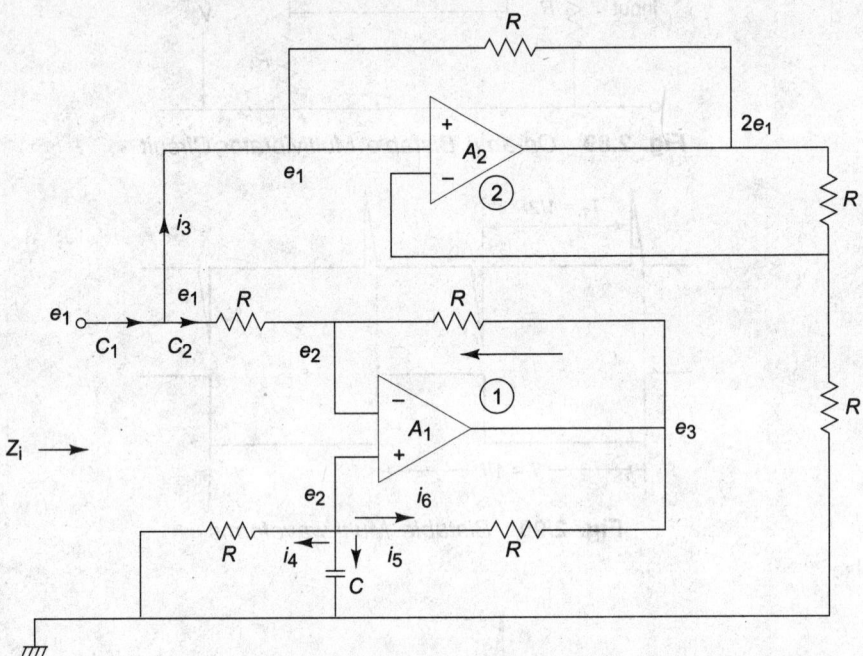

Fig. 2.92 *Gyrator Circuit*

A gyrator is a two-part network in which if a capacitor is connected across one part and grounded, the 'Z' seen looking into the other part and ground is inductance.

$$Z_i = \frac{e_1}{i_1} \qquad \qquad \qquad \dots(1)$$

$$i_1 = i_2 + i_3 \qquad \qquad \qquad \dots(2)$$

So, $$Z_i = \frac{e_1}{i_2 + i_3} \qquad \qquad \qquad \dots(3)$$

But $$i_2 = \frac{e_1 - e_2}{R} \qquad \qquad \qquad \dots(4)$$

e_1 is at higher potential. So, circuit flows into the circuit.

$$i_3 = \frac{e_1 - 2e_1}{R} = \frac{-e_1}{R} \qquad \qquad ...(5)$$

For op-amp (2), input is given to non-inverting terminal with gain of 2. It is as shown in Fig 2.93.

$$e_0' = \left(1 + \frac{R}{R}\right) e_1$$

$$= 2e_1 \quad \frac{e_1 - 0}{R} = \frac{-e_1 + e_0'}{R} \qquad \therefore e_0' = 2e_1.$$

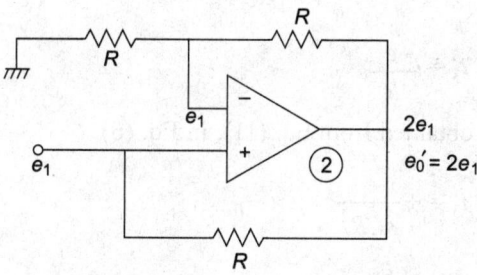

Fig. 2.93 *Part of Gyrator Circuit*

Therefore, $\qquad\qquad i_2 + i_3 = \dfrac{e_1 - e_2 - e_1}{R} = \dfrac{-e_2}{R} \qquad\qquad ...(6)$

For op-amp (1), by equating currents at the inverting input terminal (Fig. 2.92),

sum of all the circuits entering at a node = 0; $\dfrac{e_1 - e_2}{R} + \dfrac{(-e_2 + e_3)}{R} = 0 \qquad\qquad ...(7)$

So, $\qquad\qquad \dfrac{e_1 - e_2}{R} = \dfrac{e_2 - e_3}{R}$

By equating the currents at the non-inverting terminals,

$$\frac{+e_2}{R} + \frac{e_2}{1/sC} + \frac{e_2 - e_3}{R} = 0 \qquad\qquad ...(8)$$

$$\frac{-e_2}{R} = e_2 / (1/sC) + \frac{e_2 - e_3}{R}$$

$\dfrac{e_1}{R}$ has negative sign because circuit gets divided in opposite direction at the nodal point.

$$\left(\text{Total sum of circuit is zero. } \frac{-e_2 - e_3}{R} + \frac{e_2}{1/sC} + \frac{e_2}{R} = 0 \right)$$

$$i_4 + i_5 + i_6 = 0$$

Solving Eqs. (7) and (8):

from Eq. (7), $\qquad\qquad \dfrac{e_2 - e_3}{R} = \dfrac{e_1 - e_2}{R}$

Substitute this value in Eq. (8),

$$\frac{-e_2}{R} = \frac{e_2}{1/sC} + \frac{e_1 - e_2}{R} \qquad \qquad ...(9)$$

or

$$\frac{-\cancel{e_2}}{R} + \frac{\cancel{e_2}}{R} = \frac{e_2}{1/sC} + \frac{e_1}{R} \qquad \qquad ...(10)$$

So,

$$e_2 = \frac{-e_2}{sCR} \qquad \qquad ...(11)$$

But from Eq. (6),

$$i_2 + i_3 = \frac{-e_2}{R}$$

Substituting the value of e_2, obtained from Eq. (11), in Eq. (6)

We get,

$$i_2 + i_3 = +\frac{e_1}{sCR^2} \qquad \qquad (12)$$

from Eq. (1),

$$Z_i = \frac{e_1}{i_2 + i_3}$$

$$= \frac{e_1}{e_1/sCR^2} = sCR^2$$

so,

$$Z_i = sCR^2 = sL = j\omega L$$

where

$$L = CR^2$$

Therefore, the circuit acts like an inductor with a value

$$L = CR^2$$

if

$$R = 10 \text{ k}\Omega, \quad C = 0.01 \text{ μF}, \quad L = 1 \text{ H}.$$

In this circuit, magnetic field will not be induced when current passes through it, as in pure inductor.

2.35 WAVEFORM GENERATOR IC 8038

It is a 14 pur IC square wave; triangular and sinewave can be generated using this IC.
10 kΩ resistor is a pull-up resistor to obtain output voltage.

If pin 11 is given negative, then both positive and negative voltage swings will be obtained. If pin 11 is grounded above the 0 V, only excursions of the output voltage are obtained.

$$f = \frac{1}{\dfrac{5}{3} R_A C \left[1 + \dfrac{R_B}{2R_A - R_B} \right]}$$

for 50% duty cycle, $R_A = R_B = R$. Therefore, $f = 0.3/RC$.

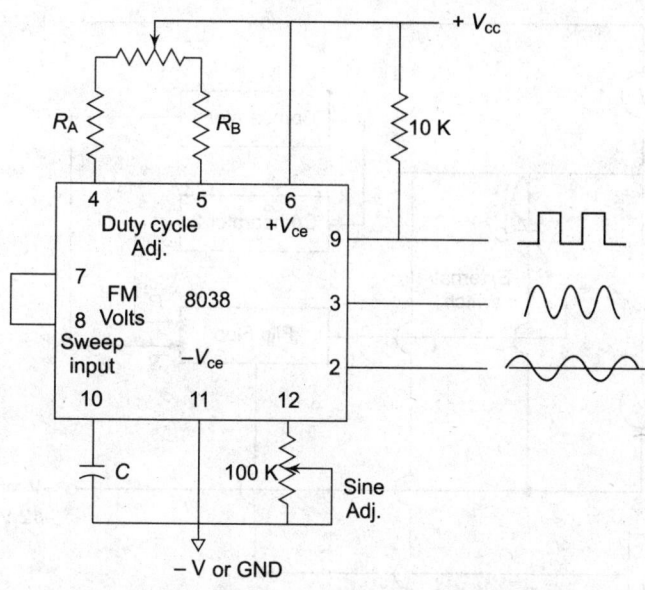

Fig. 2.94 *Waveform Generator IC 8038*

566: Waveform Generator IC. We can generate ⊓⊔ ∧∧ waveforms and also ramp waveform.

$$f = \frac{\frac{5}{2}(V - V_c)}{VRC}$$

Type 205 and 266 are also waveform generator ICs.

IC 8038 is the function generator 14 pin IC. The internal schematic is shown in Fig. 2.95.

When current source (1) drives the C voltage above the reference level, comparator 1 sets the FF. Then current source (2) is switched on. It discharges the capacitor at the same rate as charging. When the triangular wave voltage falls below the negative reference level, comparator 2 charges state and resets the FF. Therefore, current source 2 charges C in the negative direction. So, V_c, voltage across C, decreases.

Flip-flop will give positive and negative voltage (not just from 0 to negative value) because of the DC level shifter.

Output $|Z|$ from buffer circuit is 200 Ω.

Frequency range: 0.001 Hz to 500 kHz

Low distortion as 1%

Low-frequency drift with $T = 50$ ppm/°C more

By changing C, frequency can be changed. C can be varied from 0.001 μF to 10 μF. R value can be 10 kΩ.

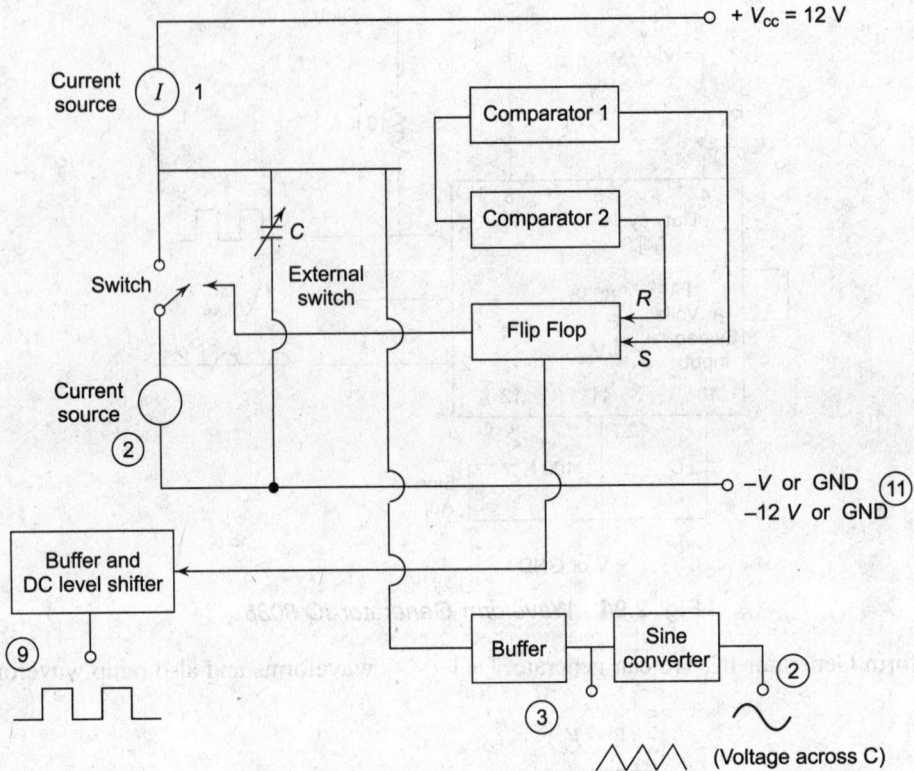

Fig. 2.95   *Internal Schematic of 8038 IC*

2.36   V/I CONVERTER

The op-amp V/I converter circuit is shown in Fig. 2.96.

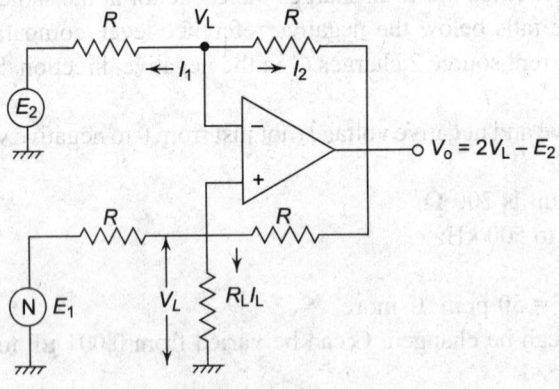

Fig. 2.96   *V/I Converter*

The current through R_L will not depend on R_L

$$I_L = \frac{E_1 - E_2}{R}$$

$$V_L = I_L R_L$$

$$I_{\rm L} \propto (E_1 - E_2), \qquad V/I \text{ conversion}$$

$$I_1 = -I_2$$

$$\frac{V_{\rm L} - E_2}{R} = \left(\frac{V_{\rm L} - V_{\rm o}}{R}\right) \qquad I_1 = \left(\frac{V_{\rm L} - E_2}{R}\right)$$

$$\frac{V_{\rm L}}{R} + \frac{V_{\rm L}}{R} = \frac{V_{\rm o}}{R} + \frac{E_2}{R} \qquad I_2 = \left(\frac{V_{\rm L} - V_{\rm o}}{R}\right)$$

or $\qquad\qquad 2V_{\rm L} = (V_{\rm o} + E_2)$

So, $\qquad\qquad V_{\rm o} = (2V_{\rm L} - E_2)$

2.37 BRIDGE AMPLIFIERS

The circuit for op-amp bridge amplifier is shown in Fig. 2.97.

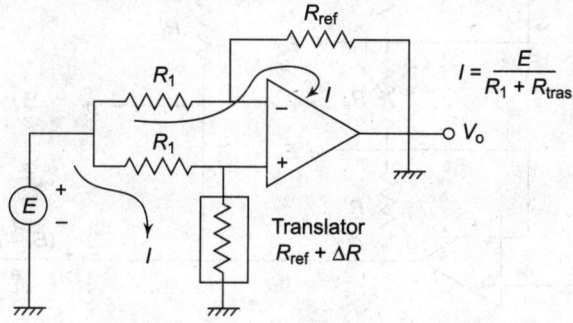

$$I = \frac{E}{R_1 + R_{\rm tras}}$$

Fig. 2.97 *Op-amp Bridge Amplifier*

$$I = \frac{E}{R_1 + R_{\rm ref} + \Delta R}$$

$$V_{\rm o} = E \frac{\Delta R}{R_1 + R_{\rm ref} + \Delta R}$$

2.38 RESISTANCE BRIDGE

Resistance bridge circuit is shown in Fig. 2.98.

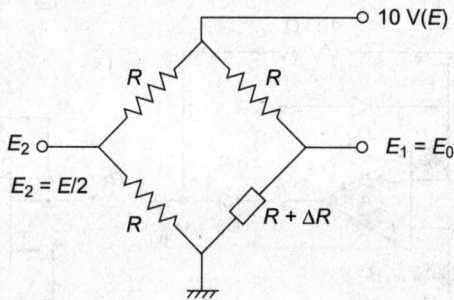

Fig. 2.98 *Resistance Bridge Circuit*

$$E_1 = E\,\frac{(R + \Delta R)}{R + R + \Delta R}$$

$$= \frac{E(R + \Delta R)}{(2R + \Delta R)}$$

$$E_2 = E/2$$

$$(E_1 - E_2) = \frac{E\Delta R}{4R}$$

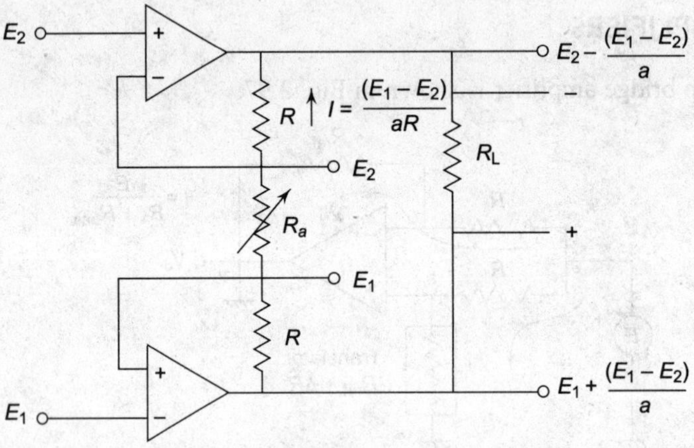

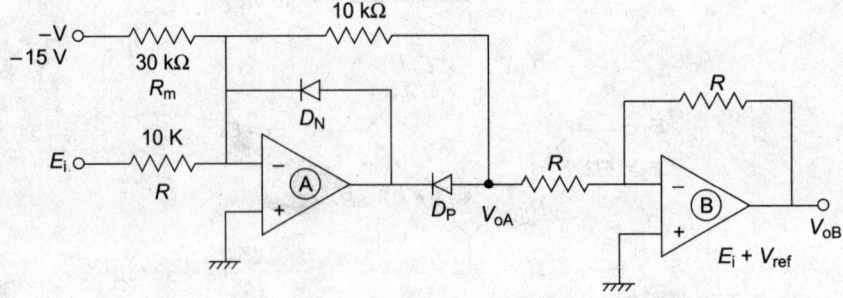

Fig. 2.99

$$V_\text{o} = (E_1 - E_2)\left(1 + \frac{2}{a}\right)$$

2.39 DEAD-ZONE WITH POSITIVE AND NEGATIVE OUTPUTS

2.39.1 Dead-zone Circuit with Positive Output

The circuit diagram is shown in Fig. 2.100. The waveforms are shown in Fig. 2.101.

Fig. 2.100 *Dead-zone Circuit with Positive Output*

$$V_{ref} = \frac{-V}{m} = -5 \text{ V}$$

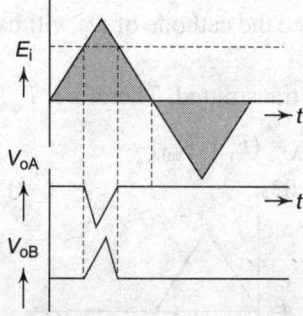

Fig. 2.101 *Waveforms*

2.39.2 Dead-zone Circuit with Negative Output

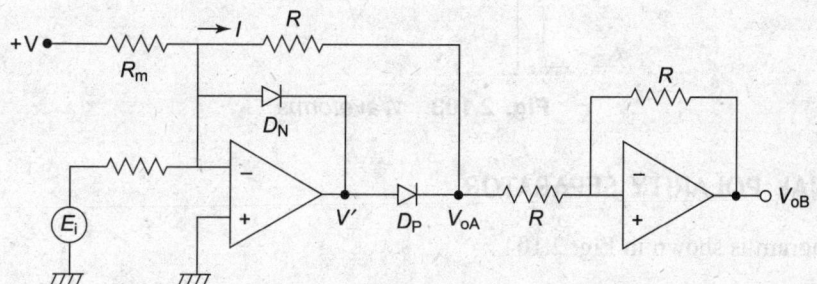

Fig. 2.102 *Dead-zone Circuit with Negative Output*

The operation of the circuit and analysis is just similar to op-amp HWR circuit.

+ V is a positive voltage source.

R_m is a resistor.

V_{ref} is such that $V_{ref} = +V/m$.

$$I = \frac{V_{ref}}{R} = \frac{V}{R_m}$$

If E_i is positive, D_N is forward biased.

Since the anode of D_N is at positive potential, V' is negative.

Therefore, D_P is reverse biased. Since its anode is at negative potential,

$$\therefore \qquad V_{oA} = 0. \quad \text{Therefore, } V_{oB} = 0$$

So, all positive E_i is blocked. Output is 0. When D_N is forward biased the current is confined to the inner loop only, as shown by the arrow mark. Current flowing through mR, R, D_P loop is zero. Therefore, $V_{oA} = 0$, $V_{oB} = 0$. All positive inputs of E_i are eliminated.

When E_i is negative

D_N is reverse biased, since its anode is at negative potential.

D_P is forward biased, when $E_i > V$ since the cathode of D_P will have D_P voltage, which is negative and hence D_P will get forward biased.

So, negative E_i greater than V will be transmitted. Therefore, $V_{oA} = + (E_v + V_{ref})$

$$\therefore \qquad V_{oB} = - V_{oA} = (E_i + V_{ref})$$

The waveforms are shown in Fig. 2.103.

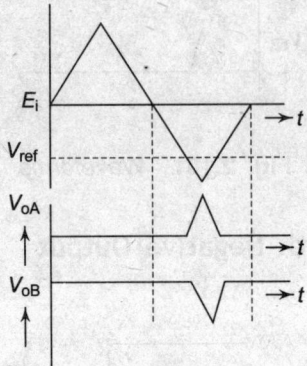

Fig. 2.103 *Waveforms*

2.4.0 SIGNAL POLARITY SEPARATOR

The circuit diagram is shown in Fig. 2.104.

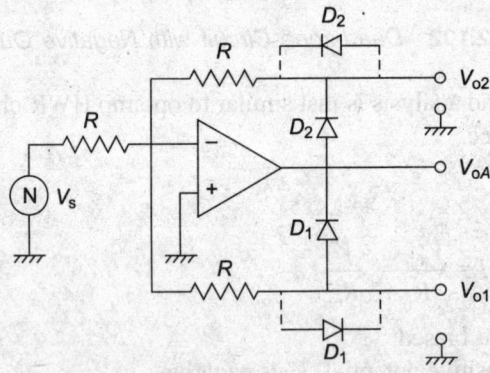

Fig. 2.104 *Signal Polarity Indicator*

D_1 can also be connected as shown ⌐---▷---⌐

D_2 can also be connected as shown ⌐---◁---⌐

that is, the positive of D_1 and D_2 can be reversed.

Using this circuit, we can find the polarity of the input V_s (which is not known).

If V_s is positive, V_{oA} is negative, V_{o1} is negative, D_1 is forward biased, and D_2 is reverse biased as in HWR circuit.

$V_{o2} = 0$ since V_{oA} is negative, and V_s is inverted since D_1 is forward biased.

If V_s is negative, V_{o1} is 0 V. V_{o2} is positive. D_1 is revese biased, D_2 is forward biased as in HWR *ckt*.

When V_s is positive, V_{oA} is negative $= V_{o1} = -0.6$ V

V_{oA} is negative, therefore, V_{o1} is negative $= -V_s$. D_2 is reverse biased. So, this loop is open, there is no current through R, hence, there is drop across R and therefore $V_{o2} = 0$ V.

When U_s is negative, D_2 conducts.

Because to the cathode of D_2, negative voltage is applied.

D_1 is off. Therefore $V_{o1} = 0$.

D_1 is off because to anode of D_1 negative voltage is applied.

$$V_{o2} = -(-V_s) = +V_s.$$

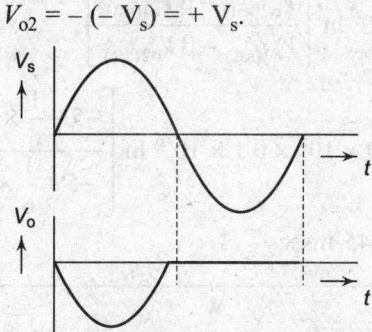

Fig. 2.105 *Output Waveforms*

Example 2.5 The following component values are used in the circuit of op-amp astable multivibrator $C = 0.1\ \mu\text{F}$, $R = 50\ \text{k}\Omega$, $R_1 = 10\ \text{k}\Omega$, $R_2 = 50\ \text{k}\Omega$. The output voltage is bounded to the limits $+10$ V and -5 V. Calculate the timing periods t_1 and t_2.

$$V_{o1\ (\text{sat})} = +10\ \text{V}$$

$$V_{o2\ (\text{sat})} = -5\ \text{V}$$

$$\beta = \frac{R_1}{R_1 + R_2}$$

So,

$$\beta = \frac{10}{10 + 50} = \frac{1}{6}$$

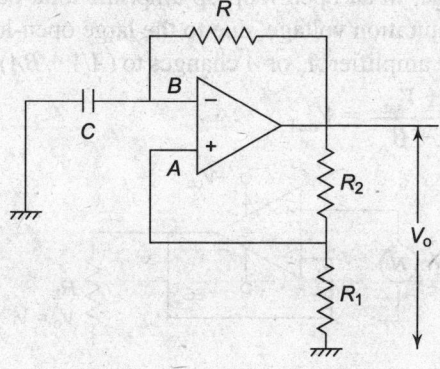

Fig. 2.106 *Circuit for Example 2.5*

$$e_c = E - (E - E_0) e^{-t/RC}$$

Since $V_{o1(sat)} \neq V_{o2\,(sat)}$

$$t_1 = RC \ln \left\{ \frac{V_{o1(sat)} - \beta V_{o2(sat)}}{V_{o1(sat)} - \beta V_{o1(sat)}} \right\}$$

$$= 50 \times 10^3 \times 0.1 \times 10^{-6} \ln \left\{ \frac{10 + \dfrac{1}{6} \times 5}{10 - \dfrac{1}{6} \times 10} \right\}$$

$$= 1.312 \text{ msec}$$

$$t_2 = RC \ln \left[\frac{V_{o2(sat)} - \beta V_{o1(sat)}}{V_{o2(sat)} - \beta V_{o2(sat)}} \right]$$

$$= 50 \times 10^3 \times 0.1 \times 10^{-6} \ln \left[\frac{-5 - \dfrac{1}{6} \times 10}{-5 + \dfrac{1}{6} \times 5} \right]$$

$$= 3.45 \text{ msec}$$

Example 2.6

For the circuit shown, determine A_f, R_{if}, R_{of}, and f_f and V_{ooT}

Given, $\qquad\qquad A_v = 200,000$ or A

$\qquad\qquad\qquad R_i = 2 \text{ M}\Omega$

$\qquad\qquad\qquad R_o = 75 \; \Omega$

$\qquad\qquad\qquad f_o \simeq 5 \text{ Hz}$

Supply voltages: ± 15 V

Output voltage supply: ± 13 V

$$f_f = \text{bandwith with feedback}$$

$$f_0 = \text{bandwith without feedback}$$

V_{ooT} is the total output offset voltage.

Due to the input offset voltage, in an open-loop op-amp, the total output offset voltage is equal to either the positive or negative saturation voltage, due to the large open-loop gain A_v or A.

With feedback, the gain of the amplifier A_v or A changes to $(A/1 + BA)$, the total output offset voltage with feedback also changes to $\dfrac{\pm V_{sat}}{1 + \beta_A} = V_{ooT}$

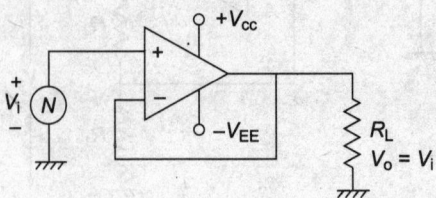

Fig. 2.107   *Circuit For Example 2.6*

since the op-amp is in voltage follower configuration, $\beta = 1$.

Therefore, $\qquad\qquad 1 + \beta A \simeq 200,000$

This is the case of voltage series feedback.

$$R_{of} = R_i (1 + \beta_A) = 2 \text{ M}\Omega \ (200,000) = 400 \ \mu\Omega$$

$$R_{of} = \frac{R_o}{1 + \beta_A} = \frac{75 \ \Omega}{200,000} = 0.375 \ \mu\Omega$$

$$f_f = (1 + \beta_A) f_o$$

$$5 \text{ Hz} \ (200,000) = 1 \text{ MHz}$$

$$V_{ooT} = \frac{\pm V_{sat}}{1 + \beta_A} = \frac{\pm 13 \text{ V}}{200,000} \pm 65 \ \mu\text{V}$$

A_f = voltage gain at frequency f
R_{if} = input resistance with feedback
R_{of} = output resistance with feedback
f_f = bandwidth with feedback.

2.41 CHARGE AMPLIFIER

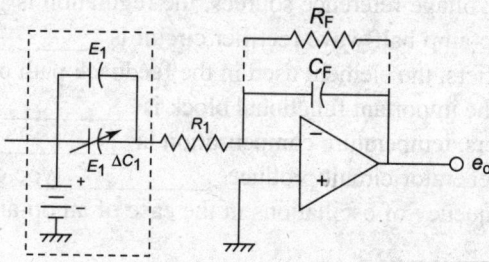

Fig. 108 *Charge Amplifier Circuit*

Some transducers such as capacitance microphases, piezoelectric transducers operate on the principle of conversion of the measurement variable into an equivalent change. The equivalent circuit of such a transducer may be represented by a battery and capacitor in series

As capacitance C_1 charges, charge becomes

$$\Delta q = \Delta C_1 E \qquad Q = CV$$

Decharge flows into the feedback capacitor CP.

Resulting charge in charge generates an output voltage.

$$e_0 = - \Delta C_1 . \frac{E}{C_F}; \text{ Therefore, } e_0 = \frac{x_{ef}}{x_{cl}}, -e_i$$

R_F will present output of op-amp going to saturation due to continuous charging.
R_1 limit the output current. $Q = CV$.
Sensitivity of charge amplifier

$$\frac{e_0}{\Delta C_1} = \frac{-E}{C_F}$$

SUMMARY

- By using an op-amp for DC or AC, V and I measurements, loading of input source can be reduced.
- Scale changing is possible when an op-amp is used as voltage reference source.
- An op-amp can be used as a V/I converter.
- By using an op-amp for rectification, half-wave or full-wave voltages less than V_γ (out in voltage) can be rectified. Therefore, such circuits are known as super diode circuits.
- The relation between V_i and rectified V_o in the case of an op-amp HWR circuit is linear. So, this circuit is, therefore, also known as a linear rectifier circuit.
- Multiplication or division of two electrical signals V_1 and V_2 is possible using op-amp logarithmic and antilogarithmic amplifiers.
- Using two op-amps, rectangular and triangular waveforms can be generated.

FILL IN THE BLANKS

1. The advantage of using an op-amp for current or voltage measurement is _____.
2. An op-amp current to voltage converter is also called _____.
3. In the case of op-amp voltage reference sources, the regulation is _____.
4. The advantage of an op-amp half-wave rectifier circuit is _____.
5. The logarithmic amplifiers, the element used in the feedback path of op-amp is _____.
6. In analog multipliers, the important functional block is _____.
7. In logarithmic amplifiers, temperature compensation is _____.
8. An op-amp function generator circuit produces _____ type of waveforms.
9. The expression for frequency of oscillations in the case of an op-amp function generator circuit with usual notation is _____.
10. An op-amp HWR circuit is also known as _____.

ANSWERS

1. loading of the input source is reduced.
2. current amplifier.
3. better
4. voltage less than cut in voltage V_γ can be rectified.
5. diode
6. logarithmic amplifier
7. necessary
8. square and triangular
9. $f = R_2/4RR_1C$
10. linear rectifier circuit, super diode

ESSAY-TYPE QUESTIONS

1. Explain how an op-amp can be used for V and I measurements. What are the advantages of this circuit?
2. Draw the circuits and explain the voltage of op-amp voltage reference sources.
3. What is a super diode circuit? What are its features? Draw the circuit and explain.
4. Draw the op-amp FWR circuit and explain its working.
5. Derive the expression for the output of an op-amp logarithmic amplifier.
6. Draw the op-amp function generator circuit and deliver the expression for f_o.

UNSOLVED PROBLEMS

1. For the circuit shown in the figure below determine the current through the meter, I_m.

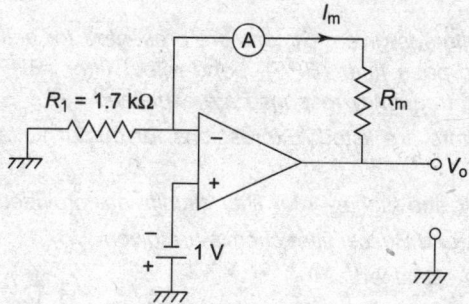

2. In the above circuit for Problem 1, if a microammeter of 100 μA full-scale range is connected in the feedback path, calculate R_1 to get full-scale output of 5 V.
3. Using op-amp, a high resistance universal voltmeter circuit is to be designed. A 50 mA ammeter is being used in the feedback path of the circuit. Determine the value of resistor to indicate full-scale deflection when the voltage to be measured to 6 V DC and 6 V peak. Give the switching network and draw the circuit.
4. Repeat the above problem if 8 V rms, 8 V peak, and 8 V p–p voltages are to be measured.
5. A zener diode is to be tested using op-amp circuit. V_o = 10 V, V_i = 5 V, R_1 = 2 kΩ. Find the zener current and zener voltage. Draw the circuit.
6. A p–n junction diode is to be tested using op-amp circuit. V_i = 2 V. R_1 = 1.2 kΩ. V_o = 0.5 V. Find the diode current and voltage drop across the diode.
7. An op-amp is being used as voltage-to-current converter. The value of resistance used in the circuit R is 6.8 kΩ, R_L = 2 kΩ, V_1 = 5 V, V_2 = 0 V. Determine the values of I_L, V_L and V_o. Draw the circuit.
8. In the above problem, if R = 4.7 kΩ, R_L = 1.2 kΩ, V_1 = 0 V, determine, I_L, V_L and V_o.
9. In an op-amp I/V converter circuit, V_o is 6 V, R_f = 470 kΩ. Determine the value of short circuit current I_{sc}.
10. For the circuit shown, a light-dependent resistor (LDR) is connected in the input circuit. The value of resistor in the feedback path is 15 kΩ. If the current through the LDR is 10 μA in darkness and 1.2 μA in sunlight, find V_o.

Active Filters and Oscillators

Objectives:

In this chapter...

- *Classification of active filters is given. Circuits are presented for active low-pass filter (LPF), high-pass Filter (HPF), band-pass filter (BPF), band-reject filter (BRF), resonant filter, and notch filter. Circuit working and frequency response are explained.*
- *Advantages of active filters are listed. Expressions for output voltage and cut-off frequency are derived.*
- *First-order, second-order, and higher-order filter circuits are provided and expressions derived.*
- *Butterworth, Chebyshev, and Bessel filter circuits are given.*
- *Design of filter circuits is explained.*
- *Various oscillator circuits using op-amps such as RC phase shift oscillator circuit, Wien bridge oscillator circuit, Hartley oscillator circuit, Colpitt oscillator circuit are described. The expression for frequency of oscillation is provided.*

3.1 CLASSIFICATION OF FILTERS

Filter circuits are used to eliminate or attenuate unwanted frequency signals. As the meaning of the term 'filter' implies, electronic filter circuits filter out unwanted signals. Filter circuits are classified as:

1. Active filters
2. Passive filters

3.2 ACTIVE FILTERS

Filter circuits using active components like op-amps are called active filters. In active filter circuits, active elements like BJTs, and ICs are used. In passive filter circuits passive components like R, L, and C are used. Network synthesis is the term applied to the systematic method of specifying components in a circuit so that the circuit's output is a prescribed function of its input function of frequency.

3.2.1 Types of Active Filter Circuits

1. Low-pass filter (LPF)   Allows low-frequency signals only. Attenuates higher-frequency signals beyond the cut-off frequency f_c

2. High-pass filter (HPF)   Allows higher-frequency signals beyond cut-off frequency f_c

3. Band-pass filter (BPF)   Allows signals in a particular frequency range called bandwidth

4. Band reject filter (BRF)   Attenuates in a particular frequency range
   or notch filter

5. Resonant filters   Allows any one particular frequency component (signal) f_o

6. Notch filters   Alternates one particular frequency component f_r

The frequency response of these circuits are shown in Fig 3.1(a).

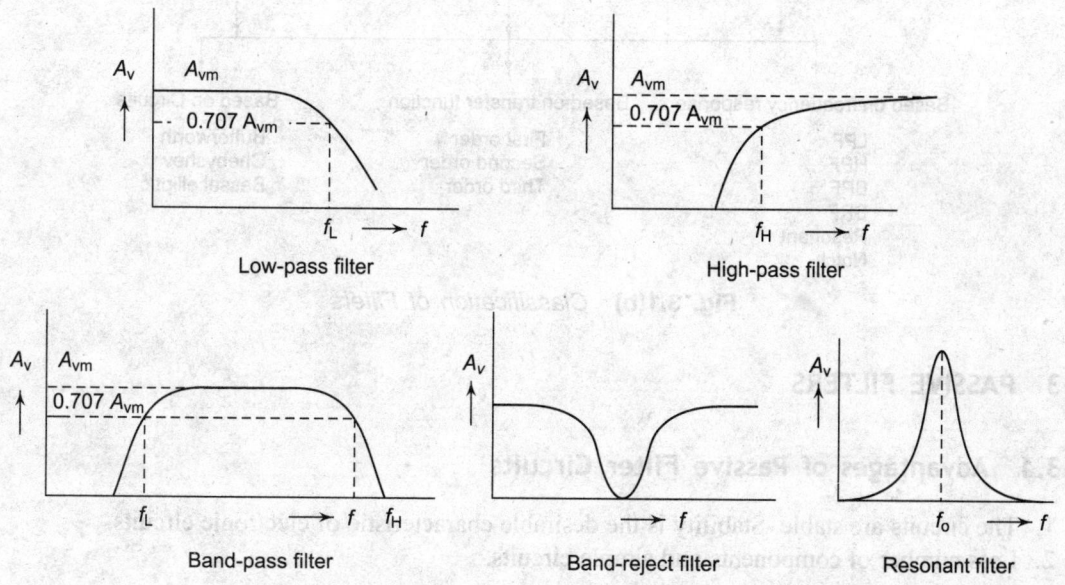

Fig. 3.1(a)   *Frequency Response Characteristics of Various Filters*

3.2.2   Advantages of Active Filter Circuits

1. Loading effect can be reduced using op-amps.
   Active filter circuits can be made to have higher input impedance and lower output impedance, using op-amps, to prevant loading effect. Active filter circuits can be made to have higher input Z and low output Z.

2. Voltage gain can be greater than 1, whereas for passive circuits it is always less than 1.

3. Cheaper circuits. Because with low values of R and C, we can realize the same transfer functions. Op-amps is cheap. Hence, filter is less costly.

4. They eliminate the need of inductors that are bulky and costly.
   1. $A_v \geq 1$ 2. No loading effect of the signal source
   3. Filter circuits in IC form   4. Large variety in standard configuration

3.2.3   Disadvantages of Active Filter Circuits

1. Active filter circuits can not perform satisfactorily, usually above 100 kHz, because of the limitation of active elements.

2. Selectivity and sensitivity are dependent on the gain of op-amp.

Filter circuits are used where the signals contain noise and so on, and only signals on a certain frequency range will be selected.

Classification of Filters: The classification of filters is shown in Fig. 3.1(b).

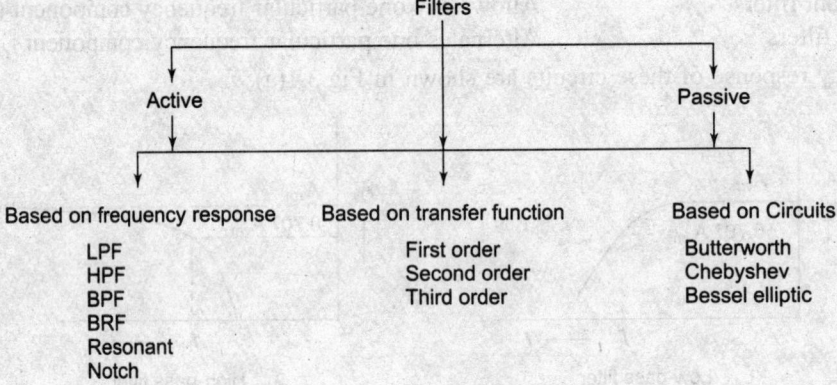

Fig. 3.1(b)   *Classification of Filters*

3.3 PASSIVE FILTERS

3.3.1 Advantages of Passive Filter Circuits

1. The circuits are stable. Stability is the desirable characteristic of electronic circuits.
2. Less number of components and simple circuits.

3.3.2 Disadvantages of Passive Filter Circuits

1. Voltage gain $\dfrac{V_o}{V_i} < 1$. Voltage gain is less than 1.
2. Large values of inductors (L), capacitors (C), and resistors (R) may be needed (which are costly) to reduce loading effect, such values of R, L and C are to be chosen.
3. Costly.
4. Loading effect will be there.

3.4 LOW-PASS NETWORK (PASSIVE FILTER CIRCUIT)

The circuit and waveforms are shown in Fig. 3.2(a). For low frequency signals, $X_c = \dfrac{1}{2\pi f_c}$ value is

large. So, the capacitor acts as an open circuit. The signal is thus transmitted without attenuation. Hence, the frequency response curve is flat as shown in Fig. 3.2(b). For higher frequencies beyond f_c, X_c is short. So, the signal is attenuated. Hence A_v falls.

Expression for voltage gain A_v:

$$V_o = I \cdot X_c$$

$$I = \frac{V_o}{R + \frac{1}{\sqrt{\omega_c}}}$$

$$V_o = \frac{V_i X_c}{R + X_c}; \quad \frac{V_o}{V_i} = \frac{(1/sC)}{R + \frac{1}{sC}}$$

$$V_o = \frac{V_{in}/j\omega C}{\sqrt{R^2 + X_c^2}}$$

$$\frac{V}{V_i} = \frac{1}{1 + sCR}$$

$$\frac{V_o}{V_{in}} = \frac{X_c}{\sqrt{R^2 + X_c^2}}$$

$$\frac{V_o}{V_{in}} = \frac{1}{\sqrt{2}} = 0.707$$

$$\therefore \quad \text{At } f = f_c, \ R = X_c \text{ or } f_c = \frac{1}{2\pi RC}$$

$$\frac{V_o}{V_i} = \frac{1}{1 + sCR}, \ f_c = \frac{1}{2\pi RC}$$

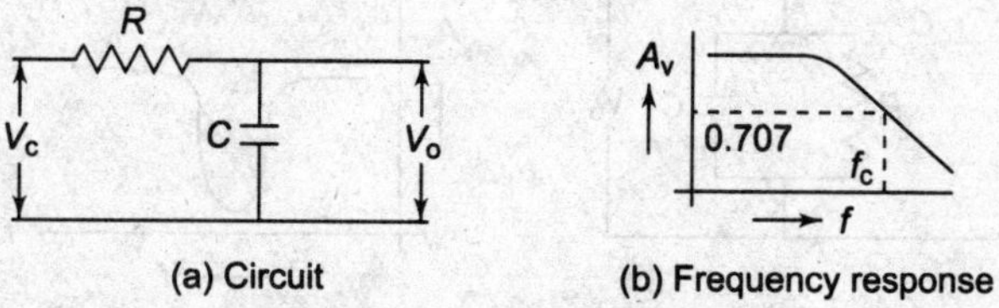

(a) Circuit (b) Frequency response

Fig. 3.2 *Low-Pass Network*

3.5 HIGH-PASS NETWORK (PASSIVE FILTER CIRCUIT)

The circuit and frequency response are shown in Fig. 3.3. When f of the input signal is small, X_c is large. So V_i is attenuated. V_o is small and A_v is less. As frequency increases, X_c decreases. So, V_o and A_v increase. Hence, the frequency response is as shown in Fig. 3.3(b).

$$I = \frac{V_{in} R}{R + \frac{1}{j\omega C}}$$

$$V_o = IR$$

$$V_o = \frac{RV_{in}}{\sqrt{R^2 + X_c^2}}$$

$$V_o = \frac{V_{in} R}{R + \frac{1}{j\omega C}} = \frac{V_o}{V_{in}} = \frac{1}{1 + \frac{1}{sCR}}$$

$$\frac{V_o}{V_{in}} = \frac{sCR}{1 + sCR}$$

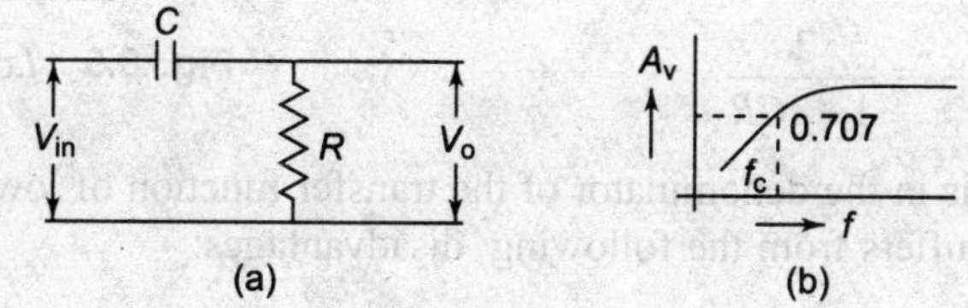

(a) (b)

Fig. 3.3 *HPF Circuit and Frequency*

$$\text{At } f = f_c, \quad \frac{V_o}{V_{in}} = \frac{1}{\sqrt{2}} = 0.707$$

$$A_v = 0.207, \text{ where } R = X_c. \qquad \frac{V_o}{V_i} = \frac{sCR}{\sqrt{1 + s^2 C^2 R^2}}$$

So, $\qquad\qquad$ at $f = f_c = R = X_c$ as $\qquad\qquad f_c = \dfrac{1}{2\pi R_c}$

$$f_c = \frac{1}{2\pi RC}$$

3.6 NOTCH FILTER OR BRF

The circuit and waveforms are shown in Fig. 3.4.

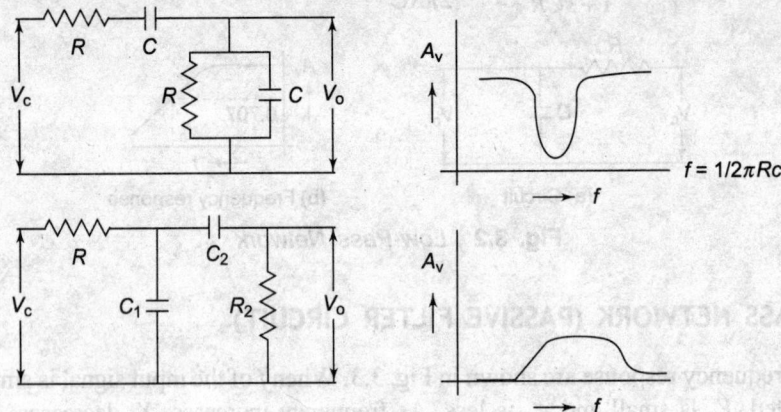

Fig. 3.4 *Notch Filter or BRF*

3.7 BAND-PASS FILTER

If f is very large, X_{c1} is short. If f is very small, X_{c2} is open. Therefore, only input signals in a certain frequency range are allowed.

The upper cut-off frequency f_H should be 10 times lower cut-off frequency f_L. Then this circuit works.

3.8 LOW-PASS FILTER

The circuit is shown in Fig. 3.5.
The transfer fraction of low pass filter is

$$\frac{V_o}{V_i} = \frac{1}{1 + sCR}$$

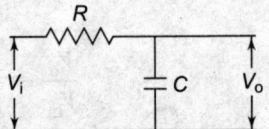

Fig. 3.5 *Low-Pass Filter Network*

Note that the operator 's' is in the denominator of the transfer function of low pass filter.
The passive LPF circuit suffers from the following disadvantages:

1. Cut-off frequency and pass band gain depend on load and source impedances. If the source supplying V_i is not ideal, its resistance will also get added to the load resistance, that is the value of R will change.
2. Another similar RC network can not be cascaded because it will change the attentuation characteristic. There will be a loading effect too.

If an op-amp is added at the input stage of the RC network and also at the output stage, the above problem can be overcome.

Thus, it becomes an active filter. And it is a first-order active filter since the transfer function contains operator 's' with power of 1 (in the term sCR).

3.8.1 First-Order Active LPF

The circuit is shown in Fig. 3.6. It is active filter since active device op-amp is used as a buffer to provide high input impedance and low output impedance.

Transfer function for first-order active LPF is $\dfrac{V_o}{V_i} = \dfrac{1}{1 + sCR}$

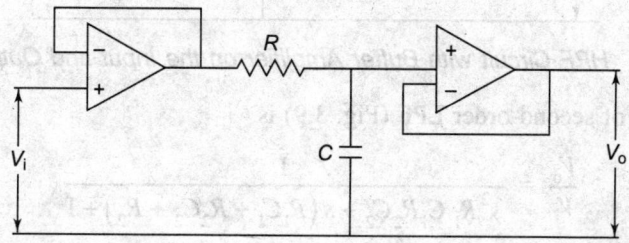

Fig. 3.6 *Active LPF with Buffer Amplifiers on Input and Output Side*

3.9 HIGH-PASS FILTER

Transfer function $\dfrac{V_o}{V_i} = \dfrac{sCR}{1 + sCR}$

Note that the operator 's' is in the numerator of the transfer function of the HPF.

3.9.1 First-Order Active HPF

The circuit is shown in Fig. 3.7.
The op-amp is used in voltage follower configuration, which has high input Z_i and low output Z_o. Thus, it provides impedance matching without affecting the characteristic resistance of the filter circuit.

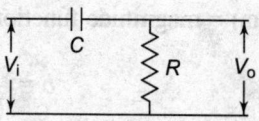

Fig. 3.7 *High-Pass Filter Network*

$$V_o = IR$$

$$I = \frac{V_i}{R + \dfrac{1}{sC}} = \frac{V_i sC}{1 + sCR}$$

$$\frac{V_o}{V_i} = \frac{sC}{1+sCR}$$

3.9.2 Second-Order LPF

A second-order LPF is obtained by cascading two first-order LPFs. Because the transfer function contains a term s^2, it is called second-order filter.

Depending upon whether the filter allows low frequencies or high frequencies, it is called second-order LPF or HPF, respectively.

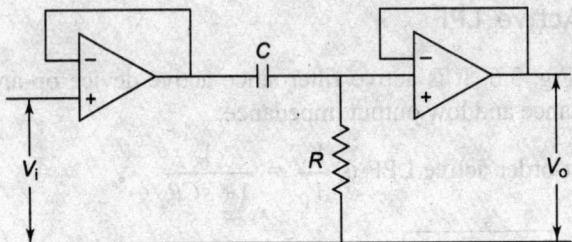

Fig. 3.8 *HPF Circuit with Buffer Amplifier on the Input and Output Side*

A transfer function of second-order LPF (Fig. 3.9) is

$$\frac{V_o}{V_i} = \frac{1}{s^2 R_1 C_1 R_2 C_2 + s\left(R_1 C_1 + R_2 C_2 + R_2\right) + 1}$$

$$R_1 = R_2, C_1 = C_2$$

$$\frac{V_o}{V_i} = \frac{1}{1 + 2\, s R_1 C_1 + s^2 R_1^2 C_1^2}$$

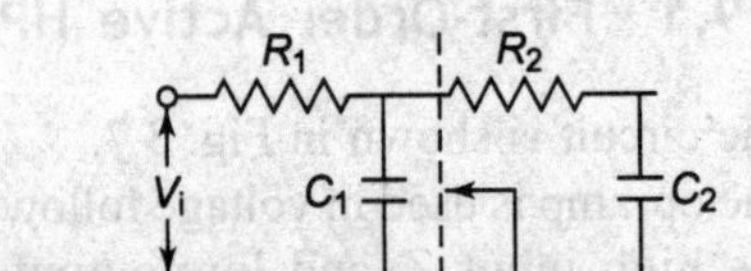

It is of the form

$$\frac{V_o}{V_i}(\omega) = \frac{1}{1 + ax + bx^2}$$

Fig. 3.9 *Second-Order LPF Network*

$H(\omega)$ = magnitude function.

$$b = \frac{1}{(\omega_n)^2}$$

$$a = \frac{2K}{\omega_n}$$

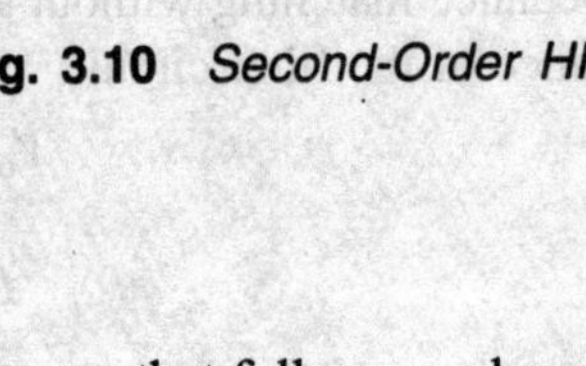

Fig. 3.10 *Second-Order HPF*

ω_n = natural resonance frequency of the circuit

K = damping coefficient

A plot of $|H(\omega)|$ with ω for second-order LPF is shown in Fig. 3.11.

If the value of the damping factor K is changed, the characteristic changes that follow are shown in the graph.

The stability of the circuit depends on the magnitude of the transfer function and the damping of the oscillations.

R_2C_2 components will load the input side (Fig. 3.9).

This will reduce the damping factor transfer function.

$$A_v = \left(1 + \frac{R_2}{R_1}\right)$$

The general form of second-order active filters can be shown as in Fig. 3.10 with unity gain.

Z_4 is grounded. Voltage across Z_4 is given as input to op-amps.

If it is a first-order active filter, there will be only Z_2 and Z_4.

If it is a second-order active filter, all the four impedances will be considered.

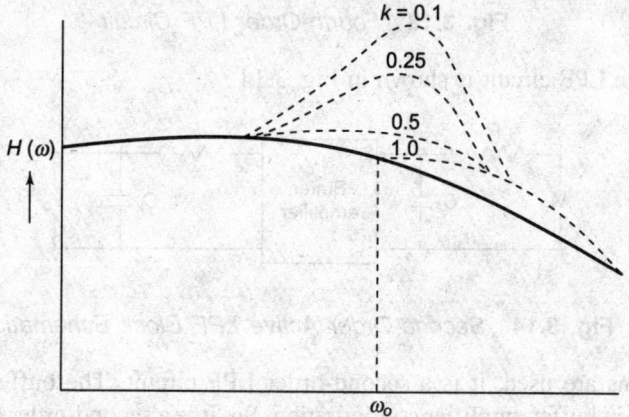

Fig. 3.11 *Frequency Response of Second-Order HPF*

If it is LPF, Z_1 and Z_2 are resistor components (Rs), and Z_3 and Z_4 are capacitor components (Cs), If the circuit is HPF, Z_1 and Z_3 are Cs, and Z_2 and Z_4 are Rs.

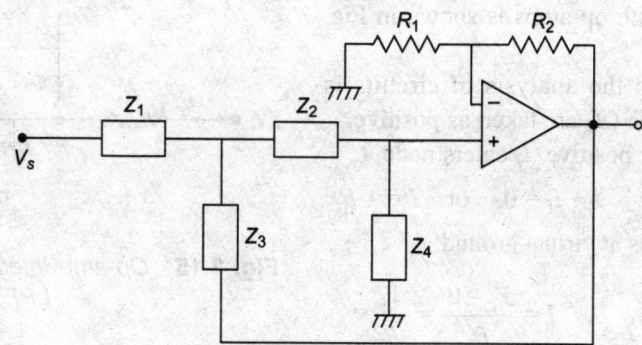

Fig. 3.12 *Fourth-Order Filter, General Circuit Configuration*

To get fourth-order filters, two second-order filters are cascaded as shown in Fig. 3.13.

3.9.3 Fourth-Order Active LPF

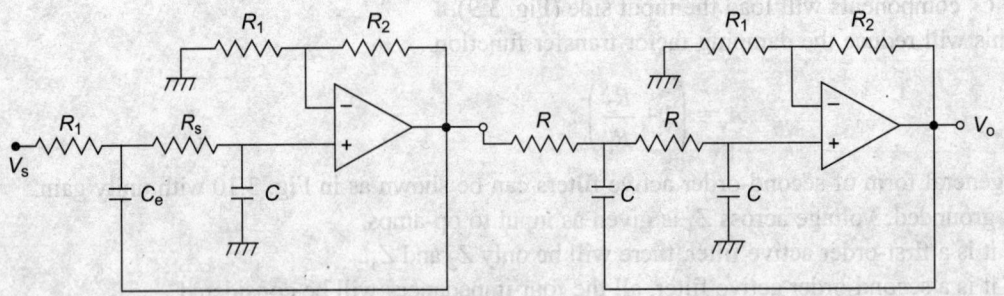

Fig. 3.13 *Fourth-Order LPF Circuit*

Second-order active LPF circuit is shown in Fig. 3.14.

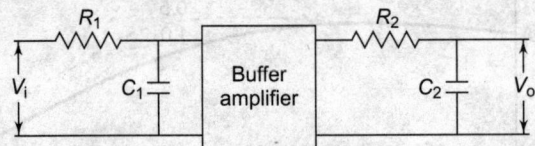

Fig. 3.14 *Second-Order Active LPF Block Schematic*

As two *RC* sections are used, it is a second-order LPF circuit. The buffer amplifier uses active elements like op-amp in buffer amplifier configuration. So, it is a second-order LPF active filter circuit.

3.9.4 First-Order LPF Circuit

An op-amp integrator circuit acts as an LPF circuit. Active LPF circuit with op-amps is shown in Fig. 3.15.

The notation used in the analysis of circuits is current entering a node (A) are taken as positive. Therefore, I and I_f are positive. I_f enters node A.

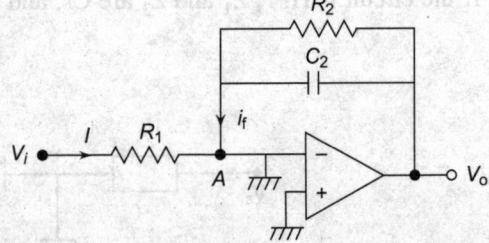

So, $I + I_f = 0 \quad \text{or} \quad I = -I_f$

Inverting terminal is at virtual ground.

Fig. 3.15 *Op-amp Integrator as First-Order LPF*

Therefore, $I = \dfrac{V_i - 0}{R_1} = \dfrac{V_i}{R_1}$

$$I_f = \frac{V_o - 0}{Z_f} = \frac{V_o}{Z_f}, \quad Z_f = \frac{R_2}{1 + j\omega C R_2}$$

$$I_f = \frac{V_o}{\left(R_2 /(1 + J\omega C R_2)\right)}$$

$$I = -I_f$$

So,
$$\frac{V_o}{V_i} = \left(\frac{R_2}{R_1}\right)\frac{1}{1 + scR_2}$$

It is a first-order filter because the order of the operator 's' is 1 (s^1).

It is LPF because as f increases, $(1 + scR_2)$ value $s = j\omega$ increases. Therefore, the $\dfrac{1}{1 + scR_2}$ value decreases. So, output V_o also decreases. So, high-frequency signals are attenuated.

Thus, low-frequency signals are transmitted and high-frequency signals are attenuated. Hence, it is an LPF circuit.

The cut-off frequency is the frequency at which the V_o value becomes $\dfrac{1}{\sqrt{2}}$ of its maximum value. This will happen when $X_c = R_2$.

$$\left|\frac{V_o}{V_i}\right| = \left(\frac{R_2}{R_1}\right)\frac{1}{\sqrt{1 + \left(\dfrac{R_2}{X_c}\right)^2}}$$

When $X_c = R_2$ $\quad \left|\dfrac{V_o}{V_i}\right| = \left(\dfrac{R_2}{R_1}\right)\dfrac{1}{\sqrt{2}}$

The corresponding frequency f at which $X_c = R_2$ is called the cut-off frequency f_c.

$$\frac{1}{2\pi f_c C_2} = R_2$$

So,
$$f_c = \frac{1}{2\pi C_2 R_2}; \quad \omega_c = \frac{1}{C_2 R_2}$$

The steady-state sinusoidal response of magnitude is

$$|A(jf\omega)| = \frac{R_2}{R_1}\frac{1}{\sqrt{1 + \left(\dfrac{\omega}{\omega_2}\right)}}$$

$$\omega_c = \frac{1}{CR} = \text{cut-off frequency}$$

Advantages of first-order active op-amp LPF

1. For this circuit, gain can be greater than 1. By making $\dfrac{R_2}{R_1} > 1$, the voltage gain can be increased.

2. Loading effect is reduced.

3.10 SALLEN AND KEY CIRCUIT

In the case of filter circuits, the sensitivity of the circuit has to be improved, that is the dependence of the transfer function on various parameters, component values, stability of the circuit, and so on. This can be done by designing higher-order-filters, like second order, fourth order, and so on. The higher the order of the filter cicuit, the better are the characteristics.

Sallen and Key active filter circuits using op-amps are commonly used. They consist of LPF or HPF *RC* networks connected in cascade with op-amps in voltage follower configuration.

3.10.1   Second-Order Active LPF (Sallen and Key Circuit)

The gain average of this circuit is 1. The circuit is shown in Fig. 3.16.

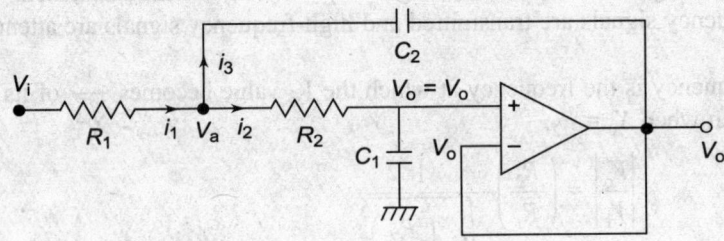

Fig. 3.16   *Sallen and Key Circuit*

Applying Kirchhoffs' current law at node V_a, $i_1 - i_2 - i_3 = 0$
Since i_1 is entering node V_a,
i_2 and i_3 are leaving node.
Therefore, $i_1 = i_2 + i_3$.
C_1 is grounded. Therefore, V_o across C_1 is given to op-amp.

3.10.1.1   Advantages

1. Voltage gain can be greater than 1.
2. Loading effect can be reduced.

If op-amp is in voltage follower configuration, the voltage at the non-inverting terminal = V_o

$$\frac{V_i - V_a}{R_1} = \frac{V_a - V_o}{R_2} + (V_a - V_o)sc_2$$

$$(i_1) = (i_2) + (i_3) \tag{3.1}$$

$$\frac{V_i - V_a}{R_1} + \frac{V_o - V_a}{R_2} + (V_o - V_a)\,sc_2 = 0 \tag{3.2}$$

$$\frac{V_a - V_o}{R_2} - \frac{V_o}{\left(\dfrac{1}{sc_1}\right)} = 0 \tag{3.3}$$

For the network $R_1\,R_2\,C_1\,C_2$, the current i_2 flows through R_2 and C_1 only because the op-amp does not draw any current $V_b = V_o$.

Since the op-amp is in voltage follower configuration, the voltage of the inverting terminal (V_b) = voltage of non-inverting terminals (V_o).

$$V_b = V_o$$

$$\therefore \qquad \left(\frac{V_o - V_a}{R_2}\right) - V_o (sc_1) = 0$$

Taking Eqs. (3.2) and (3.3) to get the transfer function

$$\frac{V_o(s)}{V_i(s)} = \frac{1}{1 + sc_1(R_1 + R_2) + s^2 R_1 R_2 C_1 C_2}$$

Currents entering node Z are taken as positive. Currents leaving node Z are taken as negative. Current flows from higher potential to lower potential. Therefore,

$V_s > V_a; V_a > V_s, V_a > V_o$

So, $i_1 - i_3 - i_2 = 0$ (at node V_a)

Therefore,
$$\frac{V_i - V_a}{R_1} - (V_a - V_o) sc_2 - \frac{V_a - V_o}{R_2} = 0 \tag{3.4}$$

or
$$\frac{V_i - V_a}{R_1} = (V_a - V_o) sc_2 + \left(\frac{V_i - V_o}{R_2}\right)$$

$$\frac{(V_a - V_o)}{R_2} - \frac{V_o}{(1/sc_1)} = 0 \tag{3.5}$$

Comparing this with the standard form,

$$\frac{V_o(s)}{V_i(s)} = \frac{1}{I_L + as + bs^2}$$

$$b = \frac{1}{\omega_n^2}, \quad \omega_n = \text{natural resonant frequency of the circuit}$$

$$a = \frac{2K}{\omega_n}, \quad K = \text{damping factor } (\zeta)$$

$$a = C_1 (R_1 + R_2)$$
$$b = R_1 R_2 C_1 C_2$$

$$\omega_n = \frac{1}{\sqrt{b}}$$

So,
$$\omega_n = \frac{1}{\sqrt{R_1 R_2 C_1 C_2}} \tag{3.6}$$

$$a = \frac{2K}{\omega_n} = C_1 (R_1 + R_2)$$

Therefore,
$$K = \frac{\omega_n a}{2} = \frac{C_1(R_1 + R_2)}{2} \frac{1}{\sqrt{R_1 R_2 C_1 C_2}}$$

or
$$K = \frac{R_1 + R_2}{2\sqrt{R_1 R_2}} \sqrt{\frac{C_1}{C_2}}$$

Therefore, the natural frequency of the circuit is

$$\omega_n = \frac{1}{\sqrt{R_1 R_2 C_1 C_2}}$$

$$K(\zeta) = \frac{R_1 + R_2}{2\sqrt{R_1 R_2}} \sqrt{\frac{C_1}{C_2}}$$

If

$$R_1 = R_2 = R$$

$$\omega_n = \frac{1}{R\sqrt{C_1 C_2}}$$

$$k(\zeta) = \sqrt{\frac{C_1}{C_2}}$$

3.10.2 Third-Order Active LPF (Sallen and Key Circuit)

By adding another R–C section to the second-order LPF, it becomes a third-order circuit as shown in Fig. 3.17.

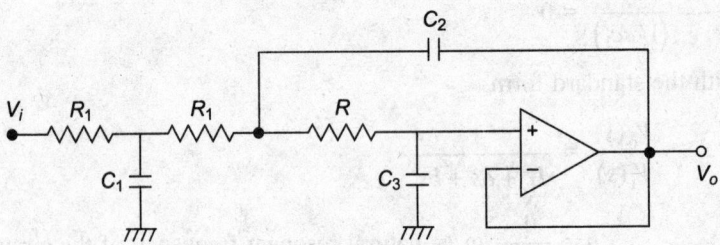

Fig. 3.17 *Third-Order Active LPF is Sallen and Key Form*

The A_v of this circuit is 1.
If equal value resistances are used, then the transfer function is

$$\frac{V_o(s)}{V_i(s)} = \frac{1}{C_1 C_2 C_3 R^3 s^3 + 2(C_1 C_3 + C_2 C_3)R^2 s^2 + (C_1 + 3C_3)R + 1}$$

nth order LPFs are designed by combining $\dfrac{n}{2}$ second-order sections in case n is even, or one-third order section with $\left(\dfrac{n}{2} - \dfrac{3}{2}\right)$ second-order sections if n is odd.

3.11 GENERAL EXPRESSION FOR SECOND-ORDER HPF

The general expression for second-order HPF for voltage gain using 's' operator $A_v(s)$ is given as

$$A_v = \frac{A_o\left(s^2/\omega_o^2\right)}{1 + \left(\dfrac{b_s}{\omega_o}\right) + \left(\dfrac{s^2}{\omega_o}\right)^2}$$

3.11.1 Second-Order Active HPFs (Sallen and Key Circuits)

The circuit diagram is shown in Fig. 3.18.
R and C positions of the LPF are interchanged.

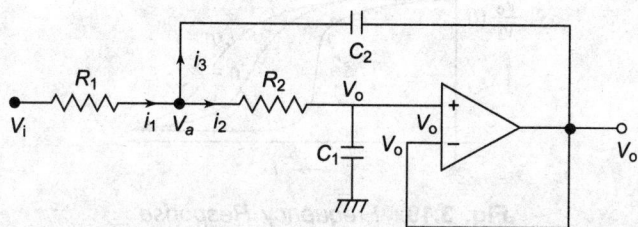

Fig. 3.18 *Second-Order Active HPF Circuit in Sallen and Key Form*

$$i_1 = i_2 + i_3$$

$$(V_i - V_a)\, sc_1 = (V_a - V_o)\, sc_2 + \frac{(V_a - V_o)}{R_2} \qquad (3.7)$$

$$(V_i - V_a) = (V_a - V_o)\, sc_2 + \frac{(V_a - V_o)}{R_2}$$

$$(i_1) = (i_2) + (i_3)$$

$$\frac{(V_i - V_a)}{(1/sc_2)} - \frac{V_o}{R_1} = 0 \qquad (3.8)$$

The transfer function is in the form f_{OH} f must be lowercase

$$\frac{V_o}{V_i}\,(s) = \frac{R_1 R_2 C_1 C_2\, s^2}{1 + s R_2\,(C_1 + C_2) + s^2 R_1 R_2 C_1 C_2}$$

This equation can be reduced to the standard form. The standard form of the transfer function is

$$\frac{V_o}{V_i}\,(s) = \frac{s^2 / \omega_n^2}{1 + 2K\left(\dfrac{s}{\omega_n}\right) + \left(\dfrac{s^2}{\omega_n^2}\right)}$$

$$= \frac{s^2 / \omega_n^2}{1 + as + bs^2}$$

$$\omega_n = \frac{1}{\sqrt{R_1 R_2 C_1 C_2}}\; ; \; K = \frac{C_1 + C_2}{2\sqrt{C_1 C_2}}\left[\sqrt{\frac{R_2}{R_1}}\right]$$

If $C_1 = C_2 = C$,

$$\omega_n = \frac{1}{C\sqrt{R_1 R_2}}$$

$$K = \sqrt{\frac{R_2}{R_1}}$$

If the order of the filter is increased, the frequency response becomes steeper as shown in Fig. 3.19.

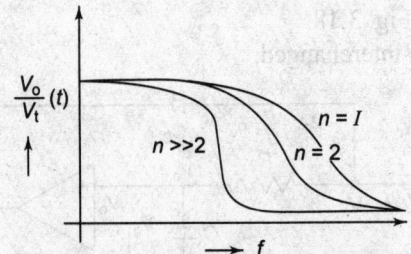

Fig. 3.19 *Frequency Response*

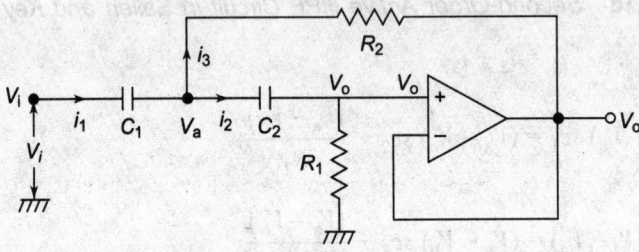

Fig. 3.20 *Second-Order HPF, Active Filter Circuit*

3.12 BAND-REJECT FILTER

A BRF can be obtained by combining an LPF with cut-off frequency f_{OH} and HPF with a cut-off frequency f_{OL}.

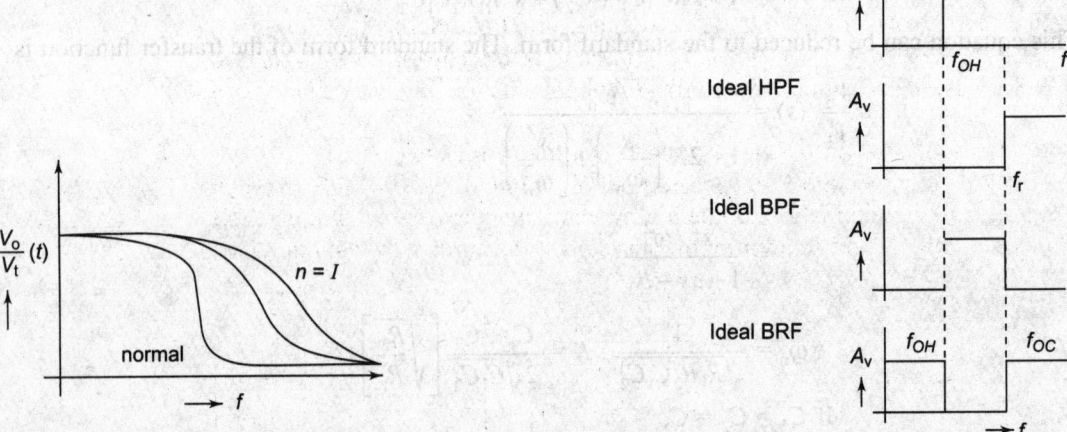

Fig. 3.21 *Frequency Response Variation with Changing Value of n*

Fig. 3.22 *Frequency Response*

Since these two values f_{OH} and f_{OL} are different and there is certain frequency range in which these two filters do not transmit the input signals, so a BRF can be obtained.

When LPF and HPF are combined, a BPF is obtained.

The transfer function for all pass filter is $\left(\dfrac{1-s}{1+s}\right)$. It is a first-order equation.

3.12.1 Band-Reject Filters

The block schematic for BRF and the frequency response are as shown in Fig. 3.23.

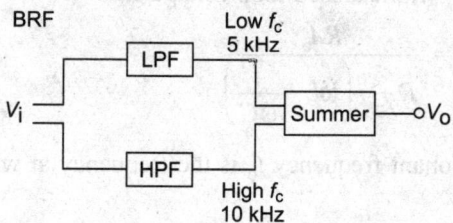

Fig. 3.23 *Block Schematic of Active Resonant BRF.*

3.13 ACTIVE RESONANT BPF

Because of the resonant circuit, signals of a particular frequency f_o will be amplified and transmitted and other signals, including noise, will be attenuated. The characteristic is as shown in Fig. 3.24.

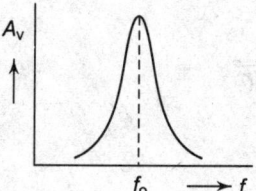

Fig. 3.24 *Characteristics of an Active Resonant BPF*

The circuit will consist of inductance L and capacitance C. It is a resonant circuit.

V_s = signal supply voltage

V_i = actual input to op-amp, there is drop across L and C and output voltage from op-amps is V_o.

$$A_o = \frac{V_o}{V_i} = \text{voltage gain w.r.t. the input of op-amps}$$

$$V_c = \text{voltage across } R$$

$$\frac{V_o}{V_i} = \frac{V_s R}{R + j\left(\omega L - \dfrac{1}{\omega c}\right)}$$

Let

$$A_v = \frac{V_a}{V_s}$$

$$A_v = \frac{V_a}{V_s} = \left(\frac{V_o}{V_i}\right)\left(\frac{V_i}{V_s}\right)$$

$$\frac{V_o}{V_i} = A_o; \quad \frac{V_i}{V_s} = \frac{R}{R + j\omega L - \dfrac{j}{\omega C}}$$

A_o is positive and constant for all frequencies because it depends on only the bandwidth of the op-amps, which is assumed to be a flat response.

A_o = open-loop voltage gain;

A_v = finite closed-loop voltage gain

So,

$$A_v = \frac{RA_o}{R + j\left(\omega L - \dfrac{j}{\omega C}\right)}$$

The centre frequency or resonant frequency f_o is the frequency at which $\omega_o L = \dfrac{1}{\omega_o C}$. Therefore, at resonance

$$\omega_o L = \frac{1}{\omega_o C}$$

So,

$$\omega_o^2 = \frac{1}{LC}$$

$$\omega_o = \frac{1}{\sqrt{LC}} \tag{3.9}$$

$$f_o = \frac{1}{2\pi\sqrt{LC}}$$

$$Q = \frac{1}{\omega_o CR} \tag{3.10}$$

Q = eliminating ω_o,

$$Q = \frac{1}{R}\sqrt{\frac{L}{C}} \tag{3.11}$$

Substituting Eq. (3.11) in Eq. (3.9),

So,

$$\omega_L = \frac{Q\omega}{\omega_o}$$

$$\frac{1}{\omega C} = Q\omega_o$$

$$|A_v(j\omega)| = \frac{A_o}{\sqrt{1 + Q^2\left(\dfrac{\omega}{\omega_o} - \dfrac{\omega_o}{\omega}\right)^2}} \tag{3.12}$$

Bandwidth $BW = (\omega_2 - \omega_1)$; $\omega_2 =$ upper cut-off frequency, $\omega_1 =$ lower cut-off frequency

$$= \frac{1}{2\pi}\left(\omega_2 - \frac{\omega_o^2}{\omega_2}\right)$$

Therefore, $\omega_o^2 = \sqrt{\omega_1 \omega_2}$

The resonant circuit is shown in Fig. 3.25.

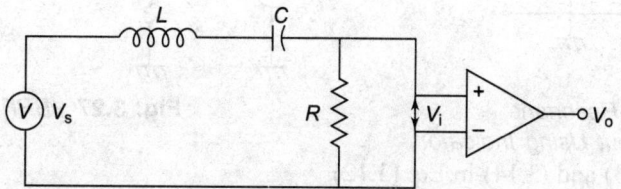

Fig. 3.25 *Active Resonant BPF*

Its transfer function is derived using operator s as given below,

$$A_o = \frac{V_o}{V_i} = \text{voltage gain of op-amps only}$$

Its transfer function is

$$A_v (j\omega) = \frac{V_i}{V_s} = \frac{V_o}{V_s}$$

$$= \frac{R_1 A_o}{R + j\left(\omega L - \dfrac{1}{\omega C}\right)} = \frac{R A_o}{R + j\omega L + \dfrac{1}{j\omega C}}$$

Substituting s for $j\omega$,

$$A_v(s) = \frac{R A_o}{R + sL + \dfrac{1}{sC}}$$

Multiplying the numerator and denominator by $\left(\dfrac{s}{L}\right)$

$$A_v(s) = \frac{s A_o\left(\dfrac{R}{L}\right)}{s^2 + s\left(\dfrac{R}{L}\right) + \dfrac{1}{LC}}$$

We know that $\qquad \omega_o^2 = \dfrac{1}{LC}$ $\qquad\qquad\qquad\qquad$ (3.13)

$$Q = \frac{\omega_o L}{R} = \frac{1}{\omega_o CR} = \frac{1}{R}\sqrt{\frac{L}{C}} \qquad\qquad\qquad (3.14)$$

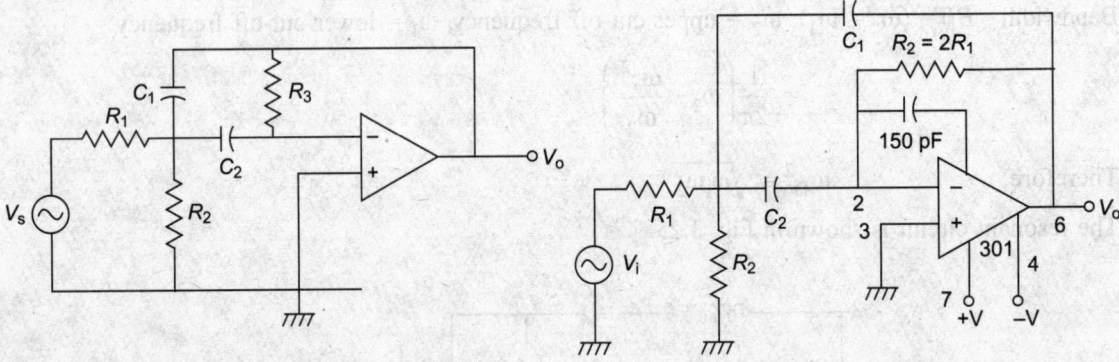

Fig. 3.26   *Circuit for Resonant*
           BPF Without Using Indicator

Fig. 3.27   *BPF Circuit*

Substituting Eqs. (3.13) and (3.14) in Eq. (3.12)

$$\frac{R}{L} = \frac{\omega_o}{Q}$$

$$\frac{1}{LC} = \omega_o^2$$

$$A_v(s) = \frac{\left(\dfrac{\omega_o}{Q}\right)A_o \cdot s}{s^2 + \left(\dfrac{\omega_o}{Q}\right)s + \omega_o^2}$$

This is the transfer function of the second-order band-pass (resonant) filter (s^2 terms are included). This can be realized without using any inductor and with only Rs and Cs as shown in the circuit. There is no inductor in the circuit.

The transfer function for this circuit is,

$$A_v(s) = \frac{V_o(s)}{V_s} = \frac{\dfrac{s}{R_1 C_1}}{s^2 + \dfrac{C_1 + C_2}{R_3 C_1 C_2}s + \dfrac{1}{R' R_3 C_1 C_2}} \tag{3.15}$$

where

$$R' = R_1 \parallel R_2 = \frac{R_1 R_2}{R_1 + R_2}\ ;\ R_1 \text{ in parallel with } R_2$$

$$A_v(s) = \frac{(R/L)A_o s}{s^2 + s(R/L) + \dfrac{1}{LC}} \tag{3.16}$$

$$A_v(s) = \frac{(\omega_o/Q)A_o s}{s^2 + (\omega_o/Q)s + \omega_o^2} \tag{3.17}$$

Comparing Eqs. (3.15), (3.16), and (3.17)

$$R_1 C_1 = \frac{L}{RA_o} = \frac{Q}{\omega_o A_o} \tag{3.18}$$

$$\frac{R_3 C_1 C_2}{C_1 + C_2} = \frac{L}{R} = \frac{Q}{\omega_o} \tag{3.19}$$

$$R'R_3C_1C_2 = LC = \frac{1}{\omega_o^2} \qquad (3.20)$$

Design equations for BPF. Circuit diagram is shown in Fig. 3.23(b)

$$\text{Bandwidth} \qquad = \frac{w_r}{Q} = \frac{1}{2\pi f_r Q}$$

$$R_2 = \frac{2}{(BW)(C)}; \quad R_1 = \frac{R_2}{2}; \quad R_3 = \frac{R_2}{4Q^2}$$

Example 3.1 *Design a second-order BPF with mid-band voltage gain $A_o = 50$, centre frequency $f_o = 160$ Hz, and bandwidth = 16 Hz.*

Solution

$$\text{Bandwidth} = \frac{f_o}{Q}$$

So,

$$Q = \frac{f_o}{\text{Bandwidth}} = \frac{160}{16} = 10$$

$$\omega_o = 2\pi f_o = 1,000 \text{ rad/sec}$$

Let

$$C_1 = C_2 = 0.1 \ \mu\text{F, for BPF, } R_1C_1 = \frac{L}{RA_o} = \frac{Q}{\omega_o A_o}$$

$$R_1 = \frac{Q}{A_o\omega_o C} \qquad \text{(from previous equations)}$$

$$R_1 = \frac{10}{50 \times 10^3 \times 0.1 \times 10^{-6}} = 20 \ \Omega$$

$$R_3 \left(\frac{C_1C_2}{C_1 + C_2} \right) = \frac{Q}{\omega_o}$$

So,

$$R_3 = \frac{Q}{\omega_o \left(\dfrac{C_1C_2}{C_1 + C_2} \right)} = \frac{10}{1000 \left(\dfrac{0.1 \times 0.1}{0.2} \right) \times 10^{-6}}$$

$$= 200 \text{ k}\Omega$$

$$R' = \frac{1}{\omega_o^2 R_3 C_1 C_2} \qquad \text{(from previous equations)}$$

$$= \frac{1}{10^6 \times 2 \times 10^5 \times 10^{-14}} = \frac{10^8}{2 \times 10^5} = \frac{10^3}{2} = 500 \ \Omega$$

Therefore,

$$R_2 = \frac{R_1 R'}{R_1 + R'} = \frac{20 \times 500}{20 + 500} = \frac{1000\cancel{0}}{52\cancel{0}} \simeq 20 \ \Omega$$

3.14 ELECTROTHERMAL (ETC) FILTERS: (ETH FILTERS) OR ELECTROTHERMAL RESISTANCE AND CAPACITANCE FILTERS

These filters are used in the frequency range <1 Hz to audio frequency range. For normal active or passive LPFs, the useful frequency range is the audio range to about 1 MHz. If these were to be used for low-frequency range, the value of C becomes very large. The structure and characteristics of electro-thermal filters are shown in Fig. 3.28.

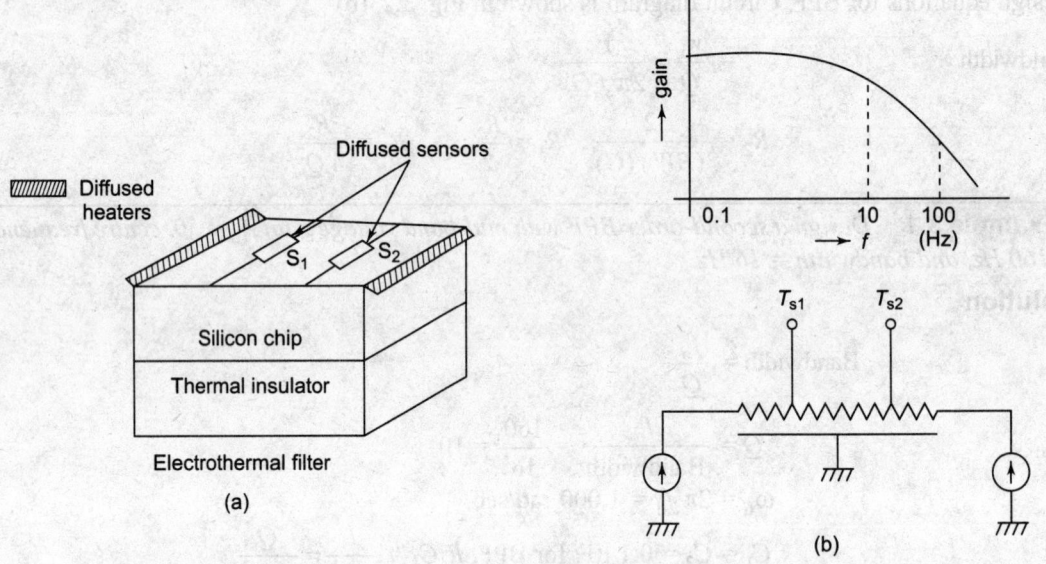

Fig. 3.28 *Frequency Response and Equivalent Circuit of ETC Filters*

So the thermal effects in silicon are used to realise filters. Such filters are called ETC filters (C for capacitance). The distributed thermal resistance and thermal capacitance of the silicon chip act as LPFs to the heat signals.

3.14.1 First-Order HPF

The circuit is shown in Fig. 3.29. A differentiator circuit used as an HPF.

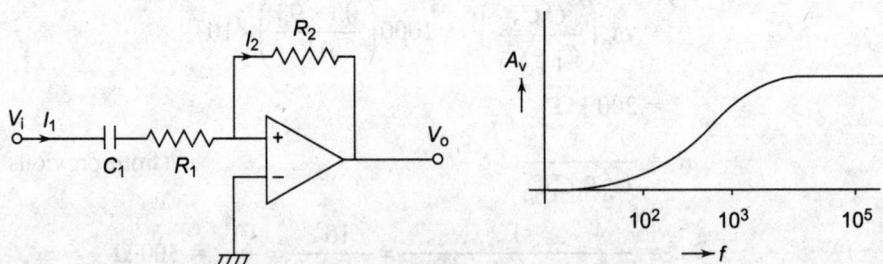

Fig. 3.29 *Circuit and Frequency Response of First-Order HPF*

$$I_1 = \frac{V_i}{Z_1} = \left(\frac{V_i}{R_1 + \frac{1}{sC_1}} \right)$$

$$I_2 = \frac{V_o}{R_2}$$

$$I_1 = I_2$$

So,
$$\frac{V_i}{R_1 + \dfrac{1}{sC_1}} = \frac{V_o}{R_2}$$

Therefore,
$$\frac{V_o}{V_i} = \frac{R_2}{R_1 + \dfrac{1}{sC_1}}$$

$$A_v(s) = \left(\frac{R_2}{R_1}\right)\frac{s}{\left(s + \dfrac{1}{C_1 R_1}\right)}$$

This is the transfer function of the first-order HPF. Note that the operator s is in the numerator of the transfer function of HPF.

3.14.2 Second-Order HPF

The circuit is shown in Fig. 3.30.

$$A_v(s) = \frac{A_o s^2}{s^2 + b s \omega_o + \omega_o^2}$$

ω_o = cut-off frequency/natural frequency of the circuit

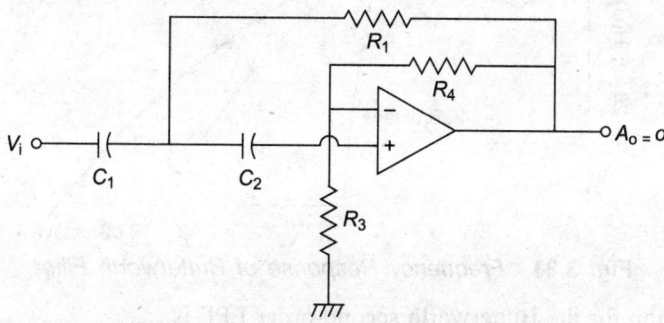

Fig. 3.30 *Circuit of Second-Order HPF*

3.15 CLASSIFICATION OF FILTERS BASED ON CHARACTERISTICS

Filters are classified as (1) low pass, (2) high pass, (3) band pass, and so on. This classification is based on frequency response.

Filters are also classified as first order, second order, third order, and so on, depending on the order of s in the transfer function. In addition, filter circuits of the same type and of the same order (e.g., first-order low-pass filter) can be further classified based on their characteristics as follows:

  (a) Butterworth filters
  (b) Chebyshev filters

(c) Bessel filters

(d) Elliptic filters

Many other combinations also exist.

3.16 BUTTERWORTH FILTERS

These filters are optimised for minimal flatness of the frequency response characteristic in the pass band. They possess the following characteristics:

1. The attenuation monotonically increases in the pass band.
2. Attenuation is 3.01 dB at the cut-off frequency.
3. It further increases at the rate of $(n \times 6)$ dB/octave, where n is the order of the filter.
4. The magnitude function $|H(\omega)|$ is given as

that is, $\dfrac{V_o}{V_i}$ (s), where $s = j\omega$, and considering its magnitude,

$$|H(\omega)| = \left\{1+\left(\frac{\omega}{\omega_o}\right)^{2n}\right\}^{-1/2}$$

The frequency response of a Butterworth LPFs is shown in Fig. 3.31.

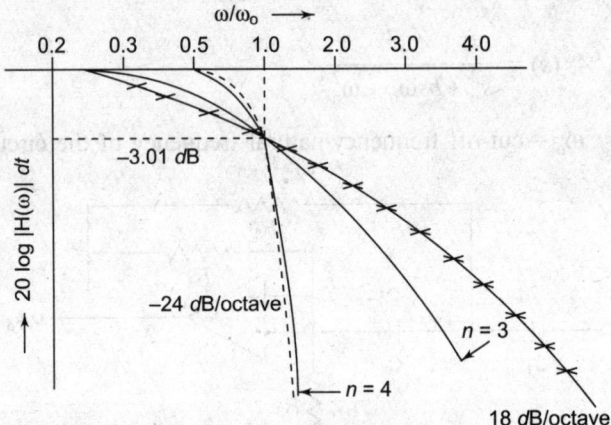

Fig. 3.31 *Frequency Response of Butterworth Filter*

The transfer function for the Butterworth second-order LPF is

$$\frac{1}{s^2+1.4142\,s+1},$$

third order $\dfrac{1}{s^3+2\,s^2+2\,s+1}$, and

fourth order $\dfrac{1}{s^4+2.6131s^3+3.4142s^2+2.613s+1}$.

A one-pole Butterworth LPF has an optimal closed-loop voltage gain 'A_{CL}'. A two-pole Butterworth LPF needs closed-loop voltage gain A_{CL} of 1.586. A three-pole Butterworth LPF requires two sections of RC networks.

3.17 CHEBYSHEV FILTERS

These filters have a higher rate of attenuation than Butterworth filters at the cut-off frequency. The pass-band regions of the frequency response of this type of filter do not show monotonic increase in filtration, as exhibited by Butterworth filters. Instead, the attenuation can increase or decrease and again increase, depending on the order. Beyond the cut-off frequency, the attenuation increases monotonically.

The rate of attenuation at the cut-off frequency is higher for Chebyshev filters. These filters also show an attenuation on the pass band, which changes non-monotonically with frequency, that is the characteristic posseses ripples. The filter with a higher ripple will also have a higher attenuation rate at the cut-off frequency.

The transfer function of third-order Chebyshev filters with ripple of 0.5 dB is

$$\left[\frac{0.83813}{s^2 + 0.53663 + 0.83813}\right]\left[\frac{0.535}{s + 0.56}\right]$$

Whether it is LPF or BPF, the transfer functions remain the same. Any circuit can be chosen such as Sallen and Key and if the transfer function used is of Butterworth polynomials, it is called Butterworth second-order LPF/BPF. The type of circuit to be used for Butterworth or Chebyshev filters is not specified. The frequency response of third-order Chebyshev LPF response is shown in Fig. 3.32.

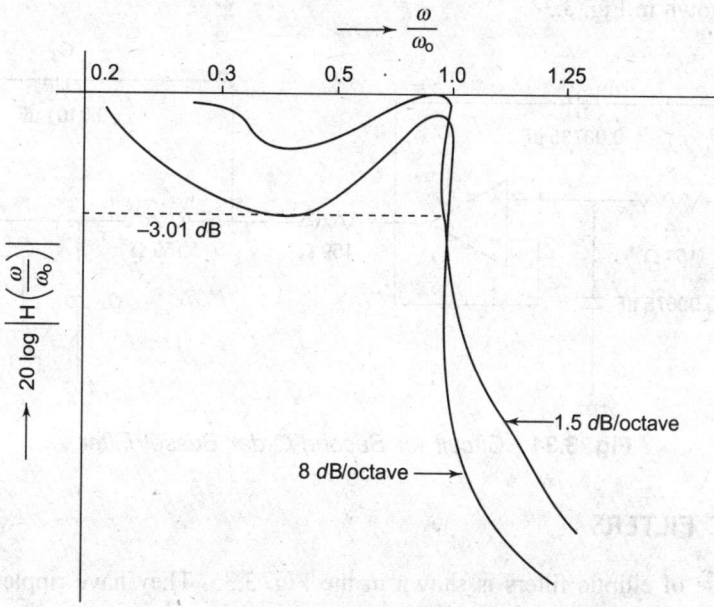

Fig. 3.32 *Third-Order Chebyshev LPF Response*

3.18 BESSEL FILTERS

Butterworth, Chebyshev, and Elliptic filters are suitable for operation with continuous signals. The translucent response of these filters show considerable overshoot. These filters are not suitable if the input is a pulse train or a signal burst. For step inputs, these filter show large overshoots.

For pulse-type input, Bessel filters are suitable. Here, the phase difference between input and output gradually increases with frequency.

Group delay: Group delay is one characteristic parameter of filter circuits. It is defined by the derivative of the gain with respect to frequency. This is flat over a wide range of frequencies. The frequency response is as shown in Fig. 3.33.

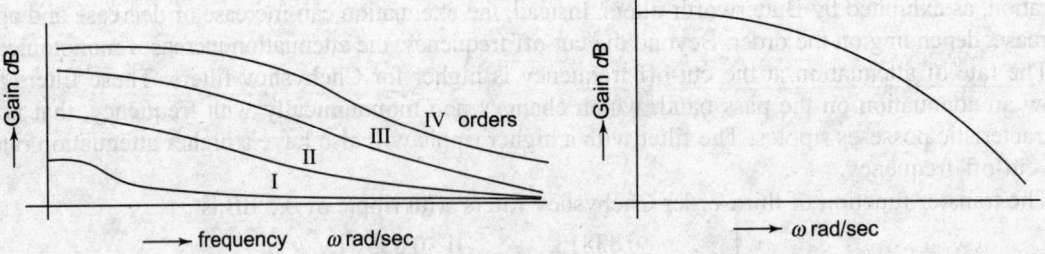

Fig. 3.33 *Frequency Response Curves for Bessel Filters*

The transfer function for a second-order Bessel filter is

$$\frac{E_o}{E_i} = \frac{5.1}{(s^2 + 2.7192s + 2.0142s + 2.9142s + 2.5)}$$

The circuit is shown in Fig. 3.34.

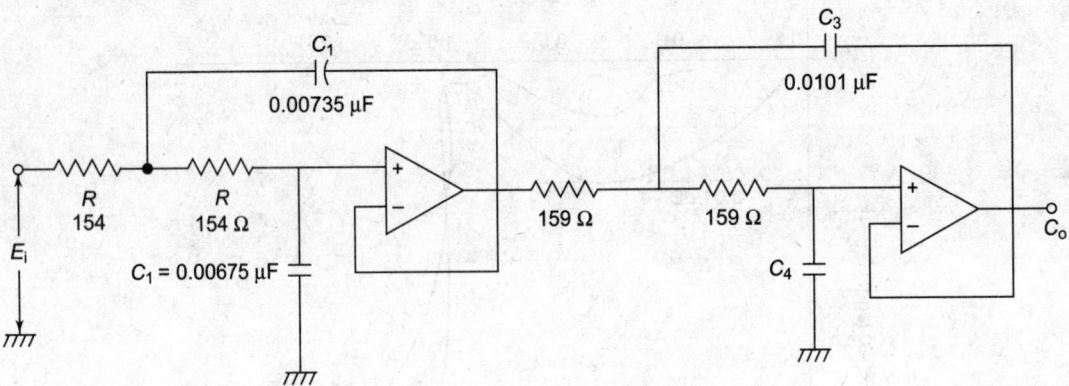

Fig. 3.34 *Circuit for Second-Order Bessel Filter*

3.19 ELLIPTIC FILTERS

Frequency response of elliptic filters is shown in the Fig. 3.35. They have ripple in the attenuation characteristic, both in pass band and stop band.

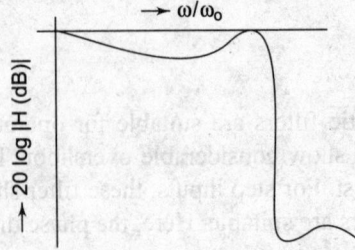

Fig. 3.35 *Frequency Response of Elliptic Filter*

3.20 ALL-PASS FILTERS

These filters are characterised by the phase shift they produce, on the signal, and the magnitude of the signal is left unaffected, that is voltage gain of the circuit $A_v = 1$. The circuit passes all frequency signals, without any change in magnitude at the output, but produces any phase shift between V_o and V_i. The first-order all-pass network has the following transfer function:

$$\frac{V_o}{V_i} = \frac{1-s}{1+s}$$

The gain of the network is constant at all frequencies. The phase shift varies from 180° lag.
The circuit and frequency response are shown in Fig. 3.36.
All-pass filter circuits are used in
1. Phase equalisers
2. Modems

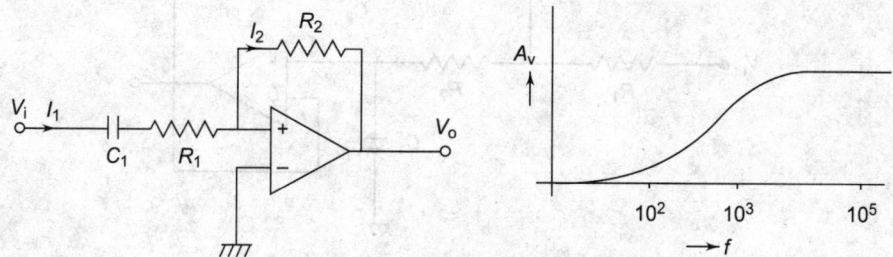

Fig. 3.36 *All-Pass Filter*

Example 3.2   *Design a second-order low-pass Butterworth filter, with unity gain, to have a cut-off frequency of 1 kHz and maximally flat response.*

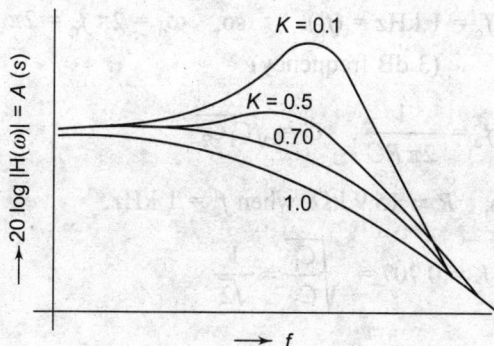

Fig. 3.37 *Frequency Response for Different Values of K*

Solution

A second-order LPF will have a frequency response, as shown, for different values of the damping coefficient K.

The damping coefficient K is also represented as ζ in some books. The frequency response is maximally flat if $K = 0.707$.

The transfer function for second-order LPF with unity gain is

$$\frac{V_o}{V_i}(K) = |H(\omega)| = \frac{1}{1 + as + bs^2}$$

$$b = \frac{1}{\omega_n^2}; \quad a = \frac{2K}{\omega}$$

The circuit is shown in Fig. 3.38.

(This is a unity gain amplifier because the op-amp is connected as a buffer.)

If $\qquad R_1 = R_2 = R, \; \omega_n = \dfrac{1}{R\sqrt{C_1 C_2}}$ $\hfill$ (3.21)

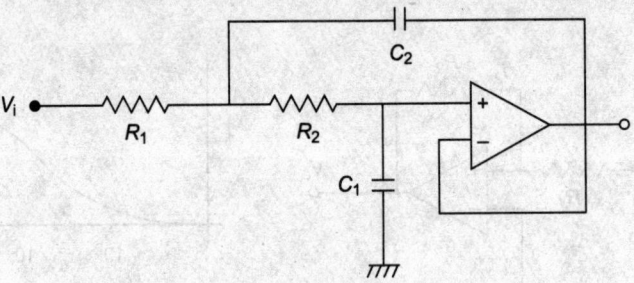

Fig. 3.38 *Second-Order LPF*

$$K = \zeta = \sqrt{\frac{C_1}{C_2}} \quad \text{or} \quad \omega_n = \frac{1}{RC}, \quad \text{where } C = \sqrt{C_1 C_2}$$

$K = 0.707$ for maximally flat response.

$$f_c = 1 \text{ kHz} = (f_n), \qquad \text{so,} \quad \omega_n = 2\pi f_n = 2\pi f_c$$
$$(3 \text{ dB frequency})$$

$$f_c = \frac{1}{2\pi RC}, \quad C = \sqrt{C_1 C_2}.$$

Choose $C = 0.01 \; \mu\text{F}$. So, $R \simeq 15.9 \text{ k}\Omega$, when $f_c = 1$ kHz.

$$K = 0.707 = \sqrt{\frac{C_1}{C_2}} = \frac{1}{\sqrt{2}}$$

Therefore, $\qquad C_1 = \sqrt{2} C = 0.01414 \; \mu\text{F}$

$$C_2 = \frac{1}{\sqrt{2}} C = 0.00707 \; \mu\text{F}$$

The circuit is shown in Fig. 3.35.

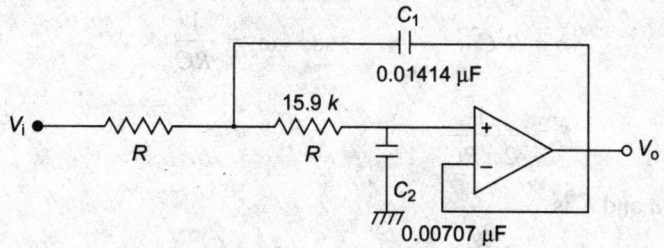

Fig. 3.39 *Second-Order Active Filter (LPF) (Unity Gain)*

Second-order LPFs can also be built with non-unity gain amplifier as shown in Fig. 3.39. A (Sallen and Key give $A_v > 1$) non-unity gain Sallen and Key filter circuit is shown in Fig. 3.40.

$$V_b = \frac{V_o R_f}{(R_1' + R_f)}$$

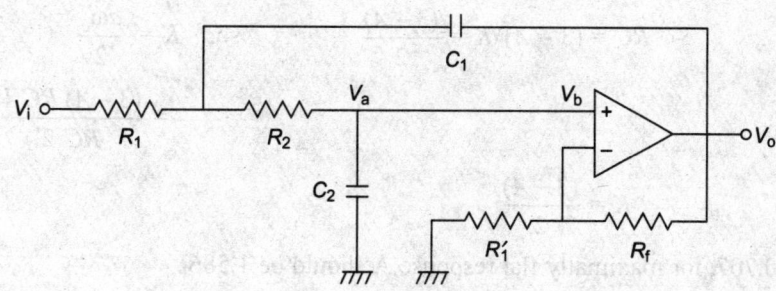

Fig. 3.40 *Second-Order LPF, in Sallen and Key Form, Non-unity Gain*

$$V_a = \frac{V_o R_1'}{(R_f + R_1')} (1 + s C_2 R_2)$$

$$A = 1 + \frac{R_f}{R_i}$$

A is the gain of the isolator.

In this case,

$$\frac{V_o}{V_i}(s) = \frac{A}{s^2 R_1 C_1 R_2 C_2 + s(R_1 C_1 + R_2 C_2 + R_1 C_2 - A R_1 C_1) + 1}$$

If $R_1 = R_2 = R$; $C_1 = C_2 = C$,

then

$$\frac{V_o}{V_i}(s) = \frac{A}{1 + (3 - A) s RC + s^2 R^2 C^2}$$

Damping factor $K = \dfrac{(3 - A)}{2}$, comparing thus with the standard expression,

$$\frac{V_o}{V_i}(s) = \frac{1}{1 + as + bs^2}, \quad \omega_c = \omega_n$$

$$b = R^2C^2 = \frac{1}{\omega_c^2} \quad \text{so,} \quad \omega_c = \frac{1}{RC}$$

$$f_c = \frac{1}{2\pi RC}$$

Relation between a and K is

$$a = \frac{2K}{\omega_c}; \quad \omega_c = \frac{1}{RC} \quad \text{comparing}$$

$$b = \frac{1}{\omega_c^2} = R^2C^2 = (3-A)\, RC \qquad \frac{1}{\omega_c^2} = b = (3-A)\, RC$$

$$K = \frac{RC}{2}, \quad K = \frac{(3-A)}{2} \qquad\qquad a = \frac{2K}{\omega_c}; \quad \omega_c = \frac{1}{RC}$$

$$RC = (3-A)\backslash K = \frac{(3-A)}{2} \qquad \text{so,} \quad K = \frac{a\omega_c}{2}$$

$$= \frac{(3-A)\, RC.1}{RC.2}$$

$$K = \frac{(3-A)}{2}$$

If K is to be 0.707, for maximally flat response A should be 1.586.

$$\left[\text{Since } \frac{1}{\sqrt{2}} = \frac{(3-A)}{2}; (3-A) = \sqrt{2}, \quad \text{so,} \quad A = 3 - \sqrt{2} = 1.586 \right]$$

Design for $f_c = 1$ kHz, choose $C = 0.01\ \mu\text{F}$.

So,

$$f_c = \frac{1}{2\pi RC}; R \simeq 15.9\ \text{k}\Omega$$

$$A = \frac{R_f}{R_c} = 1.5858.$$

Therefore,

$$\frac{V_o}{V_i} = 0.5858, \quad \text{Choose } R_f = 58.58\ \text{k}\Omega$$

$$R_i = 100\ \text{k}\Omega$$

Example 3.3 *Design a Chebyshev filter with positive, non-unity gain LPF of second order to have 3 dB frequency = 250 Hz and to have 0.5 dB ripple. Cutoff frequency f_c is 3 dB.*

Solution

For a second-order Chebyshev filter having a ripple of 0.5 dB, the transfer function is

$$\frac{V_o}{V_i}(s) = |H(\omega)| = \frac{0.83813}{s^2 + 0.5366s + 0.83813}$$

Comparing the above expression with the standard expression, $\dfrac{1}{1+as+bs^2}$

$$a = \frac{0.5366}{0.83813};$$

$$b = \frac{1}{0.83813};$$

$$\frac{V_o}{V_i}(s) = \frac{A}{1+(3-A)s\,RC+s^2R^2C^2}$$

$$a = (3-A)\,RC, \qquad b = R^2C^2 = \frac{1}{\omega_{c^2}} = 0.83813$$

So,
$$\omega_c = \frac{1}{\sqrt{b}} = \frac{1}{\sqrt{0.83813}} = 0.2931$$

$$A = 0.83813;$$

$$K = \frac{(3-A)}{2} \qquad\qquad b = R^2C^2 = \frac{1}{\omega_{c^2}} = 1$$

Since
$$\frac{V_o}{V_i}(s) = \frac{1}{1+as+bs^2} \qquad b = \frac{1}{\omega_{c^2}}; \; a = \frac{2K}{\omega_c}; \quad \text{so, } K = \left(\frac{3-A}{2}\right)$$

Here
$$b = 1; \; a = (3-A)\,1; \qquad\qquad \text{so, } K = \frac{(3-A)}{2}$$

$$a = \frac{2K}{\omega_c}$$

Therefore,
$$K = \frac{1}{2}\left[\frac{0.5366}{\sqrt{0.83813}}\right] = 0.2931$$

But
$$K = \frac{a\omega_c}{2}; \; \text{So, } a = \frac{2K}{\omega_c}$$

So,
$$a = 0.5366$$

$$\omega_c = \sqrt{0.83813} = 0.2931;$$

$$\frac{V_o}{V_i}(s) = \frac{1}{1+\dfrac{0.5366\,s}{0.83813}+\dfrac{1}{0.83813}s^2}$$

$$K = \frac{1}{2}\left[\frac{0.5366}{\sqrt{0.83813}}\right] = 0.2931$$

$$\sqrt{0.83813} = 0.9155$$

$$\frac{0.5360}{0.9155} = 0.58612$$

$$\frac{0.58612}{2} = 0.2931$$

$$\frac{V_o}{V_i} = \frac{1}{1 + as + bs^2} \qquad b = \frac{1}{\omega_c^2} = \frac{1}{0.83813} ; a = \frac{2K}{\omega_c} ; \qquad \text{So, } K = \frac{a\omega_c}{2}$$

So, $A = (3 - 2K) = 2.414$

Since $K = 0.2931$ so, $A = 3 - 2\,(0.2931) = 2.414$

$$A = 1 + \left(1 + \frac{R_f}{R_i}\right) = 2.414$$

So, $\dfrac{R_f}{R_i} = (2.414 - 1) = 1.414$

$$f_c = \frac{1}{2\pi RC} = 250 \text{ Hz (given)}$$

Therefore, $RC = 0.6366$ msec

R and C are the components in the circuit, which determine the cut-off frequency f_c of the filter circuit.

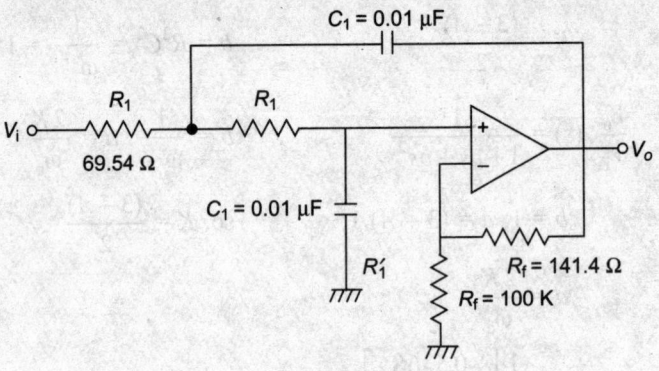

Fig. 3.41   *Second-Order LPF, Non-unity Gain*

Natural frequency $\omega_n = \sqrt{0.83813}$ since, $b = \dfrac{1}{\omega_n^2}$

So, $\omega_n = \dfrac{1}{\sqrt{b}}$

$$b = \frac{1}{0.83813}$$

$$R_1 C_1 = \frac{RC}{\sqrt{0.83813}} = 0.6954 \text{ msec}$$

Choose $C_1 = 0.01\ \mu F$ $\boxed{\text{So, } R_1 = 69.54 \text{ k}\Omega}$ Answer

---

Example 3.4   *The circuit in Fig. 3.42 uses ideal op-amp.*

(a) Find the voltage gain $A_v = \dfrac{V_o}{V_s}$, damping factor K, and the cut-off frequency ω_0.

(b) Using this circuit design, a second-order Butterworth LPF circuit $f_o = 1$ kHz and low voltage gain $= -1$.

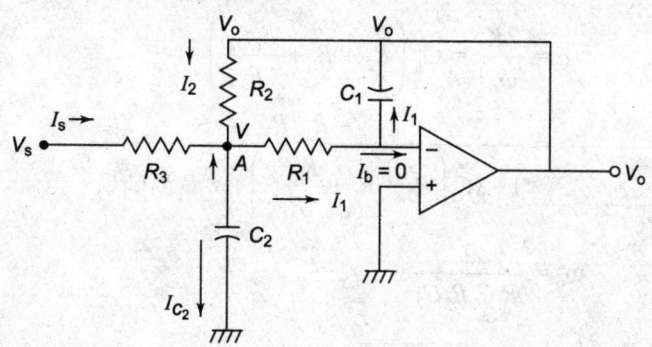

Fig. 3.42

Applying Kirchhoff's current law,

$$I_s - I_{C_2} + I_2 - I_1 = 0$$

Currents entering the node are taken as positive.
Currents leaving the node are taken as negative.

Current through $C_2 = sC_2V$ (I_{C_2})

Current through $R_3 = V_s - V/R_3$ (I_s)

Current through $R_2 = \dfrac{V_o - V}{R_2}$ (I_2)

Applying KCL to node A we find.

$$\frac{V_o - V}{R_2} + \frac{V_s - V}{R_3} - sVC_2 = I_1 \tag{3.22}$$

But $I_1 \times R_1 = V$ (3.23)

Since it is ideal op-amp substituting Eq. (3.23) in Eq. (3.22) and simplifying

$$\frac{V_o}{R_2} + \frac{V_s}{R_3} = I_1 \left(1 + \frac{R_1}{R_2} + \frac{R_1}{R_3} + sR_1C_2 \right)$$

But $V_o = I_1 \dfrac{1}{sC_1}$

So, $\dfrac{R_3}{R_2} + \dfrac{V_s}{V_o} = -R_3 \left(1 + \dfrac{R_1}{R_2} + \dfrac{R_1}{R_3} + sR_1C_2 \right) sC_1$

or $A_v = \dfrac{V_o}{V_s} = \dfrac{-1}{\dfrac{R_3}{R_2} + \left(R_3 + R_1 + \dfrac{R_3 R_1}{R_2} \right) C_1 s + R_3 R_1 C_1 s^2}$

Comparing, $\dfrac{V_o}{V_s}(s) = \dfrac{1}{1 + as + bs^2}$

Therefore, $b = \dfrac{1}{\omega_n^2} = R_1 R_2 C_1 C_2$

$$a = \frac{2K}{\omega_n} = C_1 \left(R_2 + R_1 + \frac{R_1 R_2}{R_3} \right)$$

$$= \frac{-R_2/R_3}{1 + sC_1 \left(R_2 + R_1 + \frac{R_1 R_2}{R_3} \right) + s^2 R_1 R_2 C_2 C_1}$$

Let

$$\omega_0^2 = \frac{1}{R_1 C_1 R_2 C_2} \qquad (3.24)$$

So,

$$\frac{2K}{\omega_0} = 2K \sqrt{R_1 C_1 R_2 C_2}$$

$$= (R_2 + R_1) C_1 + \frac{R_1 R_2}{R_3} C_1 \qquad (3.25)$$

Therefore,

$$K = \frac{1}{2} \left(\sqrt{\frac{R_2 C_1}{R_1 C_2}} + \sqrt{\frac{R_1 C_1}{R_2 C_2}} + \frac{1}{R_3} \sqrt{\frac{R_1 R_2 C_1}{C_2}} \right)$$

For $A_{vo} = 1$, $R_2 = R_3$. We choose $R_2 = R_3 = 5$ kΩ. Choose $C_2 = 1$ μF then from Eq. (3.22), we obtain,

$$C_1 = \frac{1}{R_1 R_2} \frac{1}{(2\pi)^2 \times 10^8 \times 10^{-8}} = \frac{1}{4\pi^2 R_1 R_2} \quad \text{farads (F)}$$

and from Eq. (3.23), we have,

$$\frac{2K}{\omega_0} = \frac{\sqrt{2}}{2\pi \times 10^3} = \frac{R_2 + R_1}{4\pi^2 R_1 R_2} + \frac{1}{4\pi^2 R_3}$$

or

$$2\pi \sqrt{2} \times 10^{-3} = \frac{1}{R_1} + \frac{1}{R_2} + \frac{1}{R_3} = \frac{1}{R_1} + 0.4 \times 10^{-3}$$

So,

$$R_1 = 0.118 \text{ k}\Omega \text{ then } C_1 = \frac{1 \times 10^{-6}}{2\pi \times 0.118 \times 5}$$

$$= 0.27 \text{ }\mu\text{F}$$

Example 3.5 *Design a BRF (second order) to have a response shown below in Fig. 3.43. LPF and HPF must be in parallel*

Solution

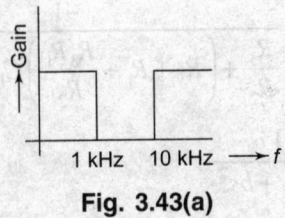

Fig. 3.43(a)

Expression for second-order Butterworth polynomial is $(s^2 \times 1.4114\, s + 1) = 0$. The block schematic for BRF with frequency rejection band between 1 kHz and 10 kHz is shown in Fig. 3.43(b).

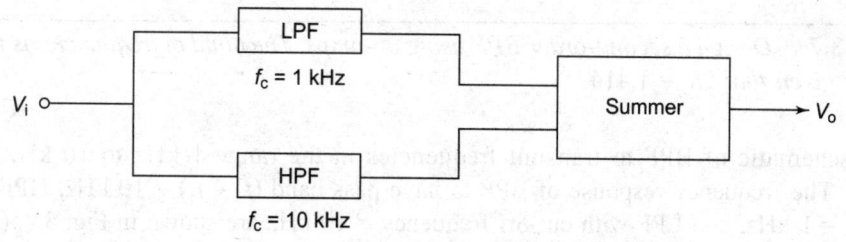

Fig. 3.43(b) *Band-Reject Filter*

Example 3.6 *Design a fourth-order Butterworth HPF with cut-off frequency of 10 kHz. It is given that*

$$2K_1 = 0.765, \quad 2K_2 = 1.848.$$

Solution

Fourth-order filter can be obtained by cascading two second-order filters

$$A_{v1} = \frac{R_1' + R_1}{R_1} = \left(1 + \frac{R_1'}{R_1}\right) = 3 - 2K_1 = 3 - 0.765 = 2.235$$

Selecting

$$R_1 = 10 \ K, \quad R_1' = 12.35 \ K$$

$$A_{v2} = \frac{R_2' + R_2}{R_2} = 3 - 2K_2 = 3 - 1.84 = 1.152$$

Choosing

$$R_2 = 10 \ K, R_2' = 1.52 \ K$$

$$f_o = 1/2\pi \ RC = 10 \ \text{kHz} \quad (\text{choosing } R = 1 \ K, C = 0.016 \ \mu F)$$

The input is being given to the non-inverting terminal/gain of op-amp $= \left(\dfrac{R_i' + R_1}{R_1}\right) = \left(1 + \dfrac{R_i'}{R_1}\right)$

The circuit is shown in Fig. 3.44.

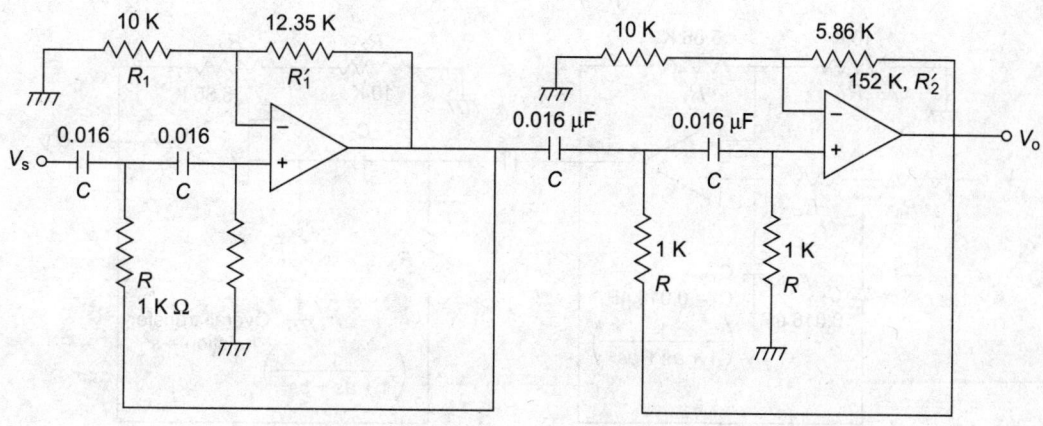

Fig. 3.44 *Circuit for Fourth-Order HPF*

Example 3.7 *Design a second-order BPF using op-amps. The band of frequencies is from 1 kHz to 10 kHz. It is given that 2K = 1.414.*

Solution

The block schematic of BPF to transmit frequencies in the range 1 kHz to 10 kHz is shown in Fig. 3.45(a). The frequency response of BPF to have pass band $(f_2 - f_1)$ = 10 kHz, HPF with cut-off frequency f_c = 1 kHz, and LPF with cut-off frequency = 10 kHz are shown in Fig. 3.45(b).

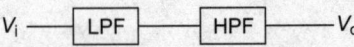

$$V_i \longrightarrow \boxed{\text{LPF}} \longrightarrow \boxed{\text{HPF}} \longrightarrow V_o$$

Fig. 3.45 (a) *Block Schematic for BPF*

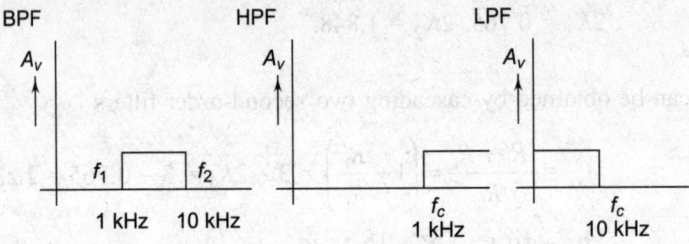

Fig. 3.45 (b) *Frequency Response of BPF*

Here, an LPF (second order) with cut-off frequency of 10 kHz and a second-order HPF with cut-off frequency of 1 kHz are to be used.

The circuit is shown in Fig. 3.46.

$$A_{v1} = A_{v2} = 3 - 2K = 3 - 1 - 0.414 = 1.586$$

$$= \frac{R_1'}{R_1} = \frac{R_2'}{R_2}, \quad \text{since} \left(1 + \frac{R_1'}{R_1}\right) = \left(1 + \frac{R_2'}{R_2}\right)$$

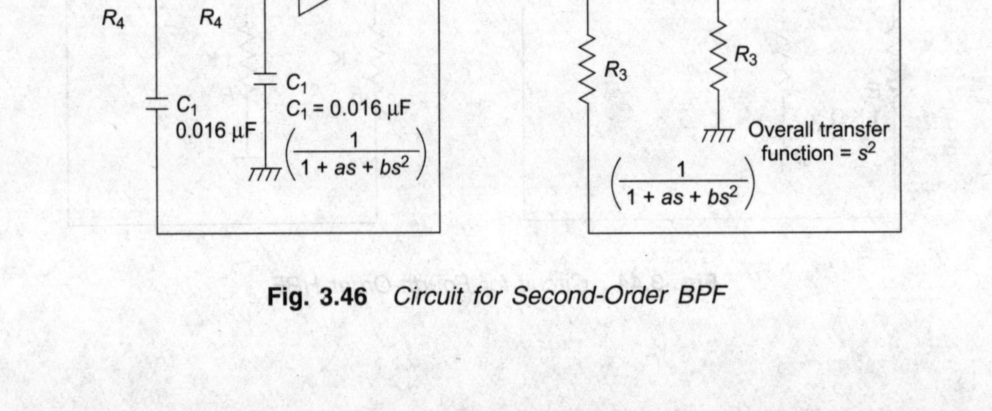

$$\text{Gain} = \left(\frac{1 + R_1'}{R}\right) \qquad\qquad \text{Gain} = \left(\frac{1 + R_2'}{R_2}\right)$$

10 K   5.86 K
R_1 R_1'

V_s R_4 R_4

C_1 0.016 μF
C_1 = 0.016 μF $\left(\dfrac{1}{1 + as + bs^2}\right)$

C_2 C_2

R_2 R_2'
10 K   5.86 K

$\longrightarrow V_o$

R_3 R_3

Overall transfer function = s^2

$\left(\dfrac{1}{1 + as + bs^2}\right)$

Fig. 3.46 *Circuit for Second-Order BPF*

Choosing
$$R_1 = 10 \text{ K}$$

$$1 + \frac{R_1'}{R} = 1.586$$

$$R_1' = 10 \times 0.586 = 5.86 \text{ K}$$

Also
$$(R_2' + R_2)/R_2 = 1.586 \text{ or } \left(1 + \frac{R_2'}{R_2}\right) = 1.586$$

$$R_2 = 5.86 \text{ K}$$

$$\frac{1}{2\pi R_4 C_1} = 10 \times 10^3$$

Selecting
$$R_4 = 1 \text{ } K$$

$$C_1 = \frac{1}{2\pi \times 1000 \times 10^4} = 0.016 \text{ } \mu F$$

$$\frac{1}{2\pi R_3 C_2} = 10^3, \text{ choosing } R_1 = 1 \text{ } K, \text{ } C_2 = \frac{2}{2\pi \times 10^3 \times 10^3} = 0.16 \text{ } \mu F$$

The overall gain $A_v = A_{v_1} A_{v_2}$ is not specified. So for each stage, gain $= \left(1 + \frac{R_2}{R_1}\right)$ must be taken.

It will not make any difference if HPF precedes LPF. It is not a fourth-order filter.

3.21 RESONANT BPF CIRCUIT WITH R_3 AND C_1 INTERCHANGED

The circuit diagram is shown in Fig. 3.47.

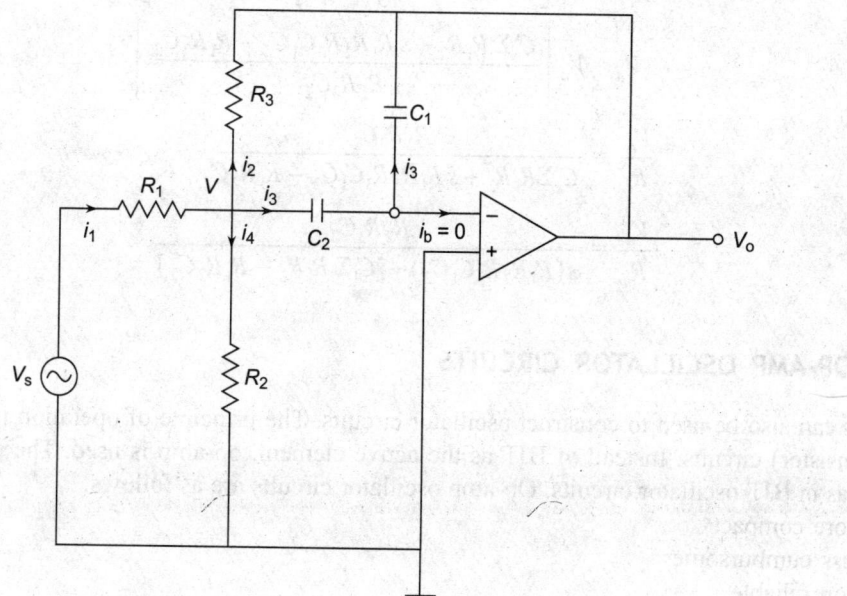

Fig. 3.47 *Resonant BPF Without Using Inductor*

Expression for $A_v = \dfrac{V_o}{V_s}$

$$\frac{V_s - V}{R_1} = i_1 \qquad\qquad \text{but } i_1 = i_2 + i_3 + i_4$$

$$\frac{V - V_o}{R_3} = i_2 \qquad\qquad \frac{V_s - V}{R_1} = \frac{V - V_o}{R_3} + V_s C_2 + \frac{V}{R_2}$$

$$\frac{V - 0}{1/sC_2} = i_3 = \frac{0 - V_o}{1/sC_1} \qquad \frac{V_s}{R_1} + \frac{V_o}{R_3} = V\left[\frac{1}{R_1} + \frac{1}{R_2} + \frac{1}{R_3} + sC_2\right]$$

$$\frac{V - 0}{R_2} = i_4 \qquad\qquad = V\left[\frac{\Sigma R_1 R_2 + s R_1 R_2 R_3 C_2}{R_1 R_2 R_3}\right]$$

also from i_3: $Vs C_2 = V_o s C_1$

So, $$V = V_o\left(\frac{C_1}{C_2}\right)$$

$$\frac{V_s}{R_1} = V_o\left(\frac{C_1}{C_2}\right)\left[\frac{\Sigma R_1 R_2 + s R_1 R_2 R_3 C_2}{R_1 R_2 R_3}\right] - \frac{V_o}{R_3}$$

$$= V_o\left[\frac{C_1 \Sigma R_1 R_2 + s R_1 R_2 R_3 C_1 C_2}{R_1 R_2 R_3 C_2} - \frac{1}{R_3}\right]$$

$$\frac{V_s}{R_1} = V_o\left[\frac{C_1 \Sigma R_1 R_2 + s R_1 R_2 R_3 C_1 C_2 - R_1 R_2 C_2}{R_1 R_2 R_3 C_2}\right]$$

$$V_s = V_o\left[\frac{C_1 \Sigma R_1 R_2 + s R_1 R_2 R_3 C_1 C_2 - R_1 R_2 C_2}{R_2 R_3 C_2}\right]$$

$$\frac{V_o}{R_s} = \frac{R_2 R_3 C_2}{C_1 \Sigma R_1 R_2 + s R_1 R_2 R_3 C_1 C_2 - R_1 R_2 C_2}$$

$$\frac{V_o}{R_s} = \frac{R_2 R_3 C_2}{s(R_1 R_2 R_3 C_1 C_2) + [C_1 \Sigma R_1 R_2 - R_1 R_2 C_2]}$$

3.22 OP-AMP OSCILLATOR CIRCUITS

Op-amps can also be used to construct oscillator circuits. The principle of operation is the same as in BJT (transistor) circuits. Instead of BJT as the active element, op-amp is used. The passive network remains as in BJT oscillator circuits. Op-amp oscillator circuits are as follows:

(1) More compact
(2) Less cumbursome
(3) More reliable
(4) Less drift problematic

3.22.1 RC Phase Shift Oscillator

The circuit is shown in Fig. 3.48. RC network provides 180° phase shift. The expression for the frequency of oscillations f_o is

$$f_o = \frac{1}{2\pi \, RC\sqrt{6}}$$

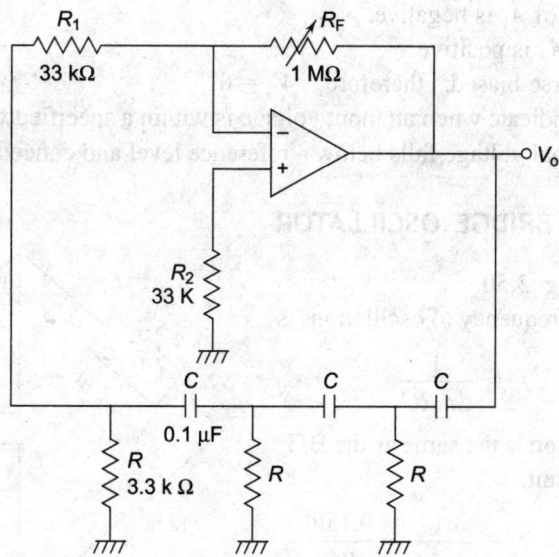

Fig. 3.48 *Op-amp RC Phase Shift Oscillator Circuit*

3.23 WINDOW DETECTOR OR DISCRIMINATOR

The circuit and frequency response are shown in Fig. 3.49.
V_1 is upper threshold voltage (UTV).
V_2 is lower threshold voltage (LTV).
A_1 and A_2 are comparators.

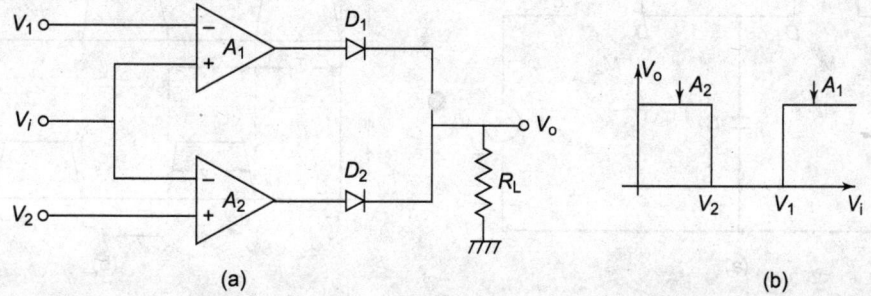

(a)   (b)

Fig. 3.49 *(a) Circuit for Op-amp Window Detector and (b) Frequency Response*

When V_i is less than LTV V_1,

When $V_i <$ LTV, D_1 is forward biased since A_2 output is positive and is equal to V_2

(V_2) D_1 is reverse biased since V_1 is negative voltage.

$V_o = V_o$ (sat)

When $V_i >$ UTV, D_2 is reverse biased.

D_1 is forward biased.

$V_o = V_o$ (sat)

If LTV $< V_i <$ UTV, V_o of A_1 is negative.

$(V_2) < V_i < (V_i)$, V_o of A_2 is positive.

So, D_1 and D_2 are reverse biased, therefore, $V_o = 0$.

This circuit is used to indicate when an input voltage is within a specified voltage range. It is used to mention when a given signal voltage falls below a reference level and concedes a reference level.

3.24 OP-AMP WIEN BRIDGE OSCILLATOR

The circuit is shown in Fig. 3.50.

The expression for the frequency of oscillations is

$$f_o = \frac{1}{2\pi RC}$$

The principle of operation is the same as the BJT Wien bridge oscillator circuit.

$$f_o = \frac{1}{2\pi RC} = \frac{0.159}{RC}$$

$$R_F = 2R_1$$

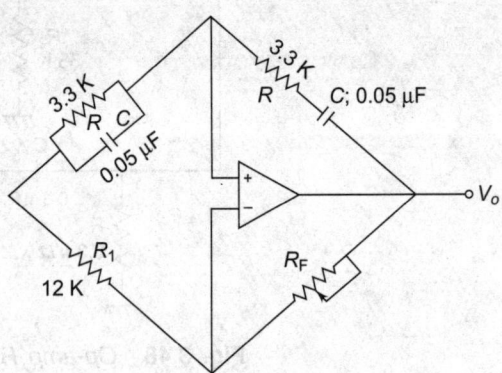

Fig. 3.50 *Op-amp Wien Bridge Oscillator*

3.25 SWITCHED CAPACITOR

Consider the circuit shown in Fig. 3.51.

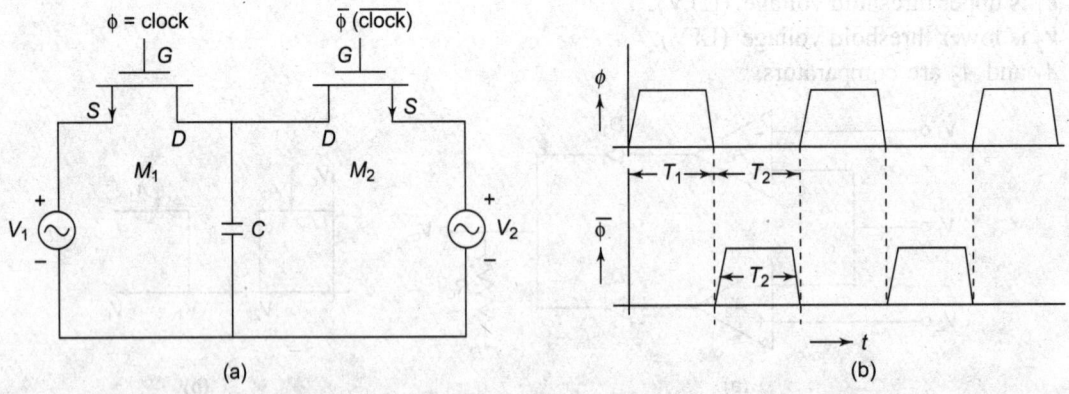

Fig. 3.51 *(a) Circuit and (b) Frequency Response of Switched Capacitor*

M_1 and M_2 are n-channel enhancement type MOSFETs. NMOS devices are used.

G = Gate D = Drain S = Source

Because it is an NMOS device, if the gate voltage V_G or clock f applied to the gate is positive and high, a channel is formed between the source and drain. So, the device will turn on.

r_{ds} (on) = 1 kΩ. That is, when the device is in the on state the resistance = 1 kΩ. If the gate voltage is low or 0 V, there is no channel between the source and drain. Therefore, $I_D \simeq 0$.

$$r_{ds} \text{ (off)} = 10^{12} \ \Omega \text{ or more}$$

So, $\dfrac{r_{ds\,(off)}}{r_{ds\,(on)}}$ ratio is higher $\dfrac{10^{12}}{10^3} = 10^9$

MOSFET devices have high on–off resistance ratio. The devices can be used as switches as they have high off resistance and low on resistence.

Consider that clock pulses as shown in Fig. 3.51 are given to the gates of the MOSFET. ϕ and $\overline{\phi}$ are as shown in Fig. 3.50. They are out of phase. When ϕ is high, $\overline{\phi}$ is low, and vice versa.

So, if the two MOSFETs are used as switches, the circuit will be similar to a single-pole, double throw (SPDT) switch as shown in Fig. 3.52.

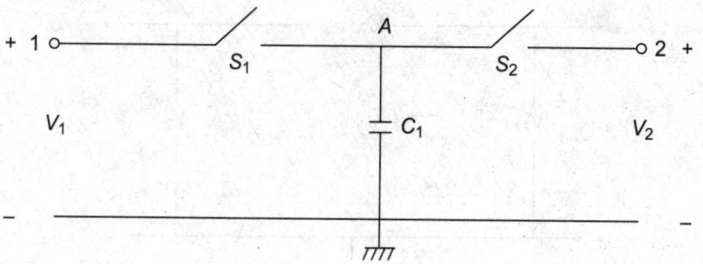

Fig. 3.52 *Single-Pole Double-Throw Switch*

S_1 and S_2 are the switches causing double throw. The pole is at point A. Both the switches supply current to point A.

There must be a phase shift between the two clock pulses and both the clocks should not occur at 'high' at the same time. Otherwise, both the switches will be closed. To ensure this, *break-before-make* type of clock or a similar operation must be made.

The switches S_1 and S_2 are complementary. The capacitor C gets charged from V_1 and V_2 alternately. S_1 is closed for T_1 seconds and open for T_2 seconds. S_2 is closed for T_2 seconds and open for T_1 seconds.

Period of one switching cycle $T_1 + T_2 = T$

Switching frequency $f_{CLK} = \dfrac{1}{T}$

Let V_1 and V_2 be ideal voltage sources,

$$V_1 > V_2$$

When S_1 is closed and S_2 is open, C gets charged to V_1 in line T_1. When S_1 is open and S_2 is closed, C_1 discharges to V_2.

The cycle is repeated at $t = T_1 + T_2$.

During one period T in such an operation, the charge Q which is transported from node 1 to node 2 is

$$Q = C_1(V_1 - V_2)$$

Since

$$Q = CV$$

$$i = \frac{Q}{T}$$

Therefore, the charge transfer Q in T seconds results in current flow i.

$$i = \frac{Q}{T} = \frac{C_1(V_1 - V_2)}{T}$$

$$\frac{1}{T} = f_{CLK}$$

So,

$$i = C_1 f_{CLK} (V_1 - V_2)$$

So, the circuit can be represented as a resistor R_1 carrying the same circuit i, as shown in Fig. 3.53.

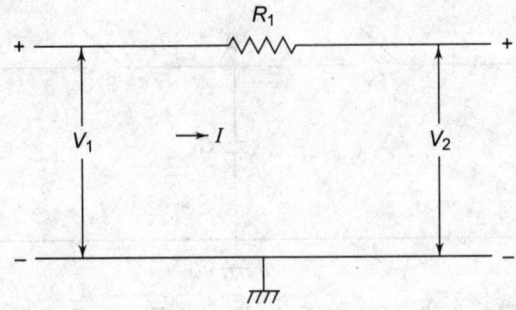

Fig. 3.53   *Equivalent Circuit*

$$\frac{V_1 - V_2}{R_1} = C f_{CLK} (V_1 - V_2)$$

or

$$R_1 = \frac{1}{C f_{CLK}}$$

The above equation indicates that switching a capacitor periodically is equivalent to a resistance R_1 connected as shown in Fig. 3.53. From the above equation, R_1 can be changed by changing f_{CLK} for a given value of C and vice versa.

The above phenomenon happens because the switch is a MOSFET device and its on and off resistances differ. On the other hand, if no clock is used, if S_1 is closed, C gets charged to V_1. If S_2 is closed, it gets charged to V_2.

Using such equivalent circuits stimulate RC combination with MOSFET as switch and normal capacitor C, active filter circuits can be built. Even integrator and op-amp circuits can be built.

Switched capacitor integrator circuits with op-amp are shown in Fig. 3.54.

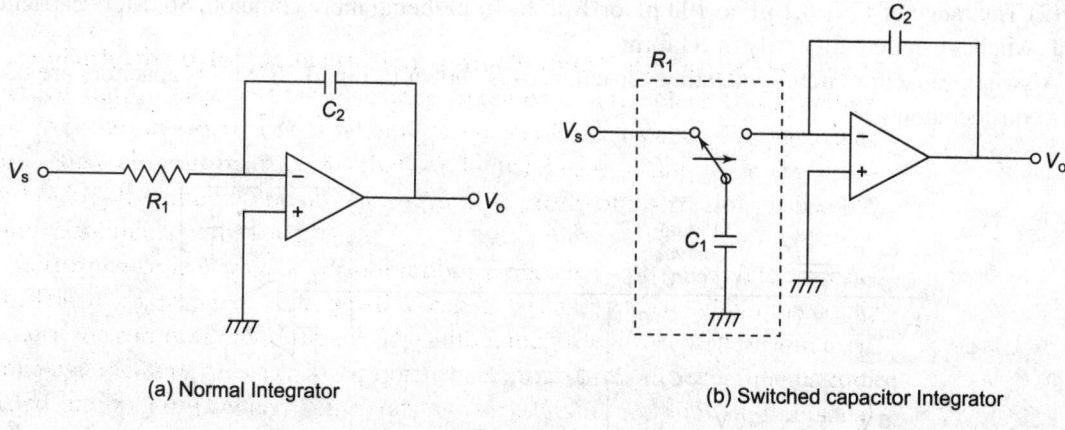

(a) Normal Integrator   (b) Switched capacitor Integrator

Fig. 3.54 *Integrator Circuits*

3.25.1 Switched Capacitor Integrator

The circuit with MOSFETS as resistors is shown in Fig. 3.55.

$$\frac{V_o}{V_s}(s) = \frac{X_{c_2}}{X_{c_1}} = \frac{C_1}{C_2} \text{ , integrator is an LPF.}$$

Its cut-off frequency $f_c = \dfrac{1}{2\pi R_1 C_2}$.

$$f_c = \frac{1}{2\pi R_1 C_2}$$

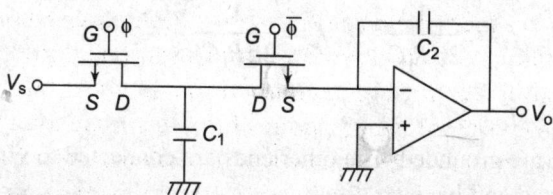

Fig. 3.55 *Switched Capacitor Integrator Circuit*

3.25.1.1 Advantages of Switched Capacitor Integrator Circuits

(1) In switched capacitors,

$$R_1 = \frac{1}{Cf_{CLK}}$$

So, the required value of R can be achieved by changing f_{CLK} or C.

In conventional active filter circuits, for very low f_c or very high f_c, the R and C values may be odd. This can be overcome by having switched capacitor circuits for R_s.

(2) The range of C_s is 0.1 pF to 100 pF or 1 pF to 10 pF being more common. So, MOS capacitors and switches can be built easily in IC form.

A switch capacitor circuit with stray capacitances is shown in Fig. 3.56. Stray capacitors are taken into consideration here.

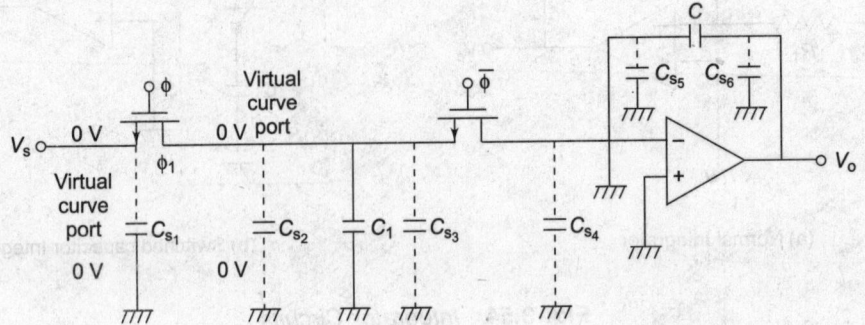

Fig. 3.56   *Switched Capacitor Circuit Considering Stray Capacitances*

C_{s_1} and C_{s_2} have no effect because they are in parallel with the voltage source. C_{s_5} and C_{s_6} have also no effect because they see virtual ground. But C_{s_1} and C_{s_2} are in parallel with C_1.

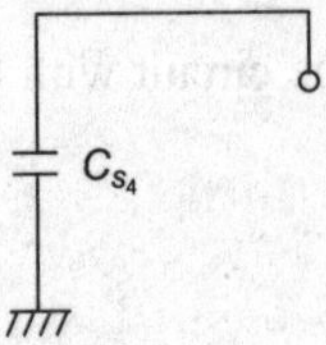

$$\text{So,} \quad C_1' = C_{s_1} + C_{s_2} + C_1$$

$$= C_1(1 + \gamma)$$

where
$$\gamma = \frac{C_{s_1} + C_{s_2}}{C_1}$$

Fig. 3.57

Therefore, the cut-off frequency or resonant form f_o will be different and there will be error.

$$f_c = \frac{1}{2\pi R_1 C_1} \qquad f_c' = \frac{1}{2\pi R_1 C_1'}$$

$$\Delta f_c = \gamma f_c$$

One end of C_{s_4} and C_{s_5} are grounded. The other ends are connected to virtual ground points. Therefore, C_{s_4} and C_{s_5} are shorted and have no effect.

3.25.2   Stray Insensitive Integrator

By connecting a MOSFET across the capacitors causing stray capacitance and making the device on, so that the stray capacitors are discharged and contribute no charge to the current flow, the effect of stray capacitances can be removed.

Therefore, the circuit is as shown in Fig. 3.58.

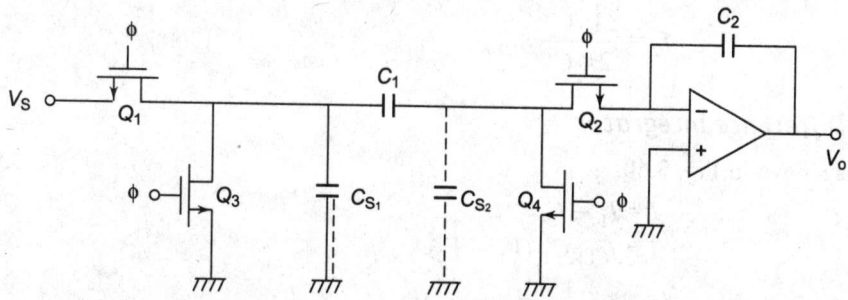

Fig. 3.58 *Stray Insensitive Integrator Circuit*

The stray capacitance problem is overcome by turning on Q_3 and Q_4 so that C_{s_1} and C_{s_2} are shorted. When Q_3 and Q_4 are turned on, they act like closed switches, shorting the stray capacitors C_{s_1} and C_{s_2}

3.25.3 First-Order Switched Capacitor Filter

3.25.3.1 Summing Integrator

This is also a first-order LPF because the integrator acts as an LPF. The circuit is shown in Fig. 3.59.

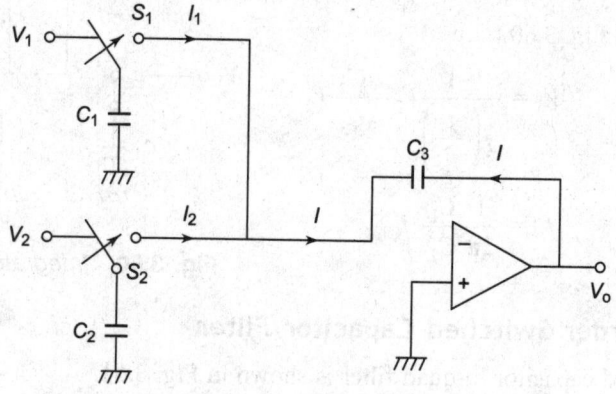

Fig. 3.59 *First-Order Switched Capacitor Filter*

$$R_1 = \frac{1}{C_1 f_{CLK}}$$

$$R_2 = \frac{1}{C_2 f_{CLK}}$$

$$I = I_1 + I_2$$

$$V_o = \frac{-1}{j\omega C_3} I = \frac{-1}{j\omega C_3} \ (C_1 f_{CLK} V_1 = C_2 f_{CLK} V_2)$$

$$V_o = \frac{1}{j(f/f_1)} V_1 - \frac{1}{j(f/f_2)} V_2$$

where

$$f_1 = \frac{1}{2\pi} \frac{C_1}{C_2} f_{CLK}$$

$$f_2 = \frac{1}{2\pi} \frac{C_2}{C_3} f_{CLK}$$

3.25.3.2 Difference Integrator

The circuit is shown in Fig. 3.60.

$$I = I_1 - I_2$$

$$I = f_{CLK} C_1 (V_1 - V_2)$$

$$V_o = \frac{-1}{j\omega C_2} C_1 (V_1 - V_2)$$

$$= \frac{-1}{j\left(\dfrac{f}{f_o}\right)} (V_2 - V_1)$$

where

$$f_o = \frac{1}{2\pi} \frac{C_1}{C_2} f_{CLK}$$

3.25.3.3 Integrator Summer

The circuit is shown in Fig. 3.60.

$$V_o = \frac{-1}{j\left(\dfrac{f}{f_o}\right)} V_1 - \frac{C_2}{C_3} V_2$$

where

$$f_o = \frac{1}{2\pi} \frac{C_1}{C_2} f_{CLK}$$

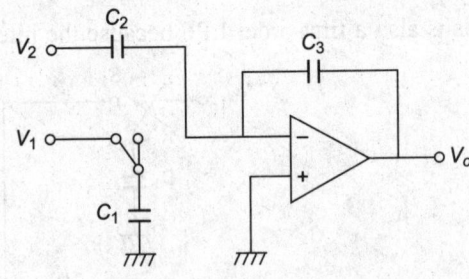

Fig. 3.60 *Integrator Summer Circuit*

3.25.4 Second-Order Switched Capacitor Filter

The circuit for switched capacitor bi-quad filter is shown in Fig. 3.61.

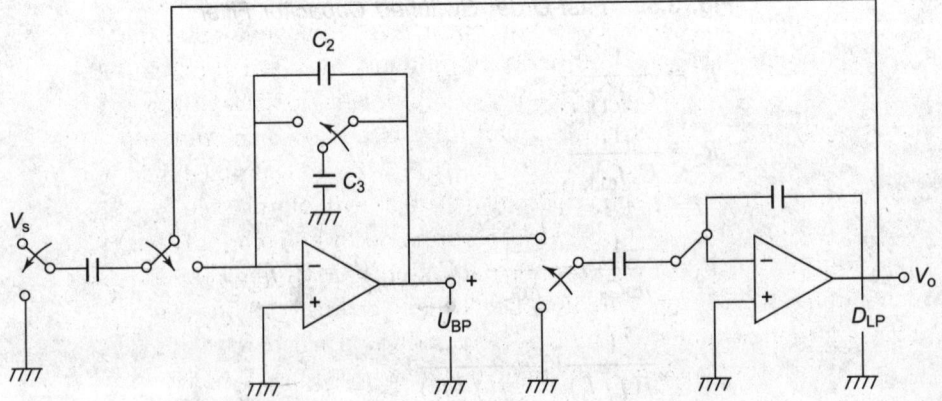

Fig. 3.61 *Second-Order Switched Capacitor*

A bi-quad section consists of a lossy inverting integrator followed by a lossless inverting integrator and a unity gain inverting amplifier.

Example 3.8 *Design a Chebyshev filter with positive non-unity gain, LPF of second order, to have 3 dB frequency equal to 250 Hz and 0.5 dB ripple.*

Solution

$H(\omega)$ equation for a second-order Chebyshev type LPF is,

$$H(\omega) = 0.83813(s^2 + 0.5366s + 0.83813)$$

Comparing with standard expression,

$$A_v(s) = \frac{A_{vo}\left(\dfrac{1}{RC}\right)^2}{s^2 + \left(\dfrac{3 - A_{vo}}{RC}\right)s + \left(\dfrac{1}{RC}\right)^2}$$

$$\left(\frac{1}{RC}\right)^2 = 0.83813 \Rightarrow \frac{1}{RC} = 0.915 \qquad \Rightarrow RC = 1.0923$$

$$\frac{3 - A_{vo}}{RC} \qquad A_{vo} = 2.41386 \text{ for second-order Chebyshev filter.}$$

$$A_{vo} = \frac{R_1 + R_1'}{R_1} = 1 + \frac{R_1'}{R_1}$$

$$\frac{R_1'}{R_1} = 2.41386 - 1 = 1.41386$$

If $\qquad\qquad R_1 = 10 \text{ K} \qquad\qquad\qquad R_1' = 14.13 \text{ K}$

$$f_o = \frac{1}{2\pi RC} \qquad\qquad\qquad f_o = 250 \text{ Hz}$$

$$RC = \frac{1}{2\pi \cdot 250}, \text{ let } C = 0.3 \text{ μF and } R = 212 \text{ kΩ}$$

The circuit diagram is as shown in Fig. 3.62.

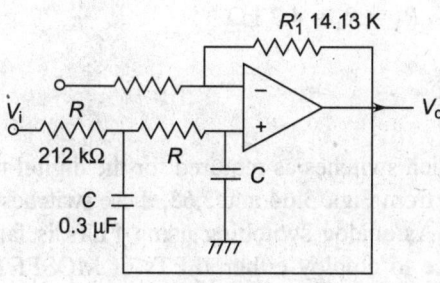

Fig. 3.62 *Circuit with Second-Order Positive Non-Unity Gain LPF*

Example 3.9 *Design a BRF to reject band in the range of 1 kHz to 10 kHz.*

Solution

$$f_L = 10 \text{ kHz and } f_H = 1 \text{ kHz}$$

$$R'_1 = \frac{1}{2\pi f_H C_2}$$

Let $C_1 = 0.05\ \mu F$

$$R'_1 = \frac{1}{2\pi \times 10 \times 10^3 \times 0.01 \times 10^{-6}} = 318\ \Omega$$

$$R_1 = \frac{1}{2\pi f_L C_2}$$

Let $C_2 = 0.01\ \mu F$

$$R = \frac{1}{2\pi \times 1 \times 10^3 \times 0.01 \times 10^{-6}} = 15.915\ k\Omega$$

The circuit diagram is shown in Fig. 3.63.

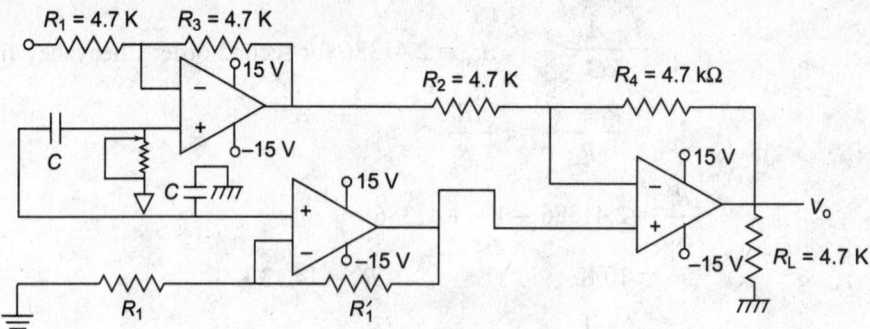

Fig. 3.63 *BRF Circuit*

There is no restriction on pass-band gain, use a gain of 1 for each section.

So, $R_1 = R'_1 = R'_F = 4.7\ K$

Further, let the gain of summing-amplifier be set at 1

So, $R_2 = R_3 = R_4 = 4.7\ k\Omega$

3.26 ANALOG SWITCHES

There are many methods by which switches as required for the digital-to-analog (D/A) converters can be implemented. As can be seen from Figs. 3.64 and 3.65, these switches are analog switches of (SPDT) single-pole double-throw type. As analog switching using FETs is far superior to bipolar transistor circuitry, it is common practice to employ either JFETs or MOSFETs for this purpose. Switching networks as required for this application using MOSFETs are directly available in IC form, whereas those using JFETs can be implemented using IC or discrete hardware. Typical circuits for switching are shown in Figs. 3.66–3.69. Fig. 3.66 gives a typical MOSFET SPDT switch along with its driver flip-flop (which is a DFF) available in IC form. Using p-MOS devices Q_1 and Q_2, the switch is shown to operate with $-V_R$. Here, *logic 1 corresponds to $-V_R$ and logic 0 to 0 V. A_1* on the bit lines sets the *FF at* $Q = 1$ and $Q = 0$, thereby turning Q_1 on. This connects R_1 to $-V_R$ while maintaining Q_2 off. Similarly, a 0 on the bit line connects R_1 to ground. The implementation of a similar circuit function using JFETs is shown in Fig. 3.66, where Q_1 and Q_2 are n-channel JFETs. These devices are of the depletion type; their bias ($=V_{GS}$) requirements

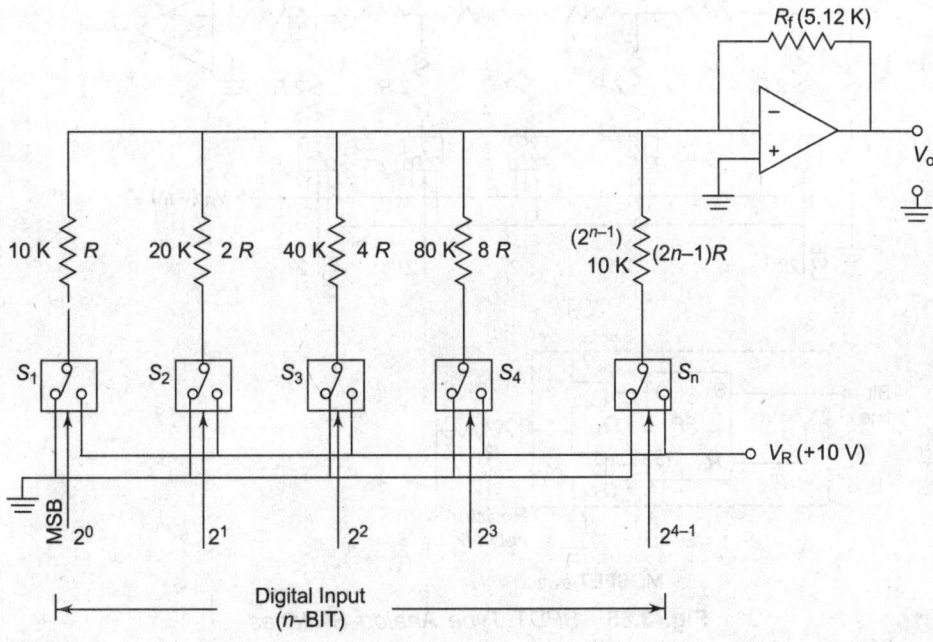

Fig. 3.64 *SPDT Type Analogue Switches*

necessitate the use of suitable driver circuitry, as shown in Fig. 3.67, which facilitates switching voltage from $-V_{EE}$ to $+V_{cc}$ at the output of Q_4 as the input to the driver switches from 0 to 1. By using $V_{EE} > V_{GS}$ (cut-off) and $V_{cc} > V_R$, it is assured that the JFET is turned off (i.e. $V_{GS} = -V_{EE}$) or on (i.e. $V_{GS} = 0\ V$) for the above two input states. This circuit is usually of discrete type. Fig. 3.65 gives the logic schematic of a typical switch in quad bilateral CMOS switch IC **(type IC 4016, 4066,** and so on) which is very popular and available from several manufacturers. This IC is known for its

1. wide supply voltage range (3–15 V),
2. high noise immunity ($\sim 45\%$ of V_{DD}),
3. wide range of digital and analog levels (~ 7.5 V)
4. low value of on resistance ($< 300\ \Omega$)
5. matched switch characteristics
6. high on/off voltage ratio (~ 65 dB for $R_L = 10$ kΩ)
7. high order of linearity ($< 0.5\%$ distortion) i.e.

it is easy to see that when the control voltage (E_n) is high (i.e. V_{DD}), the switch is on and when E_n is low (i.e. V_{SS}), the switch is off. Thus, the circuit performs the function of a single-pole single-throw (SPST) switch. The use of n-channel and p-channel MOSFETS in parallel between input–output terminals of the switch enables current flow in both directions. Two such switches can be connected together to realize an SPDT switch as required for the D/A converter.

The op-amp specifications for use in the D/A converter are governed by the clock rate of the digital input and the number of bits in the input data. An alternative approach to D/A converters makes use of binary-related constant current sources which can be switched ON or OFF to yield current output to a summation node. The control of these switches in response to digital input provides an input current, the amplitude of which is directly related to digital data.

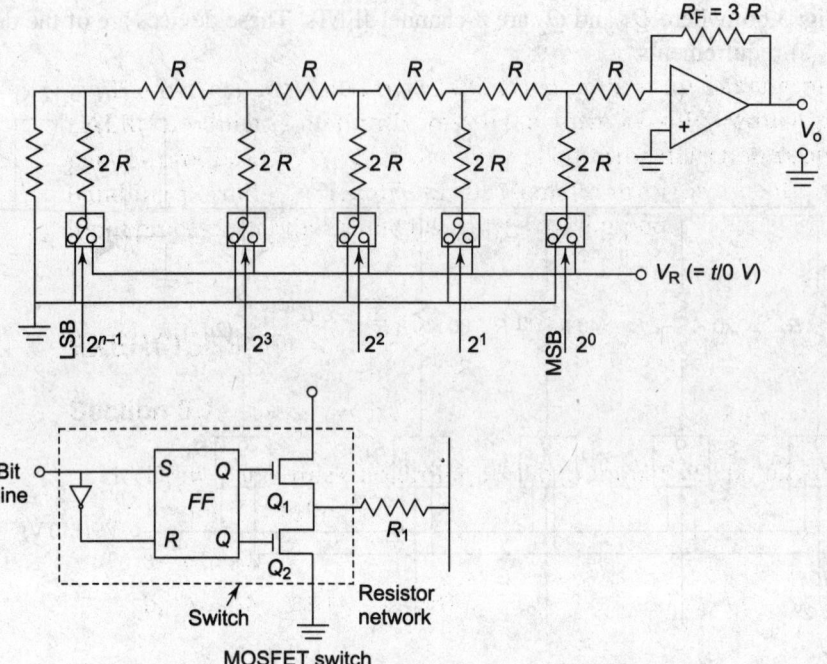

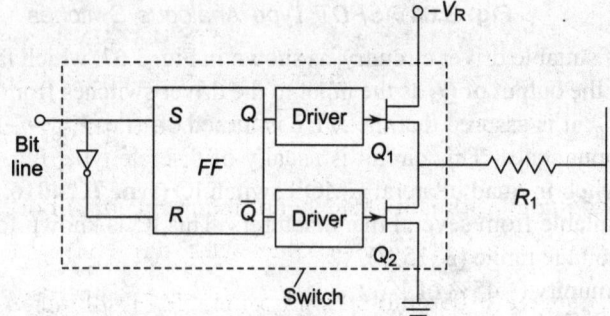

Fig. 3.65 *SPDT Type Analog Switches*

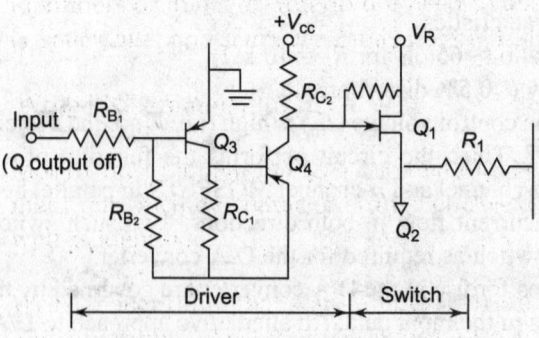

Fig. 3.66 *JFET Switches*

Fig. 3.67 *JFET Switches*

3.26.1 Multiplexers

Multiplexers are used to reduce the number of components and/or the weight required to process more than one analog signal, and to facilitate computer control of data acquisition. The number of channels in a multiplexer can vary from two to several hundred, subject to certain practical limitations. The number of samples per second that must be taken for each analog input signal is determined by the highest frequency component of the signal. At least two samples per period must be taken according to the Nyquist sampling theorem. Thus, one limitation on the maximum number of channels in a multiplexer system arises from *limited switching rate of* the multiplexer channels and the required minimum number of *samples per second*. Usually, multiplexers are designed with a fixed number of channels on a *plug-in card* or in a *plug-in module*.

The most commonly used switches in modern multiplexer designs are the JFET transistor and the MOSFET transistor. The preference for these switches is due to the excellent DC isolation between the switch driver circuitry and the analog signal path which they provide. The other desirable characteristics of such switches are as follows:

1. Zero offset voltage
2. Low leakage current
3. A very large off-to-on impedance ratio

Some types of transitor switches will be discussed in this section. In addition to discussing the operation and properties of these devices, we shall also present a treatment of differential input multiplexers and multi-tiered multiplexers.

3.26.2 Multiplexer with MOSFET Switches

The first type of multiplexer to be considered is one using MOSFET switches. Figure 3.64 shows the circuit diagram of a single-ended input multiplexer using *n*-channel depletion-mode MOSFET transistor switches. The output of each switch is tied to a common node which is the multiplexer output. Each switch driver circuit applies voltage to the gate of a MOSFET switch. This voltage controls the state of the MOSFET: a −15 V level turns it off and a +15 V level turns it on. With ±15 V power supply and an analog input range of ±10 V, the MOSFET transistor must have a gate-to-drain breakdown voltages of ±25 V minimum.

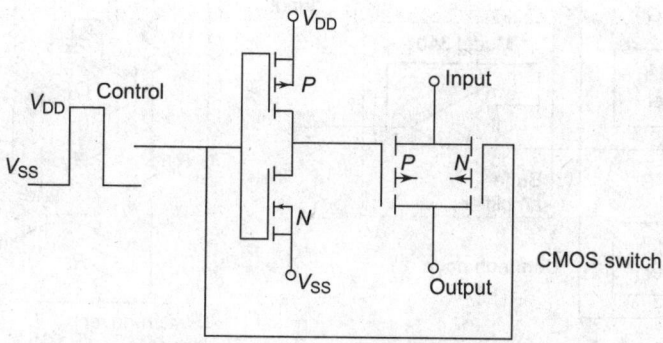

Fig. 3.68 *MOSFET Switches*

The output node of the multiplexer must be connected to a high impedance load such as a high input impedance sample–hold module to prevent part of the analog input from being dropped across the drain-to-source resistance of the on channel. If the multiplexer output must be loaded, a buffer amplifier as shown in Fig. 3.65 should be used to isolate the load from the common node. The amplifier should provide a very accurate gain of unity, which requires that the CMRR and open-loop gain be high. A gain accuracy of 0.01% requires that both parameters be in excess of 80 dB. The DC input uncertainty is caused by the inherent offset voltage of the input stage of the amplifier and by the bias current flowing from the amplifier through the impedances of the source and the on channel switch. The time required for the amplifier to settle within a desired percentage (e.g., 0.01%) of its final output voltage after the application of a step input (settling time) will determine the maximum allowable sampling rate of the multiplexer.

3.26.3 Multiplexer with JFET Switches

The second type of multiplexer to be considered is one using JFET transistors. With a few changes, the circuit of Fig. 3.68 can be modified to use N-channel JFET transistors. Since the gate-to-source voltage of a JFET must be zero when the device is turned on, the gate must somehow be made to follow the analog input. To keep the analog input isolated from the gate of the JFET and the switch driver, the gate is bootstrapped by R_B from the output of a buffer amplifier, as shown in Fig. 3.69. When Q_1 and Q_2 of the gate driver shown in Fig. 3.64 are turned on, the output of the buffer amplifier must supply any

output load current plus $\left[\dfrac{(N-1)^{25}}{R_b\,\text{k}\Omega}\right]$ mA through the $(N-1)$ bootstrap resistors to the -15 V supply

(through Q_2) when the input of the on channel is at $+10$ V. When Q_1 and Q_2 are off, the diode (D_1, for instance) is reverse biased and the FET turns on. The gate-to-source cut-off voltage of the JFET must be -5 V maximum (for inputs up to 10 V) and the buffer amplifier must have the characteristics previously described if it is connected to a signal common at the signal source, but each shield is not connected at the multiplexer input. If a guard is needed around the multiplexer or the differential amplifier, the common-mode signal can be extracted from the voltages on each differential input, namely, $1/[2\,(C_1+C_2)]$. Often this signal is readily available at a point in the differential amplifier. This method of splitting the shield reduces the complexity of the differential multiplexer by reducing the number of parts required and the number of input pins required on a module.

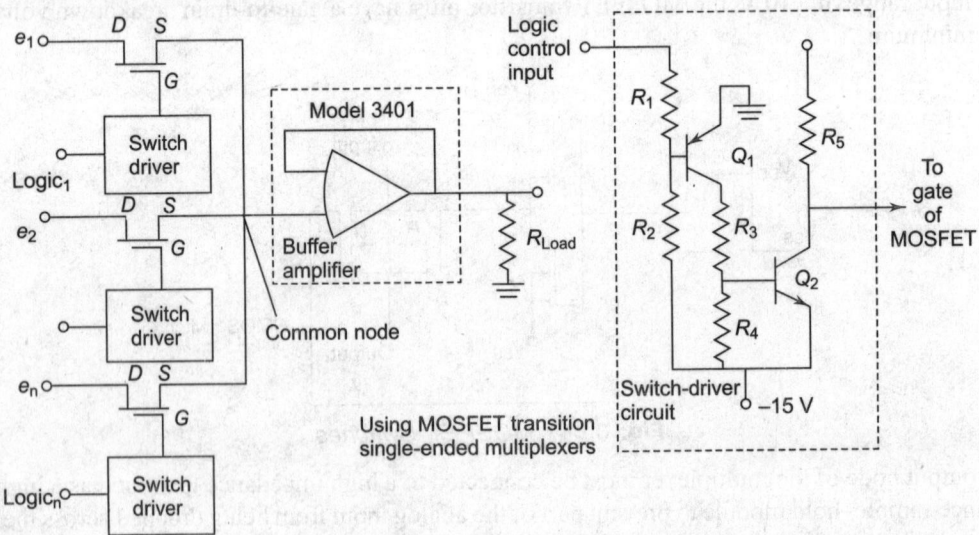

Fig. 3.69 *Switch-Driver Circuit*

The CMRR is limited by the common-mode input impedance of the differential amplifier and by signal-source impedance imbalances. Any mismatch in the R_{ON} resistance of the two multiplexer switches of a differential channel will be interpreted by the instrumentation amplifier as the signal-source impedance imbalance. Consequently, the R_{ON} of the FETs in any channel should be matched.

The same considerations and limitations apply to differential multiplexer FET switches as for the single-ended input multiplexer switches.

3.26.4 Multi-tiered Multiplexing

There are two common ways to combine more than one multiplexer module or plug-in card to expand the number of channels in a system, the method used may depend upon the type of decoding logic that is used to drive the multiplexers. If a 10-channel *BCD* to decimal decoder is used, the pyramid structure can be used, as shown in Fig. 3.70, for a 10-channel multiplexer. The same *multi-tiered* connections also apply to the differential type multiplexer, although only the single-ended type is shown in the figure. Outputs of the units counter drive the module logic inputs, all in parallel. The 10's counter drive the output accumulator. This technique greatly simplifies the logic required to drive the multiplexers. Up to 100 channels can be built using only 11 modules. The technique can be extended to over 100 channels by adding another *BCD* counter and decoder for the 100's position. When less than 100 channels are used, some scheme must be devised to reset the counters to zero when the last channel is reached. This can be accomplished by connecting gates to the four outputs of the 10's and units counters to sense the highest required count. The output of these gates then resets the counters to zero. In some systems, binary coding is used. In this case, the number of inputs per module would be eight.

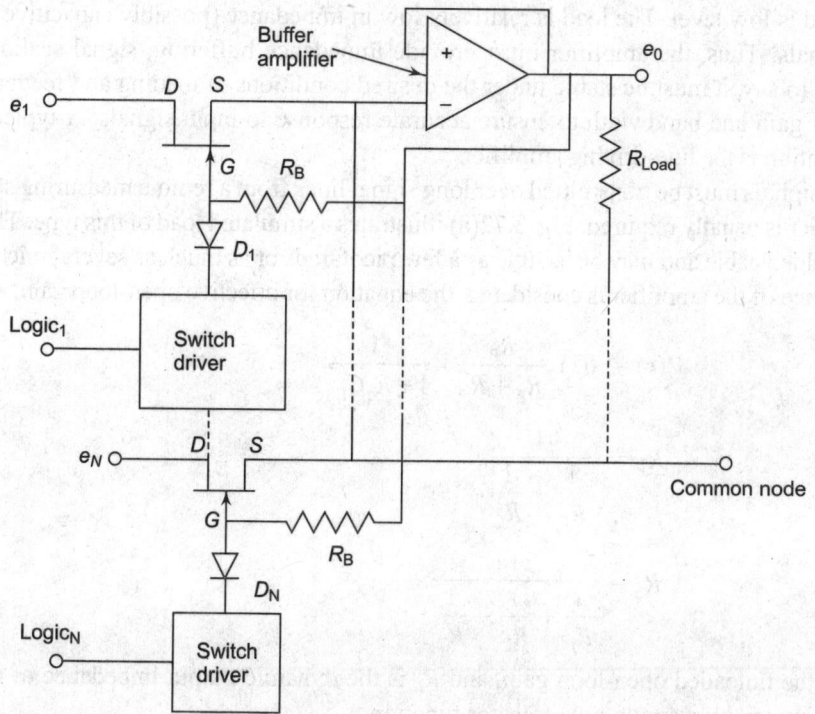

Fig. 3.70 *Single-Ended Multiplexer Using JFET*

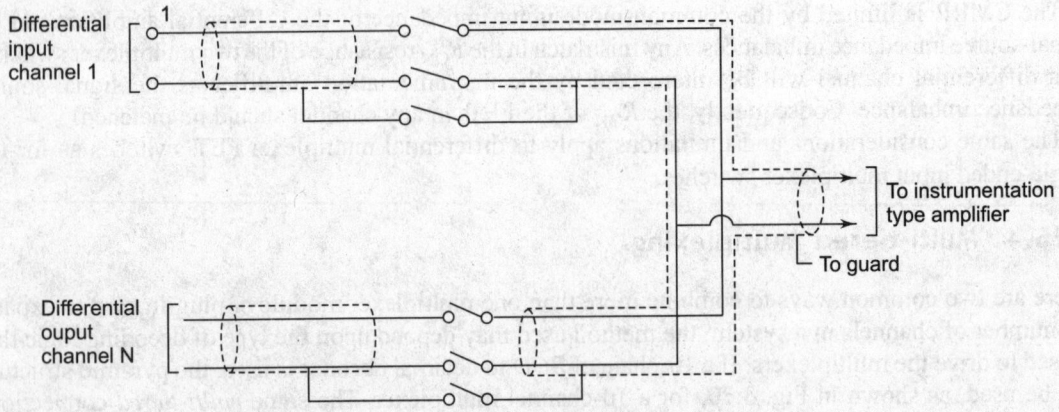

Fig. 3.71 *Differential Input Multiplexer*

3.27 LINE DRIVERS

3.27.1 Line-Driving Amplifier

One of the primary areas of application for the op-amp is that of buffering between a signal source and the desired load. Usually the signal source is very limited in powers. It has relatively high internal impedance, and is low level. The load is relatively low in impedance (possibly capacitive) and requires high-level signals. Thus, the amplifier must provide impedance buffering, signal scaling, and power gain. Needless to say, it must be stable under the desired conditions of loading and feedback, and must have sufficient gain and bandwidth to ensure accurate response to input signals. A typical example of such an application is the line-driving amplifier.

When data signals must be transmitted over long signal lines from a remote measuring station, the line driving amplifier is usually required. Fig. 3.72(a) illustrates a simulated load of this type. The capacitance is that of a shielded cable and may be as little as a few picofarads or as much as several microfarads. If the output impedance of the amplifier is considered, the equation for effective open-loop gain, $A'(s)$, becomes

$$A'(s) = A(s) \frac{R_P}{R_P + R_o} \times \frac{1}{1 + R_q C_{LS}}$$

where
$$R_P = \frac{1}{\dfrac{1}{R_F} + \dfrac{1}{R_L}};$$

$$R_q = \frac{1}{\dfrac{1}{R_F} + \dfrac{1}{R_L} - \dfrac{1}{R_o}}$$

where $A(s)$ is the unloaded open-loop gain, and R_o is the dynamic output impedance of the op-amp. If $A(s)$ is approximated by a single-pole transfer function

$$A(s) = \frac{A_o}{1 + \dfrac{s}{\omega_o}}$$

then the effective (loaded) open-loop gain becomes

$$A'(s) = \frac{R_P}{(R_P + R_o)} \frac{A_o}{1 + \dfrac{s}{\omega_o}} \frac{1}{(1 + R_q C_{LS})}$$

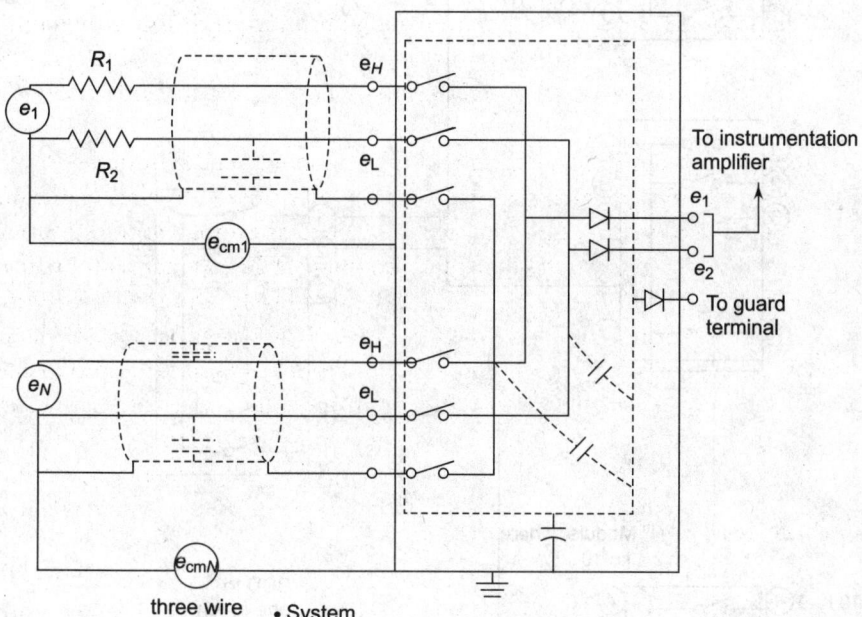

three wire • System

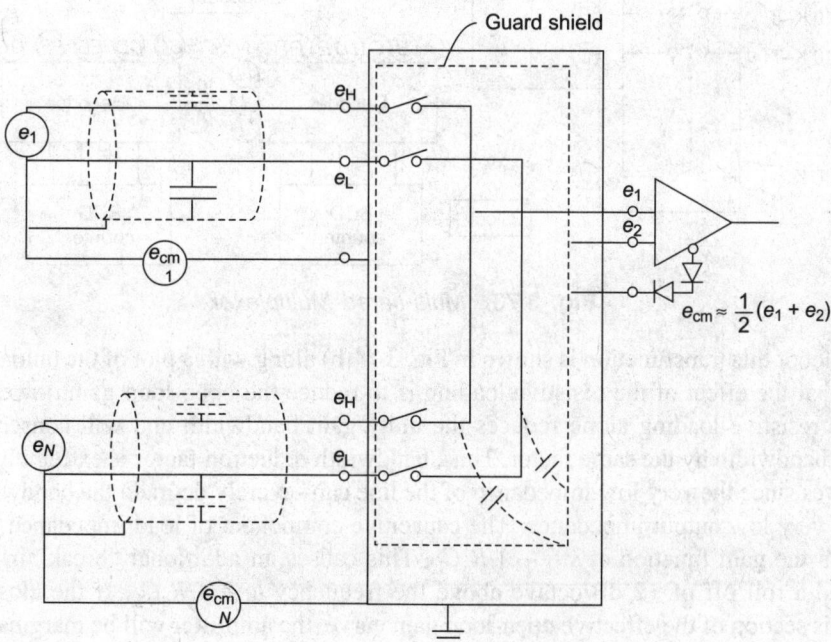

Fig. 3.72 *Multi-tiered Multiplexer*

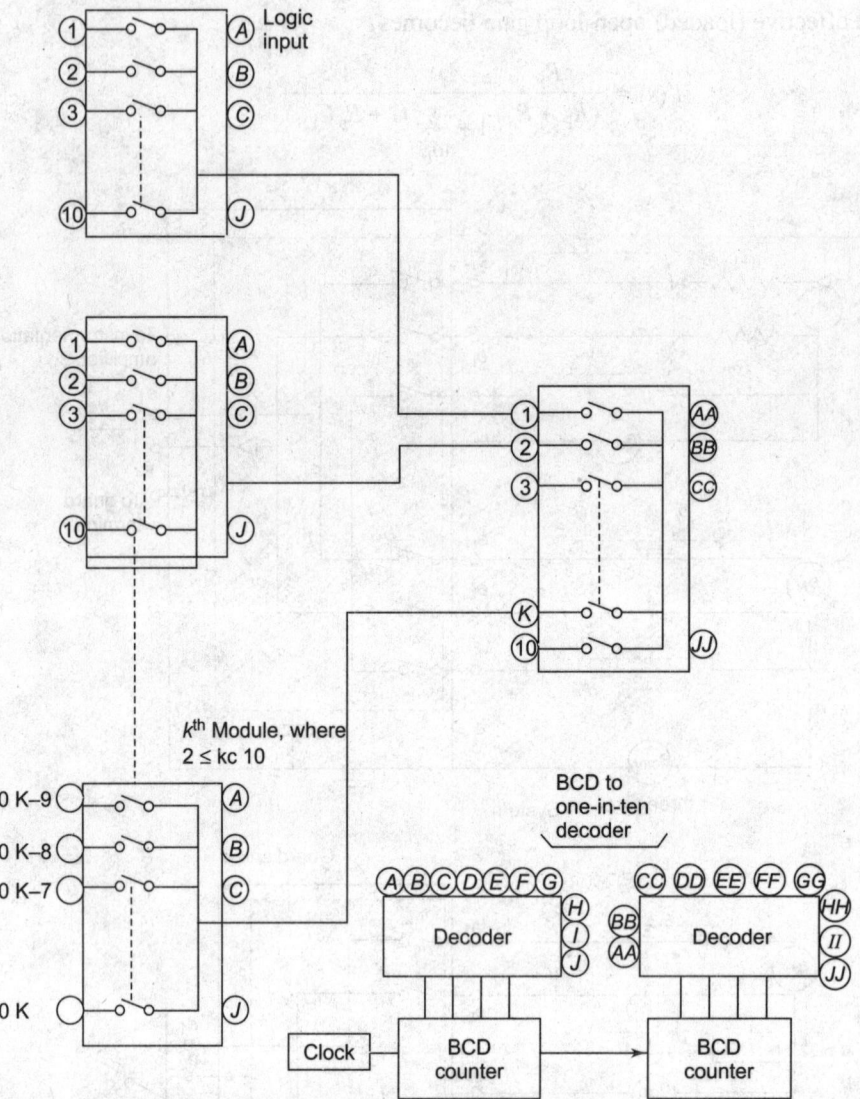

Fig. 3.73 *Multi-tiered Multiplexer*

A bode plot of this transfunction is shown in Fig. 3.74(b) along with a plot of the unloaded open-loop gain. Note that the effect of the resistive loading is to reduce the open-loop gain, lowering the entire curve. Thus resistive loading alone reduces the unity gain bandwidth and will consequently reduce closed-loop bandwidth by the same factor. Thus, bandwidth reduction factor is extremely important for fast line drives since the very low impedance of the line can severely degrade the bandwidth unless the op-amp has very low output impedance. The capacitive component of load impedance introduces another pole in the gain function at $s = -1/R_q C_L$. This causes an additional "break" in the frequency response and a roll off of +2 dB/octave above the frequency $\omega = 1/R_q C_L$. If the closed-loop curve intersects this section of the effective open-loop gain curve, the amplifier will be marginally stable with unacceptable transient responses.

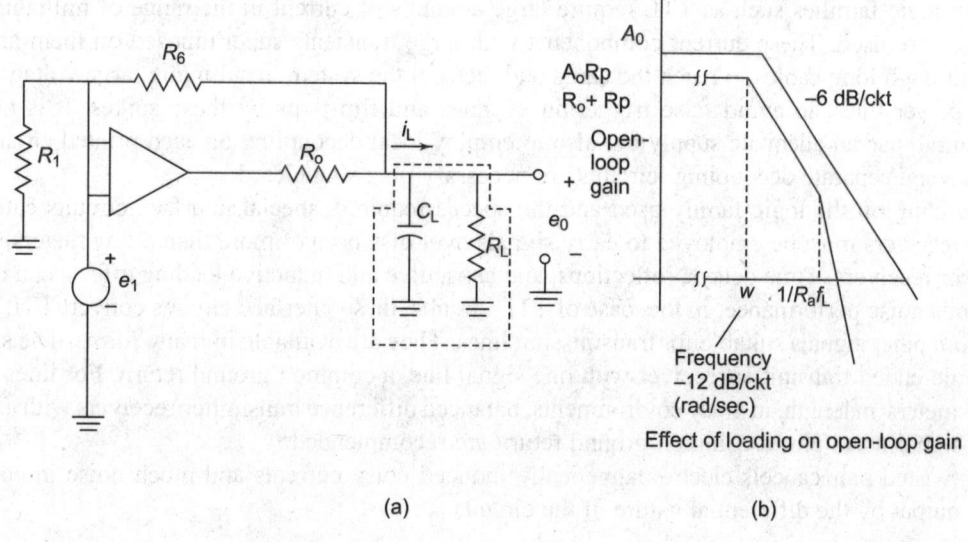

Fig. 3.74 *(a) Line-Driving Amplifier Circuit for Long Signal Lines. (b) Bode Plot*

There are a number of techniques for dealing with the problems of loading. The most satisfactory of these is to choose an amplifier with very low open-loop output impedance or to create one by adding a power booster stage to an available op-amp. This will reduce the gain and bandwidth loading factors caused by the load resistance, and will increase the frequency at which the additional pole occurs. The higher the frequency in this pole, the more stable the closed-loop response. The power output stage also supplies the current necessary to meet the condition

$$i_{L\ max} = C_i \left(\frac{de_o}{dt} \right)_{max}$$

3.27.2 Line Drivers and Receivers

As with all electronic circuits, there are certain considerations that must be taken into account if digital logic families are to be applied successfully. For example, when working with TTL, it is often stated that +5 V represents a logic 1 and +0 V a logic 0. But, in practice, "1" and "0" are represented by a range of voltage levels. These levels are different for output and input to ensure that any output is always more than adequate to drive any input under worst-case conditions. For example, for TTL, the output range is 0 to 0.4 V for a logic "0" and 2.4 to 5 V for a logic "1", while the input range is 0 to 0.8 V for logic "0" and 2.0 to 5 V for logic "1". Hence a worst-case output of +2.4 V is greater than the minimum of 2.0 V required to ensure a "1" input.

The high-value ranges define a noise margin which is the difference between the worst input and output voltages for that level. For example, for TTL levels,

the high (1) noise margin = 2.4 − 2.0 = 0.4 V and

the low (0) noise margin = 0.8 − 0.4 = 0.4 V.

Similar conditions apply to all logic families and the specific number must be extracted from the appropriate data sheet.

Many logic families such as TTL require large amounts of current in the range of milliamperes if many ICs are used. These current components with large transients superimposed on them are often routed through long cables to reach the cards and racks in the system, resulting in large voltage spikes on the power lines, to avoid false triggering of gates and flip-flops by these spikes. It is not only important to use an adequate supply but also to employ local decoupling on each printed circuit card. Often several separate decoupling circuits are necessary on a single card.

Depending on the logic family used and the speeds required, special interface circuits called line drivers/receivers must be employed to carry signals over distances of more than a few meters without drivers or receivers. Time delays, reflections, and capacitive and inductive loading effects can degrade a system's noise performance. In the case of TTL circuits, these interface circuits convert TTL signals to or from other signals suitable for transmission lines. They are available in many forms. The simplest is a single-ended transmitter/receiver with one signal line, a common ground return. For lines greater than 33 meters in length, in noisy environments, balanced difference transmitter/receivers with a twisted pair of signal wires plus a common ground return are recommended.

The twisted pair cancels electromagnetically induced noisy currents and much noise in common-moded output by the differential nature of the circuits.

3.27.3 Multiplexer Circuit Principle

The Fig. 3.75 shows the basic multiplexer principle. As shown, the multiplexer selects one of the several data sources to transmit over a single output line. Prior to the advent of semiconductor switching, such data sampling was accomplished by mechanical multi-position or stepping switches which mechanically sampled each of the inputs in a given sequence.

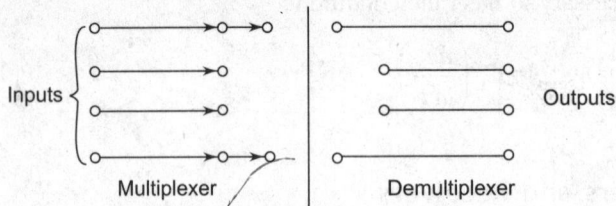

Fig. 3.75 *Principle of Multiplexer and Demultiplexer*

Typical multiplexing applications include computers, remote pressing control, and communication systems, where it is desirable to reduce the number of wires between subsystems. In such systems, multiple data inputs are multiplexed to line share a single data transmittion line to the distribution point where they are demultiplexed and are routed to their individual destinations.

In multiplexer applications such as time division multiplexing (TDM), multiple measurements are sent over a common channel by dividing the available time intervals among the measurement to form a composite pulse train.

In its computer applications, the multiplexer receives binary information from 2^n time lines and transmits on a single output line. The one output line being selected is determined from the bit combination of "n" selection lines. The multiplexer circuits are formed using AND, OR, NOT, and so on.

A multiplexer is also formed on a "data selector" since it selects one of the multiple inputs and steers the information to the output.

3.27.4 Analog Switches

Analog switches are used when continuous waveforms are applied. The function of this switch circuit is to produce an output voltage (V_C) proportional to the input analog voltage (V_i), when the control voltage $V_C = V_{on}$ and have $V_C = 0$ when $V_C = V_{off}$. The circuit can be realized using a diode quad shown in Fig. 3.76. The circuit consists of a diode quad and two isolated diodes, all manufactured on the same silicon chip.

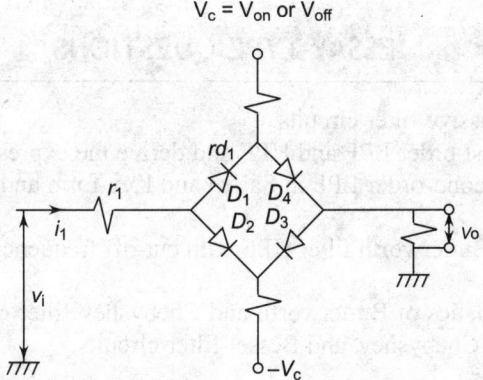

Fig. 3.76 *Analog Switch Circuit Using Diode Quad*

The diodes in the circuit are all manufactured simultaneously on an IC and therefore have identical characteristics, while the characteristics of discrete diodes selected at random may differ widely. In practice, the analog switch is driven from an op-amp that has an output impedance much less than 1 Ω.

The speed of the diode switches is less than 0.1 nsec, whereas the speed of circuits using FETs is of the order of 10 nsec. Generally, FET switches are used to switch low-frequency signals.

The IC, MC 1545 may be used as an analog switch controlled by digital logic signal applied to the gate terminal. This application is illustrated in Fig. 3.77.

If an input is applied to pin 4, this signal is amplified at the gate. It is high, corresponding to a logic state. The signal is not passed through the amplifier when the gate control signal is low (logic 0).

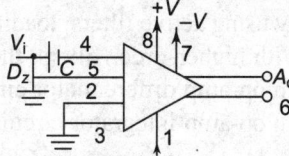

Fig. 3.77 *MC 1545 IC Analog Switch*

SUMMARY

- Active filter circuits provide voltage gain and help avoid the loading effect.
- A differentiator circuit can be used as an HPF.
- An integrator circuit can be used as an LPF, $f_c = \dfrac{1}{2\pi RC}$.
- Combining active LPF and HPF, BPF or BRF circuits can be designed.
- Higher-order filter circuits give better and sharper frequency response characteristics. But more circuit components have to be used, which increases the complexity.

- In Butterworth filters, attenuation is 3.01 dB at the cut-off frequency.
- Rate of attenuation at the cut-off frequency is higher for a Chebyshev filter.
- APFs transmit all frequencies, but provide phase shift.
- Switching a capacitor periodically is equivalent to a resistance R_1. $R_1 = \dfrac{1}{Cf_{CLK}}$. By changing clock frequency f_{CLK}, the required value of R_1 can be achieved.
- First-order and second-order switched capacitor circuits are explained.

ESSAY-TYPE QUESTIONS

1. Compare active and passive filter circuits.
2. Draw the circuit for first-order LPF and HPF and derive the expressions for cut-off frequencies.
3. Draw the circuit for second-order LPF in Sallen and Key form and derive the expression for the transfer function.
4. Design a fourth-order Butterworth filter HPF with cut-off frequency of 10 kHz, given that $2K_1 = 0.765$ and $2K_2 = 1.848$.
5. What are the characteristics of Butterworth and Chebyshev filter circuits?
6. Compare Butterworth, Chebyshev, and Bessel filter circuits.
7. Explain analog switches.
8. Write notes on analog multiplexers.
9. Draw the circuit for second-order switched capacitor filter circuit and explain its working.
10. Explain summing integrator and difference integrator.

FILL IN THE BLANKS

1. In the case of active filters, voltage gain A_v can be _____.
2. By using active filters, loading effect can be _____.
3. With higher-order filters, the sharpness of the frequency response curve _____.
4. An op-amp differentiator circuit can be used as a _____ filter.
5. An op-amp integrator circuit can be used as a _____ filter.
6. The transfer function $\dfrac{sCR}{1+sCR}$ represents a _____ filter.
7. The transfer function $\dfrac{1}{sCR}$ represents a _____ filter.
8. The transfer function of an APF circuit is _____ .
9. The kink in the frequency response of an active filter circuit will be more if the value of the damping coefficient K is _____ and typical value is _____ .
10. To obtain BRF, LPF and HPF must be connected in _____ .

ANSWERS

1. $A_v \geq 1$
2. Minimized
3. Increases

4. HPF
5. LPF
6. First-order HPF
7. First-order LPF
8. $(1 - s)/(1 + s)$
9. < 1, about 0.1
10. Parallel

UNSOLVED PROBLEMS

1. A second-order BPF is to have midband voltage gain of $60 f_o$. The centre frequency required is 50 Hz and bandwidth must be 10 Hz. Determine the component values.
2. Design a second-order LPF in Butterworth type to have unity gain and cut-off frequency of 800 Hz, and maximally flat response.
3. Determine the component values for a Chebyshev filter with positive non-unity gain second-order LPF to have $f_c = 500$ Hz and 0.5 dB ripple.
4. Determine the component values for a second-order BRF to reject the frequencies in the range of 5 kHz to 15 kHz.
5. Determine the component values for a fourth-order Butterworth HPF with $f_c = 15$ kHz, given $2K_1 = 0.765$ and $2k_2 = 1.848$.
6. Design a second-order BPF using op-amps, given $2K = 1.414$, to pass signals in the band of 2 kHz to 20 kHz.
7. Determine the component values of a Chebyshev filter with positive non-unity gain LPF of second order to have 3 dB frequency equal to 500 Hz and 0.5 dB ripple.
8. Design a BRF to have $f_L = 25$ kHz and $f_H = 1$ kHz.
9. For the circuit shown if $R = 22$ kΩ and $C = 0.01$ μF, determine the value of f_c.

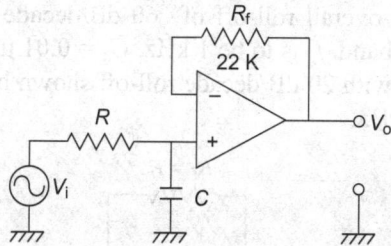

10. For the above circuit if f_c is to be 5 kHz and $C = 0.0047$ μF, determine the value of R.
11. For the op-amp LPF circuit shown in Problem 9, if the desired cut-off frequency in radian/sec is 20×10^3 and $C = 0.047$ μF, determine the value of R.
12. For the circuit shown, determine the values of R_1 and R_2 to get a cut-off frequency of 5 kHz. Assume $C_1 = 0.01$ μF.

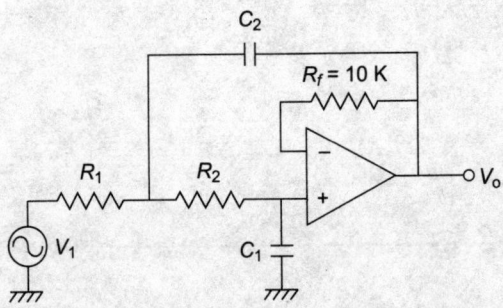

13. For the circuit shown an LPF of −40 dB/decade is cascaded with another of −20 dB/decade to give an overall roll-off of −60 dB/decade. Overall A_{CL},

$$A_{CL} = \frac{V_o}{V_i} = \frac{V_{ol}}{V_i} \times \frac{V_o}{V_{ol}}$$

For a Butterworth filter, the magnitude of A_{CL} must be 0.707 at $f = f_c$. Design the circuit to guarantee that the frequency response is flat in the pass band, for $f_c = 3$ kHz. Assume $C_3 = 0.01$ μF.

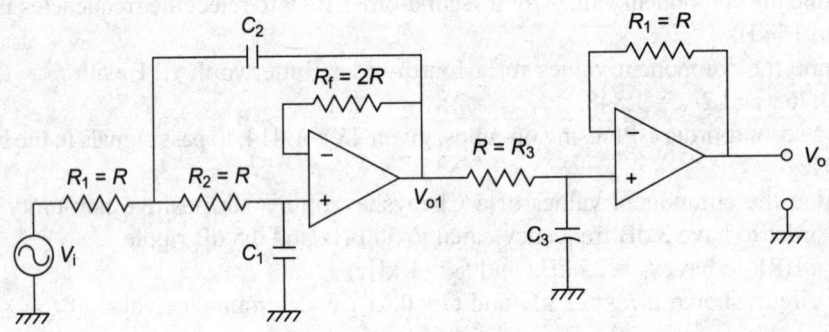

14. Design an LPF to give an overall roll-off of −60 dB/decade to guarantee that the frequency response is flat in the pass band. f_c is to be 1 kHz. $C_3 = 0.01$ μF.

15. Consider the op-amp HPF with 20 dB/decade roll-off shown below. Calculate the value of R if $C = 0.001$ μF and $f_c = 15$ kHz.

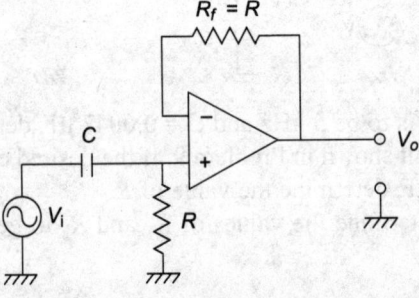

16. For a circuit of Problem 15, if $R = 47$ kΩ and $C = 0.01$ μF, calculate f_c.

17. Consider a second-order HPF with 40 dB/decade roll-off. To satisfy Butterworth criteria, the frequency response must be 0.707 at ω_c and 0 dB in the pass band. Calculate the values of R_1 and R_2 for $f_c = 3$ kHz.

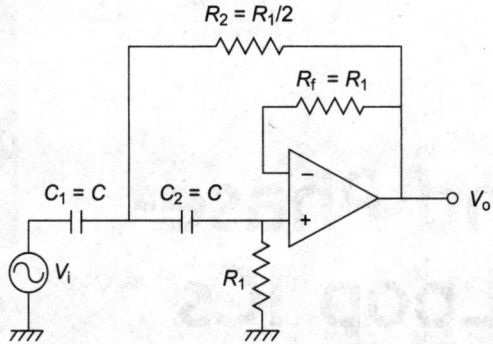

18. Determine the values of R_1 and R_2 for the above circuit if ω_c is to be 100 K rad/sec. $C_1 = C_2 = 150$ pF.
19. Give the design procedure for 60 dB/decade HPF in Butterworth configuration when the desired cut-off frequency is 200 Hz.
20. Determine the component values of 60 dB/decade HPF for $f_c = 100$ kHz. Assume $C_1 = C_2 = C_3 = C = 220$ pF with usual notation.
21. Design a BPF to have $f_c = 1800$ Hz, with Q = 10. Assume $C_1 = C_2 = C = 0.01$ μF.
22. Design a Notch filter to give $f_r = 500$ Hz and Q = 5. Assume $C_1 = C_2 = 0.01$ μF. Draw the circuit.

Timer and Phase-Locked Loop ICs

Objectives:

In this chapter...

- *555 Timer IC internal schematic is explained.*
- *555 Timer as a stable multivibrator circuit is discussed. The expression for time period T of output is given. Waveforms are explained.*
- *Timer IC in monostable mode is also given. The circuit diagram is given and the expression for pulse width T is derived. Output waveforms are explained.*
- *Applications of timer IC and specifications and parameter values are given.*
- *The principle of phase-locked loop (PLL) is explained with the help of block schematic. The circuit specifications and applications of the 565 PLL IC circuits are given.*

4.1 NE555 TIMER

NE555 is a monolithic (single crystal) IC that is made from a single-crystal silicon and this IC is used for timing operations. It operates in a stable and monostable-mode circuits with adjustable duty cycles and timings ranging from *few microseconds* to *few hours*. It can easily *drive a* resistive load, drawing load current $I_L = 200$ mA. Because of this it is called as Timer IC

The description of the IC block schematic is given below:

The functional block diagram is shown in Fig. 4.1. Discharge transistor T_1 is connected to $\overline{Q}$. Therefore, when $\overline{Q} = 1$, that is, when the flip-flop is reset, T_1 is turned on.

The timing of the output waveform can be controlled by pin (5) control voltage. If this is not being used, it is connected to ground through 0.1 µF capacitor. Reset pin 4 has overriding effect on the output of the flip-flop.

The unit is intended to be used with supply voltages from 5 to 15 V. The resistance between V_{cc} and ground provide the reference voltages for the comparator. The reference voltage for the comparator pin (2) is $\dfrac{V_{cc}}{3}$. The reference voltage for comparator 1 is $\dfrac{2V_{cc}}{3}$. Under normal operation, if no modulating voltage is being applied to the control voltage terminal (5), it is connected to the ground through a capacitor of 0.01 µF (as recommended by the manufacturer). The capacitor will block DC. For DC, it is open circuit pin (6), the threshold voltage which monitors the voltage across the external capacitor C.

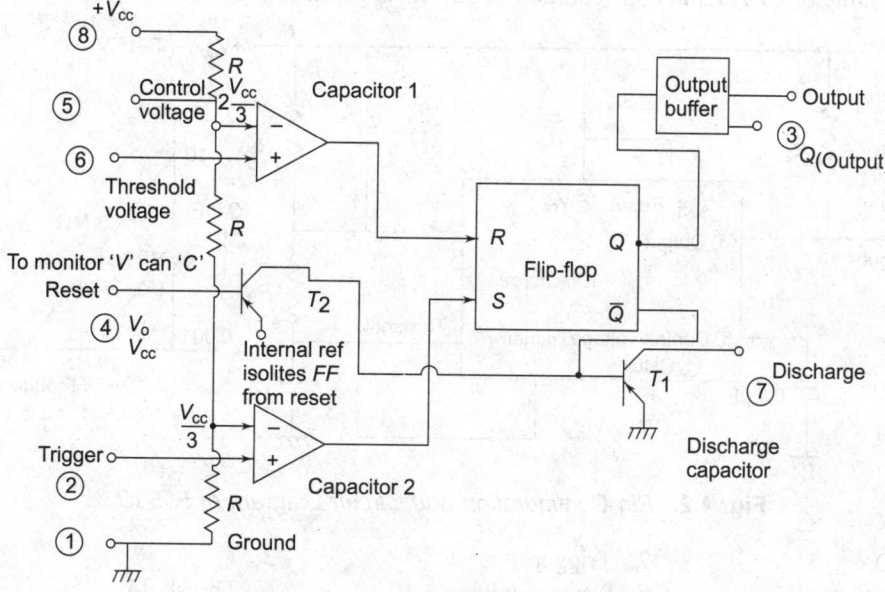

Fig. 4.1 *Block Schematic of 555 IC*

If the trigger input is negative going, then the value of the trigger input is $\frac{V_{cc}}{3}$; the output of capacitor 2 sets the flip-flop. So, the output is in logic 1 state and is considered to be high. During the positive excursion of the threshold voltage, when the threshold voltage reaches $\frac{2}{3}V_{cc}$, comparator 1 resets the flip-flop and the output changes state. Output-changing state can be controlled by the time taken for the comparator input voltage to reach to the threshold values of the comparator, that is $\frac{1}{3}V_{cc}$ and $\frac{2}{3}V_{cc}$.

The reset input (4) provides a mechanism to reset the flip-flop. Transistor T_2 serves the purpose of isolating the reset input from the flip-flop and transistor T_1. Because of the output buffer, the IC can supply an output load current up to 200 mA. The external timing capacitor C is connected to the discharge point (7). When the output of the flip-flop is in a reset state, the output to the base of the transistor T_1 will be higher, since, if the FF is reset state, $Q = 0$ and $\overline{Q} = 1$. Since $\overline{Q}$ is connected to the base of the transistor, T_1 turns on. So, the transistor T_1 will be in saturation (on), discharging the external capacitor. Thus, T_1 is the discharging capacitor. C discharges through T_1 to the ground. Since T_1 is on, it is a short circuit.

4.1.1 NE555 as Astable Multivibrator

To get 50% duty cycle, T_1 must be equal to T_2. That is R_A must be shorted by a diode so that while charging, the diode is forward biased, and while discharging, it is reverse biased. In this mode, the circuit can trigger itself and run as a free-running multivibrator or as an astable multivibrator. The external capacitor C charges through R_A and R_B. However, it discharges only through R_B (the inside circuit being made that way). Thus the duty cycle can be adjusted precisely by the ratio of R_A and R_B for a given value of C. The pin configuration of 555 IC is given in Fig. 4.2.

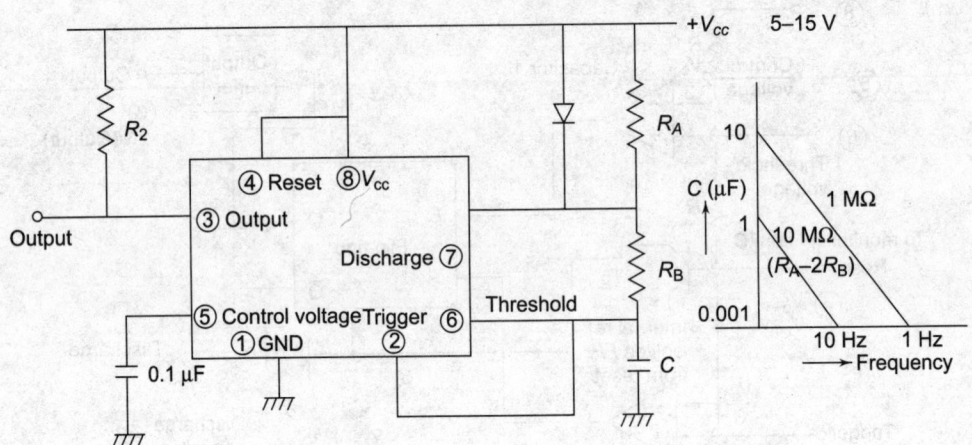

Fig. 4.2 *Pin Configuration and Circuit Diagram of 555 IC*

1. GND
4. Reset
7. Discharge

2. Trigger
5. Control voltage
8. V_{cc}

3. V_o
6. Threshold

The capacitor charges to $\left(\dfrac{2}{3}\right)V_{cc}$, through R_A and R_B and discharges to $\dfrac{1}{3}V_{cc}$ through R_B only.

The voltage across the external capacitor C is applied to the threshold input (6). When the capacitor voltage increases to $\dfrac{2}{3}V_{cc}$ (positive going), capacitor 1 responds and resets the flip-flop. The discharge transistor T_1 is turned on. So, C starts discharging through R_B. When the capacitor voltage (negative going) reaches $\dfrac{1}{3}V_{cc}$, capacitor 2 responds and sets the flip-flop. Thus, the timing interval of output waveform is varied by controlling the time taken by the capacitor to get charged to $\dfrac{2}{3}V_{cc}$ and discharged to $\dfrac{1}{3}V_{cc}$.

$$f = \frac{1}{T} = \frac{1.46}{(R_A + 2R_B)C}$$

Note that 'f' is independent of V_{cc}.

Expression for T

The capacitor gets charged exponentially to $\left(\dfrac{2}{3}\right)V_{cc}$. Initial voltage across the capacitor is $\left(\dfrac{1}{3}\right)V_{cc}$ and not zero, because C gets discharged to $\left(\dfrac{1}{3}\right)V_{cc}$ only, in the previous cycle.

The expression for charging time t of a capacitor with initial voltage E_0, charging from a source voltage E, to a voltage e_c in time t is,

$$t = RC \ln \left\{ \frac{E - E_0}{E - e_c} \right\}$$

$$E_0 = \left(\frac{1}{3} \right) V_{cc}$$

$$E = V_{cc}$$

$$e_c = \left(\frac{2}{3} \right) V_{cc}$$

$$R = (R_A + R_B)$$

C gets charged through $(R_A + R_B)$

So,

$$T_1 = (R_A + R_B) \, C \;\; \ln \left\{ \frac{V_{cc} - \frac{1}{3} V_{cc}}{V_{cc} - \frac{2}{3} V_{cc}} \right\}$$

$$= (R_A + R_B) \, C \ln (2)$$

$$\boxed{T_1 = 0.69 \, (R_A + R_B)C}$$

The capacitor discharges from $\frac{2}{3} V_{cc}$ to $\frac{1}{3} V_{cc}$ during a time interval T_2 (Fig 4.3).

The expression for the final voltage V_f of a capacitor with peak voltage charging V_p through R and C during time $(t - t_p)$ is

$$V_o = V_f = V_P \, e^{\frac{-(t - t_p)}{RC}} \qquad\qquad R = R_B$$
$$(V_1) \;\; (V_2) \qquad\qquad\qquad (t - t_p) = T_2$$

$$\frac{V_f}{V_p} = e^{\frac{-(T_2)}{R_B C}} \qquad\qquad V_f = \frac{1}{3} V_{cc} = V_1$$

$$- T_2 = R_B C \ln \left\{ \frac{V_f}{V_p} \right\} \qquad\qquad V_p = \frac{2}{3} V_{cc} = V_2$$

$$T_2 = R_B C \ln \left\{ \frac{V_P}{V_f} \right\} \qquad\qquad V_f = \text{final voltage}$$

$$= R_B C \ln \left\{ \frac{2 \times V_{cc} \times 3}{3 \times 1 \times V_{cc}} \right\} \qquad V_p = \text{peak voltage}$$

$$T_2 = 0.69 \, R_B C$$

$$T = T_1 + T_2 = 0.69 \, (R_A + 2R_B)C$$

$$f = \frac{1}{T} = \frac{1}{0.69(R_A + 2R_B)C} = \frac{1.46}{(R_A + 2R_B)C}$$

$$f = \frac{1.46}{(R_A + 2R_B)C}$$

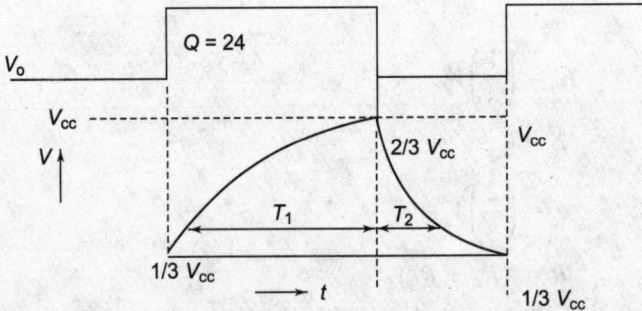

Fig. 4.3   *Output Waveforms of 555 in Astable Mode*

4.1.2   NE555 Astable with 50% Duty Cycle

For 555,
$$T_1 = 0.69 \ (R_A + R_B)C$$
$$T_2 = 0.69 \ R_B C$$

That is, the capacitor C gets charged through R_A and R_B. It discharges through R_B only. To get 50% duty cycle, T_1 must be equal to T_2. That is,
$$R_A = 0$$

But, if $R_A = 0$, the collector of discharge transistor is directly connected to V_{cc}. This will make the discharge transistor always on because the collector is n-type and collector base junction is reverse biased and transistor conducts or discharges the transistor.

Therefore, to get 50% duty cycle, make $T_2 = \dfrac{1}{2T}$.

Then, calculate the value of R_B.

Make
$$R_A = R_B$$

Connect a diode D across R_B with its anode connected to V_{cc}.

So while charging, diode D is forward biased. C gets charged through R_A and the diode. While discharging, the diode is reverse biased. Since the capacitor is charged, the cathode of diode D is at a higher potential. The anode of diode D is at comparatively less potential because of the drop across R_A. Therefore, the diode is reverse biased and C discharges through R_B only.

Therefore,
$$T_1 = T_2 \ (\text{since } R_A = R_B)$$

and a 50% duty cycle is obtained.

The circuit diagram is shown in Fig. 4.4.

4.1.3   NE555 as Monostable Multi

Consider the block schematic shown in Fig. 4.1. If $Q = 0$ and $\overline{Q} = 1$, discharge transistor is on, op-amp is in saturation, and $V_o = 0$, when voltage at the connecting terminal is greater, then it will change state.

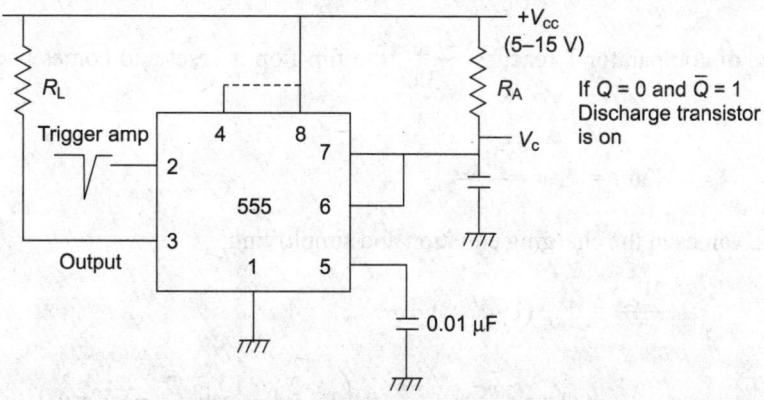

Fig. 4.4   *555 in Monostable Mode*

1. Trigger pulse should be negative because it is to set the flip-flop.
2. What should be the voltage levels for comparator 1 and 2 to change state?

For comparator 1, the reference voltage is $\frac{1}{3}V_{cc}$ and for comparator 2, it is $\frac{2}{3}V_{cc}$. Capacitor discharges instantaneously through transistor when the transistor is in ON state. Capacitor charges from 0 to $\frac{2}{3}V_{cc}$.

Initially C is held discharged by a transistor inside the timer because $\overline{Q} = 1$. So, the transistor is ON and the capacitor is shorted. Upon application of a negative trigger pulse to pin 2 (trigger input), the flip-flop is set. Comparator 2 responds, which releases the short circuit across the external capacitor and drives the output high. The voltage across the capacitor now increases exponentially with a time constant $R_A C$. When the voltage across the capacitor reaches $\frac{2}{3}V_{cc}$, the comparator responds and resets the flip-flop. So, the capacitor discharges rapidly and the output becomes low.

4.1.3.1 *Timer in Monostable Mode:* The waveforms in monostable operation are shown in Fig. 4.5 (a).

Expression for T : capacitor charging equation is
$$v = V_{cc}(1 - e^{t/RC})$$
where v is the voltage at any time t, V_{cc} is the charging voltage, and RC is the tone constant.

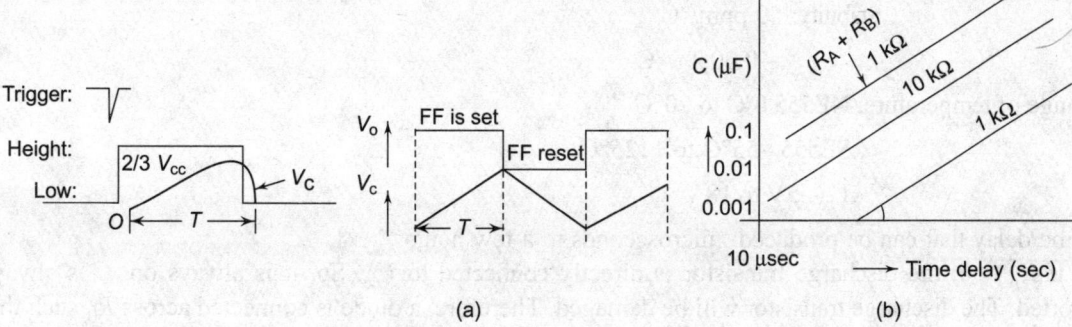

Fig. 4.5   *Waveforms for 555 Monostable Circuit*

When the voltage of comparator 1 reaches $\dfrac{2V_{cc}}{3}$, the flip-flop is reset and comes back to its original state.

So,   at $t = T$, $v = \dfrac{2V_{cc}}{3}$

Substituting these values in the charging equation and simplifying,

$$\frac{2V_{cc}}{3} = V_{cc} (1-e^{T/RC})$$

Therefore, $\dfrac{2}{3} = (1-e^{+T/RC}); \ e^{-T/RC} = \left(1-\dfrac{2}{3}\right) = \dfrac{1}{3}; +\dfrac{T}{RC} = e^{\frac{1}{3}} = 1.1$

Hence, $\boxed{T \simeq 1.1 \ RC}$

$$\frac{2V_{cc}}{3} = V_{cc} (1 - e^{T/RC})$$

Values of $(R_A + R_B)$ to be chosen and corresponding values of C to get a particular time delay are as shown in Fig. 4.5 (b):

$$\boxed{T \simeq 1.1 \ RC}$$

In the absence of a trigger pulse, the timing capacitor C is held in the discharged state. So, the output is low. The DC level at the trigger terminal should be held at a value above the threshold level of the lower comparator $\left(\dfrac{V_{cc}}{3}\right)$. The circuit triggers on negative-going slope of the signal. So, when transistor T_1 becomes off, the capacitor C charges through R exponentially towards V_{cc}. When $V = \dfrac{2V_{cc}}{3}$, the flip-flop resets.

4.1.4 Specifications of NE555

Sink or source current (maximum): 200 mA

Load current range: 2 mA to 200 mA

Stability: 50 ppm/°C

= 0.005%/°C

Range of temperature: NE555 0°C to 70°C

SE555 –55°C to +125°C

V_{cc} : 5 V to 18 V

Time delay that can be produced: microseconds to a few hours.

If $R_A = 0$, the discharge transistor is directly connected to V_{cc}. So, it is always on. C is always shorted. The discharge transistor will be damaged. Therefore, a diode is connected across R_B such that its cathode is connected to V_{cc}. When the diode is forward biased, R_B is shorted. Therefore, C gets charged through R_A and the diode. While discharging, the diode is reverse biased. Thus, C discharges through R_B only. Since

$$R_A = R_B. \ T_1 = T_2$$

we get a 50% duty cycle.

4.1.5 NE555 Monostable Multi Applications

Frequency Divider

Frequency divider is based on the property that if the trigger pulse applied has time period $< t_p$ of the monostable, it has no effect on the output waveform.

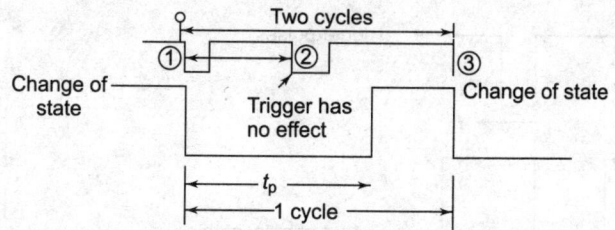

Fig. 4.6 *Time Less Than t_p Operation*

t_p = timing or gate width of the monostable

T = timing of the triggering pulse

If t_p is slightly greater than T, the second portion of the pulse, the negative-going pulse of the trigger, has no effect. The monostable changes state only when the third portion of negative trigger pulse is applied. This corresponds to two cycles of trigger input only. One cycle is obtained at the output of monostable multi. Thus the frequency is divided by two (Fig. 4.6).

Condition $t_p > T$.

If $t_p > 2T$, a divide by three operation is obtained. For three trigger input pulses, only one output is obtained at pin 3 of 555 IC.

Thus, division can be done to any value. If we use decade counters, more ICs will be removed. Here with one or two ICs in cascade, we can get more than 1000 operations. V_{cc} can be 5 to 15 V.

4.1.6 Pulse Stretcher

The pulse width given for triggering is narrow. However, the output of the monostable multi-pulse width is large, $t_p = 1.1 \ R_A \ C_A$. Thus, a narrow input (trigger) pulse is stretched to a larger pulse width (output of monostable multi) (Fig. 4.7).

This circuit is used to make LEDs glow. If the output of a circuit is a narrow pulse, the LED glow may not be visible. So, 555 monostable and larger pulses are given to LEDs to make them glow more brightly.

4.1.7 Pin Configuration Functions of 555

Pin 1: Ground

Pin 2: Trigger pulse is applied here. The magnitude of trigger pulse must be at least greater than $\frac{1}{3}V_{cc}$, to charge state of the comparator.

Pin 3: Output is obtained from this pin.

Pin 4: Reset. 555 Timer is reset, then a negative trigger pulse to the pin.

Pin 5: Control Voltage. An external voltage applied to this terminal changes the threshold voltage as well as trigger voltage.

Pin 6: Threshold. It is the non-inverting input terminal of comparator 1. It monitors the voltage across the external capacitor C.

Pin 7: Discharge. This pin is connected internally to the collector of Q_1. This acts as a switch to permit charging and discharging of external capacitor C.

Pin 8: $+V_{cc}$. V_{cc} of +5 V to +18 V can be applied here.

Waveforms for 555 monostable as divided by two circuits are shown in Fig. 4.7 (b).

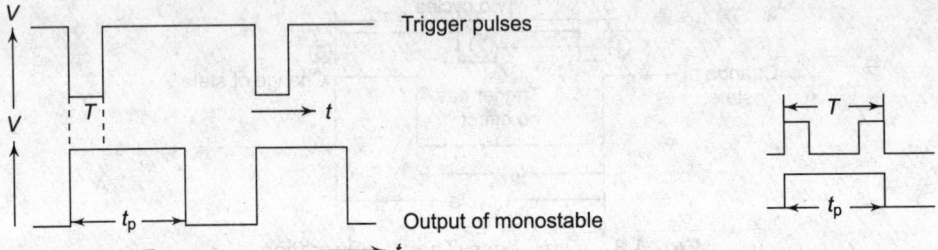

Fig. 4.7(a) *Pulse Stretcher Waveforms*   **Fig. 4.7(b)** *Waveforms for 555 monostable*

If $-V_T > \dfrac{1}{3} V_{cc}$ is applied, V_o is high.

The output is low if the voltage at this pin is $\dfrac{2}{3} V_{cc}$. But if a negative-going pulse greater than $\dfrac{1}{3} V_{cc}$ is given, comparator 2 falls low and the output of the timer rises high.

XR2556 IC is a dual timer. It consists of two 555 ICs inside. It is a 14-pin IC used in sequential timing.

4.2 PHASE-LOCKED LOOP

A phase-locked loop (PLL) is a *closed-loop system*. The function of PLL is to lock or synchronize the frequency of a voltage-controlled oscillator (VCO) to that of the incoming signal. If the frequency of the incoming signal changes, the centre frequency of the VCO also changes so that both frequencies are the same. The phase and frequencies of the two signals are compared in a phase detector. The error signal is proportional to the frequency difference of the two signals. If the difference is more, the magnitude of the error signal is more and the supply voltage to the VCO circuit changes till the two frequencies are locked. The error signal can be positive or negative, depending upon whether the centre frequency "f_0" of VCO is greater than or less than the frequency of the incoming signal. Accordingly, the supply voltage to VCO circuit will be increased or decreased.

This is an old concept and was described in 1932. But the circuit was complex and costly. With the advantages of IC technology, PLL has become popular and the PLL circuit is now available in IC form. NE565 is a PLL IC as are XR210, CA3090, MC1310B, SN61115, and LM310 PLL ICs.

4.2.1   Applications

1. *FM demodulation:* When PLL is locked on a frequency-modulated signal, the VCO frequency tracks the frequency of the signal instantaneously.
2. *AM demodulation:* The PLL locks on the *carrier of the AM signal* so that the output of VCO has the same *frequency as that of the carrier*, but no amplitude modulation.
3. *Frequency synthesis* in a PLL circuit can be modified so that it locks on to the second harmonic or third harmonic of the input signal. Thus, VCO output will be a multiple of the input signal. Harmonic locking is obtained by setting the VCO frequency to a multiple of the input frequency.
4. *Frequency translation:* Changing the frequency of output to $2 f_o$, $3 f_o$, and so on.
5. *Automatic frequency control (AFC):* When PLL is used to control the frequency of an oscillator, this circuit is used.
6. *Automotive frequency tuning* (AFT): The RF tuner in television sets is an example for frequencey tuning.

4.2.2   PLL Building Blocks

PLL consists of basically four building blocks:
1. Phase detector
2. VCO
3. LPF
4. Amplifier
The block schematic is shown in Fig. 4.8:

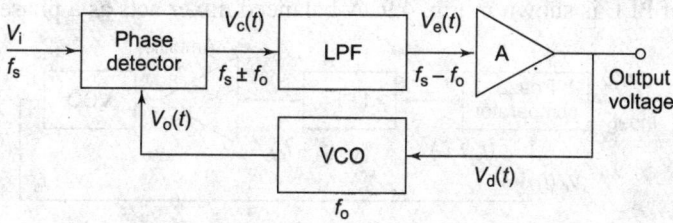

Fig. 4.8   *Block Schematic of PLL*

The phase detector is a frequency multiplier. With no input signal applied to PLL, the output from the phase detector is zero. Thus, the error voltage applied to VCO is also zero. Therefore, the VCO operates at its free-running frequency f_0. If the input signal frequency f_s is different from f_0, the phase-detector output will have frequency signals $(f_s + f_0)$ and $(f_s - f_0)$. If f_0 is very different from f_s, both frequency signals will not fall into the pass band of the LPF and hence will be attenuated. The VCO frequency then will not change. Hence, the lock will not be acquired.

If the value of the input frequency signal f_s is such that the frequency component $(f_s - f_0)$ at the output of phase detector lies within the pass band of LPF, the signal is amplified and applied as a control signal to VCO. $(f_s + f_0)$ frequency signal will be filtered. This causes the VCO frequency to vary in a direction that reduces the frequency difference between f_s and f_0. If $f_s \geq f_o$, the error signal will be positive, increasing supply to VCO and increasing f_0 as well to match f_s. If f is less, the error signal will be negative, reducing V_{cc} to VCO and also f_0 to match f_s. Once f_0 is the same as f_s, VCO is locked.

$(f_s - f_0)$ is proportional to the magnitude of the cosine of θ.

θ = phase difference between f_s and f_0.

An error signal is produced even if there is a phase difference between V_i and V_o, though f is the same. Even the phase difference must be nullified.

4.2.3 VCO Characteristics

VCO is a circuit whose frequency of oscillations change with voltage f. The characteristic of the circuit, which is variation of frequency of oscillations f with voltage V is shown in Fig. 4.9. f_0 is called as centre frequency. f_0' and f_0'' are two other frequencies below and above f_0.

$$(f''_0 - f_0): \text{lock range}$$

$$(f''_0 - f_0) = (f_0 - f'_0)$$

$$= \text{tracking range}$$

$$\text{Tracking range} = \frac{1}{2} \text{ (lock range)}$$

VCO circuits are used in:

1. Instrumentation systems
2. Space telemetry
3. Microwave application and so on where higher degree of noise immunity and narrow bandwidth is required

4.2.4 Block Diagram of PLL

The block diagram of PLL is shown in Fig. 4.9. A balanced mixer acts as a phase detector.

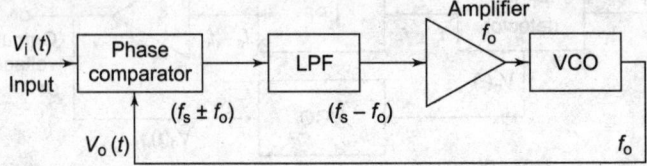

Fig. 4.9 *Block Diagram of PLL*

VCO frequency f_o changes to match the input frequency f_s. In the locked condition, any slight change in f_s first appears as a change in phase between f_s and f_o. The phase shift then acts as an error signal to change the frequency of VCO to match f_s. Basically, the VCO is a multivibrator circuit producing square waves.

4.2.5 VCO Applications

VCO circuits are used on:

1. AM demodulation
2. FM demodulation
3. Frequency translation
4. Automatic frequency control
5. Automatic fine-tuning

4.2.6 Mechanical Analogy

Consider two heavy discs with centre shafts. Let a spring connect the two shafts, that is each end of the spring is connected to one shaft. The input disc is analogous to input signal frequency. The output disc is analogous to VCO (Fig. 4.10).

Input   Output

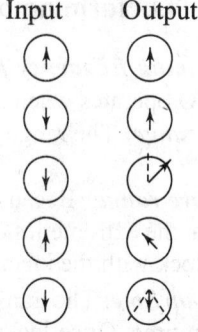

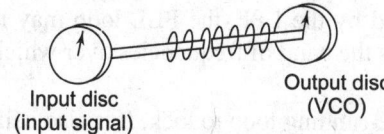

Fig. 4.10 *Mechanical Analogy for PLL*

Fig. 4.11(a) *Mechanical Analogy*

Case I: Initially in a locked condition, the two shaft centre positions are synchronous. When the input shaft moves by ϕ_1 due to the inertia, the output disc can not respond. When the input disc moves by ϕ_2, the spring gets tightened and exerts a force on the output disc causing it to rotate by ϕ_4. Thus, the error signal is proportional to $(\phi_2 - \phi_4)$.

The movement of the input disc is analogous to the change in frequency of the input signal f_s. The movement of the output disc is a change in the VCO frequency f_o.

Case II: Suppose the input disc is rotated suddenly by an angular displacement of 90° [Fig. 4.11(a)]. The output disc can not respond instantaneously. But due to the great tension on the spring and the torque produced, the output disc also starts rotating slowly and then over-shoots and returns to a position equal to the position of the input disc, that is $\phi \simeq 90°$, and oscillates and comes to rest with certain error ϕ_e.

This is an example of the response of the PLL to *step inputs*.

Case III: Suppose the input disc rotates slowly [Fig. 4.11(b)]. The output disc responds after ϕ crosses a certain minimum value due to the inertia of the output disc. Then the rate at which the output disc is moving can be the same as that of the input disc. But there will always be some error ϕ_e. PLL also works in a similar way.

Input   Output

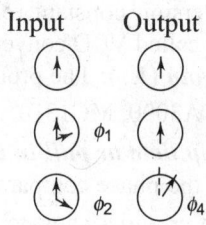

The phase comparator or a multiplier is analogous to a mixer in ratio receivers. Phase multipliers can be *analog* or *digital*. Accordingly, there can be

Fig. 4.11(b) *Mechanical Analogy*

1. Analog PLL
2. Digital PLL

In PLL, the VCO frequency f_0 should first *capture* the input signal frequency f_s and then *lock on* to it.

4.2.7 Frequency Shift Keying

PLL is used in frequency shift keying (FSK). Here, if the input is logic 1, the VCO free-running frequency f_0 is changed by some value and for input logic 0. f_o is changed by some other value. Thus, the digital input is converted to *tones* for transmission to remote points. At the receiving end, they are converted back to digital signals. This is the principle of MODEMs.

A digital PLL employs two input *exclusive* OR *gates* as the phase comparators.

4.2.8 PLL Terminology

Free-running frequency f_0: It is also called the centre frequency. This is the frequency at which the loop VCO operates when not locked to an input signal.

Lock range: The range of frequencies over which the loop will remain in locks (or can maintain the lock).

Capture range: Because of the selectivity afforded by the LPF, the PLL loop may not acquire or capture at the extremeties of the tracking range. This is the band of frequencies over which the PLL can acquire lock with the incoming signal.

Lock-up time: The transient time required for a free-running loop to lock. It is also called lock range or pull in time. Once locked, it remains in lock for higher value of frequencies. But to acquire lock it needs a lower-frequency range. Therefore, lock range is more than capture range.

Phase comparator conversion gain (K_d): The conversion constant relating the phase comparator's output voltage to the phase difference between input and VCO signals when the loop is decoded. The unit is Volt per radian.

Lock range: The range of frequencies over which the PLL can maintain lock with an input signal. It depends upon the self-correcting ability of the PLL. Once locked, changes in the frequency of the input signal should not cause unlocking.

Pull-in time: The total time taken by the PLL to establish lock. Pull-in time depends on the initial frequency and phase difference between two signals.

4.2.9 VCO Conversion Gain (K_0)

The conversion constant relating the oscillator centre-frequency shift from f_0 to the applied input voltage is called VCO conversion gain K_0.

Loop gain (K_V): The product of K_d and K_0 is called gain K_V.
NE565, CA3090, MC1310, and SN61115 are PLL ICs.

Lock-up time or pull-in time: The LPF, by rejecting the higher-frequency error components at the output of the phase comparator, enhances the interference rejection characteristics.

V_{CO}: It provides short-term memory for the PLL and ensures a rapid recapture of the signal if the system is thrown out of lock.

$$V_i(t) = V_i \sin \omega_i t$$

Mathematical equations for PLL:
V_i = input signal to phase detector

$$V_0(t) = V_0 \sin (\omega_0 t)$$

V_0 = output of VCO

$$V_e(t) = K_1 V_i V_0 [(\sin \omega_i t) [\sin (\omega_0 t)]$$

Error signal = $V_e(t)$.

Applying the trigonometric function, $\sin A \sin B = \dfrac{1}{2} [\cos (A - B) - \cos (A + B)]$
The equation becomes

$$V_c(t) = \frac{k_1 V_i V_o}{2} \left[\cos (\omega_i - \omega_o)t - \cos (\omega_i + \omega_o)t\right]$$

Phase shift ϕ cannot alter the frequency. So, it is neglected. ϕ_e gets cancelled. When $V_e(t)$ is passed through the LPF, the same frequency component is removed. Therefore,

$$V_f(t) = K_2 V_i\, V_o \cos (\omega_i - \omega_o)t$$

where K_2 is a constant.

After amplification, the control voltage for VCO appears as

$$V_d(t) = AK_2 V_i\, V_o \cos (\omega_i - \omega_o)t$$

This is converted to DC and then given to VCO.

4.2.10 AM Receiver Using PLL

The block schematic is shown in Fig. 4.12. NE561 N is one such AM recevier IC using the PLL principle.

The VCO will be locked exactly in frequency and phase to the carrier of AM signals. PLL inherently produces a 90° phase shift. The carrier is externally shifted by another 90° to compensate this. If the total phase shift is 180°, it is the same as 0°.

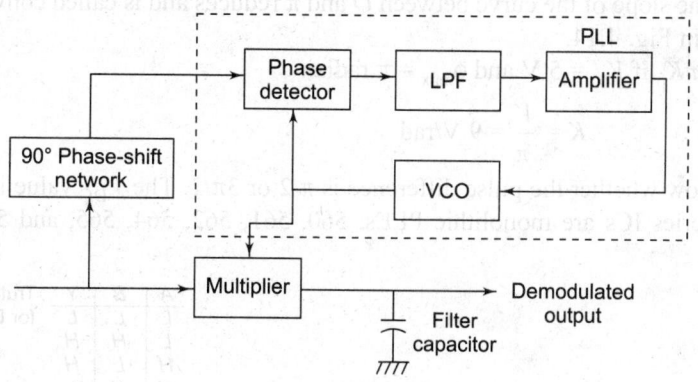

Fig. 4.12 *AM Receiver Using PLL: Block Schematic*

The circuit diagram using NE561 N IC is shown in Fig. 4.13.

$R_y\, C_y$ will shift the AM signal by 90° and feed to the phase detector input on pin 13. $R_y = 3000\ \Omega$ and $C_y = 135$ pF. C_B from pin 12 completes the return path for the shifted signal.

C_1 connected between pins 14 and 15 provides low-pass filtering of the control signal.

C_x is the LPF to remove R$_F$ components before the signal is passed to the audio amplifier. The demodulated output is obtained at pin 1.

Digital phase detectors are simple circuits compared to analog phase-detector circuits. If the input is a sine wave or pulse wave, a digital phase detector can be used.

Exclusive OR gate acts as a phase detector. CD4070 is an EX-OR gate. This IC can be used as a phase detector.

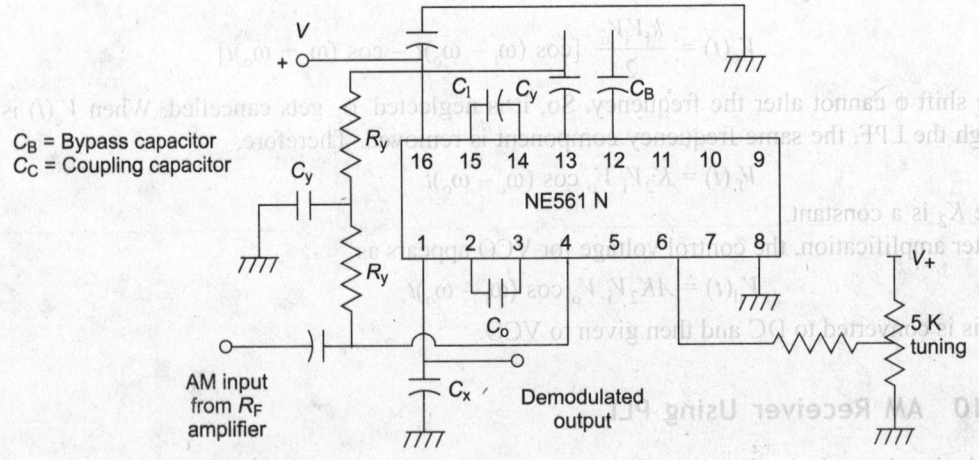

Fig. 4.13   *Circuit of NE 561 N IC PLL AM Receiver*

The average output (DC) of the EX-OR gate is proportional to the phase difference between f_i and f_o. The output of the EX-OR gate is high only when f_i is high or f_o is high. The number of pulses that occur in a given time period T will vary with the phase difference between f_i and f_o. Therefore, the average output will vary. The slope of the curve between O and π reduces and is called conversion gain K_p. The graphs are shown in Fig. 4.14.

Conversion gain K_p if $V_{cc} = 5$ V and $\phi_{max} = \pi$ radian

$$K = \frac{V}{\pi} = 9 \text{ V/rad}$$

We need not know whether the pulse difference is $\pi/2$ or $3\pi/2$. The V_{DC} value is the same for both. The SE/NE 560 series ICs are monolithic PLLs. 560, 561, 562, 564, 565, and 567 differ mainly in

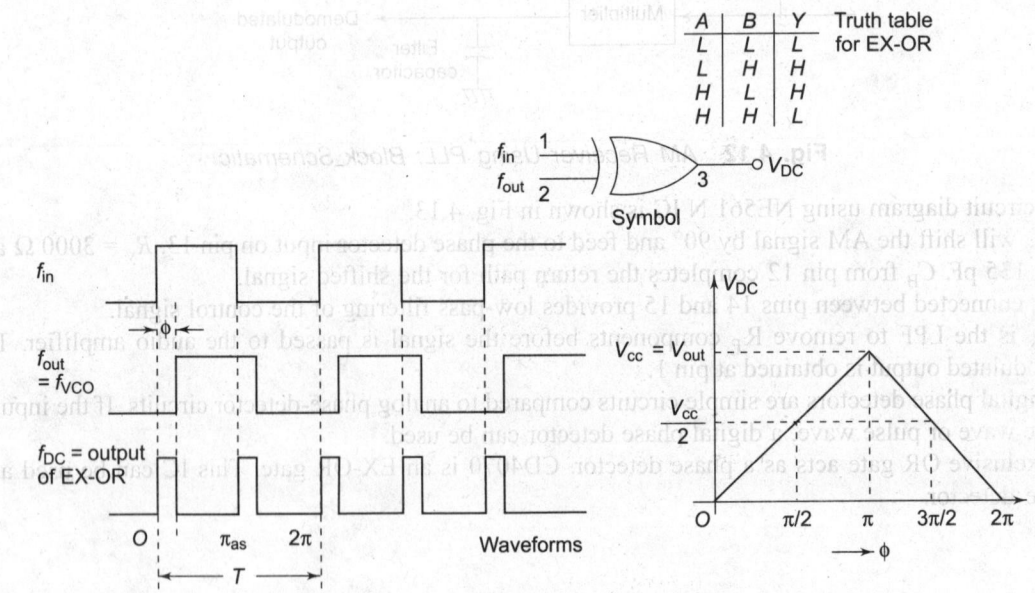

Fig. 4.14   *EX-OR Gate as Phase Detector*

operating frequency range, power supply requirements, frequency, and bandwidth ranges. The 565 PLL is very popular. NE566 is a VCO. It is used up to 500 kHz. For higher frequency ranges, MC4324/4024 and MC1648 may be used.

4.2.11 NE/SE 565 PLL Block Diagram

The block schematic of 565 PLL IC is shown in Fig. 4.15.

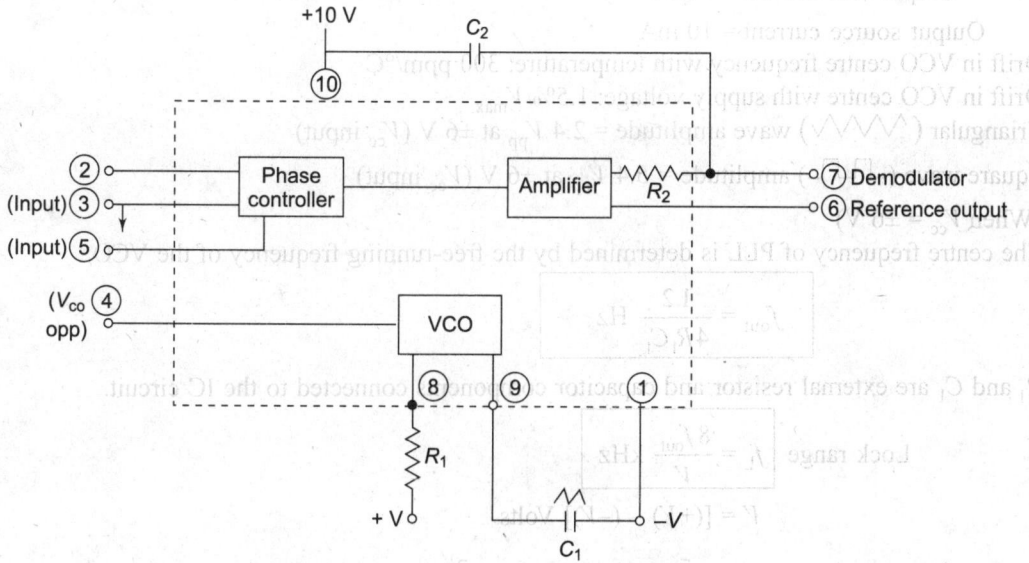

Fig. 4.15 *Block Schematic of NE565 PLL IC*

The centre frequency of the PLL is determined by the free-running frequency of the VCO. This is given by the equation

$$f_o \simeq \frac{1.2}{4 R_1 C_1} \text{Hz}$$

R_1 and C_1 are the external R and C connected to pins 8 and 9 of IC.

R_1 must have a value between 2 kΩ and 20 kΩ. C_1 can have any standard value.

The internal resistance $R_2 = 3.6$ kΩ and external capacitor C_2 acts as LPF.

The pin configuration functions of 565 PLL IC are given below:

Pin 1   Negative DC voltage is given: $- V_{cc}$.

Pin 2   Input to phase detector. External signal is given here.

Pin 3   There are two input signal pins, one pin is grounded if not being used.

Pin 4   VCO output.

Pin 5   Phase comparator VCO input. A short between pins 4 and 5 will connect the VCO output frequency f_o to the phase comparator. So f_o of VCO will be compared with f_i.

Pin 6   DC reference voltage. It is equal to the DC potential of the demodulated output at pin 7. In FSK, output at pin 6 is used as an input to the phase comparator.

Pin 7   Demodulated input or output of LPF is available.

Pin 8 $+ V_{cc}$ is given and external R_1 is connected.

Pin 9 $- V_{cc}$ is given and external C_1 is connected.

Pin 10 $+ V$ is given.

4.2.12 Specifications of 565 PLL IC

Operating frequency range: 0.001 Hz to 500 kHz

Operating voltage range: ±6 to ±12 V

Input level required for tracking: 10 mV rms minimum to 3 V p–p maximum

Input impedance $|Z|$ = 10 kΩ typical

Output sink current = 1 mA

Output source current = 10 mA

Drift in VCO centre frequency with temperature: 300 ppm/°C

Drift in VCO centre with supply voltage: 1.5%/V_{max}

Triangular ($\wedge\wedge\wedge\wedge$) wave amplitude = 2.4 V_{pp} at ±6 V (V_{cc} input)

Square wave ($\sqcap\sqcap\sqcap$) amplitude = 5.4 V_{pp} at ±6 V (V_{cc} input)

(When V_{cc} = ±6 V)

The centre frequency of PLL is determined by the free-running frequency of the VCO.

$$f_{out} = \frac{1.2}{4R_1C_1} \text{ Hz}$$

R_1 and C_1 are external resistor and capacitor components connected to the IC circuit.

Lock range $\boxed{f_L = \frac{8f_{out}}{V} \text{ kHz}}$

$$V = [(+V) - (-V)] \text{ Volts}$$

Capture range $f_c = \pm\left[\frac{f_L}{(2\pi)(3.6)(10^3)(C_1)}\right]^{1/2}$

$$R_1 = 3.6 \text{ k}\Omega$$

C_2 is in Farads.

The performance of the VCO determines the frequency stability of PLL. A good VCO must possess the following characteristics:

1. A linear voltage to frequency conversion
2. Excellent frequency stability
3. Higher-conversion gain
4. High-frequency compatibility
5. Wide-tracking range
6. Ease of tuning

4.2.12.1 Applications

Frequency multiplication: A divide by N counters is inserted between VCO and phase comparator when in lock.

$$f_o \ f_o' = Nf_s$$

When lock is achieved,

$$\frac{f_o}{N} = f_s$$

The new f_o $f_o' = N f_s$. Thus, input signal frequency f_s is multiplied N times.
The block schematic for frequency multiplication using PLL IC is shown in Fig. 4.16.

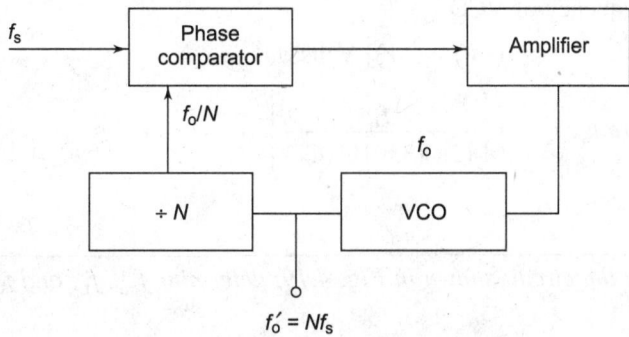

Fig. 4.16 *Frequency Multiplication*

The new $f_o f_o' = N f_s$.

4.2.13 Measurement of Capture and Lock Range

The experimental procedure to determine the lock range is explained here. Increase the input signal frequency. The f_o of VCO suddenly changes and locks to input signal frequency. This is f_1. Increase input frequency still further till it unlocks. This is $f_2 f_2'$. Now decrease frequency of input signal till lock is achieved. This is f_2. Decrease it still further till unlocking is done. This is f'_1 (see Fig. 4.17).

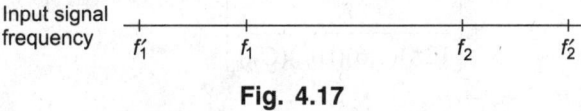

Fig. 4.17

f_2-f_1: Capture range
$f_2'-f_1'$: Lock range
f_1 = Frequency when lock is achieved while increasing f_i
f_2' = Frequency when unlock is achieved while increasing f_i
f_1' = Frequency when unlock is achieved while decreasing f_i
f_2 = Frequency when lock is achieved while decreasing f_i
f_1 = Frequency when lock is achieved while increasing input signal frequency
f_2' = Frequency when unlock is achieved while increasing input signal frequency

Relation between $f_c, f_L,$ and f_o

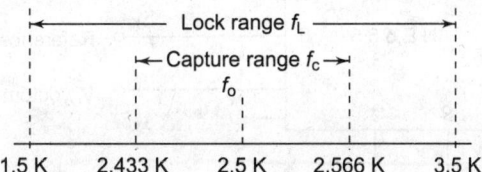

Fig. 4.18 *Relation between f_o, f_c and f_L*

565 can lock and track an input signal over ±60% w.r.t. f_o as the centre frequency.

$$Lock\ range\ (f_L) = \pm\frac{8f_o}{V}\ Hz$$

f_o = free-running frequency of VCO

$$V = [(+V) - (-V)]\ Volts$$

$$Capture\ range\ f_c = \pm\left[\frac{f_L}{(2\pi)(3.6)10^3(C_2)}\right]^{\frac{1}{2}}$$

where C_2 is in Farads.

Example 4.1   *For the circuit shown in Fig. 4.19, determine f_{out}, f_L, and f_c. f_L = lock range and f_c = capture range.*

Solution

$$f_{out} = \frac{1.2}{4R_1C_1} = \frac{1.2}{4\times12\times10^3\times10^{-8}} = 2.5\ kHz.$$

Lock range, $$f_L = \pm\frac{8f_{out}}{V}$$

$$V = [(+10) - (-10)] = 20\ V$$

So, $$f_L = \pm\frac{8(2.5)10^3}{20} = \pm1\ kHz$$

Capture range, $$f_c = \pm\left[\frac{f_L}{(2\pi)(3.6)(10^3)(C_2)}\right]^{\frac{1}{2}}$$

$$= \left[\frac{10^3}{2\pi(3.6)10^3(10)(10^{-6})}\right]^{\frac{1}{2}}$$

$$= \pm66.49\ Hz.$$

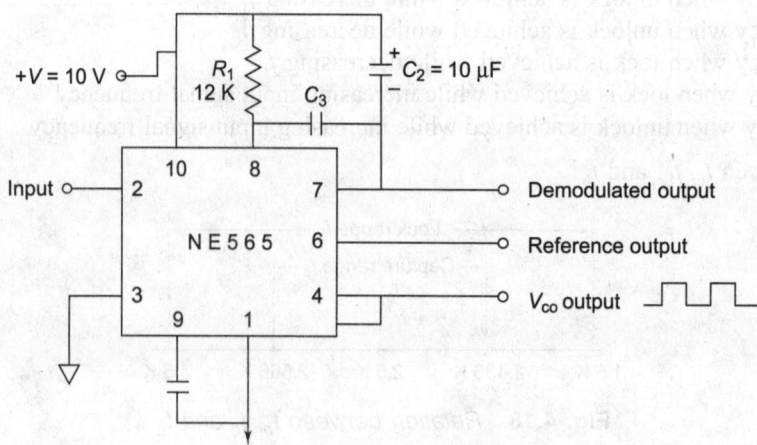

Fig. 4.19   *Circuit for Example 4.1*

Example 4.2 *A 555 Timer is being used as divided by two networks. The frequency of the input trigger signal is 2 kHz. If the value of C = 0.01 μF, what should be the value of R_A?*

Solution

For a divide by 2 networks, t_p should be larger than T. Let $t_p = 1.2T$; $T = \dfrac{1}{2\,\text{kHz}}$

So,
$$t_p = 1.2 \times \dfrac{1}{2\,\text{kHz}} = 0.6\ \text{msec}$$
$$t_p = 1.1\,R_A C$$

Hence,
$$R_A = \dfrac{0.6}{1.1 \times 0.01 \times 10^{-6}} = 54.5\ \text{k}\Omega$$

SUMMARY

- A 555 Timer IC is so called because time-interval waveforms can be generated using this IC. The time can range from few microseconds to few hours. It can easily drive a relay unit directly with $I_0 = 200$ mA.
- In the astable mode, the expression for T_1 for 555 IC is $T_1 = 0.69\,(R_A + R_B)C$ *and* $T_2 = 0.69\,R_B C$

$$f = \dfrac{1}{T_1 + T_2} = \dfrac{1.46}{[(R_A + 2R_B)C]}$$

- To get waveform with 50% duty cycle, a diode must be connected across R_B with its anode connected to V_{cc}.
- In the monostable mode, the expression for $T = 1.1\,RC$. The 555 monostable can be used as a frequency divider, pulse stretcher, and so on.
- The XR2556 is a dual timer. It consists of two 555 ICs. It is a 14-pin IC used in sequential timing.
- The 565 PLL IC can be used for FM demodulation, AM demodulation, frequency synthesis, automatic fine tuning, automotive frequency control, and so on.
- The range of frequencies over which PLL can maintain lock with input signal is called the **lock range**.
- The band of frequencies over which PLL can acquire lock is called the **capture range**. The lock range is always greater than the **capture range**. The variation from centre frequency f_0 of **VCO** is called the **tracking range**. It is half of the lock range. For a 555 IC, the expression for

$$f_0 = \dfrac{1.2}{4R_1 C_1}.$$

ESSAY-TYPE QUESTIONS

1. Draw and explain the internal schematic of a 555 timer IC.
2. Draw the 555 Timer circuit in astable mode and explain the working with the help of waveforms.
3. Draw the circuit of a 555 Timer IC in astable mode to get output waveform with 50% duty cycle.

4. Draw the 555 monostable multivibrator circuit and explain its working with the help of wave-forms.
5. Explain the applications of a timer IC.
6. What is the principle of PLL? Draw the block schematic and explain the same.
7. Explain the applications of PLL.
8. Define the terms: (1) free-running frequency f_o, (2) lock range, (3) capture range, and (4) pull-in time, pertaining to PLL.
9. Explain the principle of PLL with the help of equations.
10. Explain the pin configuration of 565 IC.
11. Draw the circuit of a 565 PLL IC and explain its working.
12. Give the pin configuration of 555 IC and explain its working.

FILL IN THE BLANKS

1. The maximum output current I_o that can be delivered from a 555 IC is _____.
2. The expression for T_1 in the case of a 555 astable multivibrator is _____.
3. The expression for f, the frequency of oscillations, in the case of 555 astable multi is _____.
4. To get 50% duty cycle in a 555 circuit, if R_A is made zero then the difficulty with discharge transistor is _____.
5. With 555 monostable circuit, if $t_p > 2T$, _____ operation is obtained.
6. For a divide by two operation, the relation between t_p and T is _____.
7. XR2556 IC is _____
8. XR210, CA3090, and MC1310B are _____ ICs.
9. The relation between lock range and capture range is _____.
10. Digital PLL employs _____ input _____ type logic gate as the phase comparator.
11. The expression for f_0 in the case of 565 PLL IC is _____.
12. The frequency range over which a 565 PLL IC can be used is _____.

ANSWERS

1. 200 μA
2. $T_1 = 0.69 (R_A + R_B) \cdot C$
3. $f = 1.46/(R_A + 2R_B)C$
4. It always remains on
5. Divide by 3 operation
6. $t_p > T$
7. Dual-timer IC
8. PLL IC
9. Lock range is greater than capture range
10. Two input, EX-OR
11. $f_0 = 1.2/4R_1C_1$.
12. 0.001 Hz to 500 kHz

UNSOLVED PROBLEMS

1. For the circuit shown below, determine f_{out}, f_L capture range, and lock range.

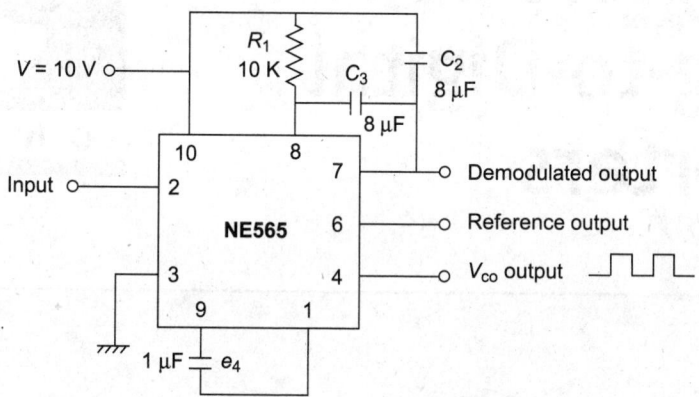

2. It is desired to get divide by 2 operation from a 555 Timer circuit. If the trigger signal frequency is 4 kHz and $C = 0.01$ μF, what should be the value of R_A?
3. Consider 555 Timer IC circuit. Given $R_A = 6.8$ kΩ, $R_B = 3.3$ kΩ, $C = 0.1$ μF, and $V_{cc} = +5$ V. Calculate the values of T_{High}, T_{Low} and free-running frequency of the timer circuit.
4. For the above problem, with the component values mentioned, determine the percentage duty cycle.
5. For the frequency-shifter circuit using 555 Timer IC, calculate the charge current I for $E = 0$ V, f_c when $E = 0$ V, the frequency shift for $E = \pm1$ V, and f_{out}.
6. Consider 555 monostable multivibrator circuit. If $R_A = 10$ kΩ, determine the value of C for output pulse duration of 1 msec.
7. For the above circuit, if $R_A = 10$ kΩ and $C = 0.1$ μF, determine the value of t_H.
8. In the case of 555 monostable circuit, if $R_A = 10$ kΩ and $C = 0.2$ μF, calculate the time interval.
9. Considering the circuit of Problem 6, 7 and 8, what value of R_A should be installed to divide a 5 kHz input signal by 3?
10. Design an astable multivibrator using 555 Timer to get output waveform at $1f = 10$ kHz with a duty cycle of 40%.

Digital-to-Analog Converters and Analog-to-Digital Converters

Objectives:

In this chapter...

- *The two types of digital-to-analog converters (DACs)—weighted resistor-network type and R–2R ladder-network type—and their principles are explained alongwith circuit schematics.*
- *Block schematic of DAC IC is given and type number of DAC ICs, parameters, and specifications are also given.*
- *The classification of analog-to-digital converters (ADCs) into direct type and indirect type is explained. In direct type ADCs, (a) simultaneous conversion type, (b) counter ramp conversion technique, and (c) successive approximation type are explained. In indirect type ADCs, (a) voltage-to-time (V/T) conversion method and (b) voltage-to-frequency (V/F) conversion method are explained. The block schematics, principle, and working are explained.*
- *Application of ADCs and DACs in digital voltmeters (DVMs) is described.*
- *Type numbers and specifications of some ADC and DAC ICs are provided.*

5.1 INTRODUCTION

There are many system problems that require connecting a digital portion of the system to an analog component. This meeting of the circuits is called an "interface". The aspects associated with this interface are digital-to-analog (D/A) and analog-to-digital (A/D) conversion. D/A and A/D converters are used in data acquisition systems for simulation and recording. Data acquisition systems measure and record signals obtained by direct measurement or signals originating from transducers wherein a physical parameter of interest is being measured. Data acquisition systems are used in a variety of industrial and scientific areas such as biomedical, aerospace, telemetry, and instrumentation.

If the signals being recorded are in digital form, they comprise digital data acquisition systems. If the signals being recorded are analog, they make for analog data acquisition systems.

D/A converters (DACs) are used when

1. The variation of the quantity with time is large or when a wide bandwidth is required.
2. Low precision is required.

A/D converters (ADCs) are used when

1. High precision is required.
2. The physical quantity being measured varies slowly (narrow bandwidth).

Any analog voltage can be expressed as a binary word by assigning voltage weights to each bit position. In a 4-bit (binary digit) word, voltage values of 8, 4, 2, and 1 could be assigned to each bit position.

For example, binary number 1011 can be explained as

MSB		LSB		
1	0	1	1	LSB: least significant bit
2^3	2^2	2^1	2^0 $= (11)_{10}$	MSB: most significant bit
(8)	(4)	(2)	(1)	

Binary number	Decimal equivalent
0000	0
1101	13
1111	15

So for a 4-bit word, the decimal value is 15. Each successive binary count represents $\dfrac{1}{15}$ th of the entire voltage. Similarly, for eight bits, the decimal count is 255. Each binary count would be equal to 1/255 or 0.392% of full scale. Therefore, $(11111111)_2 = (255)_{10}$.

$$\text{Percentage resolution} = \frac{1}{2^N - 1} \times 100$$

where N is the number of bits.

Example 5.1   *What is the percentage resolution of a 5-bit DAC given that the maximum number that can be represented using 5 bits is 11111 = 31$_{(10)}$.*

Solution

$$\text{Percentage resolution} = \frac{1}{2^N - 1} \times 100 = \frac{1}{2^5 - 1} \times 100 = 3.23\%$$

Example 5.2   *A 6-bit DAC has a maximum precision supply voltage of 20 V.*
(a) What voltage change does each LSB represent?
(b) What voltage does 100110 represent?

Solution

(a) The maximum value that can be represented by a 6-bit word is $111111_2 = 63_{10}$.

So, each bit represents $\dfrac{1}{63}$.

Hence, LSB represents a voltage of $\dfrac{1}{63} \times 20 \text{ V} = 0.317 \text{ V}$

The binary number 100110 = 38_{10}.

(b) Therefore, 100110 represents a voltage of $\dfrac{38}{63} \times 20 = 12.06 \text{ V}$

5.2 D/A CONVERTER

There are two types of DACs.
1. Weighted resistor network
2. R–2R ladder network

5.2.1 Weighted Resistor Network

The circuit is shown in Fig. 5.1. The reference supply voltage is 8 V, that is, if the bit is 1, a voltage of 8 V is applied. If the bit is zero, 0 V is applied. *A* is MSB. *D* is LSB. If all the bits are 0000, the switches are connected as shown in Fig. 5.1. If the bit *B* is 1, then the switch is connected to the +8 V reference supply. The resistor values 1 K, 2 K, 4 K, and 8 K are weighted inversely with the value of flowing current. *A* is the MSB. If $A = 1$, it corresponds to 8 in the decimal system, or the current is maximum. Therefore, the corresponding resistance is low. If *D* is 1, *D* is LSB; it corresponds to 1 in the decimal system also.

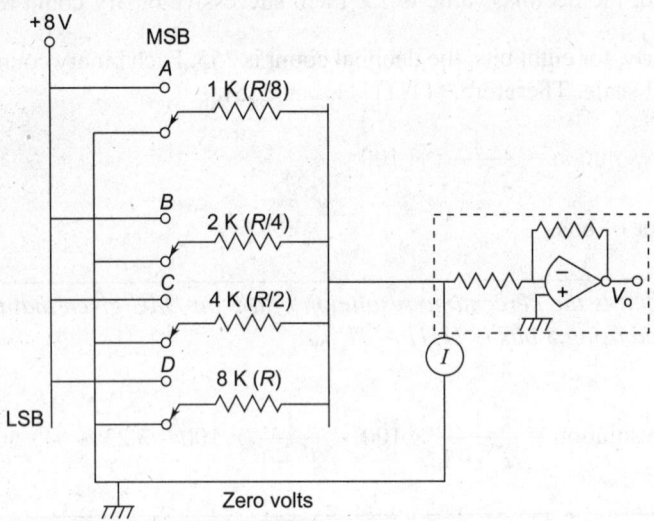

Fig. 5.1   *Weighted Resistor Network Circuit*

Therefore, the corresponding resistance is of higher value. The resistors are weighted against their decimal values. Hence, they are called weighted resistor networks. If *ABCD* were 0000, the output would be 0 mA. If it were 1010, the output current would be $8 + 0 + 2 + 0 = 10$ mA. If it were 0110, then the output would be $0 + 4 + 2 + 0 = 6$ mA. Therefore, the output current represents the decimal equivalent of the binary number. Thus, digital input is converted to analog signal because if the binary input changes with time, the output in decimal form also changes with time. The output will be as shown in Fig. 5.2. If a capacitor is chosen, the curve is smoothed out and appears as shown in Fig. 5.3 (a).

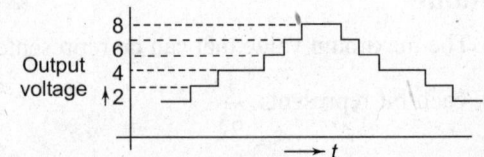

Fig. 5.2   *Output for Bit Increment of Input*

In this network, FETs are used as switches. Depending upon whether the binary input is 1 or 0, or high or low, FETs connect reference voltage or ground, as shown in Fig. 5.3 (b).

For the above circuit, voltage output is taken across R_L and $R_L \gg R$.

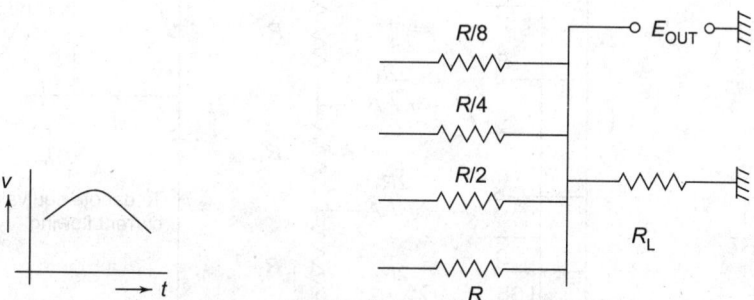

Fig. 5.3 (a) *Smoothened Output Waveform*   **Fig. 5.3 (b)** *Output Across R_L*

5.2.1.1 Advantages

1. It is simple to construct.
2. Cost is low.

5.2.1.2 Disadvantages

1. It requires many resistors of different values and of high precision. If the value of resistance changes, the analog output varies from the digital input value.
2. Supply voltage needs to be constant. DC voltage supply V_{cc} or V_{dd} of +8 V or must be exactly constant. If it varies, log output V_A also varies.

5.2.2 R-2R Ladder Network

Circuit looks like a ladder as shown in Fig. 5.4 (a). Hence, the name ladder network. The circuit is shown in Fig. 5.4 (b). *ABCD* are the input bits. *A* is MSB and *D* is LSB. If all the bits are 0000, the switches are connected to the ground. So, the current $I = 0$. Suppose *A* is 1 and *B, C, D* are 000. Hence, *A* is connected to +8 V. The equivalent circuits are shown in Fig. 5.5 (a)–(g) for simplifying and analyzing the circuit.

The total resistance of the circuit is $2R_1 + R_1 = 3R_1$.

Total current $I = \dfrac{E}{R} = \dfrac{8 \text{ V}}{3R_1}$. The circuit is shown in Fig. 5.6.

Current flowing through ammeter *A* is $I/2 = \dfrac{4}{3R_1}$.

Suppose now, *B* is connected to +8 V and other switches are grounded, that is the digital binary input given is 0100. The equivalent circuit is shown in Fig. 5.7 (a)–(d).

So, $\qquad\qquad I = 8 \text{ V}/3R_1$

I gets divided into I_1 and I_2. I_1 gets equally divided into $I_1/2$ and $I_1/2$.

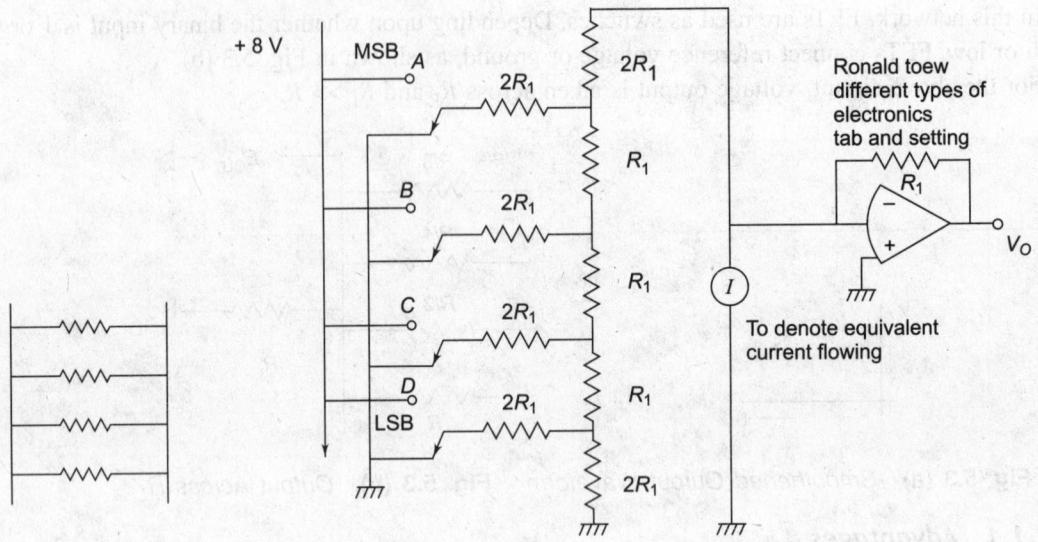

Fig. 5.4 (a)   *Ladder Resistor Network*   **Fig. 5.4 (b)**   *R–2R Ladder-Network Type DAC*

$$I = \frac{8\ V}{3R_1},\ I_1 = \frac{4\ V}{3R_1},\ I_2 = \frac{2\ V}{3R_1}$$

So, $$I = E/R = \frac{8\ V}{3R_1}$$

$$I_1 = I/2 \text{ if } I = \frac{4\ V}{3R_1}$$

Current flowing through the ammeter is $\dfrac{1}{2} I_1 = \dfrac{2\ V}{3R_1}$.

Therefore, if $A = 1$, current through the ammeter is $\dfrac{4\ V}{3R_1}$.

If $B = 1$, current through the ammeter is $\dfrac{2\ V}{3R_1}$.

If $C = 1$, current through the ammeter is $\dfrac{1\ V}{3R_1}$.

If $D = 1$, current through the ammeter is $\dfrac{1\ V}{6R_1}$.

Therefore, the current is proportional to the weightage of each bit A, B, C, and D. This digital signal is converted to analog form.

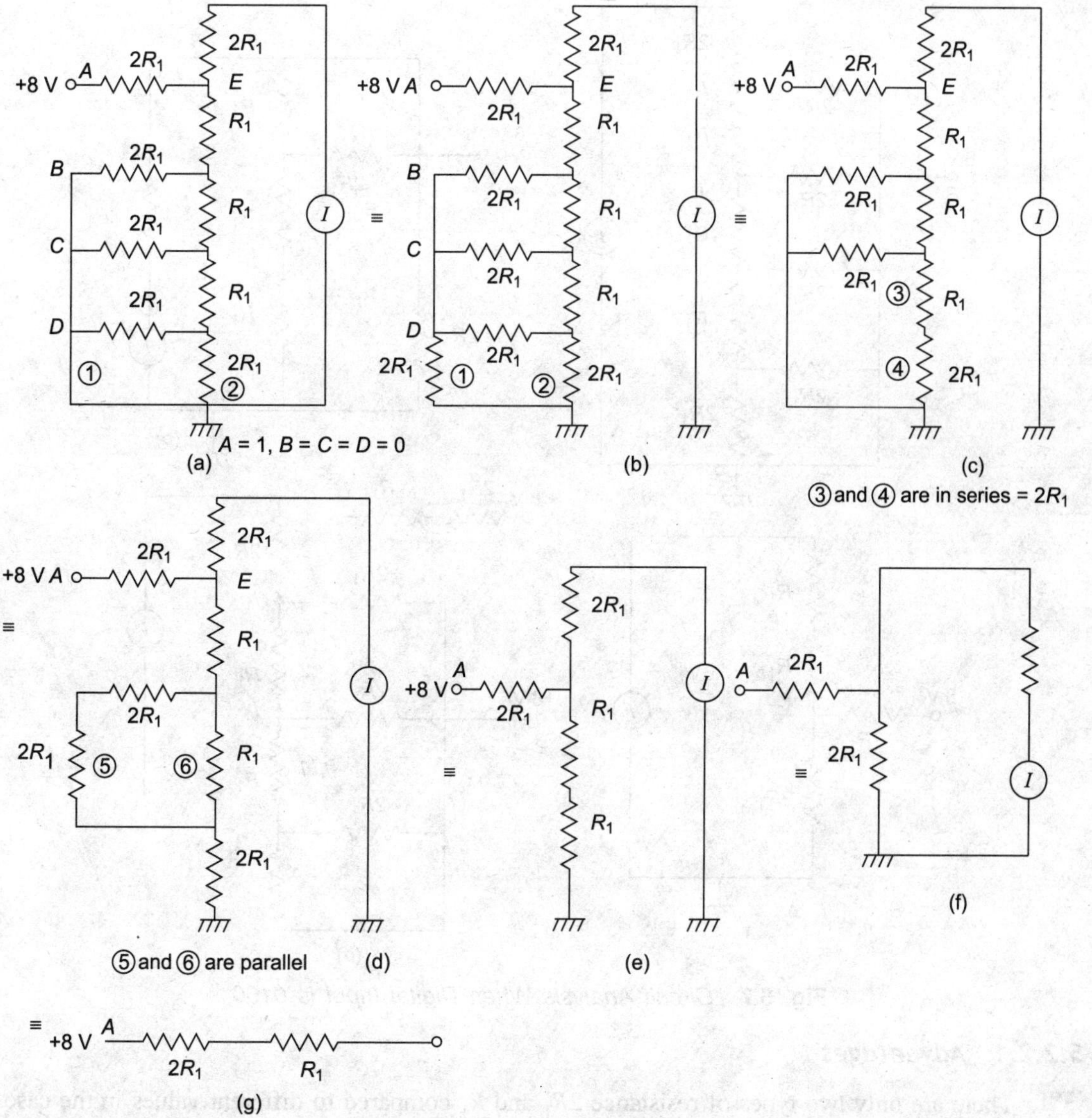

Fig. 5.5 *Equivalent Circuits*

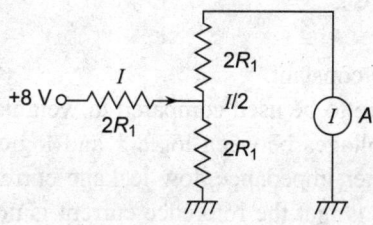

Fig. 5.6 *Simplified Equivalent Circuit*

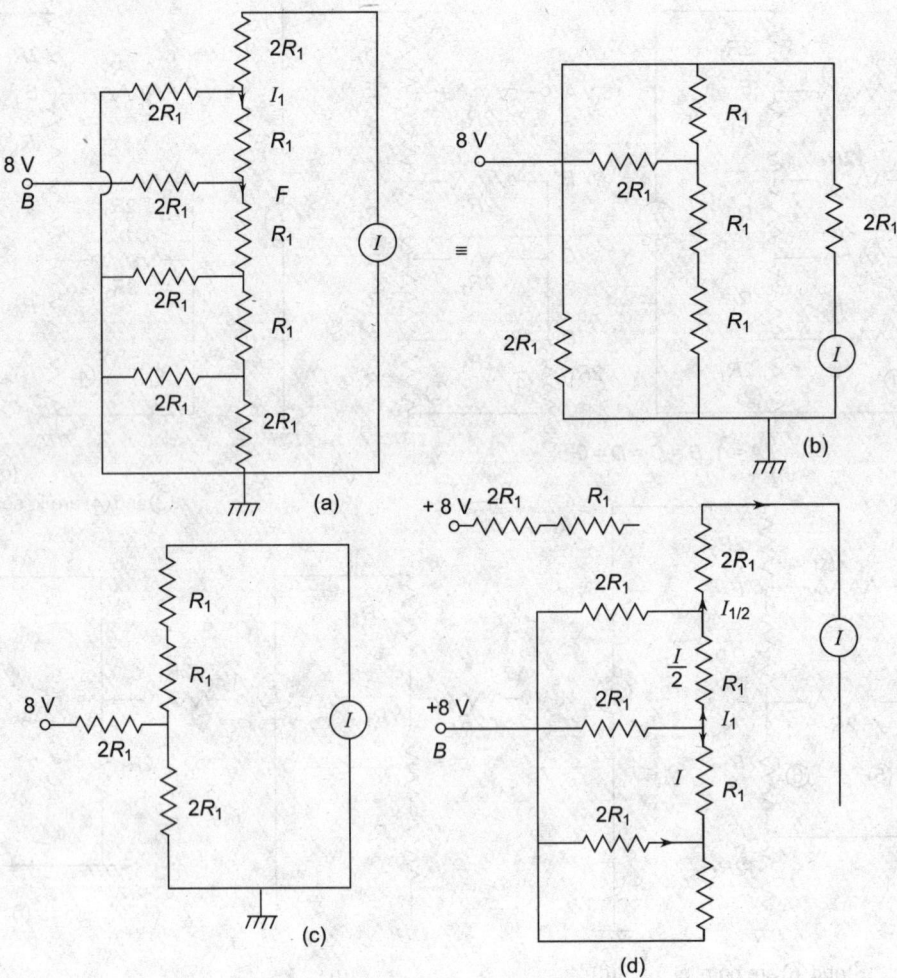

Fig. 5.7 *Circuit Analysis When Digital Input is 0100*

5.2.2.1 Advantages

1. There are only two types of resistance $2R_1$ and R_1 compared to different values in the case of weighted bridge.
2. Accuracy is better than weighted resistor network.

5.2.2.2 Disadvantages

1. Reference supply should be constant.
2. More number of resistors are to be used compared to weighted resistor type.

In DACs, for switching the voltages between logic 1 and logic 0 levels, FET switches are used because of the advantages of higher impedance, low leakage current. But for fast switching, current switches are used. The advantage is that the reference current is not interrupted as a consequence of code changing. Significant voltage change appears only at the output and not across the switches. A current switch circuit consists of zener diodes, BJTs, and op-amps. If the digital input is 1, the switching circuit connects V_{cc} or V_{dd}, +8 V supply to the corresponding resistance at A, B, C, or D, and so on.

5.3 DAC WITH MEMORY

DACs with memory buffer are also available. If a DAC is required to respond only at selected time instants to the digital input code, a memory register is incorporated in the input so that the input digital code can be stored into the memory and held at desired time instants.

Some of the DACs type numbers are AD 558, AD 7533, and AD 7110.

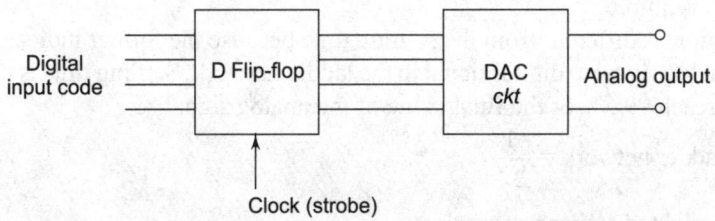

Fig. 5.8 *DAC with Memory*

In a practical DAC, the resistance network, called D/A conversion module, is connected to a flip-flop register that holds the digital number since the divider is simply a passive network. The digital input voltage (on or off levels) determines the output voltage. Digital voltage levels are not as precise as required in an analog system. So, level amplifiers are used between flip-flops and the divider network. These amplifiers switch the inputs to the divider network between ground and reference voltage supplied by precision reference supply. The practical circuit is shown in Fig. 5.9. The AND gates open only when the clock pulse is given to them. An AND gate needs two inputs—digital input and clock pulse. The digital signals are dropped into the register by a *drop in pulse* (a clock pulse). If A, B, C, and D are the bits, $A\overline{A}$, $B\overline{B}$, $C\overline{C}$, $D\overline{D}$ form the inputs for the flip-flops. The flip-flops hold the binary data for some time.

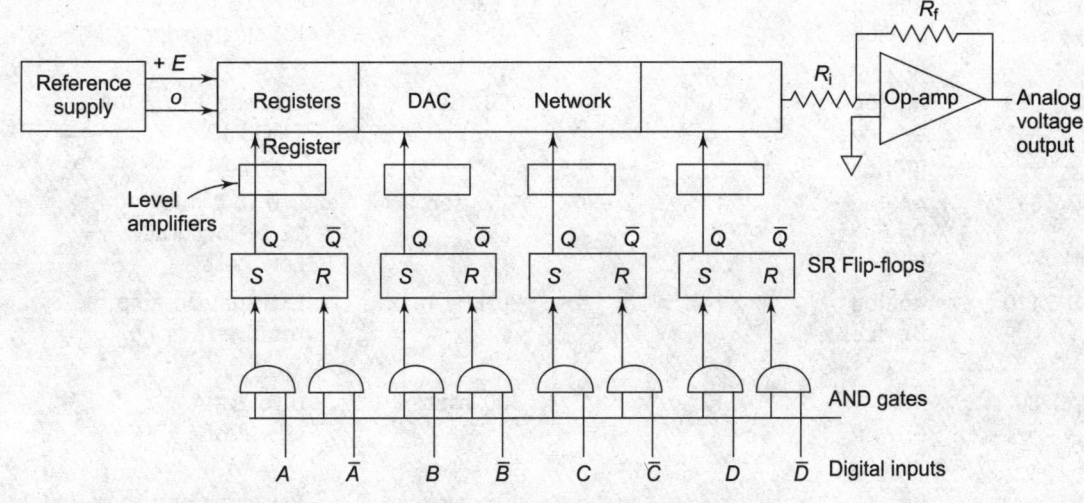

Fig. 5.9 *DAC Practical Circuit*

It always takes some time for the conversion to be completed after digital signals are dropped into the register. The settling time depends upon the flip-flops.

DACs are used to convert the computer outputs to analog signals for plotters and recorders etc. These converters form part of ADCs. (In an ADC, the analog signal is converted to a digital signal.)

Now the converted digital signal is again converted to an analog signal by the DAC and compared with the input analog signal. This is used in a counter type ADC.

The outputs of flip-flops or logic gates are not precise values like 0.0 V and 5.0 V, but vary over a given range. It is necessary to insert a precision level amplifier in between each logic input and its input resistor to the swing amplifier. The level amplifier produces precise output levels of 5 V and (or) 0–5 V depending on whether the digital inputs are high or low. A very stable, precise 5-V reference supply is required.

The conversion time is different from the settling time because the former indicates a delay involved in converting the analog signal to digital signal in the ladder network. Settling time is not the delay but the total time taken to reach 99.5% of the final value of the analog signal.

$$\text{Resolution for ladder network} = \frac{1}{2^N}$$

$$\text{Resolution for weighted resistor network} = \frac{1}{2^N - 1}$$

5.4 DAC SPECIFICATIONS

Resolution: It is the smallest change that can be detected in the analog output by a single-digit change in the digital input.

Table 5.1 Commercial DACs

Type No.	Make	Resolution $1/(2^N-1) \times 100$	Settling time	Type of output
AD 558	Analog Devices Inc.	8%	1 μsec	Voltage 0 to 2.56 V or 0 to 10 V IC, single supply
AD 7533	Analog Devices Inc.	10%	600 nsec	Current 0 to 2 mA CMOS IC Low cost
AD 563	Analog Devices Inc.	12%	1.5 μsec	*I*, 0 to 2 mA 12-bit accuracy Low cost
AD 7110	Analog Devices Inc.	14%	20 kHz max	External op-amp required
DAC 80	Burr Brown Inc.	12%	300 nsec	0 to 2 mA. Low cost
4800	Burr Brown Inc.	12%	100 nsec	Voltage ±10 V to +60 V Power DAC, providing ±200 mA circuit capability
DAC 198	Intensil	8%	20 msec	Voltage 0 to 10 V Voltage output type

For a 5-bit DAC, the maximum number that can be represented is $11111 = 31$.

Change in 1 bit represents $\dfrac{1}{31} = \dfrac{1}{2^N - 1} = \dfrac{1}{2^5 - 1}$ of the free-scale value.

So,   per cent resolution $= \dfrac{1}{31} \times 100 = 3.23\%$

Accuracy: It is the difference between the actual analog output obtained and the expected value.

Temperature sensitivity: For a fixed digital input, the analog output will vary with temperature because of the variation in reference supply voltage with temperature, resistors, summing amplifier (op-amp), and the like.

Settling time: When the digital input to a DAC changes, it takes time for the level amplifiers and other internal circuitry to respond and to produce a new analog output value. The time it takes for the output to stabilize to 99.95% of its new value is called the settling time. It will be $\simeq 500$ nsec.

For higher-speed current output DACs are used.

DACs with 12-bit code inputs have a settling time of 1–20 μsec.

5.5 A/D CONVERTERS

In instrumentation systems used for the measurement of physical parameters, the output signals are in analog form. However, there are many advantages if we can get digital signals from the instrumentation systems.

A/D converters: There are two types of ADCs—direct and indirect.

Direct ADCs can be further divided into:

1. Simultaneous ADCs
2. Counter ramp conversion
3. Successive approximation

Indirect ADCs can be further divided into:

1. Voltage-to-time (V/T) conversion and
2. Voltage-to-frequency (V/F) conversion.

These advantages are as given below:

1. Human errors in reading are eliminated. No parallax error.
2. High precision measurements, that is, the precision of reading the output of a digital instrument is unlimited.
3. Digital data can be stored and processed without the loss of accuracy.
4. The aging rate of digital instruments is slow. There is no spring tension or loosening, or similar defects as found in analog instruments.

Hence, it is always advantageous to have a digital signal. If the output of a transducer is an analog signal, we can get digital signal by means of an ADC. The block diagram in Fig. 5.10 shows the location of an ADC in an instrumentation system.

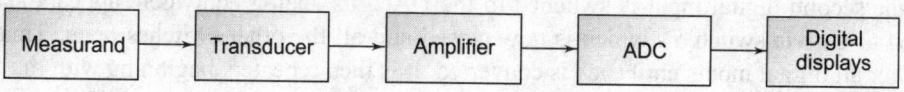

Fig. 5.10 *ADC in an Instrumentation System*

Slew rate: If the input varies at a high speed, output cannot respond at the same rate. Slew rate is the maximum rate at which output can respond faithfully to the change in input. It is expressed in $V/\mu sec$.

5.5.1 D/A Multiplexing

In many applications, more than one digital inputs have to be converted to an analog output. For example, a process-control computer may provide several digitally coded control signals to drive different control devices like valves or speed of motor. There are two methods to achieve the same. The straightforward method is to use a separate DAC for each digital input.

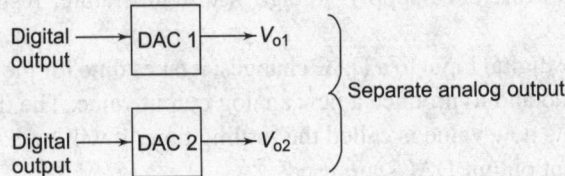

Fig. 5.11   *Multifunction D/A Conversion System*

Here, the analog output is continuously generated and there is a larger digital input. There is no need for the analog output to be stored. But as many DACs as the number of digital inputs are required.

Since DACs require precision reference supplies, resistances, level amplifiers, and so on, the cost will be more, particularly when a number of digital inputs are present. So, this method can be used where the digital inputs are few and the analog output is required immediately after the digital input is given.

A second method of performing multiple D/A conversion uses a multiplexing technique which involves time sharing. A single DAC is used to convert all the digital inputs into analog outputs in a sequential manner as shown in Fig. 5.12.

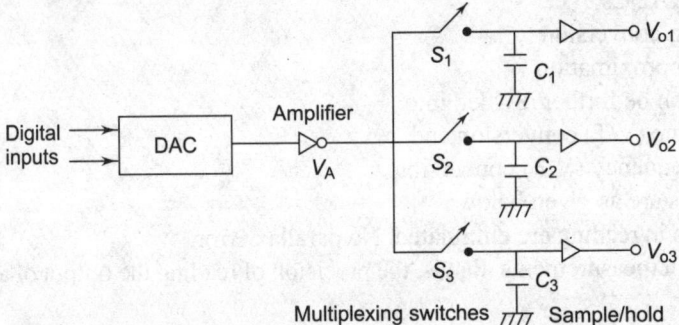

Fig. 5.12   *Multiplexing Using Single DAC*

The different sets of digital inputs are switched to the DAC in a time sequence, one at a time. For the first digital input, the DAC produces an analog voltage V_A, which is transmitted to the first sample-and-hold circuit via the closed multiplexing switch S_1. The other switches are held open. The capacitor C_1 charges to the voltage V_A while its switch is closed and holds this value when the switch is open.

When the second digital input is switched to the DAC, its analog equivalent appears at V_A and is transmitted to C_2, via switch S_2, which is now closed and all the other switches open. This sequence continues for all digital inputs until each is converted. It is then repeated, beginning with the first input, as many times as necessary until the analog output voltages are no longer required. The op-amp associated with each capacitor has a very high input impedance to prevent the capacitor from discharging while its switch is open. It also serves the purpose of load being driven and provides necessary amplification. The capacitor cannot hold the charge indefinitely. It will leak to the ground.

Multiplexing rate is the rate at which various digital inputs are switched sequentially into the DAC. One complete cycle of operation involves transferring the new digital value to the input register, converting it into analog voltage, closing the appropriate switch and opening others, and allowing the capacitor to charge fully. If electromechanical relays are used for multiplexing switches, this operation takes longer. High-speed relays can respond in 1 msec. For greater multiplexing rates, solid state CMOS bilateral switches need to be used.

Digital inputs can be given to a DAC through gating and control circuits.

5.5.2 A/D Multiplexing

In A/D multiplexing, it is convenient to multiplex the analog inputs rather than the digital outputs. Switches, either solid state or relays, are used to connect the analog input to a common bus. The bus is then connected to a single ADC, which is used for all channels.

The analog inputs are switched sequentially to the bus by a channel selection-control circuit. Sample-and-hold circuit may be used ahead of each multiplexer switch. (Sampling or sequential switches will also be there at the output because digital outputs are required one after the other.)

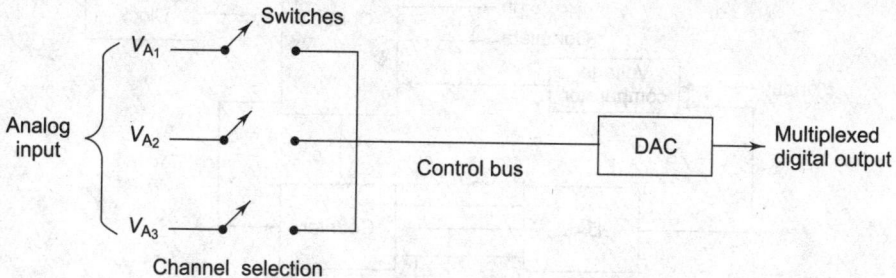

Fig. 5.13 *Multiplexing Number of Amplifier Inputs*

5.5.3 MOSFET as a Switch

When no gate voltage is given, the MOSFET is in off state. ($R = 10^{10}\ \Omega$). There is no conducting channel between the source and the drain. Now, when a certain threshold voltage is applied to the gate (typically +3 V for n-channel), there is a conducting channel between the drain and the source, and R reduces to $100\ \Omega$. So, it is on. Thus, MOSFET acts as a switch (see Fig. 5.14).

Different types of ADC techniques are described below.

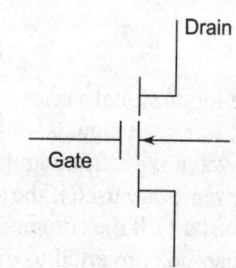

Fig. 5.14 *MOSFET as a Switch*

5.6 RAMP CONVERSION: COUNTER-RAMP CONVERSION TECHNIQUE OF ADC

This is the least expensive and slowest method of converting analog information to digital form. In a digital voltmeter, this method is used when the number of conversions required per unit time is small.

Here, the counter has to start from 0000 till V_i is reached. So, conversion time is more. How is the conversion time to be reduced? If we start from 1000, and count upwards or downwards, counting time will be reduced for 4-bit conversion.

The input voltage to be measured is fed to the voltage comparator. Upon receipt of the convert signal, the control resets the counter to 0 and then supplies clock pulses to the counter. The binary output of the counter is fed to the DAC, which outputs an analog voltage in response to its digital input. This analog voltage is then fed to the voltage comparator, which compares the output of the DAC with the analog input. As soon as the D/A output (input to the comparator to the capacitor) exceeds the input voltage, the comparator signals the control circuit, and it stops the counter.

The binary number in the counter then represents the voltage of the input signal. The control circuit will also output a polarity signal indicating when they are positive or negative, and an overflow signal, if the input signal exceeds the highest possible voltage of the DACs.

The control circuit is a flip-flop for starting and stopping the operations. When a start signal is given, all the counter flip-flops are cleared at the same time and the start–stop flip-flop is reset. This flip-flop provides a gating level to the signal gate (AND gate), allowing the clock pulses to be applied to the counter register. The clock pulses are propagated through the counter. So, the D/A output increases towards the top of the reference voltage, when the D/A output is equal to the analog input, the comparator switches (changes state) deliver an output signal to the start–stop flip-flop. The flip-flop output drops to zero, blocking the clock pulses at the signal gate.

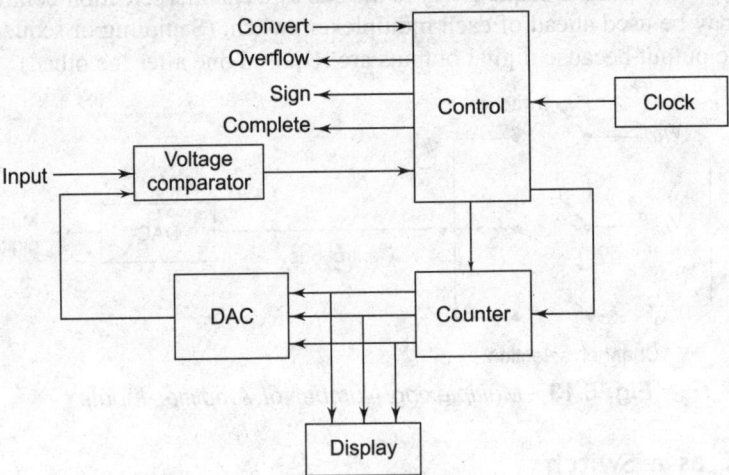

Fig. 5.15 *Counter-Ramp A/D Conversion*

If the input signal varies at a high frequency on both positive and negative sides, then this cannot respond. For such signals, delta modulation techniques have to be used. In digital voltmeters (DVMs), the sine wave is rectified and filtered, and rms value is indicated.

S/H circuits are used if the analog signal varies at a rapid rate. The S/H circuit will hold the input to the ADC constant till the conversion is completed. The conversion time is small. So, high-frequency analog signals can be converted to digital form.

Successive-approximation technique is commonly used where high speed and accuracy are needed. The principle is sequential selection of voltages leading to a successive approximation to the unknown input voltage. Standard reference voltages are generated in steps and compared with the unknown input voltage to be measured. These voltages are derived by the R–2R ladder network. The block diagram is shown in Fig. 5.16.

On receipt of the trigger command, a series of clock pulses are generated. The ring counter routes the clock pulses to the bistable multivibrators in the sequence, which in turn activate their respective switching circuits. These switching circuits turn on their respective binary-coded resistors into the circuit. The voltage derived from the resistor network forms the reference voltage for comparing with the input voltage. Suppose LSB corresponds to 1 mV and MSB to 800 mV. To start with, SW_1 is closed and hence a reference voltage of 800 mV ($V_{ref} \times \dfrac{R_f}{R}$) is generated, and is compared with the input voltage V_i, to be converted. If this voltage is greater than V_i, SW_1 is reset. If it is less than V_i, SW_1 is left in a closed position. Next, SW_2 is operated and this generates 400 mV ($V_{ref} \times \dfrac{R}{2R}$). Depending upon the condition of the previous switch, a reference voltage of 400 mV or 1200 mV is generated and compared with the input voltage. This procedure is followed until all the combinations of bits are tried. The state of the bistable multivibrator is an indication of the digital output.

Conversion speed is proportional to the number of bits. For a 12-bit conversion, the speed is 6 μsec, approximately.

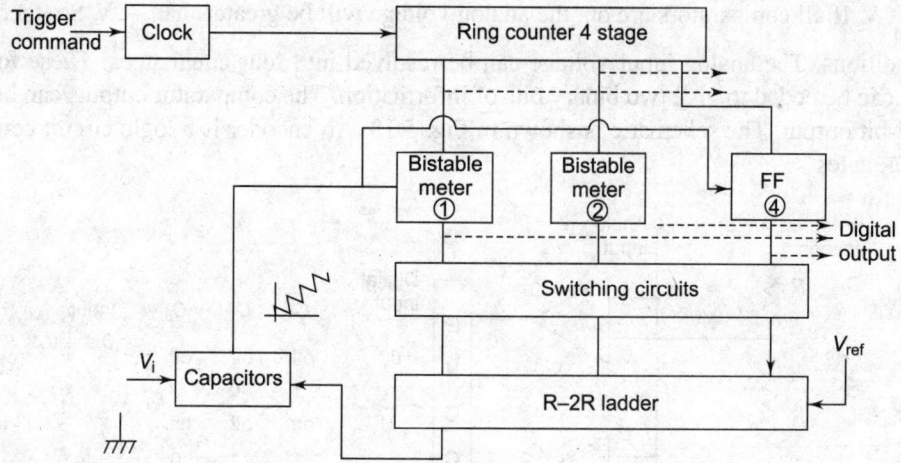

Fig. 5.16 *Successive-Approximation Technique*

5.7 BINARY-CODED RESISTANCE NETWORK FOR SUCCESSIVE-APPROXIMATION TYPE A/D CONVERSION

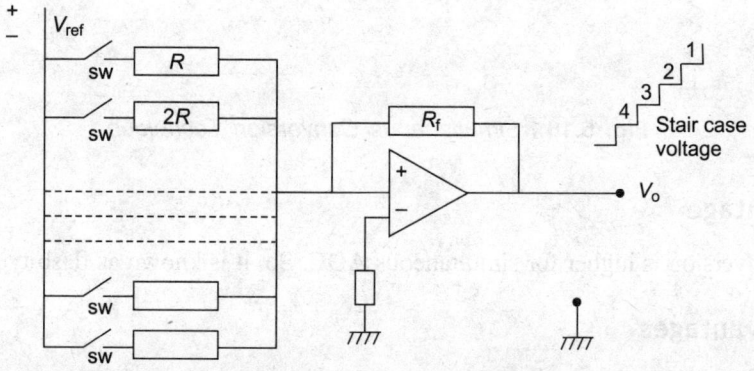

Fig. 5.17 *Resistor Network Circuit*

As many resistors R, $2R$, and so on as the number of bits are required, in this method.

5.8   SIMULTANEOUS ADCs (FLASH TYPE)

Here, three comparator circuits are used for 2-bit input. Each of the three comparators has a reference input voltage derived from a precision reference voltage source. A potential divider network provides $\frac{3}{4}$ V, $\frac{1}{2}$ V, and $\frac{1}{4}$ V. The other input terminal of each comparator is driven by the unknown analog voltage.

The comparator will be ON if the analog voltage is more than the reference voltage. If none of the comparators are on, the analog input will be $\leq \frac{1}{4}$ V. If C_1 is on, and C_2 and C_3 are off, the analog voltage will be between $\frac{1}{4}$ V and $\frac{1}{2}$ V. If C_1 and C_2 are both on, and C_3 off, the analog voltage will be between $\frac{1}{2}$ V and $\frac{3}{4}$ V. If all comparators are on, the analog voltage will be greater than $\frac{3}{4}$ V. So, there are four output conditions. The analog input voltage can be resolved into four equal steps. These four output conditions can be coded to give two binary bits of information. The comparator outputs can be encoded to give a 2-bit output. The schematic is shown in Fig. 5.18. An encoder is a logic circuit consisting of AND and OR gates.

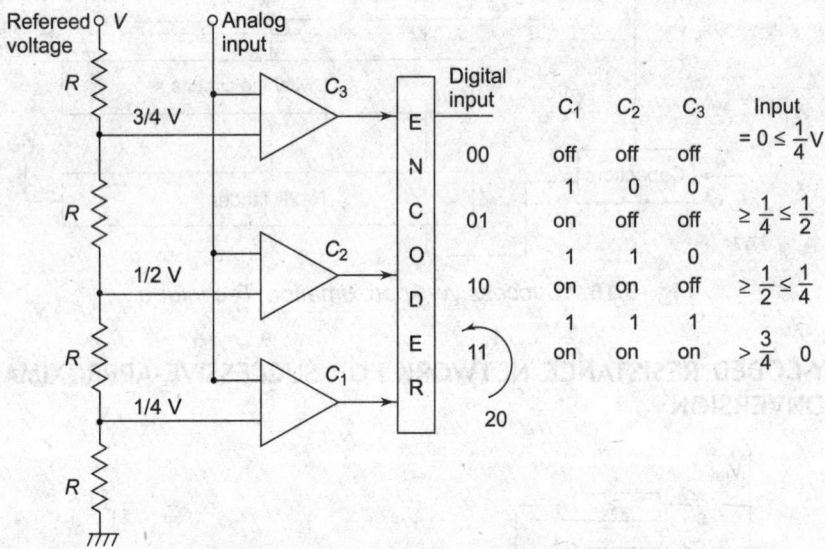

Fig. 5.18   *Simultaneous Conversion Technique*

5.8.1   Advantage

The speed of conversion is higher for simultaneous ADC. So, it is known as flash type *ABC*.

5.8.2   Disadvantages

1. $(2^n - 1)$ comparators are required for n-bit ADCs (3 comparators for 2-bit and 15 comparators for 4-bit ADC).
2. Resolution is poor. (Voltage less than V/4 cannot be detected, in the above case.)

Comparator circuit:

It consists of an op-amp used in its high-gain differential mode. Any difference in the inputs V_1 and V_2 is amplified by the gain G_v of the amplifier. It is usually 10,000 times. Thus, if V_1 is slightly greater than V_2, by an amount V_T called the threshold voltage, the comparator output saturates at +10 V. Similarly, if V_2 is greater than V_1 by an amount V_T, the comparator output saturates at –10 V. For a difference in V_1 and V_2 that is less than V_T, the comparator output is G_v times the difference.

$$V_o = \begin{cases} G_v\,(V_1 - V_2) & \text{if } |V_1 - V_2| \le V_T \\ +10\text{ V} & \text{if } V_1 - V_2 > V_T,\ \text{logic 1} \\ -10\text{ V} & \text{if } V_2 - V_1 > V_T,\ \text{logic 0} \end{cases}$$

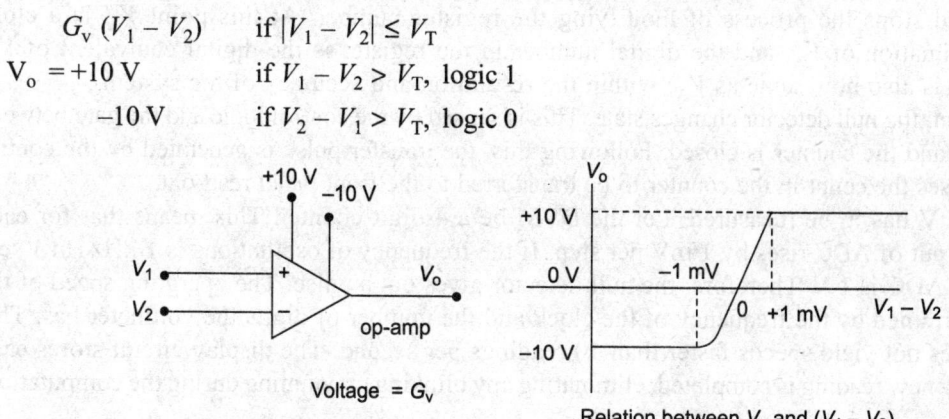

Fig. 5.19 *Comparator and Output Waveform*

In certain cases, if V_1 is greater than V_2, the output voltage is maximum and the comparator is on. If V_1 is less than V_2, the output voltage is zero or minimum and the comparator is off. Since the amplifier has a very high gain, it either saturates or cuts off at relatively low differential input levels so that it acts as a binary device.

5.9 GENERAL DESCRIPTION OF ONE CLASS OF ADC

A/D conversion is more complex and time consuming than D/A conversion. ADC of the type shown below in Fig. 5.20 uses a DAC as a part of the circuitry.

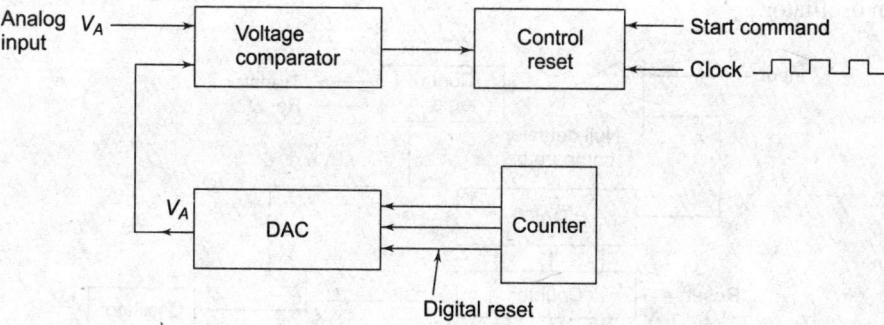

Fig. 5.20 *Schematic of ADC Using DAC*

The timing for the operation is provided by the input clock signal. The control limit indicates the conversion process. It contains logic circuits to generate the proper sequence of operations. The comparator has two analog inputs and a digital output which switches states depending on which analog input is greater.

1. The Start command goes high, starting the operation.
2. At a rate determined by the clock, the control element continually modifies the binary number, which is stored in the register (counter).
3. The binary number in the register is converted to an analog voltage V_A' by the DAC.
4. The comparator compares V_A' with the analog input V_A. As long as V_A' is less than V_A, the comparator output stays high. When V_A' equals or exceeds V_A, the comparator output goes low and stops the process of modifying the register number. At this point V_A' is a close approximation of V_A, and the digital number in the register is the digital equivalent of V_A' (which is also now same as V_A) within the resolution and accuracy of the system.

At this point, the null detector changes state. This is sensed by the control logic and the gate between the oscillator and the counter is closed. Following this, the transfer pulse is generated by the control logic and causes the count in the counter to be transferred to the front panel read-out.

Suppose 1 V has to be measured. Let the DVM be a 4-digit counter. This means that for each pulse, the output of ADC rises by 1 mV per step. If the frequency of oscillations is 1 kHz (in 1 sec) the output of ADC is 1 V. Therefore, the null detector gives out a pulse. The operating speed of the meter is determined by the frequency of the clock and the number of digits the voltmeter has. This technique does not yield speeds faster than 10 readings per second. The display circuit stores each reading until a new reading is completed, eliminating any blinking or counting during the computation.

5.9.1 Limitations

1. Conversion time is more.
2. Resolution depends on the step size of the DAC.
3. If the number of bits are more in DAC, cost will be more. Speed depends on clock frequency. The maximum speed is equal to 10 readings per second.

5.10 STAIRCASE RAMP DVM

This is also called the digital ramp technique. It is null balance or potentiometric in nature. The heart of the system is the DAC. It is driven by the digital output of an electron that is counter, which in turn is driven by an oscillator.

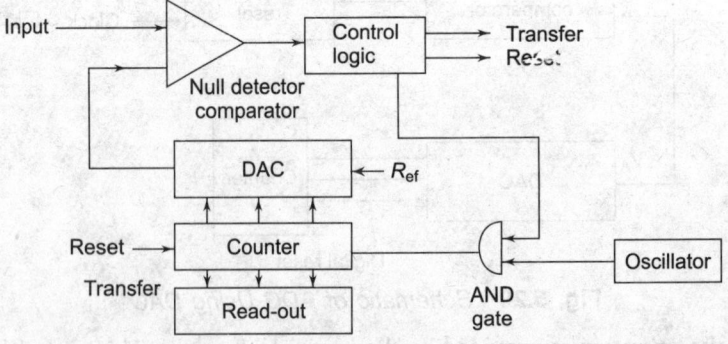

Fig. 5.21 *Staircase Ramp Type DAC*

At the beginning of a measuring sequence, the counter is reset to a zero-count condition by the control logic. Since a DAC is slaved to the counter, it also gives zero output. In the measuring sequence, the gate is opened, which is located between the oscillator and counter. The counter starts accumulating counts from the oscillator and causes the DAC to generate an output voltage equivalent to the instantaneous count. The output of the DAC which is a ramp continues to build up voltage until the null detector determines that the ramp has exceeded the input voltage with its last incremental increase.

The diagram of a ramp-type DVM (linear ramp) is shown in Fig. 5.22.

A sample rate generator triggers and resets the counters and also initiates the ramp voltage.

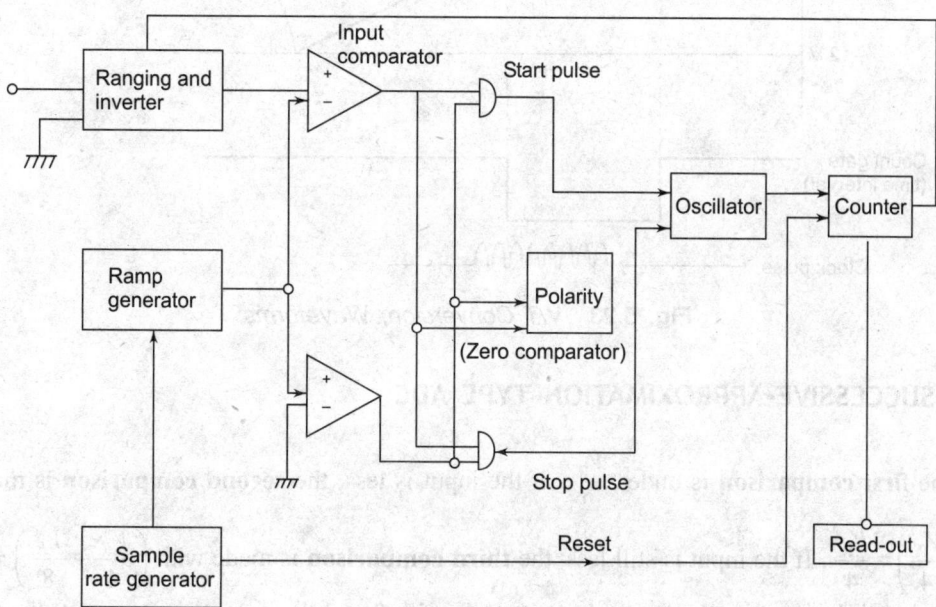

Fig. 5.22   *Linear Ramp-Type ADC Schematic*

Sample rate is the number of measurement cycles per second. Five is the typical value.

5.11 VOLTAGE-TO-TIME CONVERSION

Any DVM has a fundamental cycle sequence that involves sampling, display, and reset sequence. The application of input voltage initiates the measurement cycle. The oscillator is automatically switched with the operator and its output consists of pulses which are counted by an electronic counter. A units counter is first actuated. This units counter provides a carry pulse to the 10's counter on using the counter input pulse. In turn, the 10's counter provides its own carry pulse after it has counted 10 carry pulses from the units counter. If excess input voltage is being measured, it provides its own carry pulse which switches on a warning lamp.

Each decade counter unit in a DVM is counted as a DAC. Outputs of a DAC are connected in parallel, which build up a comparison voltage. At the instant when the comparison section senses that the input voltage and comparison voltages are equal, it produces a trigger pulse which stops the oscillator.

The sample rate function permits the DVM to follow a varying voltage. It is controlled by a simple relaxation oscillator that triggers and resets the counters to zero every half second. The display circuit stores each reading until a new sample value occurs. This eliminates blinking. Finally, when the input voltage is removed from the instrument, the reading automatically returns to zero and this completes the cycle sequence.

The measurement cycle of a DVM involves: (1) sampling, (2) display, and (3) reset.

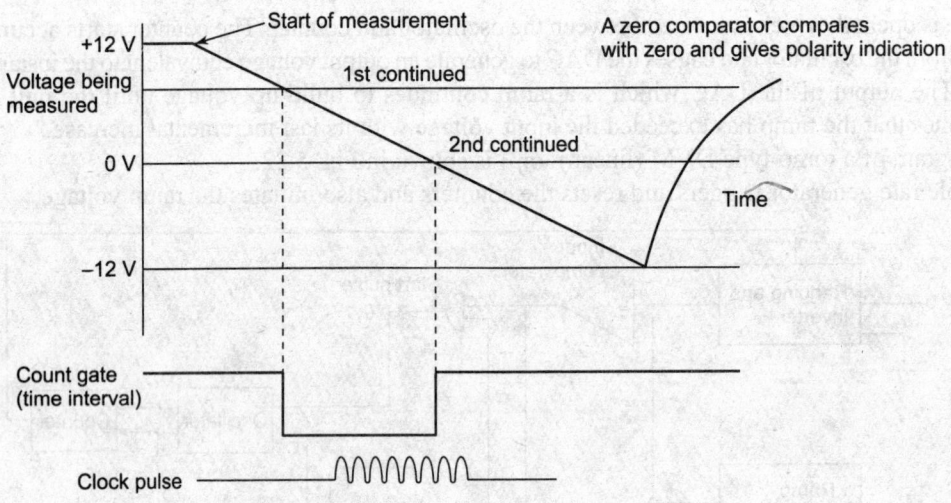

Fig. 5.23 *V/T Conversion, Waveforms*

5.12 SUCCESSIVE-APPROXIMATION TYPE ADC

When the **first comparison** is under $\dfrac{V_R}{2}$, if the input is less, the **second comparison** is made with

$\left(\dfrac{V_R}{2} - \dfrac{V_R}{4}\right) = \dfrac{V_R}{4}$. If the input is still less, the **third comparison** is made with $\left(\dfrac{V_R}{4} - \dfrac{V_R}{8}\right) = \dfrac{V_R}{8}$. If

the input is still less, a **fourth comparison** is made with 0 and the output is given as 0 (because the

comparator cannot detect voltage less than $\dfrac{V_R}{8}$. If, in the fourth comparison the analog input is greater

than 0, it is given as $\dfrac{V_R}{8}$ = 0001. Thus, for four comparisons, we get a four-digit number. For six

comparisons, we get a six-digit number. For n comparisons, we get an n-digit number. The maximum
input that can be converted is equal to V_{ref}.

The process is illustrated below in Fig 5.24.

Let us consider a 4-bit ADC. Let V_{ref} be 15 V.

So, $$\dfrac{V_R}{2} = 1000$$

The successive approximation type ADC compares the analog input to a D/A conversion module (resistor network) reference voltage, which is repeated by changing the reference voltage to half of the
previous value. A comparator is used to compare the input voltage and a sequence of voltages which are

binarily related. Initially, the input is compared with $\dfrac{V_R}{2}$, where V_R is a full-scale input voltage. If the

input is greater than $\dfrac{V_R}{2}$, then it is changed to $\left(\dfrac{V_R}{2} + \dfrac{V_R}{4}\right)$. This becomes the reference value for the new comparison. If the input is less, then it is changed to $\left(\dfrac{V_R}{2} - \dfrac{V_R}{4}\right)$. (Note the negative sign.) One more comparison is now made and the voltage is changed depending on the result of the comparison. This procedure is repeated till the desired accuracy is obtained.

Suppose the full-scale reference voltage $V_R = 1000 = (8)$.

$$\text{Then } \frac{V_R}{2} = 100\ (4) = (8/2)$$

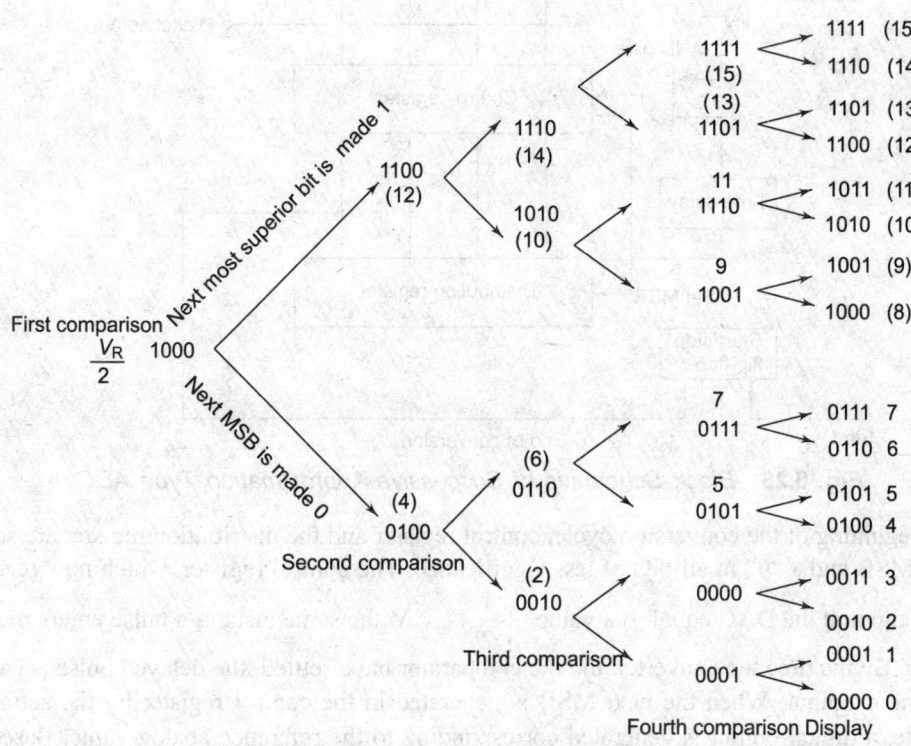

Fig. 5.24 *Principle of Successive Approximation*

So, the input to be measured is compared with $\dfrac{V_R}{2} = 100$. If the input is greater than $\dfrac{V_R}{2}$, it is compared with $\dfrac{V_R}{2} = \dfrac{3V_R}{2} = 110$. If the input is greater than $\dfrac{3V_R}{2}$, the next comparison is made with $\dfrac{3V_R}{4} + \dfrac{V_R}{8} = \dfrac{7V_R}{8} = 111$. If input is greater than $\dfrac{7V_R}{8}$, comparison is made with V_R itself.

If the input is less than $\dfrac{3V_R}{4}$, the comparison is made with $\left(\dfrac{3V_R}{4} - \dfrac{7V_R}{8}\right) = \dfrac{5V_R}{8} = 101$. If the input voltage is less than $\dfrac{5V_R}{8}$, it is compared with $\dfrac{1}{2}V_R$ and the result is given to the display unit.

$\left(\text{Conversion time is reduced by starting comparison with } \dfrac{V_R}{2}\right)$

5.13 SUCCESSIVE APPROXIMATION ADC

The block schematic is shown in Fig. 5.25.

Distribution register: It is a ring counter, circulating to bit 1 to determine which action is taking place.

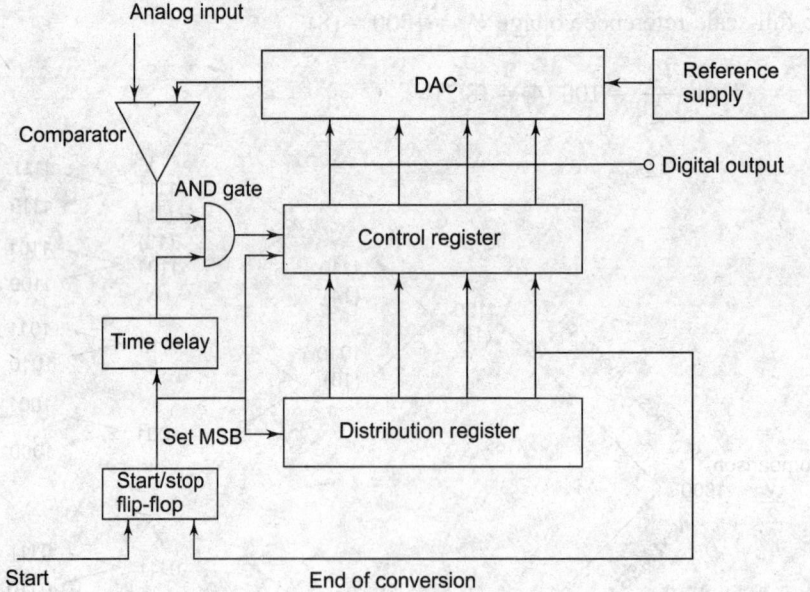

Fig. 5.25 *Block Schematic of Successive-Approximation Type ADC*

At the beginning of the conversion cycle, control register and the distribution register are set with a "1" in the MSB and a "0" in all bits of less significance. The control register, which now reads 1000, causes an output at the DAC equal to a value of $\frac{1}{2} V_{\text{ref}}$. At the same instant, a pulse enters the timing-delay chain. By the time the converter and the comparator have settled, the delayed pulse is gated with the comparator output. When the next MSB is generated in the control register by the action of the timing cycle, a digital output is generated corresponding to the reference analog value. Based on the result of the comparison, the digital quantity is changed, comparison is made, and the cycle is repeated till equivalence is reached.

The feedback path lies between the control register and distribution register. The distribution register is a ring counter. "1" is circulated. When "1" changes bit position, a particular operation takes place. It gets feedback from the control register.

Initially bit "1" is in MSB in distribution register also.

For binary 1000, $\dfrac{V_{\text{ref}}}{2}$ is the output of a DAC.

For binary 0100, reference supply is increased or decreased, depending upon the comparator output.

When the input voltage and the output of DAC coincide, an end of conversion (EOC) signal is generated in the distribution register through the control register. The conversion time is less here, compared to the counter ramp technique, since comparison starts from 1000 and not 0000.

Digital Voltmeters: ADCs and DACs are used in the construction of DVMs.

5.14 DVM TYPES

It displays measurements of DC or AC voltages as discrete numerals in the decimal numbered system. This is advantageous since it eliminates human reading error (like setting the selector switch on one scale and taking reading on the other, parallax error, and so on. It gives very precise readings.

The different types of DVMs are given below:

1. Potentiometric DVM
2. Integrating DVM and dual-slope integrating type
3. Ramp-type DVM
4. Successive approximation DVM
5. Continuous balance DVM (servobalancing type)

The block diagram for successive-approxmation DVM is the same as for successive approximation ADC.

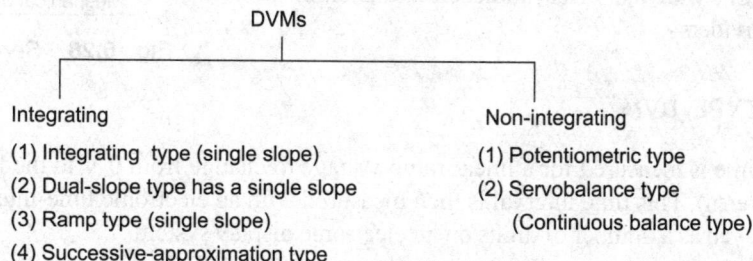

Fig. 5.26

5.14.1 Specifications of DVMs (Typical Values)

1. Input range   : ±1.0000 V to ±1000.0000 V
2. Absolute accuracy   : $\pm0.005\%$
3. Stability   :   0.002% of the reading in 24 hrs
4. Resolution   :   one part in 10^6, that is, 1 μV can be read on 1-V scale
5. Input characteristic : $R_i = 10$ MΩ, $C = 40$ pF
6. Calibration   :   Internal standard calibration from a reference supply
7. Display type   :   LED/LCD, $3\frac{1}{2}$ digit
8. Input resistance is typically 10 MΩ.

5.15 NON-INTEGRATING TYPE DVMs

5.15.1 Potentiometric Type

The principle involved in this DVM is shown in Fig. 5.27.

The linear divider (potentiometric) is adjusted until the null indicator shows equality of input voltage and output of divider potential network. It can be a differential amplifier. The range of voltage depends on V_{ref}. Resolution

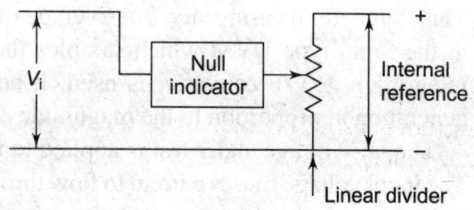

Fig. 5.27 *Potentiometric Type*

depends upon the linear divider. Sensitivity depends on the circuit of the null indicator, its noise immunity, and so on.

5.15.2 Servo Type DVM

To increase the range and to make the null technique automatic, a servo system is employed. This is an improved version of the potentiometric type DVM.

A servo system consists of a differential amplifier, motor, and linear divider. The amplifier senses the polarity of imbalance and drives the motor in such a direction so as to reduce the imbalance. This level of imbalance can be reduced by increasing gain. Attached to the shaft of the motor and divider is a mechanical read-out which in effect indicates the position of the shaft of the dividers.

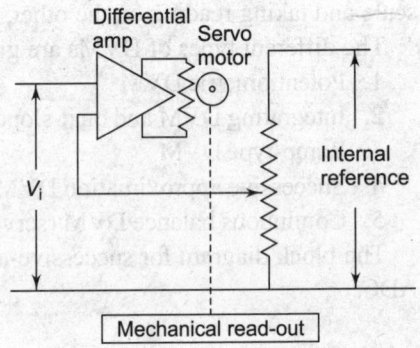

Fig. 5.28 *Servo System*

5.16 RAMP TYPE DVM

In this case, the time is measured for a linear ramp voltage to change from 0 V to the level of the input voltage (or vice versa). This time interval is then measured with an electronic time-interval counter and the count is displayed as a number of digits on an electronic display system.

At the start of a measurement cycle, a ramp voltage is initiated. This can be a positive- or negative-going ramp. The linear ramp shown in the Fig. 5.23 is compared continuously with the unknown input voltage. At the instant the ramp voltage equals the unknown voltage, a coincidence circuit on a comparator generates a pulse which opens a gate. The ramp voltage continues to decrease with time. When it finally reaches 0 V, a second comparator generates an output pulse which closes the gate.

An oscillator generates clock pulses which are allowed to pass through the gate to a number of decade-counting units which add the number of pulses that pass through the gate.

The sample rate multivibrator determines the rate at which the measurement cycle is initiated. The oscillator of this multivibrator can be adjusted by a front panel control marked "rate". It is of the order of five measuring cycles per second, a fixed value in some instruments. The sample rate circuit provides an initiating pulse for the ramp generator to start its next ramp voltage. At the same time, a reset pulse is generated which returns all the decimal counting units (DCUs) to their 0 state, removing the display momentarily from the indicator display device. The cycle repeats for the next ramp in the same manner as long as input voltage is present.

5.17 INTEGRATING TYPE DVM

This voltmeter measures the true average of the input voltage over a fixed measuring period in contrast to the ramp type DVM which samples the voltage at the end of a measuring cycle. To accomplish integration, a V/F converter is used. It acts as a feedback system which governs the rate of pulse generation in proportion to the magnitude of input voltage.

The DC voltage under test is applied to the input stage which isolates the measuring circuit.
The input voltage causes current to flow through R_1 into the summing junction of the op-amp. This current continues through C_1 and causes the output voltage of the op-amp to depart from 0 V. The output will

be a linear ramp if the input is constant. If the input is positive, the ramp will be negative going. When this voltage reaches a value equal to −V, the comparator triggers the pulse generator to inject a fixed amount of charge into the summing junction of the op-amp (that is a negative voltage step is injected to the swing junction). The sum of the input voltage and pulse voltage is negative, causing the ramp to reverse its directions. This retrace is very rapid because the pulse is large in amplitude compared to the input. When the positive-going ramp reaches 0, the comparator generates a reset trigger to the pulse generator. The negative pulse is removed and the original input is left. So, output of the op-amp is negative. The whole process is then repeated. The output of the op-amp waveform looks like a saw tooth. If the input is changed, the number of teeth per unit time of saw-tooth wave will be doubled. Coincident with each tooth is a pulse which passes through T_1 and to the input control gate. These pulses are allowed to enter the reversible counter when the gate is opened and this opening of gate is the beginning of the measurement cycle. The gate can remain open for 0.1 sec or 1 sec. During this period, the counter totals the pulses. At the end of the period, the count stored in the counter is transferred to the display which indicates the voltage being measured in decimal form.

If an input signal were to change its polarity during a measuring period, the pulses accumulated like this should be subtracted from those that have been accumulated after reversal. So, the counter will have revising capability. If the input voltage is negative, the output of op-amp is positive. Another comparator determines when the signal passes through +V and triggers another pulses generator which gives a positive pulse from the test circuit, and provides the necessary input attenuation. The attenuated input signal is applied to the V/F converter. This circuit consists of an integrating amplifier, a level detector or a comparator circuit, and a pulse generator. The integrating amplifier produces an output voltage proportional to the input voltage and related to the input feedback elements by the equation: V_{out}

$= \dfrac{1}{C} \int i\, dt = \dfrac{1}{RC} \int V_i\, dt$. If the input voltage is constant, the output is a linear ramp following the equation

$$V_0 = -V_i \dfrac{t}{RC}$$

The schematic is shown in Fig. 5.29.

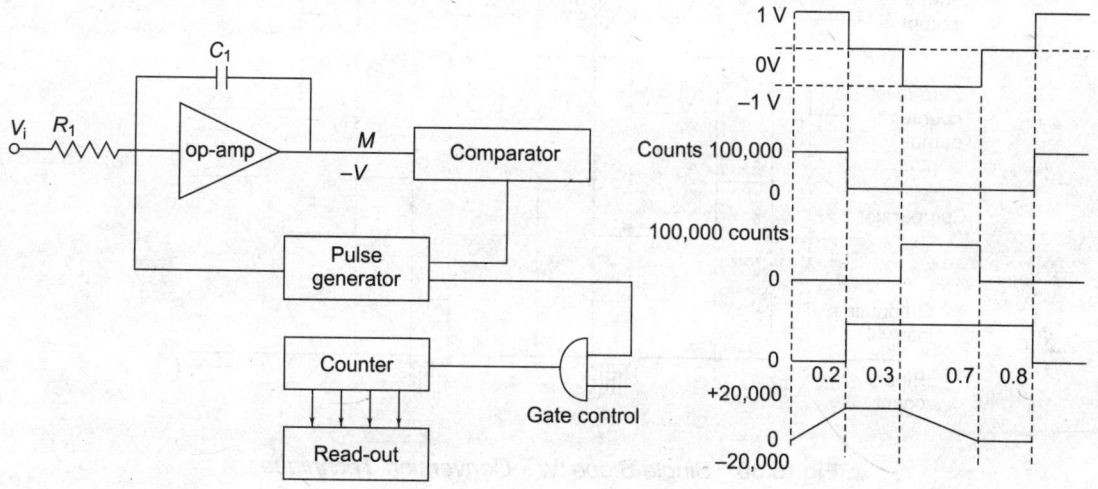

Fig. 5.29 *Schematic with Waveforms*

When a ramp reaches a negative voltage level, the level detector triggers the pulse generator which applies a negative voltage step to the summing junction of the integrating amplifier. The sum of the input voltage and the pulse voltage is negative, causing the ramp to reverse its direction. This retrace is very rapid since the pulse is large in amplitude compared to the input voltage. When the now positive-going ramp reaches 0 V, the level detector generates a reset trigger to the pulse generator. The negative pulse is removed from the summing junction of the integrating amplifier and only the original input voltage is left. The amplifier then produces a negative-going ramp and the procedure repeats itself.

A larger input voltage causes a steeper ramp and therefore a higher pulse repetition rate (PRR). Since the input is integrated, inspite of the large amount of noise present, measurement can be done accurately.

5.18   V/T CONVERSION

A/D conversion can also be activated by V/T conversion.

5.18.1   Single-Slope V/T Conversion

The block diagram and waveforms are shown in Fig. 5.30.

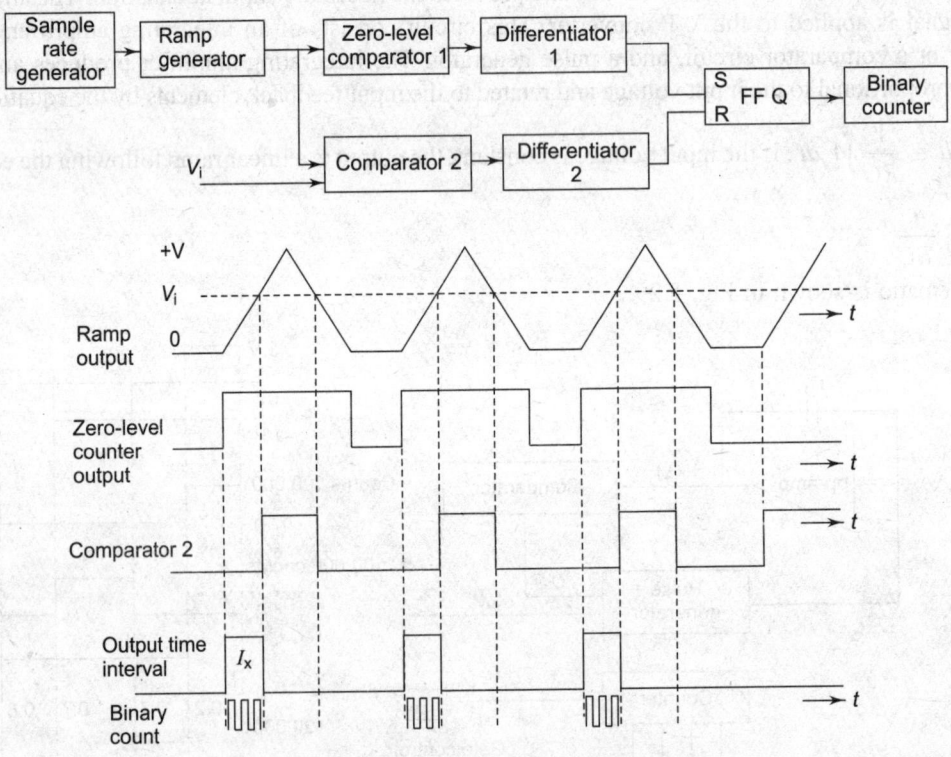

Fig. 5.30   *Single-Slope V/T Conversion Technique*

Only positive-going spikes will set or reset the flip-flop.

The ramp generator produces a repetitive ramp with highly stable amplitude and high linearity. This is applied to two comparators. One of the comparators changes state at every zero crossing and the other comparator changes state when the input voltage and the ramp are of equal magnitude.

The outputs of these are differentiated and the positive-going spikes are used to control an S–R flip-flop. The zero-level comparator sets the flip-flop and the signal comparator resets it.

The line interval so produced can be shown to be directly proportional to the input signal. The proof is as follows:

Let the expression for ramp voltage be $V(t) = \dfrac{V}{T}t = kT$

V = peak amplitude of the ramp voltage in volts
T = duration of ramp voltage in seconds
$V(t)$ = instantaneous value of the ramp in volts at time
K = slope of the ramp in V/sec

$$V_i = \left(\frac{V}{T}\right)T_x \text{ or } T_x = \frac{V_i}{V}T$$

$$T_x = \frac{V_i}{V/T} = \frac{V_i}{R} \text{ or}$$

$$\boxed{T_x \text{ is proportional to } V_i}$$

Thus, time is converted into voltage.

To convert T_x to a proportional number, a counter can be made to count a stable clock. During the period T_{x1}, at the end of this period the counting is stopped. So the counter now contains a number proportional to T_x, which is in turn proportional to the input voltage. Hence A/D conversion is achieved because the counter contains binary information proportional to T_x.

5.18.2 Limitations

1. Comparators are assumed to be ideal (gain is ∞, offset = 0). A comparator of sensitivity less than 1 μV is required.
2. It cannot take bipolar signals.
3. Noise is not rejected.

V/T or V/F method is superior to successive approximation method because in the latter method, V_i is compared to a voltage generated in steps $\dfrac{V_{ref}}{2}$, $\dfrac{V_{ref}}{4}$, and so on. So, accuracy of measurement depends on the step size.

5.19 V/T CONVERTERS FOR A/D CONVERSION

The ramp generator produces a highly stable repetitive ramp. This ramp is applied to the two comparators. Comparator 1 changes state at every zero crossing of the ramp and comparator 2 changes state when the input and ramp voltages are equal. The outputs of these are differentiated and positive-going spikes are used to control the S–R flip-flop. The zero-level comparator sets the flip-flop while the signal comparator resets the flip-flop. The time interval so produced is proportional to the input signal. To convert this time interval to a proportional number, a counter is made to count stable clock pulses during that period. At the end of the period, counting is stopped. So, the counter reading indicates the analog input. The sample rate generator drives the ramp generator to produce a linear ramp waveform of fixed duration and amplitude at every sample input. The sampling rate is greater than 16 Hz to avoid flicker in the display, usually it is 50 Hz. The wave input of the analog signal that can be measured is the peak ramp voltage.

The block schematic is shown in Fig. 5.31 for convenience.

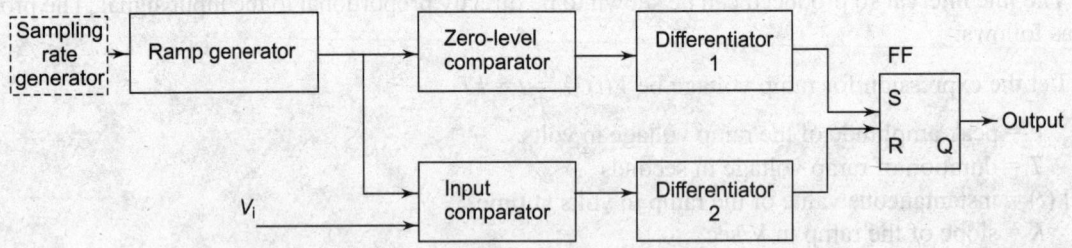

Fig. 5.31   *Block Schematic for V/T Conversion*

5.19.1   Disadvantages

1. Any noise overriding the analog signal is not rejected by the system and is taken as an analog input.
2. The ramp waveform is a function of resistors and capacitors and their temperature coefficients. It is the critical circuit in the system affected by temperature.

The system as such cannot detect negative inputs. So, a ramp waveform with centre zero has to be employed. The above disadvantages are reduced in dual-slope V/T converters.

5.20   V/F CONVERTER

In this method, the frequency of output pulses generated is proportional to analog input voltage.

When steady or slowly varying voltage is applied, the charging current $i_i = \dfrac{e_i}{R_i}$.

So, the output of the integrator $e_0 = -\dfrac{1}{C_f} \displaystyle\int_0^t i_i \, dt$

$$e_0 = \frac{-e_i}{R_i C_f} t$$

at any time t from the commencement of integration.

The voltage e_0 will continue to rise till it reaches E, the threshold voltage of each level detector. At this instant, the level detector triggers and operates a pulse generator that resets the integrator. The pulse generator injects a large negative charge at the summing junction of the op-amp. Hence, $e_0 = 0$. So, again i_i is applied and it starts integrating. This results in the generation of a repetitive ramp waveform with duration T given by

$$T = \frac{E R_i C_f}{e_i} \text{ since at } t = T, e_0 = E$$

The pulse generator output is given through a resistance R because it should provide discharging current opposite to the input-charging unit.

The frequency of this ramp waveform is the same as that of the pulse output with a frequency,

$$f = \frac{e_i}{ER_iC_f}$$

E, R_i, and C_f are constants. $f \propto e_i$ or frequency of output is proportional to input voltage.

So, the resulting pulses are counted for a fixed period of time in a binary counter, and the digital count at the end of this time is an index of the amplitude of the analog input.

5.20.1 Advantages

1. The converter can handle input signals with both polarities. There is a negative-level detector. The counter is a reversible counter. If the input changes sign, the counter resets and starts counting again.
2. Sensitivity can be easily controlled. Sensitivity is expressed in kHz/V. By changing E, R_i, and C_f, the sensitivity can be varied.
3. Linearity is good.

V/F converter provides a clock input which is proportional to the output V_c of comparator. These pulses are given to a counter through AND gate. The second input to the AND gate is sampling voltage V_s which holds the logic level for a fixed time T_H. As long as $V_s = 1$, the sampled value V_a is held at its value at the beginning of the interval T_H. (S/H circuit is used if the magnitude of the analog input changes with time). With AND gate enabled for the time T_H, the counter reading will equal the number of cycles executed by the V/F converter output in the specified time interval. The pulse input to the integrator is of a fixed value so that after the ramp output reaches the level V_R, the output voltage becomes zero.

Ramp voltage is 10 V. Slope = 1000 V/sec.

Table 5.2 Specifications of some DAC ICs.

Model Parameter	DAC 80P	DAC 703	DAC 707 JP	DAC 729
Digital input	—			
Resolution	Max. 12 bits	Max. 16 bits	Max. 16 bits	18 bits
Accuracy				
Linearity error	±1/4 LSB	+0.003% of FSR	±0.003% of FSR	±0.0015% of FSR
Drift	±10 ppm of FSR/°C	−20 to +20 ppm/°C	±10 ppm of FSR/°C	±9 ppm of FSR/°C
Conversion speed	3 µsec	4 µsec	4 µsec	5 µsec
(V_0 model)				
Ref. voltage O/P	+6.3 V	+6.3 V	+10 V	10 V
Power supply				
Requirements	±11.4 V to ±16.5 V	+15 V (+V_{cc})	+15 V (+V_{cc})	+15 V (+V_{cc})
Requirements	VDC	−15 V (−V_{cc})	−15 V (−V_{cc})	−15 V (−V_{cc})
Temp. range	0 to 70°C		0 to 70°C	0 to 70°C

Table 5.3 Specifications of some ADC ICs

Model	ADC 80 MAH-12	ADC 85H	ADC 600K
Parameter			
Resolution	Max. 12 bits	Max. 12 bits	Max. 12 bits
Accuracy must be there			
Linear error	±1/2 LSB	±0.012% of FSR	1.5 LSB
Drift (gain)	±15 ppm/°C	Max. ±15 ppm/°C	
Conversion time	22 μsec	Max. 10 μsec	140 μsec
Ref. voltage	+6.3 V	+6.3 V	
Power supply	+5, ±12, or ±15 V	+5, ±12 or ±15 V	+15 V $(+V_{cc})$
Requirements		−15 V $(-V_{cc})$	
Power supply			
Sensitivity	±0.003% of FSR/% V_{cc}	±0.004% of FSR/% V_{cc}	
Temp. range	−25 to +95°C	−25 to + 85°C	0 to +70°C

5.21 DUAL-SLOPE INTEGRATING TYPE DVM

Integrating DVMs reject inherent noise present with the input signal. The digital read-out will be the final count and not the initial count. By integrating twice, the errors due to R and C components get cancelled.

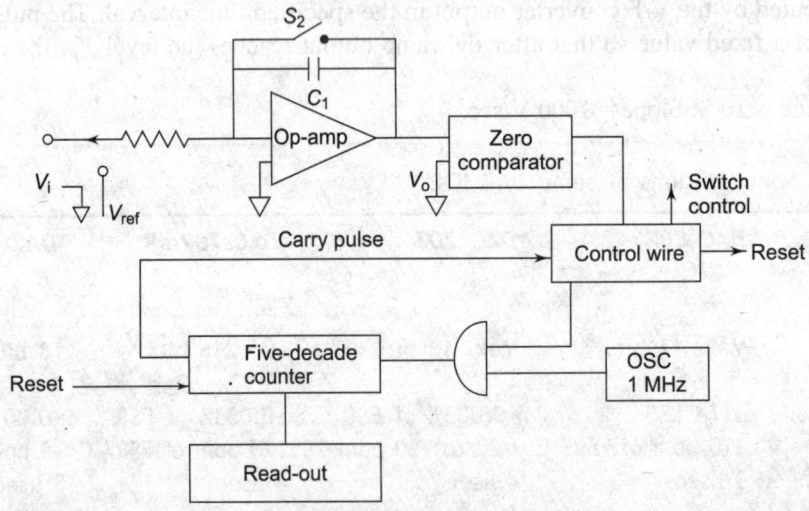

Fig. 5.32 *Block Schematic of Dual-Slope Integrating Type DVM*

In this circuit, the period of integration is determined by a 1 MHz oscillator and the counter. The first counter is reset to zero. The gate between the oscillator and counter is opened by the control logic and at the same time S_2 is opened. (Initially, the charge across C is zero because S_2 is closed.) When S_2 is open, C_1 gets charged at a rate proportional to the magnitude of the input voltage. After a fixed interval, a carry pulse is generated by the counter. The counter resets to zero when the carry pulse is generated. At the same time, S_1 will be connected to V_{ref}. V_{ref} polarity is opposite to that of V_i. Therefore, the charge accumulated at C_1 during the integration period is now reduced by the reference signal. When all the charge is removed, the output of the op-amp becomes 0 V. The zero comparator detects this. At that instant, the AND gate between oscillator and counter is closed, and the counter reading gives the digital value of analog V_i.

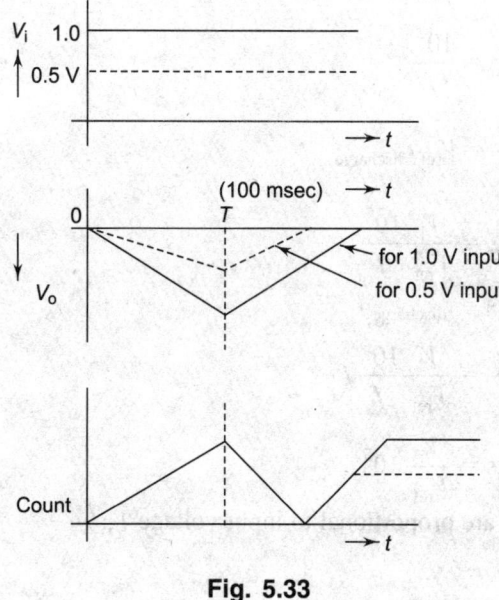

Fig. 5.33

(Integration of V_i is done till the counter reading reaches the maximum value or a predetermined time.)

In this circuit, the two slopes corresponding to when the voltage across C increases and decreases are different. Hence, the name dual slope. The first slope or rate of charging of C depends on the magnitude of V_i. But when V across e is decreasing, the slope is same because it is connected to V_{ref}. But discharging is done for a fixed time till the voltage across C becomes zero. Therefore, the count during this period is proportional to V_i.

In this circuit, the accuracy does not depend upon R_1, C_1, and the frequency of the clock. This is the advantage of the circuit.

Since
$$|V_o \text{ op-amp}| = \frac{1}{C_1}\frac{V_i}{R_1} t_{integrate}$$

$$= \frac{1}{C_1}\frac{V_{ref}}{R_1} t_{discharge}$$

For a five-decade counter, that is one which counts 100000, $t_{integration} = \dfrac{10^5}{f}$

(Five decades means $10 \times 10 \times 10 \times 10 \times 10 = 100000$.)

(Five-decade counter will count 10^5 at a frequency f.) $t_{discharge} = \dfrac{\text{Accumulated counts}}{f}$

$$\left[\text{Therefore, accumulated counts} = \frac{V_i}{V_{ref}} \times 10^5 \right]$$

$$\frac{V_i}{R_1 C_1} t_{integrate} = \frac{V_{ref}}{R_1 C_1} t_{discharge}$$

$$t_{integrate} = \frac{10^5}{f}$$

So,

$$V_i \frac{10^5}{f} = V_{ref} t_{discharge}$$

Therefore,

$$t_{discharge} = \frac{V_i}{V_{ref}} \frac{10^5}{f}$$

$$\text{Accumulated counts} = t_{discharge} f$$

$$= \frac{V_i}{V_{ref}} \frac{10^5}{f} f$$

$$\text{Accumulated counts} = \frac{V_i}{V_{ref}} 10^5$$

Thus, accumulated counts are proportional to input voltage V_i.

5.21.1 Advantage

Output reading is independent of R and C.

5.21.2 Disadvantage

1. Linearity of the ramp affects measurement.
2. In the case of a staircase ramp, $\sqrt{}$, resolution depends on the step size. Conversion time will be greater.

5.22 SPECIFICATIONS OF ADCs

1. **Analog input voltage:** This is the maximum allowable input voltage range 0–10 V, ±5 V, ±10 V, and so on.
2. **Input impedance:** Its value ranges from 1 kΩ to 1 MΩ depending upon the type of ADC. Input capacitance is in the range of tens of pF.
3. **Stability:** The temperature dependence. Even if analog input is kept constant, the digital output may change with temperature. This is called stability. It is expressed as percentage error per degree rise in temperature.
4. **Conversion time:** The time taken for the conversion of analog input to digital output. It may be 50 μsec for moderate speed ADC to 50 μsec for a very high speed ADC.
5. **Accuracy:** ADCs up to 0.001% of full scale are available. The error is contributed by digital system noise, deviation from linearity, and the like.

Example 5.3 *For a given digital ramp ADC, the clock frequency = 1 MHz, threshold voltage $V_T = 1$ mV (below which an ADC cannot detect). DAC has full-scale output of 10.23 V and a 10-bit input. Determine (a) the digital equivalent obtained for $V_A = 3.728$ V, (b) conversion time, and (c) resolution of this converter.*

Solution

(a) A DAC has a 10-bit input. Full-scale output = 10.23 V

So, total number of possible steps = $2^{10} - 1 = 1023$

Therefore, $\qquad$ step size = $\dfrac{10.23 \text{ V}}{1023} = 10$ mV

That is, the output of the DAC increases by 10 mV for every count.

Since $\qquad\qquad V_A = 3.728$ V and $V_T = 1$ mV

The output of the comparator should reach a value greater than V_A because $V_T = 1$ mV = 0.001, the output of the DAC should be 3.729 V before the comparator switches low. This will require

$\dfrac{3.727 \text{ V}}{10 \text{ mV}} = 373$ steps.

(a) At the end of conversion, the counter will hold binary equivalent of 373 = 0101110101. This is the desired digital equivalent of $V_A = 3.728$ V.

(b) Three hundred seventy three steps were required to complete the conversion. Thus 373 clock pulses occurred at the rate of one per μsec.
So, total conversion time = 373 μsec

(c) Resolution = $\dfrac{1}{2^{10} - 1} \times 100 \simeq 0.1\%$

Example 5.4 *A 5-bit DAC produces $V_o = 0.2$ V for a digital input of 00001. Find the value of V_o for a 11111 input.*

Solution

$\qquad\qquad\qquad$ LSB = 0.2 V because 00001 is LSB, and $V_A = 0.2$ V.

So, $\qquad\qquad\qquad$ 11111 = 3.2 + 1.6 + 0.8 + 0.4 + 0.2 = 6.2 V

5.23 AUTOMATIC POLARITY INDICATION FOR DVMs

An automatic polarity indicator which can be used with the DVM is shown in Fig. 5.34. The circuit comprises of two voltage comparators, both with zero reference and connected to handle signals of opposite polarity. The input terminals of the comparators can be connected to the outputs of the buffer amplifier.

The output of comparator 1 will be in "1" state for positive input signals whereas comparator 2 will go to "1" state for negative inputs. Thus, either a positive or negative indicator signal results at the output of these voltage comparators depending on the input signal polarity. The high sensitivity of IC voltage comparators (μA 710) or op-amps used for the purpose (μA 741) ensures excellent performance.

The comparator outputs are fed to lamp driver circuits which facilitate automatic polarity indication. Alternatively, LEDs or any other single-element display device can be used for this purpose.

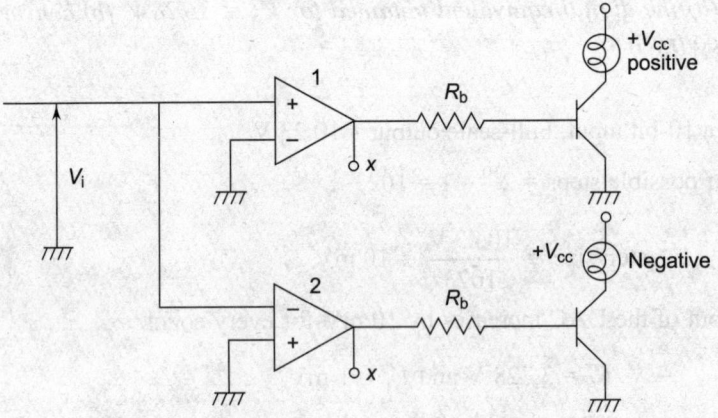

Fig. 5.34 *Auto Ranging for DVM*

The basic DVM has a fixed voltage range of 1 V. The performance of this DVM can be improved by the addition of an input attenuator which can be controlled to change the range of the instrument.

A typical attenuator suitable for use at the input to terminals of the DVM, that is preceding the buffer amplifier, is shown in Fig. 5.36.

The attenuator ensures a constant R_i for the DVM, for example 10 MΩ in this case, so long as the R_i of the buffer is much higher than that value. In addition, the voltage input to the buffer can be controlled in decade steps by the setting of the switches $S_1 - S_4$, which will be on one at a time. Thus, when S_1 is on, the input voltage range is the basic range of the DVM, that is, 1 V. The input voltage range is 1000 V when S_4 is on.

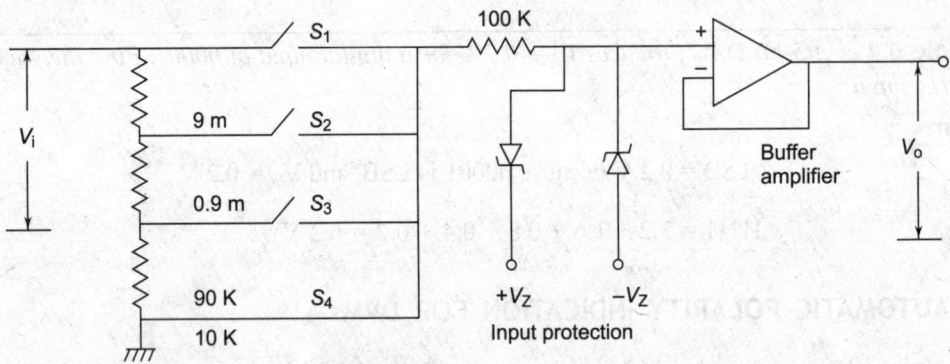

Fig. 5.35 *Switching Circuit for DVM*

For ensuring accurate division of V_i, it is necessary to have precision resistors, for example, metal-film type, low-temperature coefficient, and high-stability resistors for this application.

The switches $S_1 - S_4$ may be controlled manually or automatically. The latter offers advantages for the user, especially from the point of view of operating convenience.

5.23.1 Auto Ranging

Auto ranging involves the generation of suitable signals for controlling the switch matrix as in Fig. 5.35.

5.23.2 Typical Case

Let the DVM is set in its proper voltage range if the input to the basic DVM block is 0.1 to 1.2 V. This corresponds to a range of 0.5 msec to 6 msec or to the counts accumulated in this range, that is, between 100 and 1200. If the counts accumulated in any sample are outside this range, the DVM has to be switched to other ranges.

For example, if the counts are less than 100 at the end of a sample, the instrument should select the next lower range (that is, down-ranging), provided the instrument is not already in this range.

Similarly, if the counts are greater than 1200, the instrument should go to its next higher range (up-ranging), provided it is not already in this range. This range-changing operation has to continue until the instrument locks on to the proper range of measurement.

Since four ranges are involved in the present case, a maximum of four sampling intervals are necessary for the instrument to choose the proper range under the worst case. Therefore, the transfer of counts from the counters to auxiliary storage flip-flops has to be done only after the range switching is complete. Up-range or down-range signals have to be generated using the information of overflow as well as the state of the MSB in the decade counter as shown in Fig. 5.36.

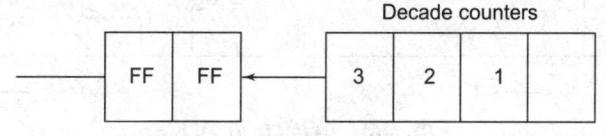

Fig. 5.36 *DVM Display*

The greater than 100 counts information is obtained by reading the state of the overflow flip-flop (first *T*-flip-flop at the output of the decade counter chain) designated as Q and *BCD* output of the *MSD* counter $Q\bar{A}\bar{B}C\bar{D} = 1$.

Whenever the above equation is satisfied, along with the information that the DVM is in its lower range (which is obtained from the switch driver of S_1), the point X changes from $1 \rightarrow 0$. This information resets the down-range flip-flop and the up–down counter counts up by one digit.

The less than 1200 counts information is also obtained similarly. In this case, the MSD counter should read 2, and overflow or Q must be in "1" state. This logic is generated by considering all the combinations followed by simplification. The following relationship is the final result:

$$Q(B + C)\bar{D} = 1$$

Whenever the equation is satisfied along with the information that the DVM is not in its highest range (which is obtained from the switch driver of S_4), the point Y changes from 1 to 0. This information resets the up-range flip-flop and up–down counter counts down by one digit.

5.24 ADC TERMINOLOGY

1. **Absolute accuracy:**   It is the maximum possible error in the full-scale digit reading (maximum voltage rating of the ADC) of the ADC compared to the true analog voltage value for that reading.

2. **Acquisition time:** It is the time taken by the S/H circuit to acquire the input signal within a stated accuracy. This also includes the settling time of the S/H amplifier, that is, the time taken by the capacitors and the amplifier so that the sampled value becomes equal to the actual input voltage value (operative time).

3. **Conversion time:** It is the time required for a complete measurement by an ADC. (This includes the sampling time, settling time, ramp generator and conversion time, all delays, and so on.)

4. **Droop rate:** When an S/H circuit using a capacitor for storage is in the "hold" mode, the information, that is, analog voltage will not remain constant during that whole period. The rate at which the output voltage changes is termed the droop rate.

5. **Feed through:** This is the undesirable signal leakage around switches or other devices expected to provide isolation (that is, when the switch is off $I = 0$. But it may not be zero in the case of CMOS or any other type of switch, but a slow leak.) It is expressed in nA.

6. **Glitch:** One bit may change earlier than the other due to non-uniform delays. So, the resulting number will be momentarily different. Due to unmatched switching delays from on to off and vice versa, in the various bits of a DAC, the resulting staircase waveform will have undesirable spikes. This is referred to as a glitch.

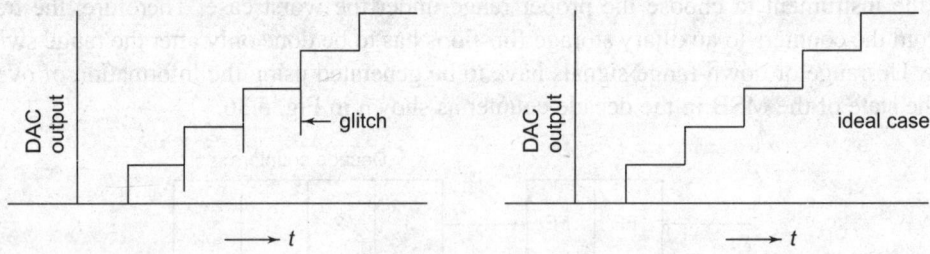

Fig. 5.37 *Glitch in ADCs*

7. **Monotonicity:** Suppose the input signal is continuously increasing. Then the output of ADC or DAC should be such that the output voltage value at any instant should not be less than the previous value or code. This is called monotonic behaviour. It requires that differential non-linearity should be less than 1 LSB. It is stated whether the DAC or ADC is monotonic. It is tested for continuously increasing ramp input signal. The output can be less than the previous value while in the case of a DAC the digital input is changing from 01111 to 10000, the rate of rise of ramp input should not be greater than the settling time, acquisition time of the DAC or ADC.

8. **Settling time:** For DAC, it is the time taken to settle for a full-scale code change, usually to within $\pm\dfrac{1}{2}$ LSB.

9. **Zero setting:** The zero level is set to zero Volts at the code corresponding to 0 V in a unipolar DAC.

$$\text{Aperture time of ADC} = \frac{\Delta V}{V}\,\frac{1}{2\pi f}$$

$$\frac{\Delta V}{V} = \text{resolution of ADC}$$

Example 5.5 *An analog voltage signal whose highest significant frequency is 1 kHz is to be digitally coded with a resolution of 0.01% covering the voltage range 0–10 V. To avoid loss of information, determine*

(a) The minimum sampling rate
(b) Minimum number of bits in the digital code
(c) Analog value of the least significant bit
(d) rms value of the quantization noise
(e) Aperture time required for ADCs (conversion time) sampling time
(f) Dynamic range of the A/D conversion in dB
(g) Suggest a suitable method of A/D conversion.

Solution

(a) According to sampling theorem, the frequency of sampling should be at least $2f$. Therefore, 2 kHz can be chosen although 5 kHz is needed in practice.

(b) Resolution should be 0.01%, that is, $\dfrac{1}{10,000} \times 100 = 0.01\%$. If we choose 14 bits, LSB =

$\dfrac{1}{2^{14}} \times 100$ $\dfrac{1}{2^n - 1} \times 100 \ (-1)$ can be neglected. $2^{10} = 1024$.

14 bits will give more than 10,000 counts.

(c) $2^{14} = 16,384$. Full-scale reading = 10 V

So, analog value of LSB = $\dfrac{10}{16,384} = 610.2 \ \mu V$

(d) $\dfrac{610.2}{2\sqrt{3}} = 176 \ \mu V$ (rms)

(e) $\dfrac{1}{16,384 \times 2\pi \times 10^3} = 9.74$ nsec $t_a = \dfrac{\Delta V}{V} \dfrac{1}{2\pi f}$; $\dfrac{\Delta V}{V} = $ resolution

Aperture time: the time taken for sampling to be done.

(f) $14 \times 6.02 = 84.3$ dB

(g) Potentiometric ADCs

Example 5.6   *A 4-bit DAC is used with reference voltage $V_R = 5 \ V$ and $R = 5 \ k\Omega$. What are the values of the output voltage V_o for the following digital inputs?*
(a) 1111   (b) 1001   (c) 0101   (d) 0001

$$V_o = - V_R \sum_{n=1}^{4} \dfrac{a_n}{2^n} \text{, } n \text{ should be from 1 to 4 and not 0 to 3 because } V_o \text{ cannot be greater than } V_{ref}.$$

Solution

(a) $$V_o = -5 \ V \left(\dfrac{1}{2} + \dfrac{1}{4} + \dfrac{1}{8} + \dfrac{1}{16} \right) = \dfrac{15}{16} \times (-50)$$

$$= 4.69 \ V$$

(b) $$V_o = -5 \ V \left(\dfrac{1}{2} + \dfrac{1}{16} \right) = -2.81 \ V$$

(c)
$$V_0 = -5 \text{ V} \left(\frac{1}{4} + \frac{1}{16} \right) = \frac{5}{16} (-5 \text{ V}) = 1.56 \text{ V}$$

(d)
$$V_0 = -5 \text{ V} \left(\frac{1}{16} \right) = -0.312 \text{ V}$$

Example 5.7 *For a dual-slope ADC, it is often desirable to make the integration period exactly equal to one period of the 50 Hz AC power line to reduce the effects of power line frequency noise on the conversion accuracy. If a $3\frac{1}{2}$ digit BCD counter (full-scale count 1999) is used and the signal is integrated until the two most significant bits of the counter are 1, what value of f_c (clock frequency) should be used?*

Solution

When the two MSBs of the counter are first 1, the count is 1800. Thus n_i (total counts) = 1800

The integrator period is $\dfrac{n_i}{f_c}$, and it is desired that $\dfrac{n_i}{f_c} = \dfrac{1}{50}$ sec.

So,
$$f_c = \frac{n_i}{f_c} = \frac{1}{50}$$

$$f_c = 50 \, n_i = 50 \times 1800 = 90000 = 90 \text{ kHz}$$

5.25 COMPARISON OF ADCs

5.25.1 Advantages of V/F Converter (Integrating Type)

1. It can handle input signals with both polarites—positive as well as negative.

2. The sensitivity of the conversion can be controlled easily. Sensitivity is defined as $\dfrac{f}{e_i}$ in kHz/V.

   Since $\quad f = \dfrac{e_{in}}{ER_{in}C_f}, \dfrac{f}{e_i} = \text{sensitivity} = \dfrac{1}{ER_{in}C_f}$.

   So, the sensitivity depends on R_i C_f and time constant, and R_i is input resistance and C_f the feedback capacitance in the integrator.

3. *V/f* characteristic is linear (Fig. 5.38).

4. The circuit can handle input signals of mV range directly. Therefore, the transducer output can be directly connected to the ADC. (It is a few hundred mV. If V_i is 1 mV of this order, the integrator output cannot be distinguished from the noise.)

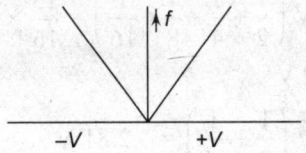

Fig. 5.38 *V–f Characteristic of V/F Converter*

5.25.2 Disadvantages

1. Accuracy is limited. If the accuracy of the instrument were 0.01%, the linearity of the converter should be better than 0.001%.
2. *R* and *C* values and their temperature coefficients affect the ramp waveform.

5.25.3 V/T Converter

5.25.3.1 Advantages

1. It is simple and easy to realize.
2. Analog signal is converted to time information. So, it can be transferred from one place to the other.

5.25.3.2 Disadvantages

1. Conversion process is slow.
2. Accuracy better than 0.01% cannot be achieved.
3. High stability components are required or otherwise slope will not be linear.
4. Noise immunity is less. Hum in the power supply will affect the performance of the circuit and output. Hum-rejection filters will have to be employed.

5.25.4 Simultaneous ADC

5.25.4.1 Advantages

1. It is faster.
2. Simple circuit.

5.25.4.2 Disadvantage

Large number of comparators are required.

5.25.5 Successive Approximation ADC

Counter ramp-conversion techniques employ DACs in their feedback path. So, the accuracy that can be achieved is limited by the accuracy of the DAC.

5.26 COMMERCIAL ADCs

Table 5.4 AD: Analog Devices Incorporated

No.	Principle	No. of bits	Conversion	Input
AD 570	Successive approximation	8	25 μsec	±5 V 0–10 V
AD 574	Successive approximation	12	25 μsec	±5 V 0–10 V

AD 7555	Integrating type	$5\frac{1}{2}$ digit	200 μsec	±5 V
				CMOS
ADC 60-08 Burr Brown	Successive approximation	8 bit	0.88 μsec	±10 V ±5 V High speed
ADCH Data Intersil (H: High speed)	Successive approximation	12 bit	20 μsec	High speed

Table 5.5 ADCs Comparison

Type	Accuracy	Input Z	Series-mode-rejection (SMR)	Speed
1. Staircase ramp (compensation)	High	Not constant (Low to higher)	Low	Medium
2. Successive approximation	High	Not constant Low to high	Low	Higher
3. Variable frequency	Medium	Limited	High	Medium
4. Dual-slope integration	High	Very high	High	Medium
5. Delta pulse modulation	High	High	High	Medium

5.27 SERIES-MODE REJECTION

In some cases, ADCs should convert the DC signal to digital form. The input is DC. But in some cases, the DC may not be pure. It may contain AC components.

So, before the input is given to ADC, it should be filtered. But a filter may slowdown the speed of the ADC. So, another solution is to integrate the input over a long period of time. Then the average AC signal will be zero, and a good series-mode rejection will be obtained.

Series-mode rejection is nothing but rejecting AC components in DC signals.

5.28 SIMULTANEOUS A/D CONVERSION USING TUNNEL DIODES

The switching time required for tunnel diodes to change state is 1 nsec and responds to a pulse of very small energy. A pulse of 10^{-15} joules can take the tunnel diode past a critical point, and initiates transition. For a transistor, 10 times higher energy of pulse is required, and the speed is less. A tunnel diode has two stable states, whereas two transistors are required to get the same.

A 2-bit tunnel diode ADC with its $V - I$ circuit is as shown in Fig. 5.39. The two tunnel diodes TD_1 and TD_2 are chosen such that $IP_1 < IP_2$, $IV_1 > IV_2$, and $V_{F_2} \simeq 2V_{F_1}$.

The peak forward voltage V_F is almost the same for all tunnel diodes made from the same semiconductor material. TD_1 and TD_2 could be of different semiconductors, germanium and gallium arsenide. The analog voltage to be measured V_i is connected in series with the bias voltage V.

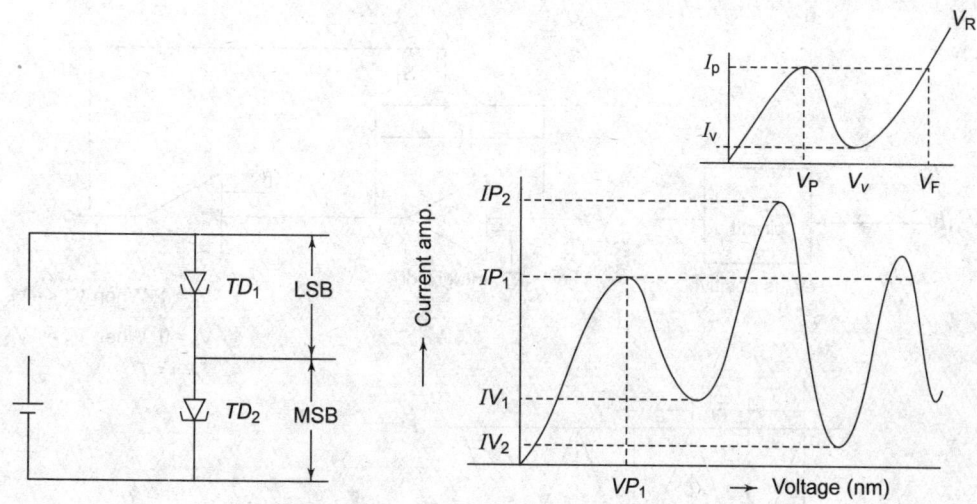

Fig. 5.39 *Tunnel Diode ADC with Waveforms*

Table 5.6

Input voltage	Tunnel diode states		Digital output	
	TD_1	TD_2		
	Low	Low	0	0
	Low	High	0	1
			1	0
	High	High	1	1

5.29 V/F CONVERSION (INTEGRATING TYPE ADC)

The input voltage to be converted into digital form is made proportional to the frequency of a clock waveform. The frequency measured is an indication of the analog input. Thus, the analog signal can be converted into a digital form.

An ADC can be built using a V/F converter. The principle of operation can be explained with the help of the schematic diagram shown in Fig. 5.40.

A pulse generator injects an opposing voltage at the summing junction of the op-amp to cancel the input.

$$T_d = \text{discharge time}$$

The analog input voltage to be converted to digital voltage is sampled and held at a level V_A. This voltage is applied to an integrator, which is followed by a comparator. The other input to the comparator is a reference voltage $-V_R$. ($-V_R$ because the input V_A has a positive voltage. So the integrator output is negative.) Initially, the switch S bridging the integrator is open. While integrating, the output of the integrator decreases linearly with time. If $\tau = RC$,

$$V_o = -\frac{V_A t}{\tau}$$

When $V_o \geq -V_R$ after a time $t = T$, the comparator changes state. V_c becomes positive for a small time T_d because $-V_o > -V_R$ and inverted.

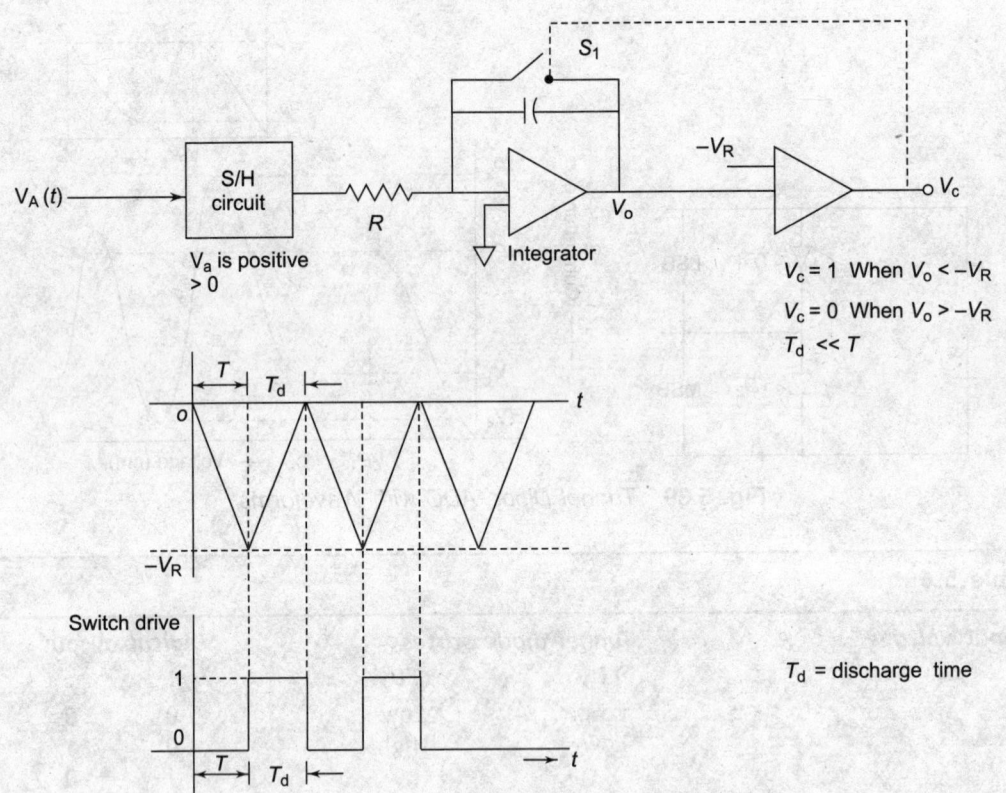

Fig. 5.40 *V/F Conversion Technique*

During this time, the switch S closes, thereby discharging the capacitor C and returning the integrator output $V_o \simeq 0$ V. In actual system, a monostable multi may be used to discharge the capacitor completely.

After time interval T_d, the comparator voltage drops to "0" state. Switch S will open and V_o starts decreasing. The counter is a reversible one. Another comparator set is used for negative input voltages. If $T_d \ll T$, the integration time,

$$f = \frac{1}{T + T_d} \simeq \frac{1}{T}$$

$$V_o = -\frac{V_A t}{\tau} \quad \text{when } t = T, \ V_o = -V_R$$

So, $$V_R = \frac{V_A T}{\tau}; \quad \text{therefore, } T = \frac{V_R \tau}{V_A}$$

or $$\boxed{f = \frac{V_A}{V_R \tau}}$$

Thus, frequency of the clock waveform f is proportional to analog voltage V_A.

In an ADC, using V/F converter, the circuit of the latter provides the clock pulses to a counter through an AND gate. The second input to the AND gate is the sampling voltage V_s, which holds the logic level for a fixed time T_H. When $V_s = 1$, the analog voltage V_A is held constant. The AND gate is enabled for time T_H, and the counter reading will equal the number of cycles executed by the V/F converter output in the specified time interval.

5.30 S/H AMPLIFIERS IN ADCs

S/H amplifiers and ADCs are operated synchronously. The ADC status indicates to the S/H amplifier when to sample and when to hold. The schematic diagram is shown in Fig. 5.41. The input to the timing circuit is the sampling pulse train. The timing circuit provides all the clock pulses needed for the ADC to convert the analog sample into an n-bit output signal. The timing circuit also generates a timing waveform called the EOC waveform, which tells the S/H circuit when to sample and when to hold. The sampling pulse train is applied to the ADC and not the S/H amplifier. The control employed by the S/H circuit is the EOC waveform generated by the ADC. So, the S/H circuit holds the analog voltage constant till the conversion is completed in the ADC circuit. When the converted value is being displayed in the digital read-out, the S/H output is allowed to change.

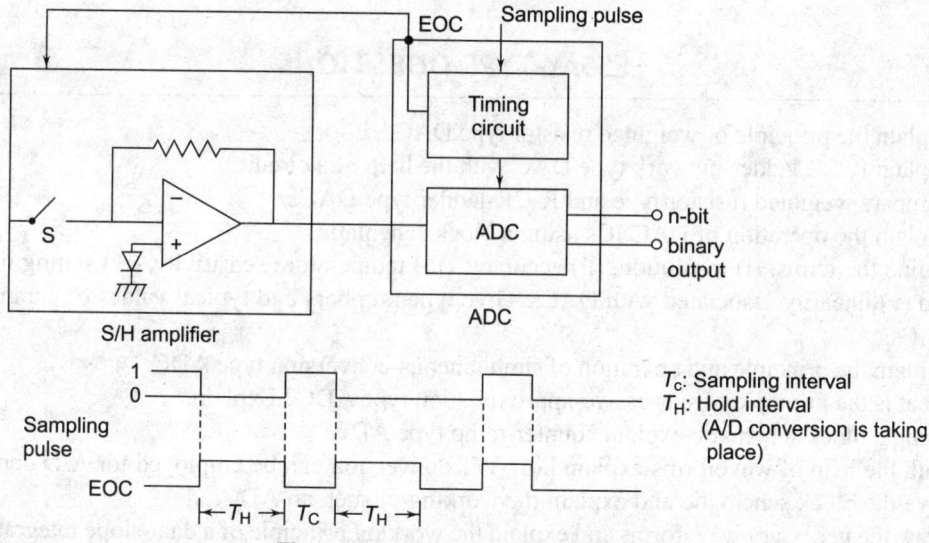

Fig. 5.41 *S/H Amplifier in ADCs*

T_c: Sampling intervals
T_H: Hold interval (A/D conversion is taking place)

SUMMARY

- There are two types of DACs namely: (i) weighted resistor type and (ii) R–2R ladder network type. In the weighted resistor type, different values of resistors are used. Therefore, due to variation in the precision of resistors, there can be more error in voltage conversion. This is reduced in R–2R ladder network as only two types of resistor values are used. But in this type, more resistors are required.

- DACs are also used in some ADCs. DACs are used where the output display has to be in analog form. For processing of signals, they are converted to digital form by using ADCs and the digital output is converted back to analog form by DACs. DACs are used in biomedical, aerospace, telemetry, and instrumentation applications.
- ADCs are classified broadly as
  1. Direct conversion type
  2. Indirect conversion type
- In the direct type, there are three types of ADCs: (a) simultaneous or flash type ADCs, (b) counter-ramp conversion technique, and (c) successive approximation-type ADC.
- In the indirect-type: (i) V/T conversion and (ii) V/F conversion methods are used. The dual-slope integrating-type ADC employs the V/F conversion method.
- Among all ADCs, the flash type or simultaneous conversion type is the fastest. In dual-slope integrating type, the conversion result is independent of the tolerance of the passive components, as charging and discharging is done with the same time constant RC, so that error due to tolerance of passive components gets cancelled.
- Conversion time, accuracy, input impedance, acquisition time, droop rate, feed through, and monotonicity are parameters of ADCs and DACs.

ESSAY-TYPE QUESTIONS

1. Explain the principle of weighted resistor type DAC.
2. Explain R–2R ladder-network type DAC with the help of a circuit.
3. Compare weighted resistor type and R–2R-ladder type DACs.
4. Explain the operation of DAC ICs using a block schematic.
5. Define the terms: (i) resolution, (ii) accuracy, (iii) temperature sensitivity, (iv) setting line time, and (v) linearity associated with DACs. Give type numbers and typical values of parameters of DACs.
6. Explain the principle and operation of simultaneous-conversion type ADC.
7. What is the principle of successive-approximation type ADC? Explain.
8. Using a block schematic explain counter-ramp type ADC.
9. With the help of waveforms explain how V/T conversion can be employed for A/D conversion. Give the block schematic and explain the working of such an ADC.
10. Draw the necessary waveforms and explain the working principle of a dual-slope integrating type ADC. Describe the necessary expression for accumulated counts.
11. Compare different types of ADCs in all respects.
12. Define the terms: (i) stability, (ii) conversion time, (iii) accuracy, (iv) acquisition time, (v) glitch, (vi) monotonicity, (vi) droop rate, (vii) feed through, and (viii) settling time associated with ADCs.
13. Give the type numbers and typical values of the parameters of ADC ICs.
14. Explain the various applications of ADCs.
15. Compare direct conversion and indirect conversion techniques of ADCs.

FILL IN THE BLANKS

1. The disadvantage of a weighted resistor-network type DAC is _____.
2. The disadvantage of an R–2R ladder-network type DAC is _____.

3. To account for variations in logic levels, DACs employ _____.
4. For an *n*-bit DAC, the expression for resolution is _____.
5. The range of settling time for a 12-bit DAC is _____.
6. The slew rate in DACs is defined as _____.
7. Between counter-ramp type and successive-approximation type ADCs, the one having lesser conversion time for a given input is _____.
8. The major advantage of a dual-slope-integrating type ADC is _____.
9. The number of comparators required in the case of a simultaneous-conversion type ADC for *n*-bit conversion is _____.

ANSWERS

1. Different values of resistors have to be used.
2. Greater number of resistors have to be used for a given bit compared to the weighted resistor-network type DAC.
3. Level amplifiers
4. $\dfrac{1}{2^n - 1} \times 100\%$
5. 1 μsec to 20 μsec
6. Maximum rate of change of output voltage.
7. Successive-approximation type.
8. The output is independent of the tolerance of the passive components R and C.
9. $(2^n - 1)$

UNSOLVED PROBLEMS

1. What is the percentage resolution of a 4-bit DAC given that the maximum number that can be represented using 4 bits is 15.
2. A 4-bit DAC has a maximum precision supply voltage of 15 V. What is the voltage change for each LSB.
3. For the DAC mentioned above, what is the analog output voltage for a digital input of 1100.
4. For a digital-ramp type ADC, $f_c = 2$ MHz, $V_T = 2$ mV, DAC has full-scale output of 12.70 V with 8-bit input. Determine the resolution of the ADC.
5. For the above problem, for an input of 6.24 V, determine the conversion time and the digital equivalent output obtained.
6. A 4-bit DAC produces output voltage of 0.1 V for a digital input of 0001 V. Find the value of V_o for maximum input.
7. An analog signal is to be digitally coded with a resolution of 0.01%. The highest significant frequency is 1.6 kHz. The voltage range is 0–8 V. Determine
   (a) The minimum sampling rate
   (b) Minimum number of bits in the digital code
   (c) Analog value of the least significant bit
   (d) rms value of quantization noise

    (e) Aperture time required for ADC sampling time

    (f) Dynamic range of ADC in dB.

8. A 4-bit DAC is used with a reference voltage $V_R = 8$ V and $R = 4.7$ kΩ. If the digital inputs are as given, determine the equivalent analog output voltage.

9. Dual-slope ADC is designed such that the integration period is exactly equal to one period of 50 Hz AC power-line frequency. A $4\frac{1}{2}$-digit *BCD* counter with a full-scale count of 19999 is used. The signal is integrated until the two most significant bits of the counter are 1. What value of clock frequency must be used?

10. An 8-bit ADC can accept voltages from +0 V to +12 V. What is the minimum value of input voltage to cause digital output change of 1 LSB. If the applied input voltage is 4.8 V, what is the digital output?

11. In the above Problem 10, if the digital output is 1111 1111, what is the analog input voltage?

12. The 8-bit ADC of Problem 10 has an offset error of $\pm\frac{1}{2}$ LSB. What should be the value of analog voltage to get full-scale digital output of all 1s.

13. The ADC of Problem 9 has a gain error of 0.5% of FSR. What should be the value of V_i to get all 1s in digital output.

14. Consider a dual-slope-integrating type ADC. For the integrator, $R_{int} = 100$ kΩ and $C_{int} = 0.47$ μF. Calculate T_2, the time to discharge the capacitor if $V_{in} = \pm150$ mV.

14. For the Problem of 14, if $V_{in} = \pm250$ mV, what is the Time T_2 for discharging capacitor. If $V_{in} = +150$ mV, find the digital output.

16. The clock frequency of an 8-bit successive-approximation type ADC is 2 MHz. Determine its conversion time.

17. In a dual-slope ADC, a $3\frac{1}{2}$-digit *BCD* counter is used and the signal is integrated until the two most significant bits of the counter are 1. What is the decimal count?

Voltage Regulators

Objectives:

In this chapter...

- *The principle of voltage regulator circuit is explained.*
- *Different types of voltage regulator ICs are given.*
- *Voltage doubler, quadrupler, and voltage multiplexer circuits are described.*
- *The reader is expected to know the principles, types, specifications, and circuits of voltage regulator ICs and other configurations like voltage doubler and quadrupler.*

6.1 INTRODUCTION

IC voltage regulators give constant output DC voltage irrespective of variations in DC input voltage. Electronic voltage regulator circuits like series voltage regulator circuits are made in the form of integrated circuits. The working principle of the internal circuit is similar to electronic circuits. In IC form, different features are given to the internal schematics. If the input is AC, it has to be rectified, filtered, and then applied to the IC within voltage limits. Depending on the specifications of the IC, the circuit will be able to maintain constant output voltage irrespective of the current drawn by the load. The various parameters that affect the output voltage are as follows:

1. Input voltage
2. Temperature
3. Load current

If the input voltage is 230 V AC, and a constant DC output voltage is required, a stepdown transformer must be used and rectifier and filter circuits are to be employed. The block diagrams are shown in Figs. 6.1 and 6.2.

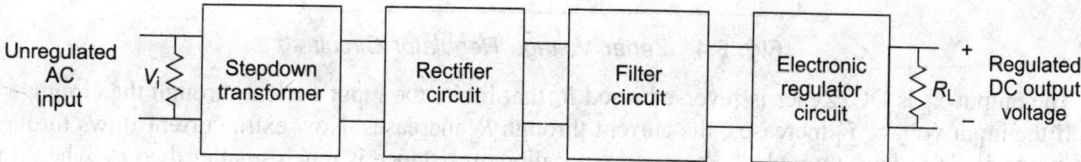

Fig. 6.1 *Block Diagram of Voltage Regulator Circuit—AC Input*

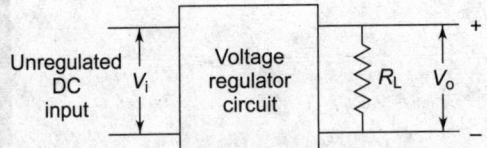

Fig. 6.2 *Block Diagram of Voltage Regulator Circuit—DC Input*

Voltage stabilizer circuits provide constant AC within a range from the AC input itself. The circuit employed for **AC voltage stabilizers** is different.

The commonly used voltage regulator ICs are given below:

1. μA 723
2. LM 309
3. LM 105
4. CA 3085 A
5. 78XX series: 7805, 7806, 7808, 7812, 7815—three terminal-fixed positive voltage regulator ICs.
6. 79XX series: 7905, 7906, 7908, 7912, 7915—three terminal-fixed negative voltage regulator ICs.

6.2 VOLTAGE REGULATORS

The different types of voltage regulator circuits without using ICs are as follows:

1. Zener voltage regulator circuit
2. Shunt voltage regulator circuit
3. Series voltage regulator circuit

6.2.1 Zener Voltage Regulator Circuit

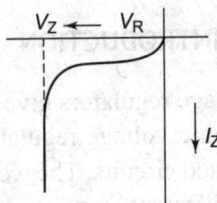

A simple circuit without using any transistor is with a zener diode. In the reverse characteristic of the zener diode, the voltage in the breakdown region remains constant irrespective of the current that is flowing through the diode (Fig. 6.3). Therefore, the zener can be used as a voltage regulator in this region. If the output voltage is taken across the zener, even if the input voltage increases, the output voltage remains constant. The circuit is shown in Fig. 6.4.

Fig. 6.3 *Reverse Characteristic of Zener Diode*

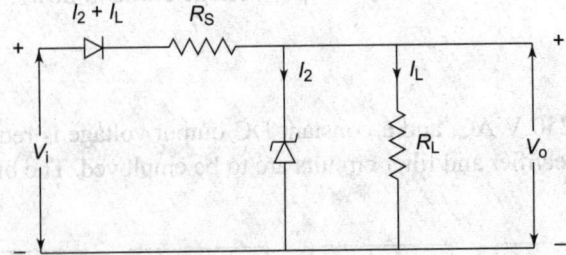

Fig. 6.4 *Zener Voltage Regulator Circuit*

The output V_o is DC. Zener is reverse biased R_s that limits the input current through the circuit.

If the input voltage V_i increases, the current through R_s increases. Now extra current flows through the zener diode and not through R_L, because zener diode resistance is much smaller than R_L when it is conducting, because I_L remains constant and so V_o remains constant.

6.2.2.1 Limitations of the Circuit

The output voltage remains constant only when the input voltage is sufficiently large so that the voltage across the zener is V_Z. Again, there is a limit to the minimum current that can pass through the zener. If V_i is increased enormously, I_Z increases and hence breakdown will occur. Therefore, voltage regulation is maintained only between these limits, minimum current and the maximum permissible current through the zener diode. Typical values are from 10 mA to 1 A. If I_L is greater than this value, the voltage regulation is lost.

6.2.2 Shunt Voltage Regulator Circuit

The shunt regulator uses a transistor to amplify the zener diode current and thus extends the zener's current range by a factor equal to transistor h_{FE}.

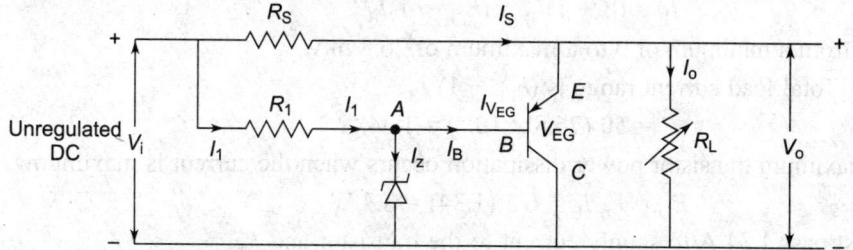

Fig. 6.5 *Shunt Voltage Regulator Circuit*

Zener current I_Z passes through R_1.

Nominal output voltage of the regulator circuit = V_Z/V_{EB}.

The current that gets branched as I_B is amplified by the transistor. Therefore, the total current $I_o = (B + 1)I_B$ flows through the load resistance R_L. Thus, for a small current through the zener, large current flows through R_L and the voltage remains constant. In other words, for large current through R_L, V_o remains constant. Voltage V_o does not change with current I_o within certain limits.

Example 6.1   *For the shunt regulation shown in Fig. 6.5, determine:*
1. The nominal voltage
2. Value of R_1
3. Load current range
4. Maximum transistor power dissipation
5. The value of R_s and its power dissipation.

Solution

Zener diode 6.3 V, 200 mW, requires 5.0 mA minimum current.
Transistor: $V_{EB} = 0.2$ V, $h_{FE} = 49$, $I_{CBO} = 0$.

1. The nominal output voltage is the sum of the transistor V_{EB} and zener voltage.
$$0.2 + 6.3 = 6.5$$

2. R_1 must supply 5 mA to the zener diode.

So,
$$R_1 = \frac{8\,V - 6.3\,V}{5 \times 10^{-3}} = \frac{1.7\,V}{5 \times 10^{-3}} = 340\ \Omega$$

3. The maximum allowable zener current is

$$\frac{\text{Power rating}}{\text{Voltage rating}} = \frac{0.2}{6.3} = 31.8 \text{ mA}$$

The load current range is the difference between minimum and maximum current through the shunt path provided by the transistor. At junction A, we can write, $I_B = I_Z - I_1$ (Fig. 6.5).

I_1 is constant at 5 mA

So, $I_B = (I_Z - I_1)$, I_1 is constant at 5 mA

Therefore, $I_{B(\text{Min.})} = (5 \times 10^{-3}) - (5 \times 10^{-3}) = 0$

$$I_{B(\text{Max.})} = (I_{Z\text{Max.}} - I_2); \ I_{Z\text{Max.}} = 31.8 \text{ mA};$$
$$= (31.8 \times 10^{-3}) - (5 \times 10^{-3}) = 26.8 \text{ mA.}$$

The transistor emitter current $I_E = I_B + I_C$

$$I_C = \beta I_B = h_{FE} I_B$$

So, $I_E = (\beta + 1)I_B = (h_{FE} + 1)I_B$

I_B ranges from a minimum of 0 to a maximum of 26.8 mA.

Therefore, Total load current range is $(h_{FE} + 1) I_B$

$$= 50 (26.8 \times 10^{-3}) = 1.34 \text{ A}$$

4. The maximum transistor power dissipation occurs when the current is maximum. $I_E \simeq I_C$

$$P_D = V_o I_E = 6.5 (1.34) = 8.7 \text{ W}$$

5. R_s must pass 1.34 A to supply current to the transistor and R_L.

Power dissipation $R_s = \dfrac{V_i - V_o}{1.34} = \dfrac{8 - 6.5}{1.34} = 1.12 \ \Omega$

(i) The competent by $R_s = I_s^2 R_s = (1.34)^2 (1.125) = 2.4 \text{ W}$

6.3 UNREGULATED POWER SUPPLY

An unregulated power supply consists of a transformer, a rectifier, and a filter. For such a circuit, regulation will be very poor. Load means load current. No load means no load current or open circuit. Full load means full load current or short circuit. As the load varies, the output voltage must remain constant for good regulation. But this will not be so for unregulated power supply circuit. The shortcomings of unregulated power supply current are as follows:

1. Poor regulation
2. DC output varies directly as the AC input varies
3. In simple rectifiers and filter circuits, the DC output voltage varies with temperature also if semiconductor devices are used.

An electronic feedback circuit is used in conjunction with an unregulated power supply to overcome the above three shortcomings. Such a system is called a regulated power supply.

6.3.1 Stabilization

The output voltage depends upon the following factors in a power supply:

1. Load voltage V_i
2. Load current I_L
3. Temperature

Therefore, change in the output voltage ΔV_o can be expressed as

$$\Delta V_o = \frac{\partial V_o}{\partial V_i} \Delta V_i + \frac{\partial V_o}{\partial I_c} \Delta I + \frac{\partial V_o}{\partial T} \Delta T$$

$$\Delta V_1 = \Delta V_i + R_o \, \Delta I_L + S_T \, \Delta T$$

where the three coefficients are defined as

Stability factor $\qquad S_V = \dfrac{\Delta V_o}{\Delta V_i}\bigg|_{\Delta I_L - 0, \, \Delta T = 0}$

This should be as small as possible, ideally 0 because V_o should not change even if V_i changes.

Output resistance $\qquad r_0 = \dfrac{\Delta V_o}{\Delta I_L}\bigg|_{\Delta V_i = 0, \, \Delta T = 0}$

Temperature coefficient $S_T = \dfrac{\Delta V_o}{\Delta T}\bigg|_{\Delta V_i = 0, \, \Delta I_c = 0}$

The smaller the values of the three coefficients, the better is the voltage regulator circuit.

6.3.2 Series Voltage Regulation

The voltage regulation (i.e. change in the output voltage as load voltage varies or output voltage varies) can be improved, if a large part of the increase in input voltage appears across the control transistor, so that output voltage tries to remain constant. That is, increase in V_i results in increased V_{CE} so that output almost remains constant. But when the input increases, there may be some increase in the output, but to a very small extent. This increase in output acts to bias to control transistor. This additional bias causes an increase in collector-to-emitter voltage which will compensate for the increased input.

If the change in output is more amplified before being applied to the control transistor, better stabilization will result.

6.3.3 Series Voltage Regulator Circuit

Q_1 is the series-pass element on the series regulator. Q_2 acts as the difference amplifier. D is the reference zener diode. A fraction of the output voltage V_o (b is a fraction, which is taken across R_2 and the potentiometer) is compared with the reference voltage V_R. The difference ($bV_o - V_R$) is amplified by the transistor Q_2. Because the emitter of Q_2 is not at ground potential, there is constant voltage V_R. Therefore, the net voltage to the base-emitter of the transistor Q_2 is ($bV_o - V_R$). As V_o increases, ($bV_o - V_R$) increases. When input voltage increases by ΔV_i, the base-emitter voltage of Q_2 increases. Therefore, collector current of Q_2 increases and hence there will be large current change in R_3. Thus, all the change in V_i will appear across R_3 itself. V_{BE} of the transistor Q_1 is small.

Therefore, the drop across $R_3 = V_{CB}$ of $Q_1 \simeq V_{CE}$ of Q_1 since V_{BE} is small and so can be neglected. Therefore, the increase in the voltage appears essentially across Q_1 only. This type of circuit takes care of the increase in input voltages only. If the input decreases, output will also decrease. (If the output were to remain constant at a specified value, even when V_i decreases, buck and boost should be there. The tapping of a transformer should be changed by a relay when V_i changes). If input voltage V_i decreases, output voltage V_o must be increased to maintain at the desired constant value. If V_i increases, V_o must be reduced to maintain the same constant value again. This is known as buck and boost.

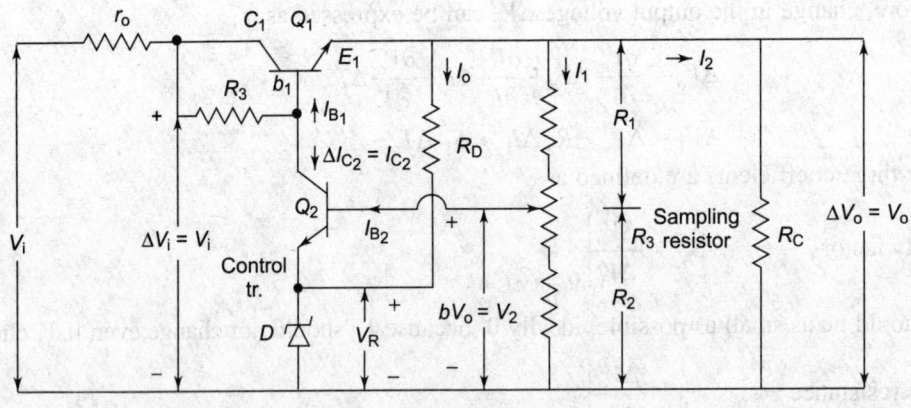

Fig. 6.6

r_o is the output resistance of the unregulated power supply which precedes the regulator circuit. r_o is the output resistance of the rectifier, filter circuit or it can be taken as the resistance of the DC supply in the laboratory experiment.

The expression for S_V (stability factor) $= \dfrac{\Delta V_o}{\Delta V_i}$

$$= \left[\frac{R_1 + R_2}{R_2}\right] \frac{(R_1 h R_2) + h_{ie_2} + (1 + h_{fe_2}) R_Z}{h_{fe_2} R_3}$$

R_Z = zener diode resistance (typical area)

$$R_o \simeq \frac{r_o + \dfrac{R_3 + h_{ie_1}}{1 + h_{fe_1}}}{1 + Gm(R_3 + r_o)}; \quad G_m = \frac{\Delta I_{c2}}{\Delta V_o}$$

The preset potential or trim potential R_3 in the circuit is called as sampling resistance because it controls or samples the amount of feedback.

Preregulator: It provides constant current to the collector of the DC amplifier and to the base of control element.

If R_3 is increased, the quantity $\dfrac{R_3 + h_{ie_1}}{1 + h_{fe}}$ does not, but this term is very small, since it is being divided by h_{fe}.

The value of the stability factor S_v of a voltage regulator should be small. S_v can be improved if R_3 is increased (from the general expression) since $R_3 \simeq \dfrac{(V_c - V_o)}{I}$. We can increase R_3 by decreasing I through R_3. The current I through R_3 can be decreased by using a Darlington pair instead of Q_1. To get even better values of S_v, R_3 is replaced by a constant current source circuit so that R_3 tends to ∞. This constant current-source circuit is often called a transistor preregulator. V_i is the maximum value of input that can be given.

Short-circuit overload protection: Overload means overload current (or short circuit). A power supply must be protected from further damage through overload. In a simple circuit, protection is provided by using a fuse so that when current of the rated value flows, the fuse wire will blow off, thus, protecting the components. This fuse wire is provided before r_o. Another method of protecting the circuit is by using diodes. The circuit is shown in Fig. 6.7.

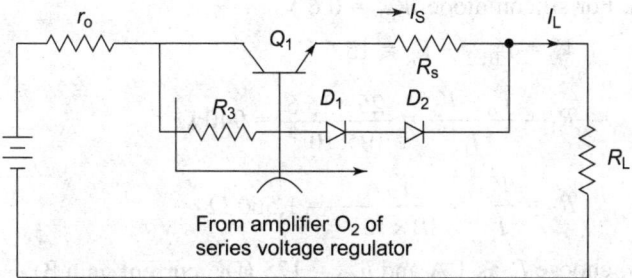

Fig. 6.7 *Short-Circuit Protection Using Diodes*

Zener diodes can also be employed, but they are relatively costly.

The diodes D_1 and D_2 will start conducting only when the voltage drop across R_s increases to the cut in voltage of both the diodes D_1 and D_2. In the case of a short circuit, the current I_s will increase up to a limiting point determined by

$$I_s = \frac{V_{v_1} + V_{v_2} - V_{BE1}}{R_s}$$

When the output is short circuited, the collector current of Q_2 will be high. $I_s R_s$ will also be large. Therefore, the two diodes D_1 and D_2 start conducting. The large collector current of Q_2 passes through the diodes D_1 and D_2 and not through the transistor Q_1. Transistor Q_1 will be safe. D_1 and D_2 will be generally silicon diodes because cut in voltage is 0.6 V. So $I_s R_s$ drop can be large.

Example 6.2 *Design a series-regulated power supply to provide a nominal output voltage of 25 V and supply load current $I_L \leq 1$ A. The unregulated power supply has the following specifications $V_i = 50 \pm 5$ V, and $r_o = 10\ \Omega$. Given $R_Z = 12\ \Omega$ at $I_Z = 10$ mA; at $I_{C2} = 10$ mA, $\beta = 220$, $h_{ie_2} = 800\ \Omega$, $h_{fe_2} = 200$, $I_1 = 10$ mA.*

Solution

The reference diodes are chosen such that $V_R \simeq \dfrac{V_o}{2}$.

$\dfrac{V_o}{2} = 12.5$ V. Therefore, two zener diodes with breakdown voltage of 7.5 V in series may be connected.

$$R_z = 12\ \Omega \text{ at } I_Z = 20 \text{ mA}$$

Choose $\qquad I_{C2} \simeq I_{E2} = 10$ mA.

At $I_{C2} = 10$ mA, the h-parameter for the transistor are measured as

$$\beta = 220, \ h_{ie_2} = 800\ \Omega. \ h_{fe_2} = 200$$

Choose $\qquad I_D = 10$ mA so that $I_Z = I_{D1} + I_{D2} = 20$ mA

$$R_D = \frac{V_o - V_R}{I_D} = \frac{25 - 15}{10} = 1 \text{ k}\Omega$$

$$I_{B2} = \frac{I_{c2}}{\beta} = \frac{10 \text{ mA}}{220} = 45 \text{ }\mu\text{A}$$

Choose I_1 as 10 mA. For silicon diode, $V_{BE} = 0.6$ V.

$$V_2 = V_{BE2} + V_R = 15.6 \text{ V}$$

So,

$$R_1 = \frac{V_o - V_2}{I_1} = \frac{25 - 15.6}{10 \times 10^{-3}} = 940 \text{ }\Omega$$

$$R_2 = \frac{V_2}{I_1} = \frac{15.6}{10 \times 10^{-3}} = 1{,}560 \text{ }\Omega$$

For the transistor Q_1, choose I_L as 1 A and $h_{fe1} = 125$ (DC current gain β)

$$I_{B1} = \frac{I_L + I_1 + I_D}{h_{fe1}(\beta)} \text{ since } I_{C1} \simeq I_{F1} = I_L + I_1 + I_n$$

$$= \frac{1000 + 10 + 10}{125} = 8 \text{ mA}$$

The circuit through resistor R_3 is $I = I_{B1} + I_{C2} = 3 + 10 = 18$ mA. The value of R_3 corresponding to $V_i = 45$ V and $I_L = 1$ A is (because these are given in the problem), $V_i = 50 \pm 5$, 45 V):

$$R_3 = \frac{V_i - (V_{BE1} + V_o)}{V_i - I} = \frac{50 - 25.6}{18 \times 10^{-3}} = 1{,}360 \text{ }\Omega$$

6.4 VOLTAGE REGULATORS: TERMINOLOGY

A voltage regulator is a circuit that maintains constant output voltage irrespective of the changes in the input voltage or the current.

Stabilizer: If the input is AC and output is also AC, such a circuit is called as voltage stabilizer circuit.

Load Regulation: It is defined as the percentage change in regulated output voltage for a change in load current from the minimum to the maximum value.

$$E_1 = \text{output voltage when } I_L \text{ is minimum (rated value)}$$
$$E_2 = \text{output voltage when } I_L \text{ is maximum}$$

Percentage load regulation $= \dfrac{E_1 - E_2}{E_1} \times 100\%$

$$E_1 > E_2$$

This value should be small.

Line regulation: It is the per cent change in V_o for a change in V_i

Percentage line regulation $= \dfrac{\Delta V_o}{\Delta V_i} \times 100\%$

In the ideal case $\Delta V_o = 0$ when V_o remains constant. This value should be minimum.

Load regulation is with respect to change in I_L.

Line regulation is with respect to change in V_i.

Ripple rejection: It is the ratio of peak-to-peak output ripple voltage to the peak-to-peak input ripple voltage.

$$\frac{V'_{o(p-p)}}{V'_{i(p-p)}}$$

V'_o = output ripple voltage

V'_i = input ripple voltage

Standby current drain: It is the current drawn by the regulator circuit when $I_L = 0$ (current drawn by the circuit components only and not by the load).

Short-circuit current limit: The output current of the regulator (I_L) when the output terminals are shorted.

Sense voltage: It is the voltage between current sense and current limit terminals.

Temperature stability or average temperature coefficient: Change in V_o per degree Celsius change in temperature (mV/ °C).

Basic regulator circuit:

A monolithic voltage regulator circuit mainly consists of three parts:

1. Reference voltage circuit
2. Error amplifier
3. Series-pass element

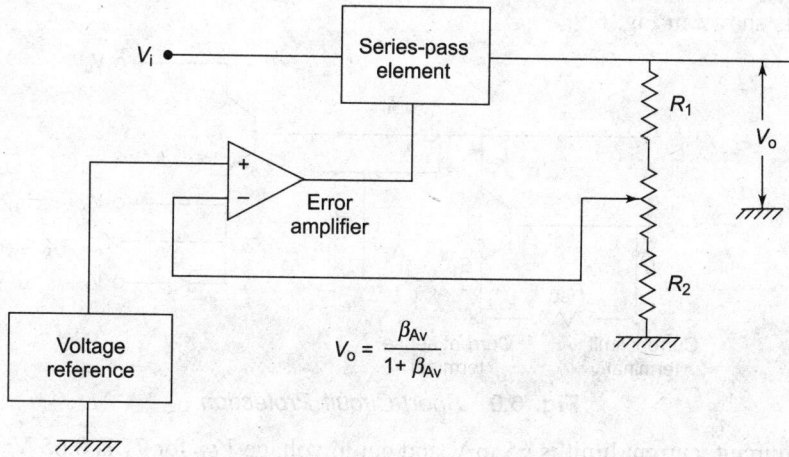

Fig. 6.8 *Basic IC Voltage Regulator Circuit*

$$V_o = \frac{\beta A_v}{1 + \beta A_v} \qquad \beta = \frac{R_2}{R_{Ve}R_2}$$

Voltage reference circuit generates a constant voltage level. A fraction of the output voltage is derived by the potential divider network R_1, R_L and this voltage is compared with the reference. The difference in these voltages is converted into an error signal, which controls the voltage drop across the series-pass element to keep V_o constant.

If V_o is less, the drop across the series-pass element is reduced so that V_o increases. If V_o is more, the drop across series pass element increases so that V_o decreases to the normal value.

The error amplifier controls the base current to the series-pass element, which is a transistor and the drop across it (V_{CE}) varies, as its I_c changes, in proportion to I_B.

The characteristics of a good regulator circuit are as follows:

1. It should have low line regulation.
2. It should have low load regulation.
3. It must have a high degree of ripple rejection.

6.5 PROTECTION CIRCUITS FOR VOLTAGE REGULATOR ICs

Voltage regulator ICs are provided with the following three kinds of protection:

1. Short-circuit protection
   (a) Active current limiting
   (b) Passive current limiting or foldback current limiting
2. Over-voltage protection
3. Thermal overload protection

6.5.1 Short-Circuit Protection

(a) Active current limiting

Normally T_1 is off. A resistor R_{SC} is connected between the current limit and current sense terminals. When V_o is shorted, the drop across R_{SC} is such that T_1 turns on. When T_1 turns on, it draws current (I_C flows) and reduces the base drive I_B to the power device T_2. Therefore, I_o is the output current reduced to zero.

The circuit is shown in Fig. 6.9.

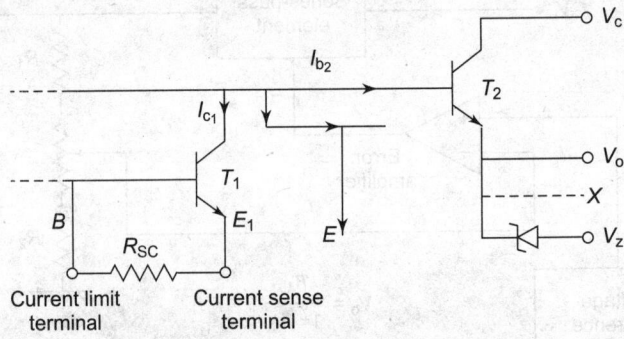

Fig. 6.9 *Short-Circuit Protection*

In the short circuit, current limit is 65 mA, and cut in voltage V_{BE} for T_1 is 0.65 V. R_{SC} must be such that when 65 mA is reached T_1 should turn ON.

$$\therefore \qquad R_{SC} = \frac{0.65\ V}{65\ mA} = 10\ \Omega$$

Current sense will detect the short-circuit. Short-circuit current will flow through this terminal.

(b) Foldback current limiting

If the short circuit is not detected, and the condition exists for a long time due to large current flow, increased heating will take place, damaging the IC. In such cases, foldback current limiting is employed. When a short circuit occurs, I_L decreases and has a minimum value.

The circuit is shown in Fig. 6.10.

T_1 **and** T_2 **are normally off.** When I_1 increases to a higher value, the voltage drop across R_{SC} increases and turns T_1 on. T_1 then turns on T_2 because I_C of $T_1 = I_B$ of T_2. When T_2 turns on, the base drive for T_3 is reduced; therefore, I_C of T_3, that is I_L, reduces.

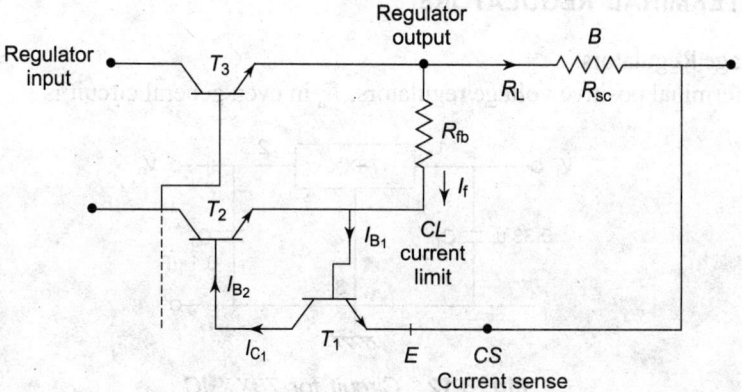

Fig. 6.10 *Circuit for Foldback Limiting*

6.5.2 Over-Voltage Protection

Here, a (avalanche/zener) diode is connected between output and ground terminals. When V_o decreases, the diode starts conducting, V_b is constant at the breakdown voltage of the zener diode.

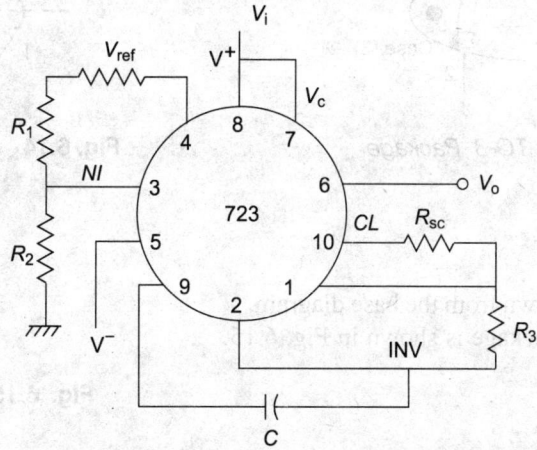

Fig. 6.11 *IC 723 Voltage Regulator*

6.6 IC 723 VOLTAGE REGULATOR

V-(pin 5) can be grounded.
Pins 8 and 7 are shorted and DC voltage is given.
Pin details:
1. Current sense
3. Non-inverted input

2. Inverted input
4. V_{ref}

5. $-V_{cc}$ 6. V_{out}
7. V_c 8. $+V_{cc}$
9. Frequency compensation   10. Current limit

6.7 THREE-TERMINAL REGULATORS

7800 Series Voltage Regulators:

These are three-terminal positive voltage regulators, V_a in even general circuit is

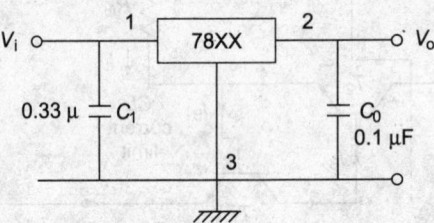

Fig. 6.12   *Circuit for 78XX IC*

XX indicates the numbers that will follow like 00, 01, 02, and so on.

These are available in TO-3 type metal package shown in Fig. 6.13. Base diagram is shown in Fig. 6.14.

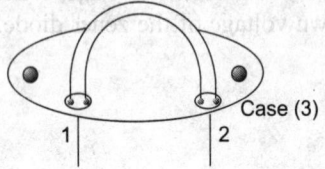

Fig. 6.13   *TO-3 Package*

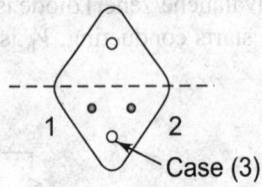

Fig. 6.14   *Base Diagram*

Pin 1: Input
Pin 2: Output
Case: Ground
Pins 1 and 2 can be known from the base diagram.
TO-20 Type: Plastic package is shown in Fig. 6.15.
Pin 1: Input
Pin 2: Ground
Pin 3: Output

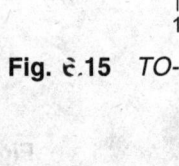

Fig. 6.15   *TO-20 Plastic Package*

Device type	Output voltage V_o	Max. input voltage (V)
7805	5.0	
7806	6.0	
7808	8.0	
7812	12.0	35 V
7815	15.0	
7818	18.0	
7824	24.0	40 V

7805 can be used as a 0.5-A current source. The current supplied to the load is

$$I_L = \frac{V_R}{R} I_Q$$

I_Q = Quiescent current and I_L = Load current
= 4.3 mA for 7805

Example 6.3 *Design a current source to supply load current of 0.5 A to draw a load of 10 Ω. Determine the maximum value of V_i. Use 7805 IC.*

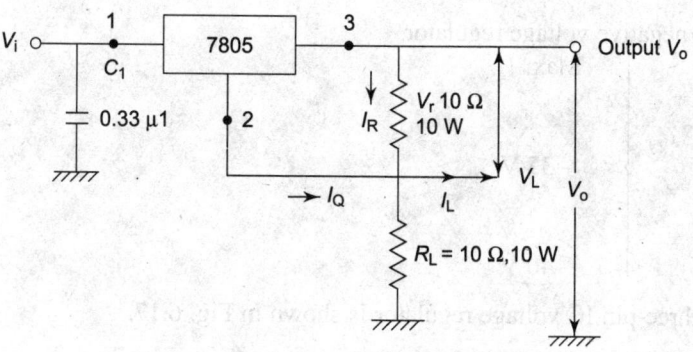

Fig. 6.16 *Circuit for Example 6.3*

$$V_R = V_{23} = 5 \text{ V}$$
$$R = 10 \text{ Ω}. \quad I_L \simeq 0.5 \text{ A}$$
$$V_o = V_R + V_L$$
$$V_L = I_L R_L$$
$$R_L = 10 \text{ Ω}$$

So, $\qquad V_L = 0.5 \text{ A} = 10 \text{ Ω} = 5 \text{ V}$

Therefore, $\qquad V_o = V_R + V_L = 5V + 5V = 10 \text{ V}$

The voltage drop across 7805 is 2 V.

Therefore, the minimum input voltage required is $V_i = V_o +$ Ampere voltage.

$$V_i = 12 \text{ V}$$

So, a current source circuit using a voltage regulator I_C can be designed for a desired value of I_L by choosing an appropriate value of R.

Example 6.4 *Using 7805C voltage regulator, design a current source that will deliver 0.25 A current to the 48 Ω, 10 W load.*

Solution

$$I_L = \frac{V_R}{R} + I_Q$$

Neglecting $\qquad I_Q, R \simeq \dfrac{V_R}{I_L} \simeq \dfrac{5\ \text{V}}{0.25\ \text{A}} = 20\ \Omega$

$$V_o = V_R + V_L = 5\ \text{V} + (48\ \Omega)(0.25) = 17\ \text{V}$$

$$V_i = V_o + \text{drop across IC}$$

$$V_i = 17 + 2 = 19\ \text{V}$$

6.8 7900 SERIES

7900 is a series of a negative voltage regulator.

	V_o	Max. V_i
7902	-2 V	
7905	-5 V	
7906	-6 V	-35 V
7908	-8 V	
7918	-18 V	
7924	-24 V	-40 V

Bottom view of three-pin IC voltage regulator is shown in Fig. 6.17.

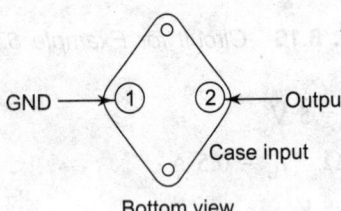

Fig. 6.17 *Bottom View of Three-Pin Voltage Regulator ICs*

6.9 PROTECTION CIRCUITS

Current limiting: This means short-circuit protection, that is, even if the output terminals of the voltage regulator are shorted, the current should be limited and should not exceed a particular value. This can be done in two ways:

1. Simple limiting circuit
2. Foldback limiting circuit

6.10 SIMPLE LIMITING CIRCUIT

The circuit is shown in Fig. 6.18. R_4 is called a current-sensing resistor.

If $I_L < 600$ mA, the voltage drop across R_4 is < 0.6 V (because $R_4 = 1\ \Omega$, $R \times I = 1 \times 600$ mA = 0.6 V). The drop across R_4 is the V_{BE} for transistor Q_3.

Therefore, Q_3 is cut off. So, the regulator circuit works as a normal circuit without any limiting provision. When I_L is between 600 mA and 700 mA, the voltage across R_4 is between 0.6 V and 0.7 V. So, Q_3 will turn on. Thus, the collector current of Q_3 will flow through R_3. So, the base voltage V_{BE} to Q_2 will decrease. Therefore, the output voltage V_o will decrease. Hence, the load current will decrease.

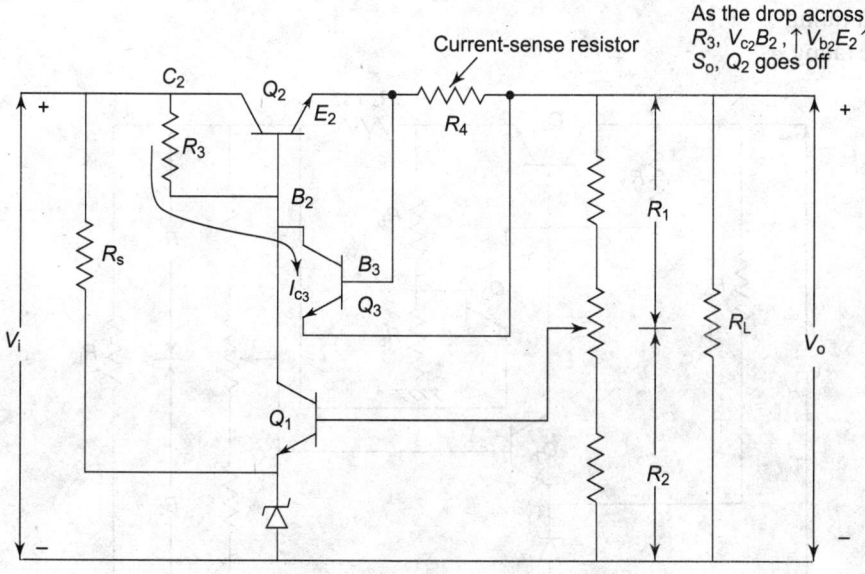

Fig. 6.18 *Current-Limiting Circuit*

If I_s = short-circuit current when output terminals are shorted, voltage across R_4 is $V_{BE} = I_s R_4$

So, $$I_s = V_{BE}/R_4$$

By choosing the value of R_4, we can change the level of current limiting.

The disadvantage with this circuit is that power dissipation across transistor Q_2 will be very large.

6.11 FOLDBACK LIMITING

The circuit is shown in Fig. 6.19. As I_B of a transistor decreases, I_C decreases. Therefore, I_o also decreases, instead of being cut off fully. This is the principle of foldback limiting.

Using this circuit, we can reduce the power dissipation in the series-pass transistor. The load current flows through R_4. So, the voltage fed to the potential divider circuit R_5 and R_6 is $(I_L R_4 + V_o)$. This voltage controls Q_3. The feedback fraction is $K = \dfrac{R_6}{R_5 + R_6}$.

The maximum load current will be higher than short-circuit current.

The specifications, pin configurations, and other details of some ICs are given in Appendix. The reader is advised to study the same.

6.12 SPECIFICATIONS OF VOLTAGE REGULATOR CIRCUITS

1. Regulation percentage
2. Input impedance
3. Output impedance
4. Ripple rejection

5. Current rating
6. Voltage rating

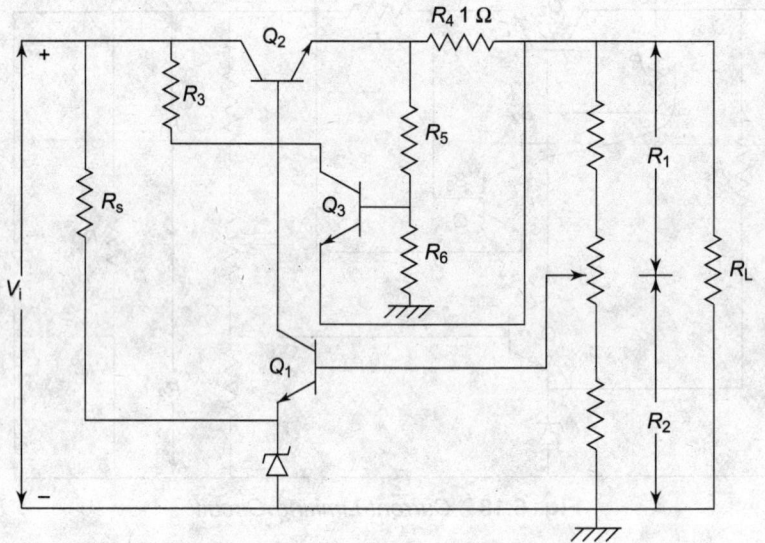

Fig. 6.19 *Foldback-Limiting Circuit*

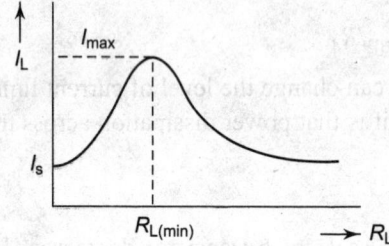

Fig. 6.20 *I–R$_L$ Characteristic in Foldback Limiting*

6.13 DC TO DC CONVERSION

These are used when large DC output voltage is required from small DC input voltage. That is, if DC 5 V is available, we can make it 15 V. The DC 5 V is used to drive an oscillator circuit. The AC output is amplified with the help of a transformer. Then it is converted to DC to get a large DC output voltage. The block schematic is shown in Fig. 6.21.

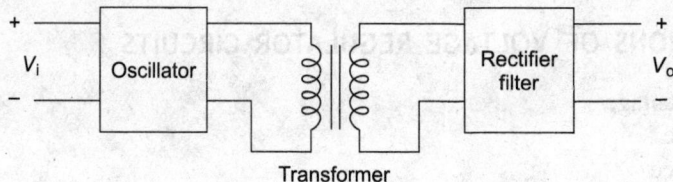

Fig. 6.21 *Block Schematic of DC-to-DC Converter*

6.14   SWITCHING REGULATORS

In the series regulator, the mean voltage is dropped across the series-pass element. So, power dissipation is greater. In order to reduce that, switching regulators are used. Here, a switch connects the input to the regulator intermittently so that the average current is passed to the load. When the switch is closed, energy is stored in an inductor. This energy is transferred to the load. When the load voltage decreases, this is sensed by the comparator and when the energy in the storage element is dissipated, the switch is closed by the comparator and the input is connected to the regulator circuit. So, the storage device gets charged again.

Switch is a transistor which is turned on and off by the voltage of the threshold detector.

When the switch is open, diode D conducts. The energy stored in L forward biases the diode. This maintains current when S is closed, D is reverse biased, and C charges. It supplies energy to the load. When V_0 decreases, detector changes state and closes the switch.

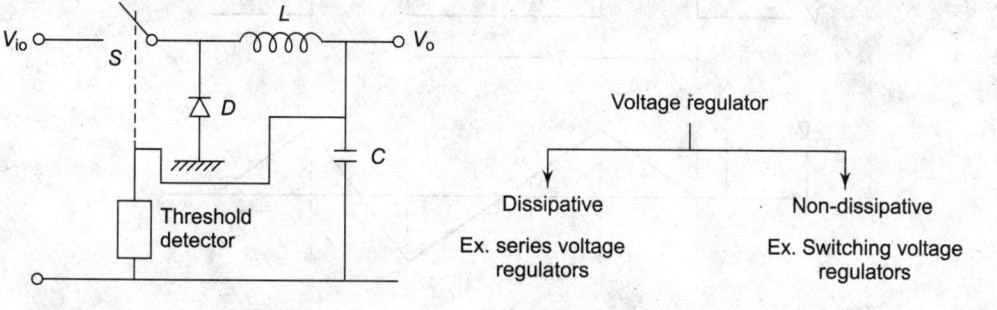

Fig. 6.22   *Switching Regulator*   **Fig. 6.23**   *Classification of Voltage Regulators*

6.15   CLASSIFICATION

Dissipative: Excess voltage when V_i increases is dissipated across a series-pass element. Efficiency of the circuit is less.
Non-dissipative: Switching type. Input is not permanently connected. Efficiency is more.
The block schematic of a switching regulator is given in Fig. 6.24.

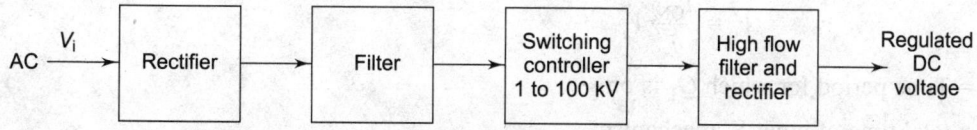

Fig. 6.24   *Block Schematic of a Switching Regulator*

The voltage-regulating transistor is operated in cut-off or saturation region. So, the current flowing through the transistor is small. Hence, power dissipation in the circuit is less and efficiency is greater. The chopped AC voltage is filtered by a high-frequency filter and rectified to get DC.

Switching regulators are (1) lighter, (2) smaller, and (3) more efficient than a series-pass type.

6.16   STEP-DOWN (BUCK) SWITCHING REGULATOR

The circuit is shown in Fig. 6.25. The output waveforms are shown in Fig. 6.26. The principle of operation of the circuit is shown in Fig. 6.27 (a) and (b).

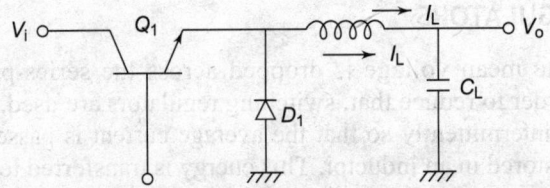

Fig. 6.25 *Bucking Type Voltage Regulator*

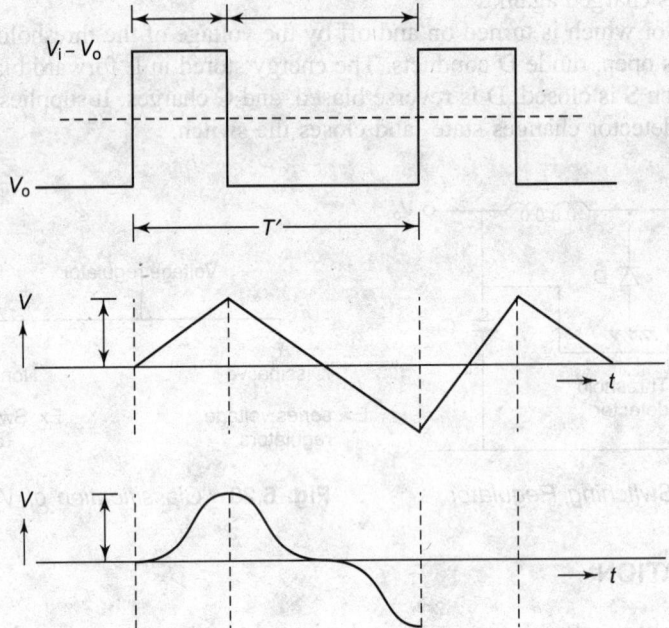

Fig. 6.26 *Waveforms for Bucking Type Voltage Regulator*

Control voltage (to turn the transistor Q_1 on/off)

$$V_o = \frac{t_{ON}}{T} V_i$$

t_{ON} = Time period for which Q_1 is on.

T = total time of input S_T mechanism

$$T = \text{Duty cycle} = \frac{t_{ON}}{t_{ON} + t_{OFF}} = \frac{t_{ON}}{\text{Total line}} = \frac{t_{ON}}{T}$$

D_1 is the catch diode in the circuit. It provides a continuous path for the inductor current when Q_1 turns off. When S_1 is closed, inductor current I_L passes from the input voltage V_i to the load. ($V_i - V_o$) appears across inductor. So I_L increases.

When S_1 is open, stored energy in the inductor forces I_L to continue passing through the load, and returns through the diode. The inductor voltage now reverses and becomes approximately equal to V_o. So, I_L decreases.

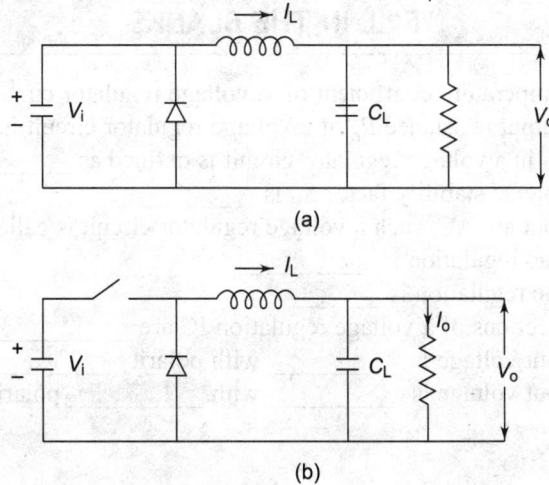

Fig. 6.27 (a) Circuit when S_1 is Closed
(b) Circuit when S_2 is Open

SUMMARY

- Voltage regulator circuits are available in IC form. Commonly used voltage regulator ICs are μAT23, LN309, LM105, CA3085A, 7805, 7905, etc.
- The three important sections of IC Voltage regulators are (i) reference voltage circuit (ii) error amplifier and (iii) series-pass element.
- The ICs will be built in short-circuit protection.
- Foldback limiting is one type of protection circuit.
- 78XX series ICs are three-pin voltage regulator ICs. They supply fixed positive output voltages of 5 V, 6 V, 8 V, 12 V etc. depending on the type No. 7805, 7806, 7808 etc.
- 79XX series ICs are also three-pin voltage regulator ICs which can supply negative output voltages like –5 V, –6 V, –8 V, –12 V etc.
- Constant current-source circuits can be built using these ICs.
- Switching voltage regulator circuits are more efficient than series-pass element type regulator circuits. Power dissipation is less in switching type circuits.

ESSAY-TYPE QUESTIONS

1. With the help of a block schematic explain the working of voltage regulator ICs.
2. Give the specifications and typical values for a +8 V constant output voltage regulator IC.
3. Draw the internal schematic of a 723 voltage regulator IC and explain the pin configuration.
4. Draw the circuit for a 7805 voltage regulation IC and explain its working.
5. Draw the circuit for a 7905 voltage regulator IC and explain its working.

FILL IN THE BLANKS

1. The expression for temperature coefficient of a voltage regulator circuit is _____.
2. The expression for output resistance R_o of a voltage regulator circuit is _____.
3. The stability factor S_V in a voltage regulator circuit is defined as _____.
4. The expression for voltage stability factor S_V is _____.
5. If both input and output are AC, such a voltage regulator circuit is called _____.
6. The expression for load regulation is _____.
7. The expression for line regulation is _____.
8. The three important sections of a voltage regulation IC are _____.
9. For IC 7806, the output voltage is _____ with polarity _____.
10. For IC 7905, the output voltage is _____ with _____ polarity.

ANSWERS

1. $S_T = \dfrac{\Delta V_o}{\Delta V_T}\bigg|_{\Delta V_i = 0, \Delta I_L = 0}$

2. $R_o = \dfrac{\Delta V_o}{\Delta V_i}\bigg|_{\Delta V_i = 0, \Delta T = 0}$

3. The ratio of change in output voltage ΔV_o corresponding to change in input voltage ΔV_i when load current I_L and temperature T are kept constant.

4. $S_v = \dfrac{\Delta V_o}{\Delta V_i}\bigg|_{\Delta I_L = 0, \Delta T = 0}$

5. Voltage stabilizer circuit

6. $\left(\dfrac{V_1 - V_2}{V_1}\right) \times 100\%$

7. $\left(\dfrac{\Delta V_o}{V_i}\right) \times 100\%$

8. (i) Reference voltage circuit
   (ii) Error amplifier
   (iii) Series-pass element.

9. The output voltage is 6 V with positive polarity (+6 V)

10. 5 V with negative polarity (–5 V).

UNSOLVED PROBLEMS

1. For the shunt regulator circuit shown, determine
   (a) Value of R
   (b) Load current range
   (c) Maximum transistor power dissipation.

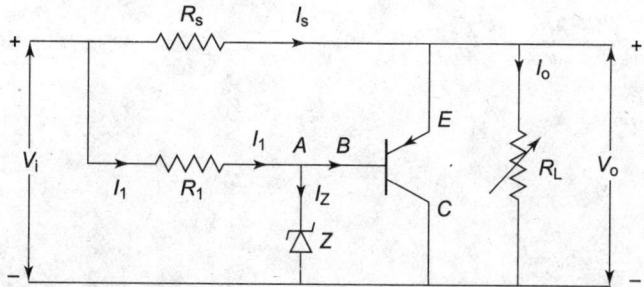

2. For the above circuit, determine
   (a) Nominal voltage
   (b) The value of R_s
   (c) Power rating of R_s

3. Design a series-regulated power supply to provide a nominal output voltage of 20 V, at $I_L = 1.2$ A. Given $V_i = 40 \pm 5$ V, $r_o = 15\ \Omega$, $R_Z = 10\ \Omega$, $I_Z = 15$ mA, $I_{C2} = 12$ mA, $h_{ie_2} = 600\ \Omega$. $h_{fe_2} = 300$. $I_L = 10$ mA (see Fig. 6.6 in the text).

4. Design a current source to supply load current of 0.4 A to drive a load of 15 Ω using 7805 IC.

5. Design a current source to deliver 0.2 A of current to a 70 W, 10 Ω load using 7805 C.

6. For a given rectifier circuit, V_{DC} (no load) = 40 V and full-load voltage = 25 V. $I_L = 0.8$ A. Determine the value of R_o. What is the output voltage when $I_L = 0.4$ A.

7. With given data of Problem 6 determine percentage regulator.

8. Consider a full-wave bridge rectifier circuit which supplies full-load current of 800 mA at full-load voltage of 25 V. The circuit uses a filter capacitor of 1000 μF. Determine (i) peak-to-peak voltage (ii) rms value of ripple voltage and (iii) minimum instantaneous output voltage.

9. With given data in Problem 8 calculate the percentage ripple.

10. Design a regulated power supply, using a 25-V transformer, FW bridge rectifier and 200 μF filter capacitor.

11. What is the value of output ripple voltage from a three-terminal voltage regulator LM 317 with regulated output voltages of 5 V and 8 V precentage ripple.

12. Design a regulated power supply using three-terminal IC to give $V_o = +5$ V. $I_o = 0.6$ A, $V_i = 12$ V_{DC}, $T_A = 60°C$.

13. Design a power supply using three-terminal IC regulator to give +5 V output at 400 μA at 30°C.

14. Using ICs, design a regulated power supply to give output voltage, which is adjustable from 1.2 V to 12 V. I_L is to be 0.5 A, $T_A = 35°C$.

15. Design a current source using 7805 to supply a load current of 0.6 A to drive a load of 20 Ω.

Appendix A
Configurations of ICs

A.1 DESCRIPTION

The μA741 is a high-performance monolithic op-amp constructed using the Fairchild Planar epitaxial process. It is intended for a wide range of analog applications. High common-mode voltage range and absence of latch-up tendencies make the μA741 ideal for use as a voltage follower. The high gain and wide range of operating voltage provide superior performance in integrator, summing amplifier, and general feedback applications.

Features of μA741 IC

- No frequency compensation required
- Short-circuit protection
- Offset voltage null capability
- Large common-mode and differential voltage ranges
- Low power consumption
- No latch-up

Connection Diagram

10-pin flatpak

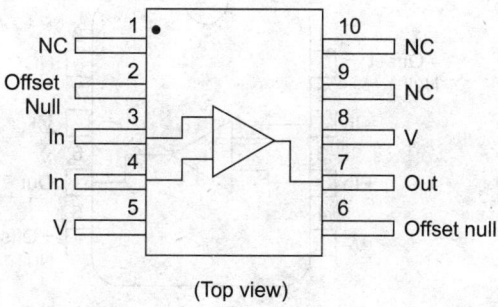

(Top view)

Fig. A.1 *μA741 Pin Configuration*

Order Information

Type	Package	Code	Part No.
μA741	Flatpak	3F	μA741FM
μA741A	Faltpak	3F	μA741AFM

Absolute Maximum Ratings

Supply voltage

μA741A, μA741, μA741E,	±22 V
μA741C	±18 V

Internal power dissipation
(Note 1)

Metal package	500 mW
DIP	310 mW
Flatpak	570 mW
Differential input voltage	±30 V
Input voltage (Note 2)	±15 V

Storage temperature range

Metal package and flatpak	−65°C to +150°C
DIP	−55°C to +125°C

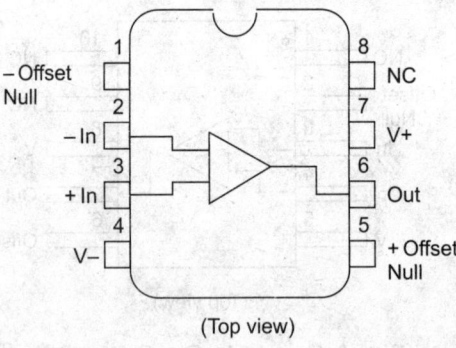

(Top view)

Fig. A.2 *μA741 Metal Package*

Connection Diagram
8-pin metal package

Pin 4 connected to case

Order Information

Type	Package	Code	Part No.
μA741	Metal	5W	μA741HM
μA741A	Metal	5W	μA741AHM
μA741C	Metal	5W	μA741HC
μA741E	Metal	5W	μA741EHC

Connection Diagram
8-pin DIP

(Top view)

Fig. A.3 *DIP-Package Pin Configuration*

Order Information

Type	Package	Code	Part No.
μA741C	Molded DIP	9T	μA741TC
μA741C	Ceramic DIP	6T	μA741RC

Operating temperature range
    Military (μA741A, μA741) $-55°C$ to $+125°C$
    Commercial (μA741E, μA741C) $0°C$ to $+70°C$
Pin temperature (soldering 60 sec)
    Metal package, flatpak, and ceramic DIP $300°C$
    Molded DIP (10 sec) $260°C$
Output short-circuit duration
    (Note 3)   Indefinite

μA741 and μA741C

Electrical Characteristics $V_s = \pm15$ V and $T_A = 25°C$ unless otherwise specified

Table A.1 Electrical Characteristics

Characteristic	Conditions	mA741			mA741C			Unit
		Min.	Typ.	Max.	Min.	Typ.	Max.	
Input offset voltage	$R_s \leq 10$ kΩ		1.0	5.0		2.0	6.0	mV
Input offset current			20	200		20	200	nA
Input bias current			80	500		80	500	nA
Power-supply rejection ratio	$V_s = +10, -20$ $V_s = +20, -10$ V, $R_s = 50$ Ω		30	150		30	150	μV/V
Input resistance		.3	2.0		.3	2.0		MΩ
Input capacitance			1.4			1.4		pF
Offset voltage adjustment range			±15			±15		mV
Input voltage range					±12	±13		V
Common-mode rejection ratio	$R_s \leq 10$ kΩ					70	90	dB
Output short-circuit current			25			25		mA
Large signal-voltage gain	$R_L \geq 2$ kΩ, $V_{OUT} = \pm10$ V	50 k	200 K		20 K	200 K		
Output resistance			75			75		Ω
Output voltage swing	$R_L \geq 10$ kΩ				±12	±14		V
	$R_L \geq 2$ kΩ				±10	±13		V
Supply current			1.7	2.8		1.7	2.8	mA
Power consumption			50	85		50	85	mW
Transient response (unity gain) — Rise time	$V_{IN} = 20$ mV, $R_L = 2$ kΩ, $C_L \leq 100$ pF		.3			.3		μsec
Transient response (unity gain) — Overshoot			5.0			5.0		%
Bandwidth (Note 4)			1.0			1.0		MHz
Slew rate	$R_L \geq 2$ kΩ		.5			.5		V/μsec

Notes

4. Calculated value from BW(MHz) = $\dfrac{0.35}{\text{Rise Time (}\mu\text{sec)}}$

5. All V_{cc} = 15 V for μA741 and μA741C.

6. Maximum supply current for all devices

   25°C = 2.8 mA

   125°C = 2.5 mA

   –55°C = 3.3 mA

A.2 GENERAL DESCRIPTION OF LM101A SERIES OP-AMPS

The LM101A series are general-purpose op-amps which feature improved performance over industry standards like the LM709. Advanced processing techniques make possible an order of magnitude reduction in input currents, and a redesign of the biasing circuitry reduces the temperature drift of input current. Improved specifications include:

- Offset voltage 3 mV maximum over temperature (LM101A/LM201A)
- Input current 100 nA maximum over temperature (LM101A/LM201A)
- Offset current 20 nA maximum over temperature (LM101A/LM201A)
- Guaranteed drift characteristics
- Offsets guaranteed over entire common mode and supply voltage ranges
- Slew rate of 10V/μsec as a summing amplifier

This amplifier offers many features which make its application nearly foolproof: overload protection on the input and output, no latch-up when the common-mode range is exceeded, freedom from oscillations, and compensation with a single 30 pF capacitor. It has advantages over internally compensated amplifiers in that the frequency compensation can be tailored to the particular application. For example, in low-frequency circuits, it can be over-compensated for increased stability margin. Or the compensation can be optimized to give more than a factor of 10 improvement in high-frequency performance for most applications.

In addition, the device provides better accuracy and lower noise in high-impedance circuitry. The low input currents also make it particularly well suited for long interval integrators or timers, sample-and-hold circuits, and low-frequency waveform generators. Further, replacing circuits where matched transistor pairs buffer the inputs of conventional IC op-amps, it can give lower offset voltage and drift at a lower cost.

The LM101A is guaranteed over a temperature range of –55°C to +125°C, the LM201A from –25°C to +85°C, and the LM301A from 0°C to 70°C.

Schematic** and Connection Diagrams (Top Views) LM101A Op-amp

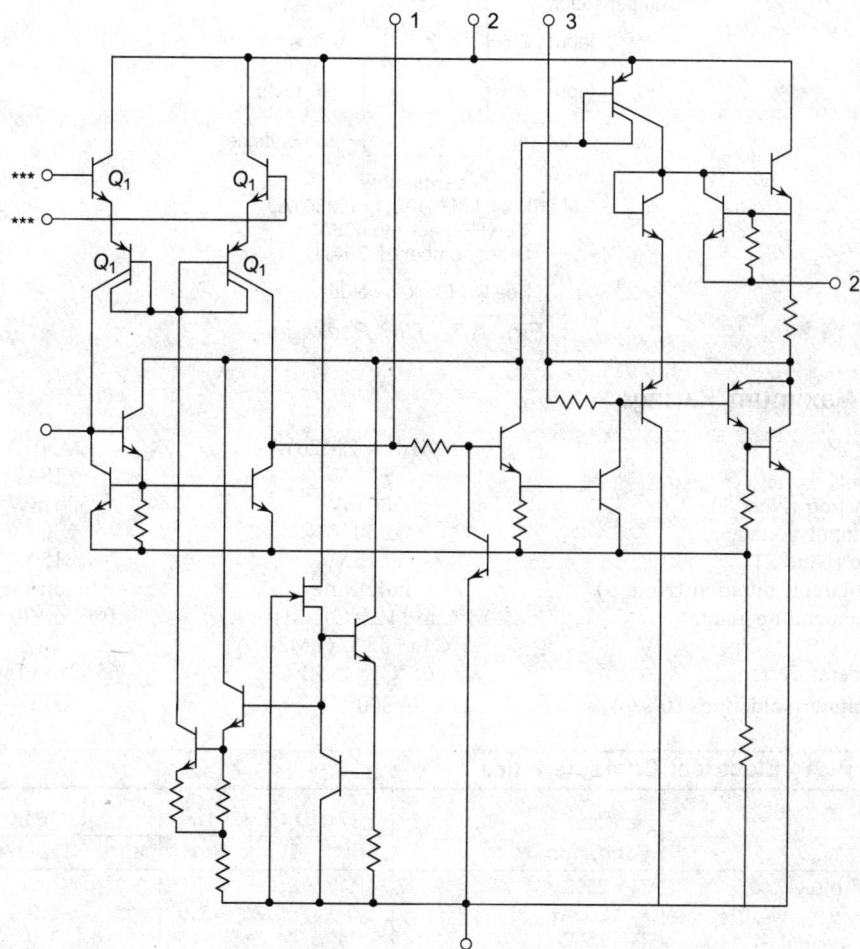

Fig. A.4 *Motorola LM101A OP-amp Connection Diagrams*

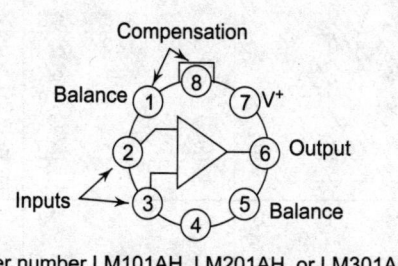

Order number LM101AH, LM201AH, or LM301AH
See NS package H08C

Fig. A.5 *Metal Can Package*

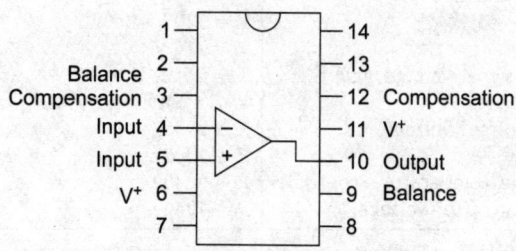

Note: Pin 6 connected to bottom of package
Order number LM101AJ-14,
LM201AJ-14, or LM301AJ-14
See NS package J14A

Fig. A.6 *DIP Package*

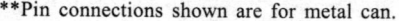

**Pin connections shown are for metal can.

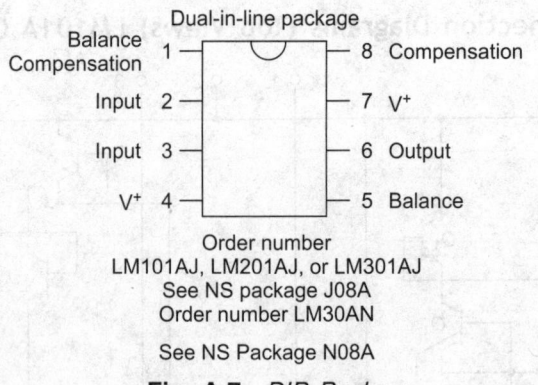

Fig. A.7 DIP Package

Absolute Maximum Ratings

	LM101A/LM201A	LM301A
Supply voltage	± 22 V	± 18 V
Power dissipation (Note 1)	500 mW	500 mW
Differential input voltage	± 30 V	± 30 V
Input voltage (Note 2)	± 15 V	± 15 V
Output short-circuit duration (Note 3)	Indefinite	Indefinite
Operating temperature range	$-55°C$ to $+125°C$ (LM101A) $-25°C$ to $+85°C$ (LM201A)	$0°C$ to $+70°C$
Storage temperature range	$-65°C$ to $+150°C$	$-65°C$ to $+150°C$
Lead temperature (soldering, 10 sec)	$300°C$	$300°C$

Table A.2 Electrical Characteristics

Parameter	Condition	LM101A/LM201A Min.	LM101A/LM201A Typ.	LM101A/LM201A Max.	LM301A Min.	LM301A Typ.	LM301A Max.	Unit
Input offset voltage LM101A, LM201A, LM301A	$T_A = 25°C$ $R_s \leq 50$ kΩ		0.7	2.0		2.0	7.5	mV
Input offset current	$T_A = 25°C$		1.5	10		3.0	50	nA
Input bias current	$T_A = 25°C$		30	75		70	250	nA
Input resistance	$T_A = 25°C$	1.5	4.0		0.5	2.0		MΩ
Supply current	$T_A = 25°C$ $V_s = \pm 20$ V		1.8	3.0				mA
	$V_s = \pm 15$ V					1.8	3.0	mA
Large signal voltage gain	$T_A = 25°C$, $V_s = \pm 15$ V $V_{OUT} = \pm 10$V, $R_L \geq 2$ kΩ	50	160		25	160		V/mV
Input offset voltage	$R_s \leq 50$ kΩ			3.0			10	mV
	$R_s \leq 10$ kΩ							mV
Average temperature coefficient of input offset voltage	$R_s \leq 50$ kΩ		3.0	15		6.0	30	μV/°C
	$R_s \leq 10$ kΩ							μV/°C
Input offset current				20			70	nA
	$T_A = T_{MAX}$							nA
	$T_A = T_{MIN}$							nA
Average temperature coefficient of input offset current	$25°C \leq T_A \leq T_{MAX}$		0.01	0.1		0.01	0.3	mA/°C
	$T_{MIN} \leq T_A \leq 25°C$		0.02	0.2		0.02	0.6	mA/°C

(Contd.)

Table A.2 *Contd.*

Parameter	Condition	LM101A/LM201A			LM301A			Unit
		Min.	Typ.	Max.	Min.	Typ.	Max.	
Input bias current				0.1		0.5		μA
Supply current	$T_A = T_{MAX}$, $V_s = \pm 20$V		1.2	2.5				μA
Large signal-voltage gain	$V_s = \pm 15$V, $V_{OUT} = \pm 10$ V	25			15			V/mV
	$R_L \leq 2$ k							
Output voltage swing	$V_s = \pm 15$ V							
	$R_L = 10$ kΩ	± 12	± 14		± 12	± 14		V
	$R_L = 2$ kΩ	± 10	± 13		± 0	± 13		V
Input voltage range	$V_s = \pm 20$ V	± 15						V
	$V_s = \pm 15$ V	+15,	−13		± 12,	± 15,	−13	V
Common-mode rejection ratio	$R_s \leq 50$ kΩ	80	96		70	90		dB
	$R_s \leq 10$ kΩ							dB
Supply-voltage rejection ratio	$R_s \leq 50$ kΩ	80	96		70	96		dB
	$R_s \leq 10$ kΩ							dB

Notes

1. The maximum junction temperature of the LM101A is 150°C and that of the LM201A/LM301A is 100°C. For operating at elevated temperatures, devices in the TO-5 package must be derated based on a thermal resistance of 150°C/W junction to ambient or 45°C/W junction to case. The thermal resistance of the dual-in-line package is 187°C/W junction to ambient.

2. For supply voltages less than ±15 V, the absolute maximum input voltage is equal to the supply voltage.

3. Continuous short circuit is allowed for case temperatures to 125°C and ambient temperatures to 75°C for LM101A/LM201A, and 70°C and 55°C, respectively for LM301A.

4. Unless otherwise specified, these specifications apply for C1 = 30 pF, ± 5 V $\leq V_s \leq \pm 20$ V and $-55°C \leq T_A \leq +125°C$ (LM101A), ± 5 V $\leq V_s \leq \pm 20$ V and $-25°C \leq T_A \leq +85°C$ (LM201A), and ± 5V $\leq V_s \leq 15$ V and $0°C \leq T_A + 70°C$ (LM301A).

A.2.1 General Description

The LM311 is a voltage comparator that has input currents more than a hundred times lower than devices like the LM306 or LM710C. It is also designed to operate over a wider range of supply voltages: from standard ±15-V op-amp supplies down to the single 5-V supply used for IC logic. Its output is compatible with RTL, DTL, and TTL as well as MOS circuits. Further, it can drive lamps or relays, switching voltages up to 40 V at currents as high as 50 mA.

A.2.2 Features

- Operates from single 5-V supply
- Maximum input current: 250 nA
- Maximum offset current: 50 nA
- Differential input voltage range: ±30 V
- Power consumption: 135 mW at ±15 V

Both the input and the output of the LM311 can be isolated from system ground, and the output can drive loads referred to ground, the positive supply or the negative supply. Offset balancing and strobe capability are provided and outputs can be wire-OR ed. Although slower than the LM306 and LM710C (200 nsec response time vs 40 nsec), the device is also much less prone to spurious oscillations. The LM311 has the same pin configuration as the LM306 and LM710C. See the "application hints" of the LM311 for application help.

Auxiliary circuits for TO-6 Package

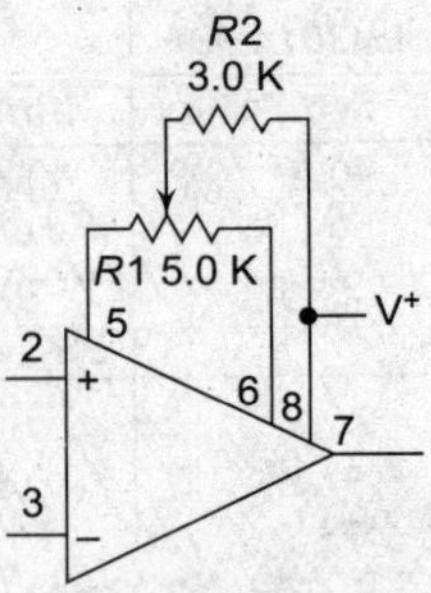

Offset Balancing
Fig. A.8

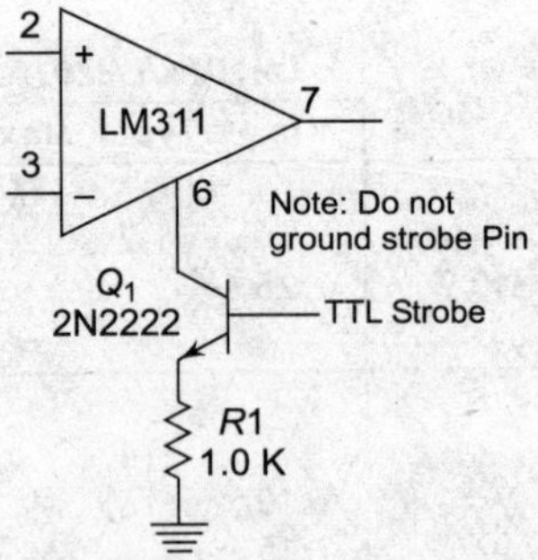

Strobing
Fig. A.9

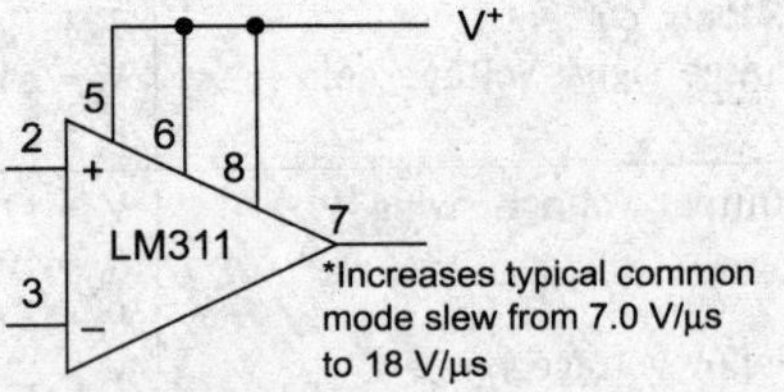

Note: Do not ground strobe Pin

TTL Strobe

Increasing Input Stage Current*
Fig. A.10

*Increases typical common mode slew from 7.0 V/μs to 18 V/μs

Typical Applications**

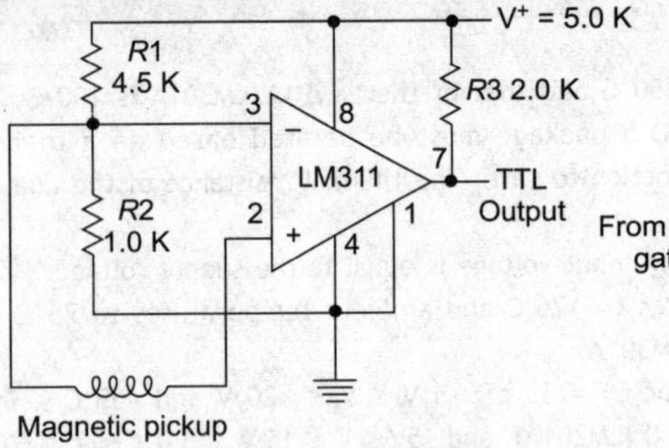

Detector for Magnetic Transducer
Magnetic pickup
Fig. A.11

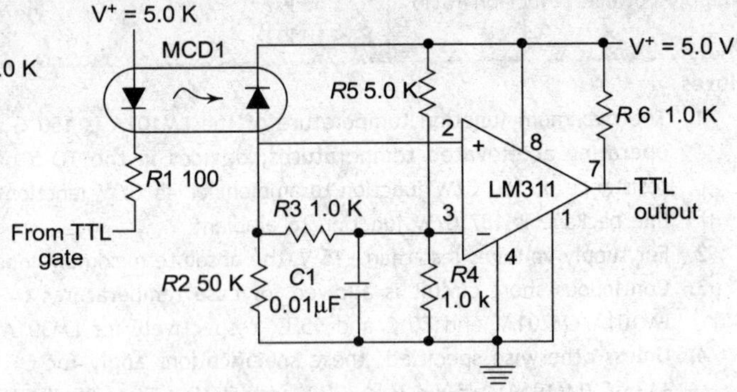

Digital Transmission Isolator
Fig. A.12

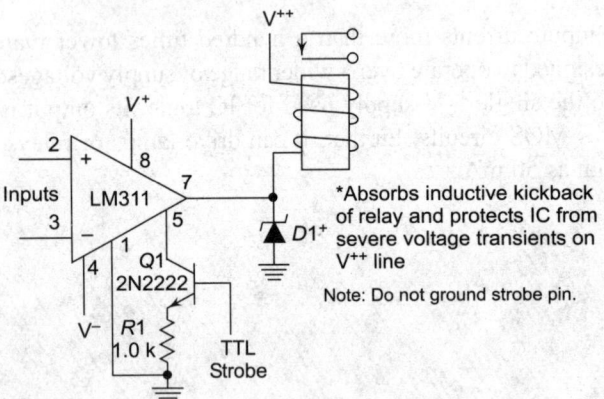

Fig. A.13 Relay Driver with Strobe

*Absorbs inductive kickback of relay and protects IC from severe voltage transients on V++ line

Note: Do not ground strobe pin.

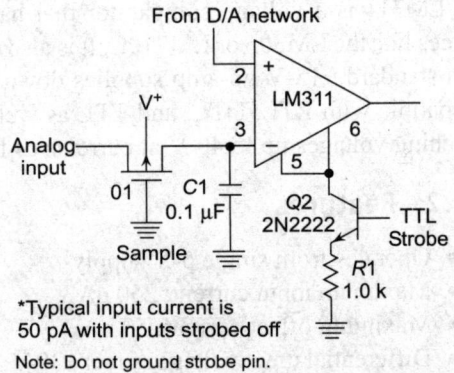

Fig. A.14 Strobing off Both Input* and Output Stages

From D/A network

Analog input

Sample

*Typical input current is 50 pA with inputs strobed off

Note: Do not ground strobe pin.

A.2.3 Absolute Maximum Ratings of LM311 Voltage Comparator

Total supply voltage ($V_{8\,4}$)	36 V
Output-to-negative supply voltage ($V_{7\,4}$)	40 V
Ground-to-negative supply voltage ($V_{1\,4}$)	30 V
Differential input voltage	±30 V
Input voltage (Note 1)	±15 V
Power dissipation (Note 2)	500 mW
Output short-circuit duration	10 sec
Operating temperature range	0°C to 70°C
Storage temperature range	−65°C to 150°C
Lead temperature (soldering, 10 sec)	300°C
Voltage at strobe pin	V^+ −5 V

Table A.3 Electrical Characteristics (Note 3)

Parameter	Conditions	Min.	Typ.	Max.	Unit
Input offset voltage (Note 4)	$T_A = 25°C$, $R_s \leq 50$ K		2.0	7.5	mV
Input offset current (Note 4)	$T_A = 25°C$		6.0	50	nA
Input bias current	$T_A = 25°C$		100	250	nA
Voltage gain	$T_A = 25°C$	40	200		V/mV
Response time (Note 5)	$T_A = 25°C$		200		nsec
Saturation voltage	$V_{IN} \leq -10$ mV, $I_{OUT} = 50$ mA $T_A = 25°C$		0.75	1.5	V
Strobe ON current	$T_A = 25°C$		3.0		mA
Output leakage current	$V_{IN} \geq 10$ mV, $V_{OUT} = 35$ V $T_A = 25°C$, $I_{STROBE} = 3$ mA		0.2	50	nA
Input offset voltage (Note 4)	$R_s \leq 50$ K			10	mV
Input offset current (Note 4)				70	nA
Input bias current				300	nA
Input voltage range		−14.5	13.8, −14.7	13.0	V
Saturation voltage	$V^+ \geq 4.5$ V, $V^- = 0$ $V_{IN} \leq -10$ mV, $I_{SINK} \leq 8$ mA		0.23	0.4	V
Positive supply current	$T_A = 25°C$		5.1	7.5	mA
Negative supply current	$T_A = 25°C$		4.1	5.0	mA

Notes

1. This rating applies for ±15-V supplies. The positive input voltage limit is 30 V above the negative supply. The negative input voltage limit is equal to the negative supply voltage or 30 V below the positive supply, whichever is less.
2. The maximum junction temperature of the LM311 is 110°C. For operating at elevated temperatures, devices in the TO-5 package must be derated based on a thermal resistance of 150°C/W junction to ambient or 45°C/W junction to case. The thermal resistance of the dual-in-line package is 100°C/W junction to ambient.
3. These specifications apply for $V_s = \pm 15$ V and the ground pin at ground, and $0°C < T_A < +70°C$ unless otherwise specified. The offset voltage, offset current, and bias current specifications apply for any supply voltage from a single 5-V supply up to ±15 V supplies.
4. The offset voltages and offset currents given are the maximum values required to drive the output within a volt of either supply with 1-mA load. Thus, these parameters define an error band and take into account the worst-case effects of voltage gain and input impedance.
5. The response time specified (see definitions) is for a 100-mV input step with 5-mV overdrive.
6. Do not short the strobe pin to ground; it should be current driven at 3 to 5 mA.

A.3 DESCRIPTION OF NE/SE 555 TIMER IC

The NE/SE 555 monolithic timing circuit is a highly stable controller capable of producing accurate time delays or oscillation. Additional terminals are provided for triggering or resetting if desired. In the time delay mode of operation, the time is precisely controlled by one external resistor and capacitor. For a stable operation as an oscillator, the free-running frequency and the duty cycle are both accurately controlled with two external resistors and one capacitor. The circuit may be triggered and reset on falling waveforms, and the output structure can source or sink up to 200 mA or drive TTL circuits.

A.3.1 Features

- Timing from microseconds through hours
- Operates in both astable and monostable modes
- Adjustable duty cycle
- High current output can source or sink 200 mA
- Output can drive TTL
- Temperature stability of 0.005% per degree celsius
- Normally on and normally off output

A.3.2 Applications

- Precision timing
- Pulse generation
- Sequential timing
- Time delay generation
- Pulse-width modulation
- Pulse-position modulation
- Missing pulse detector

A.4 LINEAR INTEGRATED CIRCUITS

Pin Configurations (Top View)

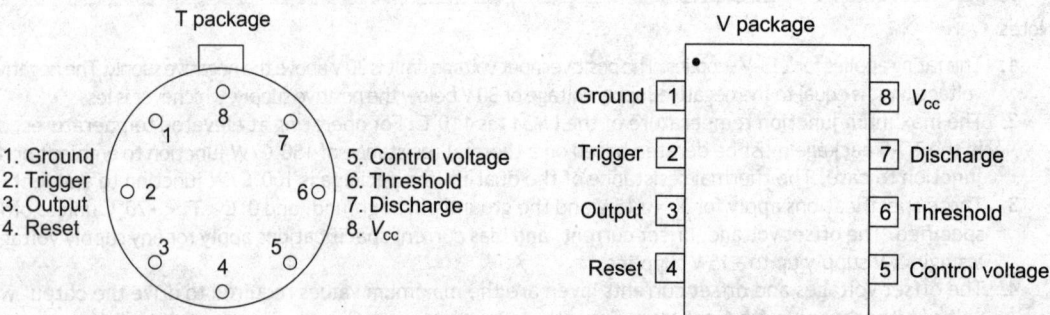

Fig. A.15 *Order Part Numbers SE555T/NE555T*   **Fig. A.16** *Order Part Numbers SE555V/NE555V*

A.4.1 Absolute Maximum Ratings of 555 Timer IC

Supply voltage $\qquad$ +18 V
Power dissipation $\qquad$ 600 mW
Operating temperature range
$\quad$ NE555 $\qquad$ 0°C to +70°C
$\quad$ SE555 $\qquad$ −55°C to +125°C
Storage temperature range $\qquad$ −65°C to +150°C
Lead temperature (soldering, 60 sec) $\qquad$ +300°C

A.5 BLOCK DIAGRAM

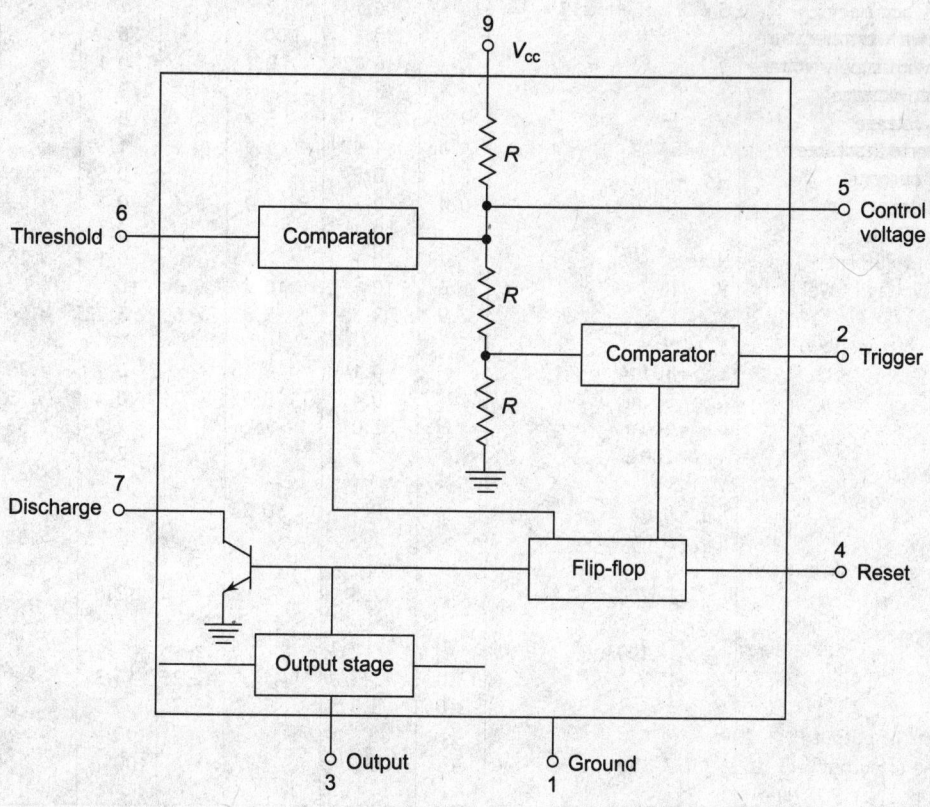

Fig A.17 *Block Diagram of 555 IC*

Electrical Characteristics of 555 IC T_A = 25°C and, V_{cc} = +5 V to +15 V unless otherwise specified.

Table A.4 Electrical Characteristics of 555 IC

Parameter	Test conditions	SE555			NE555			Unit
		Min.	Typ.	Max.	Min.	Typ.	Max.	
Supply voltage		4.5		18	4.5		16	V
Supply current	V_{cc} = 5 V, $R_L = \infty$		3	5		3	6	mA
	V_{cc} = 15 V, $R_L = \infty$		10	12		10	15	mA
	Low state, Note 1							
Timing error (monostable)	R_A, R_B = 1 kΩ to 100 kΩ							
Initial accuracy	C = 0.1 μF, Note 2		0.5	2		1		%
Drift with temperature			30	100		50		ppm/°C
Drift with supply voltage			0.05	0.2		0.1		%/V
Threshold voltage			2/3			2/3		X V_{cc}
Trigger voltage	V_{cc} = 15 V	4.8	5	5.2		5		V
Timing error (astable)	V_{cc} = 5 V	1.45	1.67	1.9		1.67		V
Trigger current			0.5			0.5		μA
Reset voltage		0.4	0.7	1.0	0.4	0.7	1.0	V
Reset current			0.1			0.1		mA
Threshold current	Note 3		0.1	.25		0.1	.25	μA
Control voltage level	V_{cc} = 15 V	9.6	10	10.4	9.0	10	11	V
	V_{cc} = 5 V	2.9	3.33	3.8	2.6	3.33	4	V
Output voltage (low)	V_{cc} = 15 V							
	I_{SINK} = 10 mA		0.1	0.15		0.1	.25	V
	I_{SINK} = 50 mA		0.4	0.5		0.4	.75	V
	I_{SINK} = 100 mA		2.0	2.2		2.0	2.5	V
	I_{SINK} = 200 mA		2.5			2.5		
	V_{cc} = 5 V							
	I_{SINK} = 8 mA		0.1	0.25				V
	I_{SINK} = 5 mA					.25	.35	
Output voltage drop (low)								
	I_{SOURCE} = 200 mA		12.5			12.5		
	V_{cc} = 15 V							
	I_{SOURCE} = 100 mA							
	V_{cc} = 15 V	13.0	13.3		12.25	13.3		V
	V_{cc} = 5 V	3.0	3.3		2.75	3.3		V
Rise time of output			100			100		nsec
Fall time of output			100			100		nsec

Notes

1. Supply Current when output high typically 1 mA less.
2. Tested at V_{cc} = 5 V and V_{cc} = 15 V.
3. This will determine the maximum value of $R_A + R_B$. For 15-V operation, the max total R = 20 MΩ.

A.6   EQUIVALENT CIRCUIT 555 IC

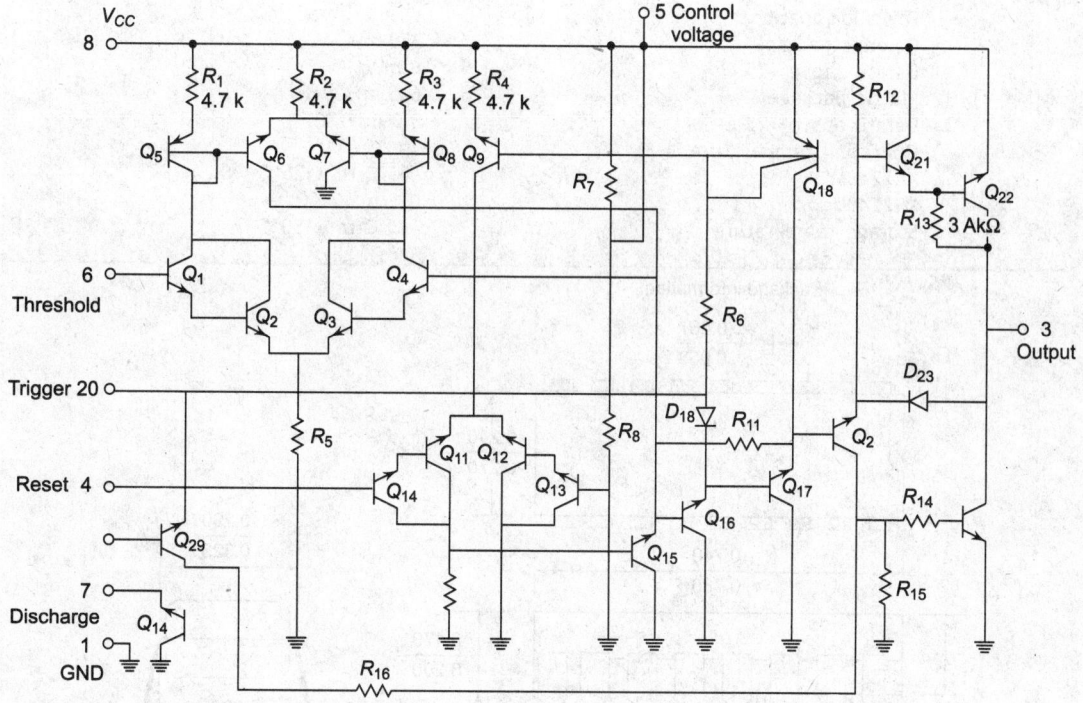

Fig. A.18   *Equivalent Circuit of 555 IC (Shown for One Side Only)*

A.6.1   The XR-2240 Programmable Timer/Counter

It is a monolithic controller capable of producing ultra-long time delays without sacrificing accuracy. In most applications, it provides a direct replacement for mechanical or electromechanical timing devices and generates programmable time delays from microseconds up to five days. Two timing circuits can be cascaded to generate time delays up to three years.

As shown in Figure A.5, the circuit is composed of an internal time-base oscillator, a programmable 8-bit counter and a control flip-flop. The time delay is set by an external RC network and can be programmed to any value from $1\,RC$ to $255\,RC$.

In astable operation, the circuit can generate 256 separate frequencies or pulse patterns from a single RC setting and can be synchronized with external clock signals. Both the control inputs and the outputs are compatible with TTL and DTL logic levels.

A. 6.1	
Features	**Applications**
Timing from microseconds to days	Precision timing
Programmable delays: $1\,RC$ to $255\,RC$	Long delay generation
Wide supply range: 4 V to 15 V	Sequential timing
TTL and DTL compatible outputs	Binary pattern generation
High accuracy: 0.5%	Frequency synthesis
External sync and modulation capability	Pulse counting/summing
Excellent supply rejection: 0.2%/V	A/D conversion
	Digital sample and hold

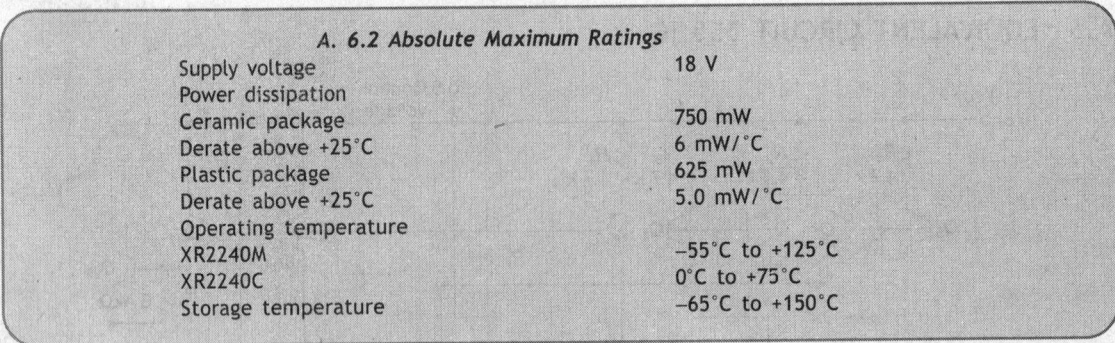

A. 6.2 Absolute Maximum Ratings

Supply voltage	18 V
Power dissipation	
Ceramic package	750 mW
Derate above +25°C	6 mW/°C
Plastic package	625 mW
Derate above +25°C	5.0 mW/°C
Operating temperature	
XR2240M	–55°C to +125°C
XR2240C	0°C to +75°C
Storage temperature	–65°C to +150°C

Package information

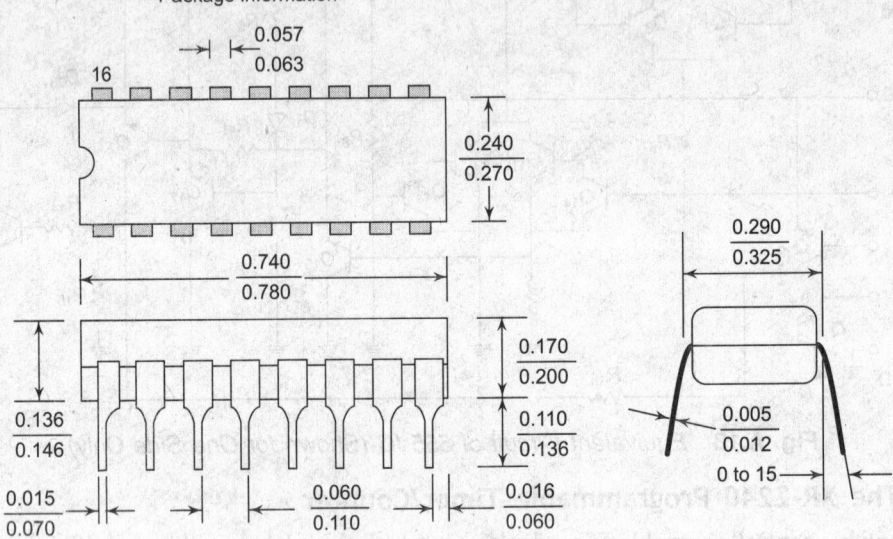

Fig. A. 19 *XR2240 Programmable Timer/Counter*

Functional block diagram

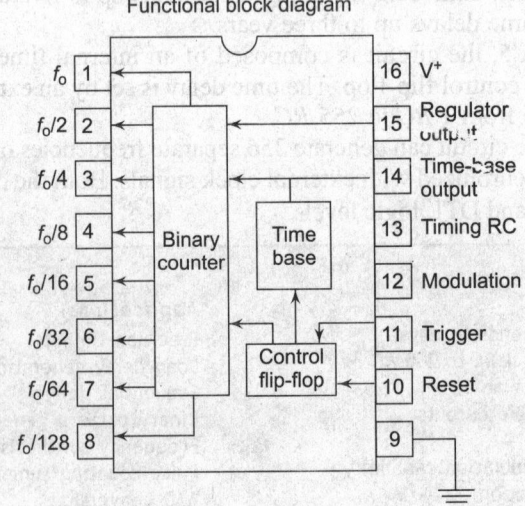

Fig. A.20 *XR2240 of Functional Block Diagram*

Electrical Characteristics of XR2240

Test Conditions: $V^+ = 5$ V, $T_A = 25°C$, $R = 10$ kΩ, $C = 0.1$ μF, unless otherwise noted.

Table A.5   Electrical Characteristics of XR2240

Parameters	XR2240			XR2240			Unit	Conditions
	Min.	Typ.	Max.	Min.	Typ.	Max.		
General Characteristics								
Supply voltage	4		15	4		15	V	For $V^+ < 4.5$ V, Short Pin 15 to pin 16
Supply current								
Total circuit		3.5	6		4	7	mA	$V^+ = 5$ V, $V_{TR} = 0$, $V_{RS} = 5$ V
		12	16		13	18	mA	$V^+ = 15$ V, $V_{TR} = 0$, $V_{RS} = 5$ V
Counter only		1			1.5		mA	
Regulator output, V_R	4.1	4.4		3.9	4.4		V	Measured at Pin 15, $V^+ = 5$ V
	6.0	6.3	6.6	5.8	6.3	6.8	V	$V^+ = 15$ V
Time-Base Section								
Timing accuracy*		0.5	2.0		0.5	5	%	$V_{RS} = 0$, $V_{TR} = 5$ V
Temperature drift		150	300		200		ppm/°C	$V^+ = 5$ V $0°C \leq T \leq 75°C$
		80			80		ppm/°C	$V^+ = 15$ V
Supply drift		0.05	0.2		0.08	0.3	%/V	$V^+ \geq 8$ V
Max. frequency	100	130			130		kHz	$R = 1$ kΩ, $C = 0.007$ μF
Modulation voltage level								Measured at pin 12
	3.00	3.50	4.0	2.80	3.50	4.20	V	$V^+ = 5$ V
		10.5			10.5		V	$V^+ = 15$ V
Recommended range of timing components								
timing resistor, R	0.001		10	0.001		10	MΩ	
timing capacitor, C	0.007		1000	0.01		1000	μF	
Trigger/Reset Controls								
Trigger								Measure at pin 11, $V_{RS} = 0$
Trigger threshold		1.4	2.0		1.4	2.0	V	
Trigger current		8			10		μA	$V_{RS} = 0$, $V_{TR} = 2$ V
Impedance		25			25		kΩ	
Response time**		1			1		μsec	
Reset								
Reset threshold		1.4	2.0		1.4	2.0	V	
Reset current		8			10		μA	$V_{TR} = 0$, $V_{RS} = 2$ V
Impedance		25			25		kΩ	
Response time**		0.8			0.8		μsec	
Counter Section								$V^+ = 5$ V
Max. toggle rate	0.8	1.5			1.5		MHz	$V_{RS} = 0$, $V_{TR} = 5$ V
Input:								Measured at pin 14
Impedance		20			20		kΩ	
Threshold	1.0	1.4		1.0	1.4		V	
Output:								Measured at pins 1 through 8
Rise time		180			180		nsec	$R_L = 3$ K, $C_L = 10$ pF
Fall time		180			180		nsec	
Sink current	3	5		2	4		μA	$V_{OL} \leq 0.4$ V
Leakage current		0.01	8		0.01	15	μA	$V_{OH} = 15$ V

*Timing error solely introduced by XR2240, measured as percentage of ideal time-base period of $T = 1.00$ RC.

**Propagation delay from application of trigger (or reset) input to corresponding state change in counter output at pin 1.

A.7   PRINCIPLE OF OPERATION OF XR2240

The timing cycle for the XR2240 is initiated by applying a positive-going trigger pulse to pin 11. The trigger input actuates the time-base oscillator, enables the counter section, and sets all the counter outputs to "low" state. The time-base oscillator generates timing pulses with its period, T, equal to 1 RC. These clock pulses are counted by the binary counter section. The timing cycle is completed when a positive-going reset pulse is applied to pin 10.

Figure A.24 gives the timing sequence of output waveforms at various circuit terminals, subsequent to a trigger input. When the circuit is at reset state, both the time base and the counter sections are disabled and all the counter outputs are at "high" state.

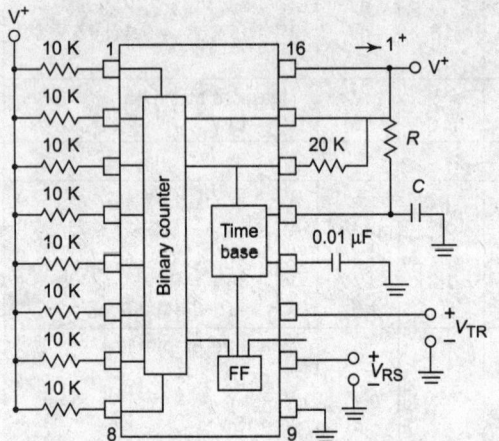

Fig. A.21   *Generalized Test Circuit*

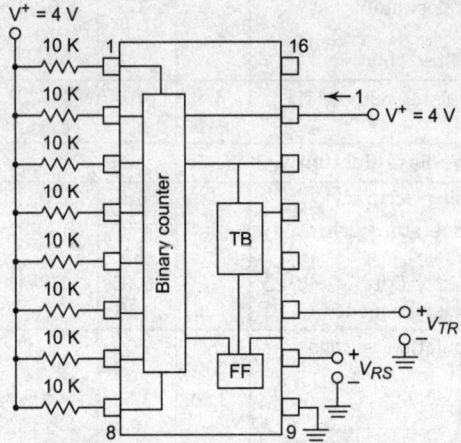

Fig. A.22   *Test Circuit for Low-Power Operation (Time-Base Powered Down)*

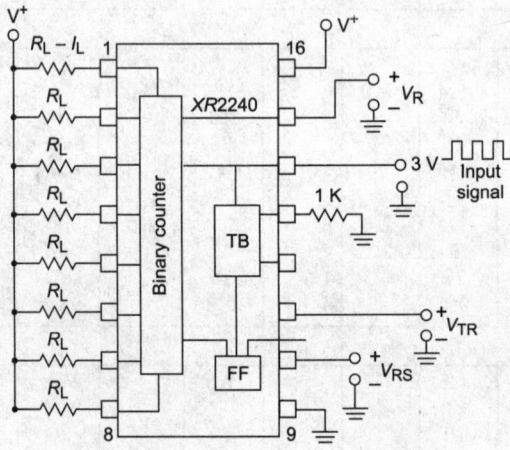

Fig. A.23   *Test Circuit for Counter Section*

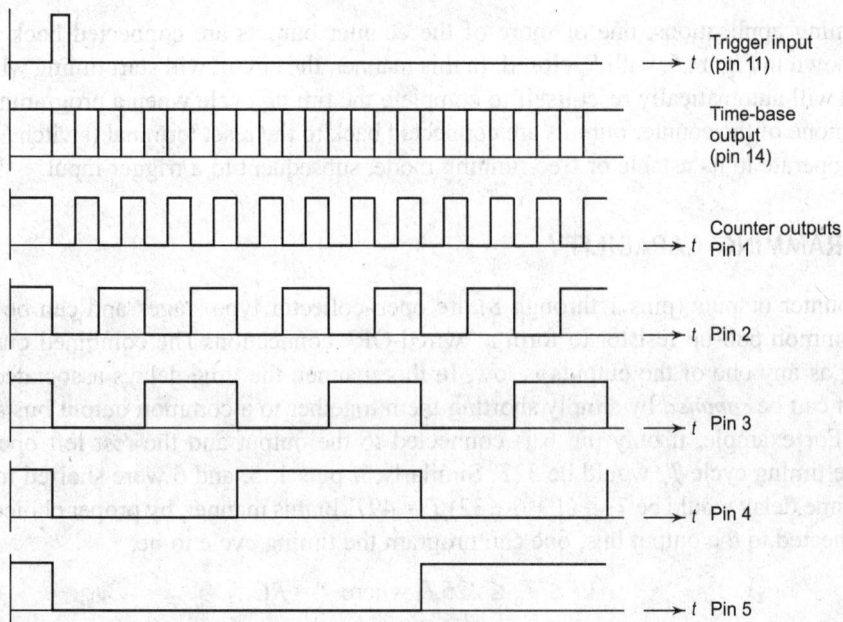

Fig. A.24   *Timing Diagram of Output Waveforms*

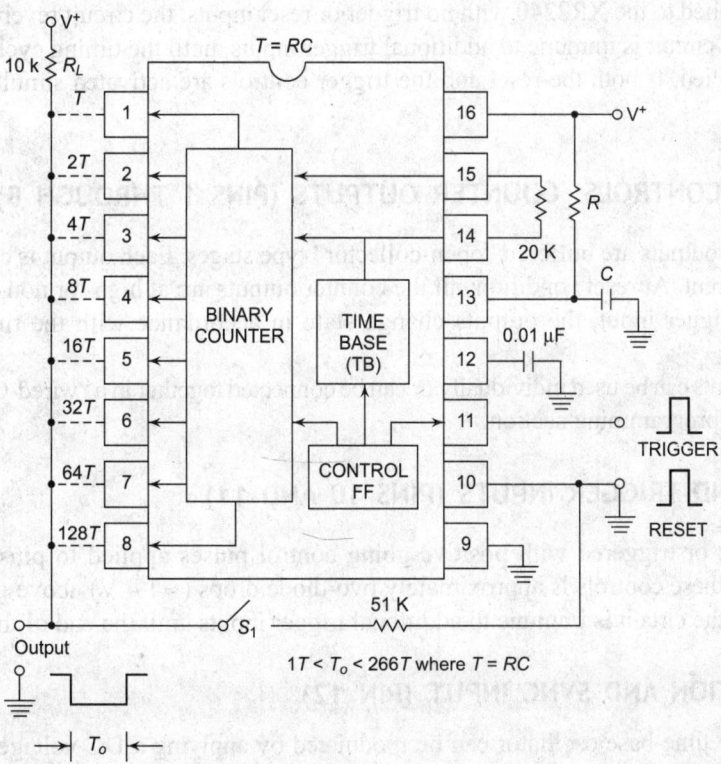

Fig. A.25   *Generalized Circuit Connection for Timing Applications (Switch S₁ Opens for Astable Operations, and Closes for Monostable Operations)*

In most timing applications, one or more of the counter outputs are connected back to the reset terminal, as shown in Figure 6, with S_1 closed. In this manner, the circuit will start timing when a trigger is applied and will automatically reset itself to complete the timing cycle when a programmed count is completed. If none of the counter outputs are connected back to the reset terminal (switch S_1 open), the circuit would operate in its astable or free-running mode, subsequent to a trigger input.

A.8 PROGRAMMING CAPABILITY

The binary counter outputs (pins 1 through 8) are open-collector type stages and can be shorted together to a common pull-up resistor to form a "wired-OR" connection. The combined output will be "low" as long as any one of the outputs is low. In this manner, the time delays associated with each counter output can be *summed* by simply shorting them together to a common output bus as shown in Figure A.25. For example, if only pin 6 is connected to the output and the rest left open, the total duration of the timing cycle T_o would be $32T$. Similarly, if pins 1, 5, and 6 were shorted to the output bus, the total time delay would be $T_o = (1 + 6 + 32) T = 49T$. In this manner, by proper choice of counter terminals connected to the output bus, one can program the timing cycle to be:

$$1T \leq T_o \leq 225T, \text{ where } T = RC.$$

A.9 TRIGGER AND RESET CONDITIONS

When power is applied to the XR2240 with no trigger or reset inputs, the circuit reverts to "reset" state. Once triggered, the circuit is immune to additional trigger inputs, until the timing cycle is completed or a reset input is applied. If both the reset and the trigger controls are activated simultaneously, trigger overrides reset.

A.10 CIRCUIT CONTROLS: COUNTER OUTPUTS (PINS 1 THROUGH 8)

The binary counter outputs are buffered "open-collector" type stages. Each output is capable of sinking ≈ 5 mA of load current. At reset condition, all the counter outputs are at high- or non-conducting state. Subsequent to a trigger input, the outputs change state in accordance with the timing diagram of Figure A.24.

The counter outputs can be used individually or can be connected together in a "wired-OR" configuration as described in the programming section.

A.11 RESET AND TRIGGER INPUTS (PINS 10 AND 11)

The circuit is reset or triggered with positive-going control pulses applied to pins 10 and 11. The threshold level for these controls is approximately two-diode drops (≈ 1.4 V) above ground.

Once triggered, the circuit is immune to additional trigger inputs until the end of the timing cycle.

A.12 MODULATION AND SYNC INPUT (PIN 12)

The period T of the time-base oscillator can be modulated by applying a DC voltage to this terminal. (as shown in the figure). The time-base oscillator can be synchronized to an external clock by applying a sync pulse to pin 12, as shown in the figure. Recommended sync pulse widths and amplitudes are also given in the figure.

A.13 HARMONIC SYNCHRONIZATION

Timebase can be synchronized with *integer multiples or harmonics* of input sync frequency by setting the time-base period T to be an integer multiple of the sync pulse period T_s. This can be done by choosing the timing components R and C at pin 13 such that:

$$T = RC = T_s/m, \text{ where } m \text{ is an integer}, 1 \leq m \leq 10.$$

Figure 17 gives the typical pull-in range for harmonic synchronization for various values of harmonic modulus m. For $m < 10$, typical pull-in range is greater than ±4% of time-base frequency.

A.14 ABSOLUTE MAXIMUM RATINGS

Power dissipation	Internally limited
Input–output voltage differential	40 V
Operating-junction temperature range	
LM117	−55°C to +150°C
LM217	−25°C to +150°C
LM317	0°C to +125°C
Storage temperature	−65°C to +150°C
Lead temperature (soldering, 10 sec)	300°C

A.15 PRECONDITIONING

Burn-in in thermal limit	100%, All devices

Table A.6 Electrical Characteristics

Parameter	Conditions	LM117/217			LM317			Unit
		Min.	Typ.	Max.	Min.	Typ.	Max.	
Line regulation	$T_A = 25°C$, $3\,V \leq V_{IN} - V_{OUT} \leq 40\,V$ (Note 2)		0.01	0.02		0.01	0.04	%/V
Load regulation	$T_A = 25°C$, $10\,mA \leq I_{OUT} \leq I_{MAX}$							
	$V_{OUT} \leq 5V$ (Note 2)		5	15		5	25	mV
	$V_{OUT} \geq 5V$ (Note 2)		0.1	0.3		0.1	0.5	%
Thermal regulation	$T_A = 25°C$, 20 msec pulse		0.03	0.07		0.04	0.07	%/W
Adjustment pin current			50	100		50	100	μA
Adjustment pin current Change	$10\,mA \leq I_L \leq I_{MAX}$ $3V \leq (V_{IN} - V_{OUT}) \leq 40V$		0.2	5		0.2	5	μA
Reference voltage	$3V \leq (V_{IN} - V_{OUT} \leq 40V$ (Note 3) $10\,mA \leq I_{OUT} \leq I_{MAX}, P \leq P_{MAX}$	1.20	1.25	1.30	1.20	1.25	1.30	V
Line regulation	$3V \leq V_{IN} - V_{OUT} \leq 40V$ (Note 2)		0.02	0.05		0.02	0.07	%/V

(Contd.)

Table A.6   *(Contd.)*

Parameter	Conditions	LM117/217			LM317			Units
		Min.	Typ.	Max.	Min.	Typ.	Max.	
Load regulation	$10\ mA \leq I_{OUT} \leq I_{MAX}$ (Note 2)							
	$V_{OUT} \leq 5V$		20	50		20	70	mV
	$V_{OUT} \geq 5V$		0.3	1		0.3	1.5	%
Temperature stability	$T_{MIN} \leq T_j \leq T_{MAX}$		1			1		%
Minimum load current	$V_{IN} - V_{OUT} = 40V$		3.5	5		3.5	10	mA
Current limit	$V_{IN} - V_{OUT} \leq 15V$							
	K and T Package	1.5	2.2		1.5	2.2		A
	H and P Package	0.5	0.8		0.5	0.8		A
	$V_{IN} - V_{OUT} = 40V$, $T_j = 25°C$							
	K and T Package	0.30	0.4		0.15	0.4		A
	H and P Package	0.15	0.07		0.075	0.07		A
RMS output noise, percentage of V_{OUT}	$T_A = 25°C$, $10\ Hz \leq f \leq 10\ kHz$		0.003			0.003		%
Ripple rejection ratio	$V_{OUT} = 10V$, $f = 120\ Hz$		65			65		dB
	$C_{ADJ} = 10\ \mu F$	66	80		66	80		dB
Long-term stability	$T_A = 125°C$		0.3	1		0.3	1	%
Thermal resistance, Junction to case	H Package		12	15		12	15	°C/W
	K Package		2.3	3		2.3	3	°C/W
	T Package					4		°C/W
	P Package					12		°C/W

Notes

1. Unless otherwise specified, these specifications apply $-55°C \leq T_j \leq +150°C$ for the LM117, $-25°C \leq T_j \leq +150°C$ for the LM217, and $0°C \leq T_j \leq +125°C$ for the LM317; $V_{IN} - V_{OUT} = 5$ V; and $I_{OUT} = 0.1$ A for the TO-39 and TO-202 packages and $I_{OUT} = 0.5$ A for the TO-3 and TO-220 packages. Although power dissipation is internally limited, these specifications are applicable for power dissipation of 2 W for the TO-39 and TO-202, and 20 W for the TO-3 and TO-220. I_{MAX} is 1.5 A for the TO-3 and TO-220 packages and 0.5 A for the TO-39 and TO-202 packages.

2. Regulation is measured at constant junction temperature using pulse testing with a low duty cycle. Changes in output voltage due to heating effects are covered under the specification for thermal regulation.

3. elected devices with tightened tolerance reference voltage available.

Typical Applications of LM311 and LM317

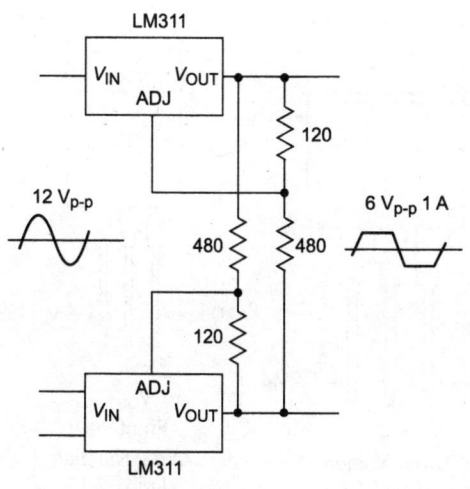

12 V Battery charger

Fig. A.26

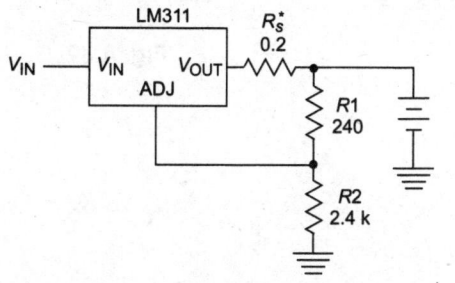

*R_S-sets output impedance of charge $Z_{OUT} = R_S \left(1 + \dfrac{R2}{R1}\right)$
Use of R_S allows low charging rates with fully
charged battery.

50 mA constant current battery charger

Fig. A.28

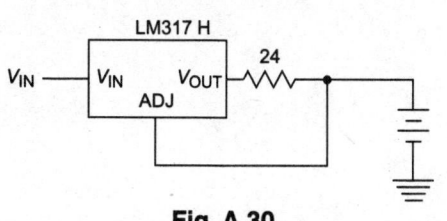

Fig. A.30

Adjustable 4A Regular

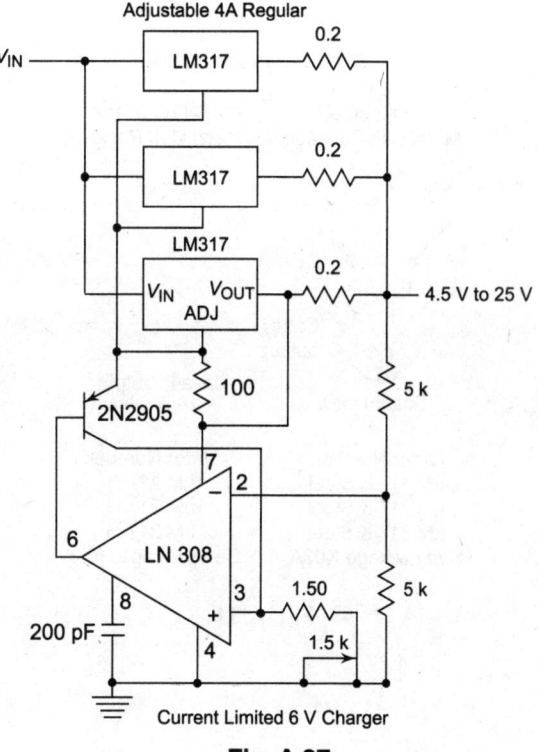

Current Limited 6 V Charger

Fig. A.27

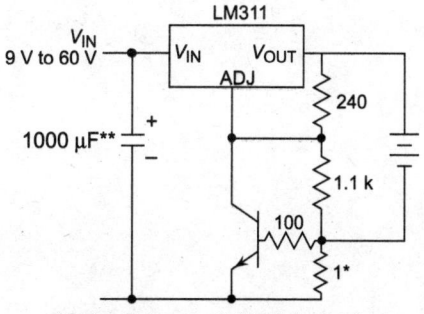

*Sets peak current (0.6A for 1 Ω)
**The 1000 µF is recommended to filter out
input transients

(TO-220)   (TO-202)
Plastic Package   Plastic Package

Fig. A.29

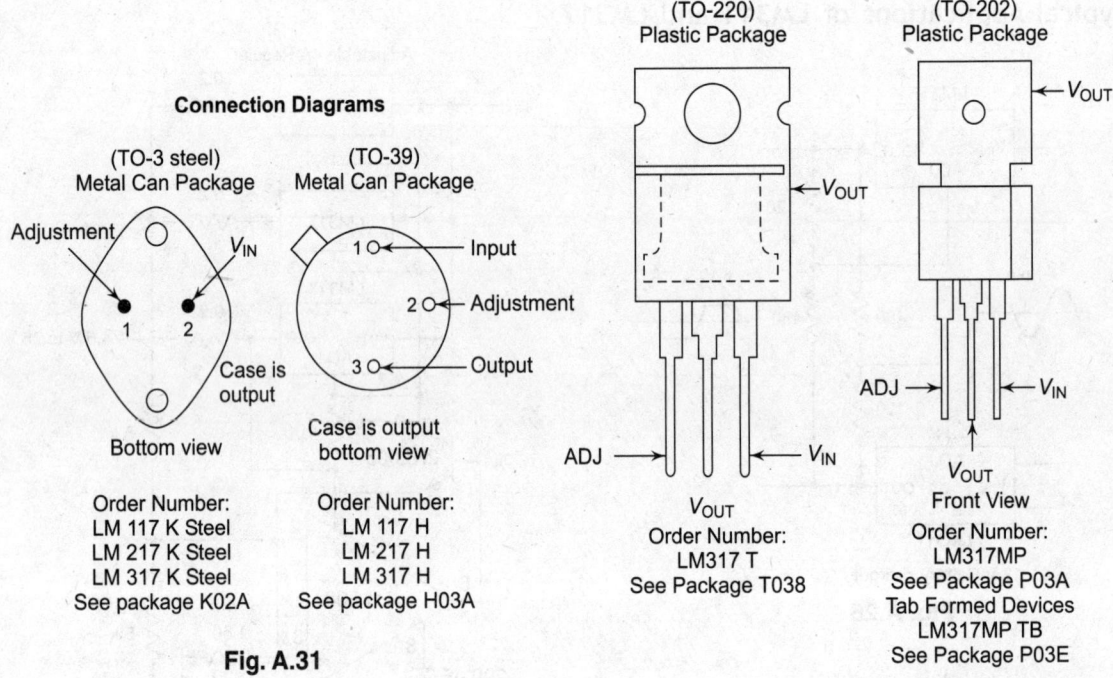

Connection Diagrams

(TO-3 steel)
Metal Can Package

Adjustment V_{IN}

Case is output

Bottom view

Order Number:
LM 117 K Steel
LM 217 K Steel
LM 317 K Steel
See package K02A

(TO-39)
Metal Can Package

1 — Input

2 — Adjustment

3 — Output

Case is output
bottom view

Order Number:
LM 117 H
LM 217 H
LM 317 H
See package H03A

Fig. A.31

(TO-220)
Plastic Package

V_{OUT}

ADJ V_{IN}

V_{OUT}
Order Number:
LM317 T
See Package T038

(TO-202)
Plastic Package

V_{OUT}

ADJ V_{IN}

V_{OUT}
Front View
Order Number:
LM317MP
See Package P03A
Tab Formed Devices
LM317MP TB
See Package P03E

Fig. A.32

Appendix B
Additional Solved Problems

CHAPTER 1

Operational Amplifiers—Basics

Ex. 1.1. For a given op-amp $PSRR = 70$ dB (min), $CMRR = 10^5$ $A_d = 10^5$. If the output voltage changes by 20 V in 4 msec, calculate
(a) $PSRR$ in numerical value
(b) A_c
(c) Slew rate

Solution: Given, $PSRR = 70$ dB (min)

$$CMRR = 10^5$$

Differential-mode gain $A_d = 10^5$

V_o changes by 20 V in 4 μsec
(a) Numerical value of $PSRR$: $20 \log (PSRR) = 70$ dB

$$\log_{10} (PSRR) = \frac{70}{20} = \frac{7}{2} = 3.5$$

So, $\qquad PSRR = 10^{3.5} = 3162.3$

Numerical value of $PSRR = 3162.3$

(b) Common-mode gain $CMRR = \dfrac{A_d}{A_{cm}}$

∴ $\qquad A_{cm} = \dfrac{A_d}{CMRR} = \dfrac{10^5}{10^5} = 1$

(c) $\qquad$ Slew rate $= \dfrac{20 \text{ V}}{4\mu \sec} = 5$ V/μsec

Ex. 1.2. Determine the output voltage produced by the cascaded integrator at $t = 0.5$ sec. Given $R_1 = 50$ kΩ; $C_1 = 0.5$ μF; $R_2 = 50$ kΩ. $C_2 = 1$ μF; and $V_i = 100$ mV.

Given: Duration of input pulse $= 0.5$ sec

$$T_1 = R_1 C_1 = 50 \times 10^3 \times 0.5 \times 10^{-6} = 25 \text{ msec}$$
$$T_2 = R_2 C_2 = 50 \times 10^3 \times 1 \times 10^{-6} = 50 \text{ msec}$$

$$V_1 = \frac{-V_i t}{R_1 C_1} = \frac{100 \times 10^{-3} \times 0.5}{50 \times 10^3 \times 0.5 \times 10^{-6}} = -2 \text{ V}$$

$$V_2 = \frac{-V_i t}{R_2 C_2} = \frac{100 \times 10^{-3} \times 0.5}{50 \times 10^3 \times 1 \times 10^{-6}} = -20 \text{ V}$$

Ex. 1.3. For a given op-amp, $CMRR = 80$ dB. Determine the numerical value of the same.

Solution: $\qquad CMRR = 80 \text{ dB}$ or $20 \log \left(\dfrac{1}{PSRR} \right) = 80 \text{ dB}$

$$\log_{10} \left(\frac{1}{PSRR} \right) = \frac{80}{20} = 4$$

$$\frac{1}{PSRR} = 10^4$$

$$PSRR = \frac{1}{10^4} = 100 \text{ } \mu\text{V/V}$$

Ex. 1.4. The change in op-amps input offset voltage V_{io} with change in supply voltage is called the supply voltage-rejection ratio $(SVRR)$. For a given op-amp, its value is 150 μV|V. Calculate change in V_{io}, if the supply voltages are varied from ±10 V to ±12 V.

Solution: $\qquad SVRR = \dfrac{\Delta V_{io}}{\Delta V}$ $\quad (SVRR$ is same as $PSRR)$

$$\Delta V_{io} = SVRR \Delta V; \quad \Delta V = (12 - 10) - (-12 - (-10))$$

$$= \left(150 \frac{\mu V}{V} \right) (4\text{V}) \text{ } \Delta V = 2 + 2 = 4 \text{ V}$$

$$= 0.6 \text{ mV}$$

Ex. 1.5. For the circuit shown in figure, determine the value of V_o, if $V_{in} = -6 \text{ } \mu$V DC. $V_{cc} = \pm 15$ V. $A_{OL} = 10^5$. Assume ideal op-amp.

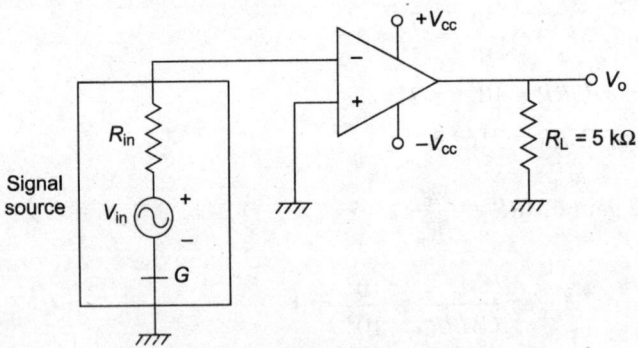

Fig. B1.1 *Circuit Diagram for Ex. 1.5*

Solution: $\qquad V_1 = V_{in}; V_2 = 0; V_o = -_{OL}(V_1 - V_2)$

$$V_o = -10^5 \text{ } (6 \text{ } \mu\text{V}) = 0.6 \text{ V}$$

Ex. 1.6. (a) Design an op-amp differentiator that differentiate an *I/P* signal with $t_{max} = 500$ Hz.
(b) Draw the output waveform for a sine wave of 1-V peak at 500 Hz applied to the differentiator.
(c) Draw the output waveform for a square wave input.

(a) $f_s = f_{max} = 500 \text{ Hz} = \dfrac{1}{2\pi R_f C_1}$

$\qquad C_1 = 0.1 \ \mu F$

$\qquad R_f = \dfrac{1}{2\pi(10^2) \times 5(10^{-7})} = 0.318 \text{ k}\Omega$

Let $f_b = 10$ and $f_a = 1 \text{ kHz}$.

$$= \dfrac{1}{2\pi R_1 C_1}$$

$$R_1 = \dfrac{1}{2\pi(10^3)(10^{-7})} = 1.59 \text{ k}\Omega$$

(b) $V_i = 1 \sin 2\pi C$

$$V_o = -R_f C_1 \dfrac{dV_i}{dt}$$
$$= -(0.3 \times 10^3)(10^{-7}) \cos [2\pi (100) \ t]$$
$$= 0.318 \times 10^{-3} \cos (2\pi (100)t)$$

(c) For square wave *I/P*, the *O/P* waveform will consist of positive and negative spikes of magnitude.

Ex. 1.7 Justify the statement:

Use of negative feedback will give stable, linear circuit gains that are independent of op-amp. The gain is determined solely by feedback resistors. So, $A_{OL}\beta \gg 1$.

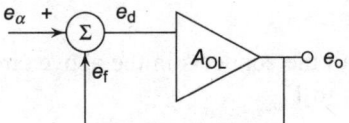

Fig. B1.2 *Circuit Diagram for Ex. 1.7*

$\qquad e_d = e_i - e_f$ \hfill (1)

$\qquad e_o = A_{OL} \ e_d$ \hfill (2)

$\qquad e_f = \beta \times e_o$ \hfill (3)

$\qquad \beta < 1$

When A_{OL} increases in Eq. (2), *O/P* e_o also increases, thereby e_f also increases; when e_d decreases e_o also decreases.

Closed-loop gain is defined as

$$A_v = \dfrac{e_o}{e_i}$$ \hfill (4)

Eq. (1) can be written as

$$e_i = e_d + e_f \tag{5}$$

$$A_v = \frac{A_{OL}e_d}{e_d + d_f} \tag{6}$$

Substitute Eq. (3) in Eq. (6),

$$A_v = \frac{A_{OL}e_d}{e_d + \beta_e} = \frac{A_{OL}e_d}{e_d + \beta A_{OL}e_d} = \frac{A_{OL}e_d}{e_d(1+\beta A_{OL})} = \frac{A_{OL}}{1+\beta A_{OL}}$$

$\beta A_{OL} \gg 1$, A_{OL} is very large in op-amp

Therefore, $\qquad A_v = \dfrac{A_{OL}}{\beta A_{OL}} = \dfrac{1}{\beta}$, so, $\boxed{\beta A_v \gg 1}$

Condition is proved.

Ex. 1.8. Bias compensation in op-amp

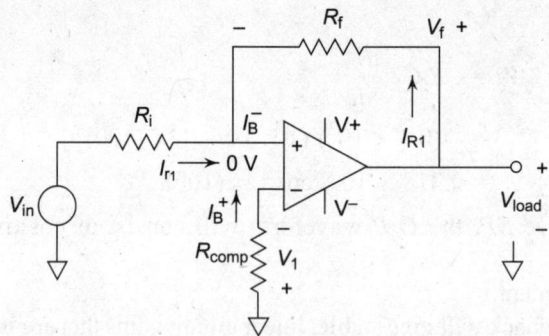

Fig. B1.3 *Circuit Diagram for Ex. 1.8*

Bias connect compensation.

Prove R_{com} is $R_f \parallel R_1$. Find all the connects in the above circuit.

(1) Apply K.V. Law from V_1 to V_{load}

$$-V_1 + 0 + V_f - V_{load} = 0$$

$$V_{load} = V_f - V_1$$

$$V_1 = T_B + R_{comp}$$

$$I_B^+ = V_1/R_{comp}$$

$$I_{R_i} = V_1/R_1$$

$$I_{R_f} = V^+/R_f$$

For compensation,

$$V_f = V_1, \; I_{R_f} = V_1/R_f,$$

$$I_B^- = I_{R_f} + I_{R_i}; \text{ Assume} = I_B^- = I_B^+$$

$$= \frac{V_1}{R_f} + \frac{V_1}{R_1}, \qquad V_i\left(\frac{R_1 + R_f}{R_1 R_f}\right) = \frac{V_i}{R_{comp}}$$

$$R_{comp} = R_1 \parallel R_f$$

Ex. 1.9. In the given circuit, find V_o for $a = 10^2$, 10^4, 10^6. If $v_1 = 1$ V, $R_1 = 2$ kΩ, and $R_2 = 18$ kΩ. Give the relationship between a and V_o of above values of a.

$$V_o = \frac{10}{1 + 10/a}$$

$$V_o = \frac{10}{1 + \dfrac{10}{10^2}} = 9.091 \text{ V, if } a = 10^2$$

$$V_o = \frac{10}{1 + \dfrac{10}{10^4}} = 9.990 \text{ V, if } a = 10^4$$

$$V_o = \frac{10}{1 + \dfrac{10}{10^6}} = 9.999 \text{ V, if } a = 10^6$$

$V_o(V_1) - a$
$9.091 - 10^2$
$9.99 - 10^4$
$9.999 - 10^6$

As V_o increases a also increases; for higher values of a, V_o resembles to 10 V accurately.

Ex. 1.10. Gain sensitivity of op-amp

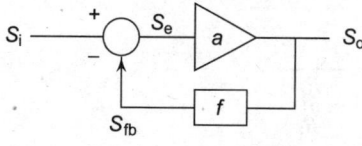

Feedback network

Fig. B1.4 *Circuit Diagram for Ex. 1.10*

$$S_{fb} = f_{so} \tag{1}$$
$$S_e = S_i - S_{fb} \tag{2}$$
$$\therefore \qquad S_e = S_i - f_{so} \tag{3}$$
$$S_o = a S_e \tag{4}$$
$$S_o = aS_i - af S_o$$
$$\frac{S_o}{S_i} = A = \frac{a}{1 + af} \tag{5}$$

$T \rightarrow$ loop gain $= af$
Gain sensitivity $\rightarrow$ Diff. (5)

$$\frac{dA}{da} = \frac{(1 + af) - af}{(1 + af)^2}$$

$$= 1/(1 + af)^2, \quad a \rightarrow \delta a$$

$$A \rightarrow \delta A$$

$$\delta A = \delta a / (1 + af)^2$$

$$\frac{\delta A}{A} = \frac{1 + af}{a} \frac{\delta a}{(1 + af)^2}$$

$$\frac{\delta A}{A} = \frac{\delta a / a}{1 + af} \tag{6}$$

Equation (6) gives gain sentivity of $A_1 a$.

Ex. 1.11   Using IC 741 op-amp, find offset voltage measurement, slew rate, and differential gain.

(1) Offset voltage measurement

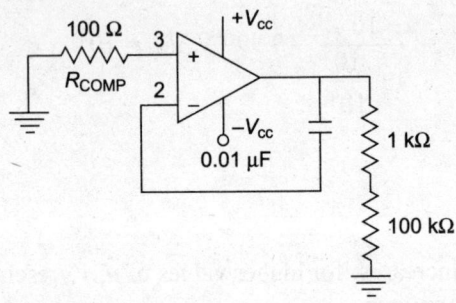

Fig. B1.5   *Circuit Diagram for Ex. 1.11 Offset Voltage Measurement*

$$V_{io} = \frac{V_o}{1 + \dfrac{R_f}{R_i}}$$

$$V_o = 40 \text{ mV}$$

$$V_{io} = \frac{40 \times 10^{-3}}{1 + (10 \times 10^3)/100} = 3.96 \times 10^{-6} \text{ V}$$

2. To measure slew rate

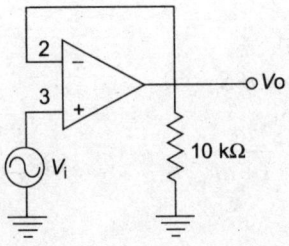

Fig. B1.6   *Measurement of Slew Rate*

$$V_{max} = 0.5 \text{ V}$$

$$f_{max} = 350 \text{ kHz}$$

$$S_R = \frac{2\pi f_{max} V_{max}}{10^6}$$

$$S_R = \frac{2 \times 3.14 \times 350 \times 10^3 \times 0.5}{10^6}$$

$$= 1.019$$

3. Differential gain

$$V_1 = G_1 = 2 \text{ V}$$
$$V_2 = G_2 = 1 \text{ V}$$
$$V_o = A(V_1 - V_2), \ V_o = 10 \text{ V}$$
$$A = V_o/V_1 - V_2$$
$$A = 10$$

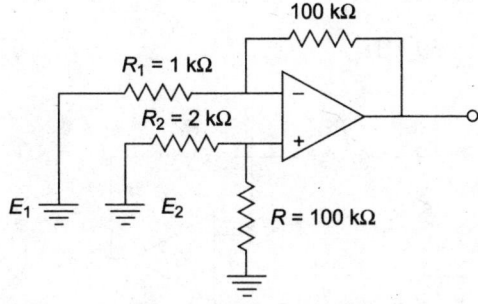

Fig. B1.7 *Measurement of Differential Gain*

Ex. 1.12 For the circuit shown below find output voltage.

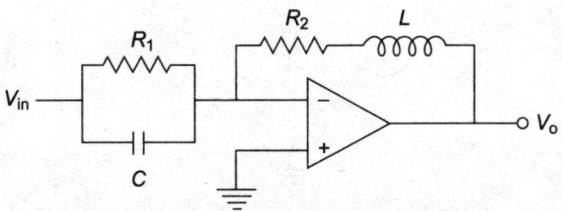

Fig. B1.8 *Circuit Diagram for Ex. 1.12*

$$V_o = V_1 + V_2 + V_3$$

$$V_1 = \frac{R_2}{R_1} V_{in}$$

$$V_2 = \left(R_2 C + \frac{L}{R_1} \right) \frac{dV_{in}}{dt}$$

$$V_3 = LC \frac{d^2 V_{in}}{dt^2}$$

$$-V_o = \frac{R_2}{R_1} V_{in} + \left(R_2 C + \frac{L}{R_1}\right)\frac{dV_{in}}{dt} + LC \frac{d^2 V_{in}}{dt^2}$$

Ex. 1.13 The following specifications are given for the figure given below. $R_4 = 3.9$ kΩ, $R_5 = 5$-K pot, $R_1 = R_2 = 1$ kΩ, $R_F = R_3 = 4.7$ kΩ, and the supply voltage is ± 15 V. If R_5 is set at 3 kΩ. Calculate (a) voltage gain, (b) *I/P* resistance, (c) output resistance, and (d) bandwidth of amplifier.

1. $A_d = -\left[1 + \frac{2(3.9 \times 10^3)}{3 \times 10^3}\right]\left(\frac{4.7 \text{ K}}{1 \text{ K}}\right) = -16.92$

2. $R_{if} = (2 \times 10^6)\left[1 + \frac{(200,000)(3.9 \text{ K} + 3 \text{ K})}{2(3.9 \text{ K}) + 3 \text{ K}}\right] = 256$ mΩ

3. $R_{of} = \dfrac{75}{1 + 200,000/16.9} = 6.3$ mΩ

4. $f_f = \dfrac{(200,000)(J)}{16.92} = 59$ kHz

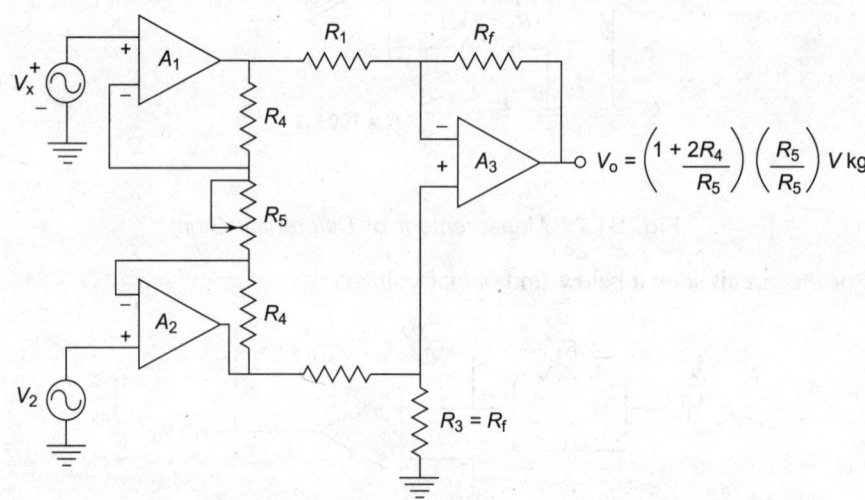

Fig. B1.9 *Circuit Diagram for Ex. 1.13*

Ex. 1.14 For the circuit shown below determine I_{C_1}, I_{C_2}, I_{C_3}, $\beta = 125$.

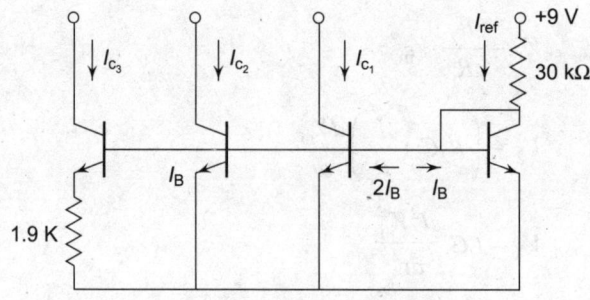

Fig. B1.10 *Circuit Diagram for Ex. 1.14*

$$I_{ref} = \frac{9 - 0.7}{30 \text{ K}} = 0.277 \text{ mA}$$

$$I_{ref} = I_C + 3I_B$$

$$= I_C \left(1 + \frac{3}{\beta}\right)$$

$$I_C = I_{ref} \frac{\beta}{1 + \beta}$$

$$I_{C_1} = I_{C_2} = 0.271 \text{ mA}$$

$$R_E = \frac{V_T}{\left(1 + \dfrac{1}{\beta}\right)I_{C_3}} \ln \frac{I_{C_1}}{I_{C_2}}$$

$$1.94 \text{ K} = \frac{0.025}{\left(1 + \dfrac{1}{125}\right)I_{C_3}} \ln \frac{0.231}{I_{C_3}}; \text{ Therefore, } I_{C3} = 0.0287 \text{ mA}$$

Ex. 1.15 For the circuit given below, find I_{C_1} and I_{C_2}, $V_o = 6$ V and $\beta = 200$. Also find R_C.

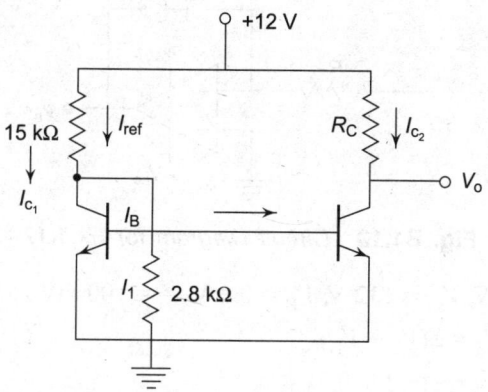

Fig. B1.11 *Circuit Diagram for Ex. 1.15*

$$I_{ref} = I_{C_1} + 2I_B + I_1$$

$$= I_{C_1} \left(1 + \frac{2}{\beta}\right) + I_1$$

$$I_{C_1} = 0.495 \text{ mA}$$

$$= 0.5 \text{ mA}$$

$$I_{C_2} = I_{C_1} \text{ (due to mirror effect)}$$

Apply Kirchhoff law

$$12 \text{ V} = I_{C_2} R_C + V_o$$

$$R_C = \frac{12 \text{ V} - 6 \text{ V}}{0.5 \text{ mA}} = 12 \text{ k}\Omega$$

Ex. 1.16   Find V_o in the circuit if $R_f = 10$ kΩ, $R_1 = 2$ kΩ, and $R_2 = 5$ kΩ.

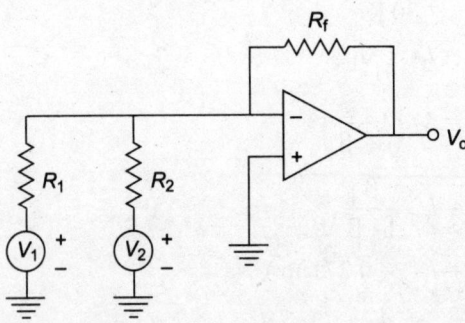

Fig. B1.12   *Circuit Diagram for Ex. 1.16*

$$V_o = -\frac{10}{2} V_1 + \frac{10}{5} V_2$$

$$V_o = -5 V_1 + 2V_2$$

Ex. 1.17   In the circuit given below, find *O/P* V_o and draw the output waveform.

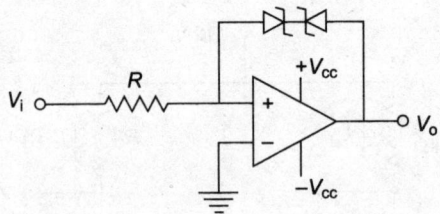

Fig. B1.13   *Circuit Diagram for Ex.1.17*

$R = 1$ kΩ, $V_{21} = V_{22} = 6.2$ V, $V_{cc} = \pm12$ V, $V_D = 0.3$ V, $V_i = 100$ mV *I/P* sine, 1 kHz frequency

$$V_o = \pm(V_2 + V_D)$$

$$= \pm(6.2 + 0.7)$$

$$V_o = \pm6.9 \text{ V}$$

O/P waveform

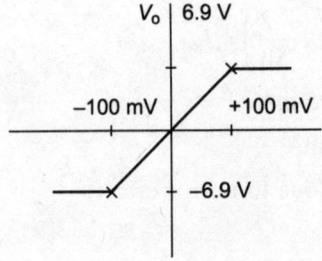

Fig. B1.14   *Output Waveform for Ex.1.17*

Ex. 1.18 For the circuit shown below, calculate V_{ref}, V_{sat}, and R_1, if $V_H = 0.1$ V, $V_{UT} = V_{ref}$, $A_{OL} = 100,000$, loop gain = 1000, $R_2 = 1$ kΩ.

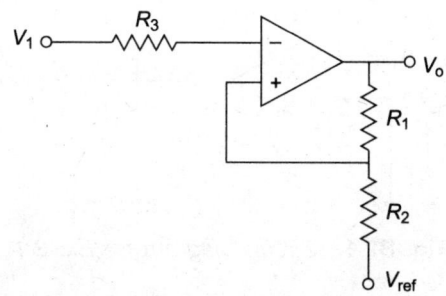

Fig. B1.15 *Output Waveform for Ex. 1.18*

$$V_{ref} + \frac{R_2}{R_1 + R_2}\ (V_{sat} - V_{ref}) = V_{UT}$$

if $V_{UT} = V_{ref}$

$$\frac{R_2}{R_1 + R_2}\ V_{sat} = \frac{R_2}{R_1 + R_2}\ V_{ref}, \text{ so, } V_{sat} = V_{ref}$$

From A_{OH} A, $V_{sat} = 5$ V

$$-V_{sat} = -5 \text{ V},$$

$$0.1 = \frac{21000}{R_1 + 1000}\ 5, R_1 = 99 \text{ k}\Omega$$

CHAPTER 2

Op-amp Applications

Ex. 2.1 Design a circuit to convert a 0 mA to 20 mA input current to a 0 V to 10 V output voltage. The circuit is powered from ±15-V regulated power supply.

Solution: $$V_o = -\left(\frac{V_i}{R_1}\right) R_F$$

If V_{in} and R_1 combination is replaced by a current source I_{in}, the output voltage becomes proportional to I_{in}. So, the circuit acts as a current to voltage converter.

$$V_o = 10 \text{ V}, \ I_{in} = 20 \text{ mA}, \ V_o = -I_{in} R_F$$

So, $$R_F = \frac{V_o}{I_{in}} = \frac{10 \text{ V}}{20 \text{ mA}} = 500 \ \Omega$$

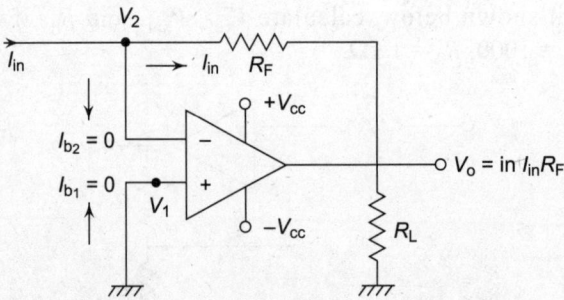

Fig. B2.1   *Circuit Diagram for Ex. 2.1*

Ex. 2.2   The input current to an op-amp current to voltage converter varies from 600 µA to 1 mA. If $R_F = 6.8$ kΩ, determine the variation of output voltage.

Solution:
$$V_o = -I_{in}R_F$$
$$V_{o1} = -(600 \times 10^{-6})\,(6.8 \times 10^3) = -4080 \text{ mV}$$
$$= -4.08 \text{ V}$$
$$V_{o2} = -(1 \times 10^{-3})\,(6.8 \times 10^3) = -6.8 \text{ V}$$

Thus, output voltage varies from –4.08 to –6.8 V.

Note: The input current is applied to the inverting terminal.

Ex. 2.3   Identify the circuit shown below and determine the value of closed-loop voltage gain A_{CL}. Calculate the input resistance $R_{if_1}{}'$ and $R_{if_2}{}'$.

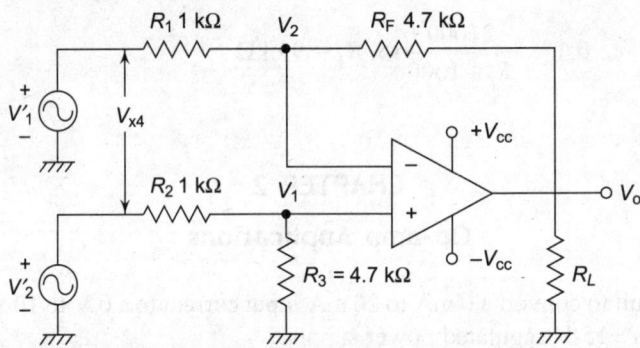

Fig. B2.2   *Circuit Diagram for Ex. 2.3*

Solution: The circuit is differential amplifier or difference amplifier.

$$V_o = V_{o1'} + V_{o2'}$$

$$V_o = -\frac{R_F}{R_1}\,(V_{1'} - V_{2'})$$

$$= -\frac{R_F}{R_1}\,(V_{1'2'})$$

Differential voltage gain $A_d = \dfrac{V_o}{V_{1'2'}} = -\dfrac{R_F}{R_1}$

$$A_d = -\frac{4.7 \text{ k}\Omega}{1 \text{ k}\Omega} = -4.7$$

$$R_{if_1}' \simeq R_1 = 1 \text{ k}\Omega$$

$$R_{if_2}' \simeq (R_2 + R_3) = (1 \text{ k}\Omega + 4.7 \text{ k}\Omega) = 5.7 \text{ k}\Omega$$

Ex. 2.4 For the circuit of differential amplifier using two op-amps shown, determine (i) A_d, (ii) bandwidth (B.W.), (iii) R_{ifx}, (iv) R_{OF}, (v) V_o. Given $R_2 = R_F = 4.7 \text{ k}\Omega$, $R_3 = R_1 = 1 \text{ k}\Omega$, $R_i = 2 \text{ M}\Omega$, $V_x = 600 \text{ mV}$ *p-p*; $V_y = 400 \text{ mV}$ *p-p* sine waves at : $R_o = 60 \Omega$, 500 Hz, $f_o = 1$ MHz, $A = 200{,}000$.

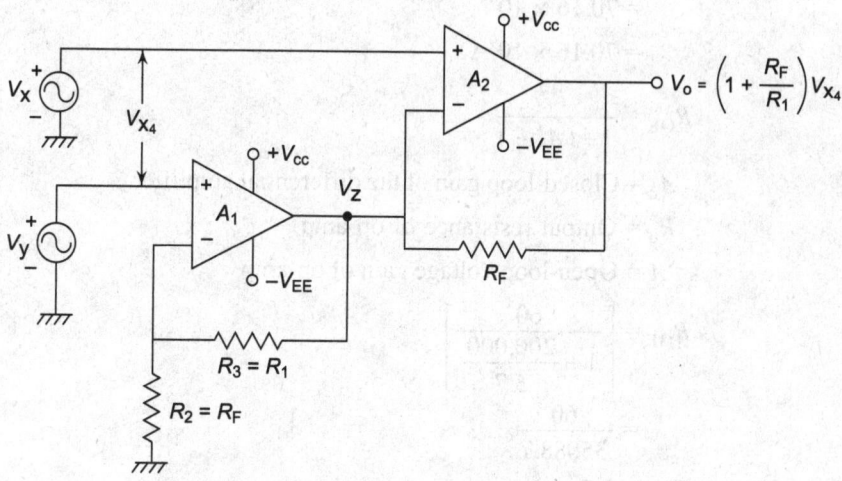

Fig. B2.3 *Circuit Diagram for Ex. 2.4*

(i)
$$A_d = \frac{V_o}{V_{xy}} = \left(1 + \frac{R_F}{R_1}\right) = \left(1 + \frac{4.7 \text{ k}\Omega}{1 \text{ k}\Omega}\right) = 5.7$$

(ii)
$$\text{B.W.} = f_F = \frac{\text{Unity-gain bandwidth}}{\text{Closed-loop gain } A_d}$$

$$= \frac{1 \text{ MHz}}{5.7} = 175.44 \text{ kHz}$$

(iii)
$R_{ify}(y)$ = The input resistance seen by signal source V_y

$$= R_i (1 + AB)$$

$$B = \frac{R_2}{R_2 + R_3}$$

$$R_{ify} = 2 \times 10^6 \left(1 + \frac{200{,}000 \times 4.7}{4.7 \text{ k}\Omega + 1 \text{ k}\Omega}\right)$$

$$= 2 \times 10^6 \left(1 + \frac{200 \times 10^3 \times 4.7}{5.7 \times 10^3}\right)$$

$$R_{ify} = 329.89 \text{ M}\Omega$$

(iv)
R_{ifx} = The input resistance seen by the voltage source V_x

$$= R_i (1 + AB)$$

$$B = \frac{R_1}{R_1 + R_F}$$

$$B = \frac{1 \times 10^3}{1 \times 10^3 + 4.7 \times 10^3} = \frac{10^3}{5.7 \times 10^3} = 0.1754$$

$$R_{ifx} = 2 \times 10^6 (1 + 2 \times 10^5 \times 0.1754)$$

$$= 0.7016 \times 10^{11}$$

$$= 70.16 \times 10^9$$

$$= 70.16 \times 10^9 \ \Omega$$

(v) $$R_{OF} = \frac{R_o}{1 + (A/A_d)}$$

A_d = Closed-loop gain of the differential amplifier

R_o = Output resistance of op-amp

A = Open-loop voltage gain of op-amp

$$R_{OF} = \left[\frac{60}{1 + \dfrac{200{,}000}{5.7}} \right]$$

$$= \frac{60}{35088.7}$$

$$R_{OF} = 1.7 \ m\Omega$$

Ex. 2.5 A 741C op-amp is being used as an inverting amplifier, what is the maximum gain that can be used keeping the amplifier response flat to 1 kHz.

Solution: $$f_o = \frac{vaB}{A}$$

vaB = unity-gain bandwidth

f_o = Break frequency

$A = A_{OL}$ = open-loop voltage gain of op-amp

A_F = closed-loop voltage gain

$$A_F = \frac{A_{OL}}{1 + (A_{OL})B} \ ; \quad B = \text{Gain of feedback circuit}$$

$$\frac{A_F}{K} = \frac{vaB}{f_F} \quad \text{or} \quad A_F = \frac{(vaB)(k)}{f_F}$$

$$A_F = \frac{(1 \, M)(A_{F+1})}{1 \, k\Omega} \quad K = (A_{F+1})$$

So, $$A_F = 999$$

Ex. 2.6 Design an amplifier of +15 using op-amp

(a) $$R_1 = 10 \ k\Omega, \ \text{Find} \ R_f$$

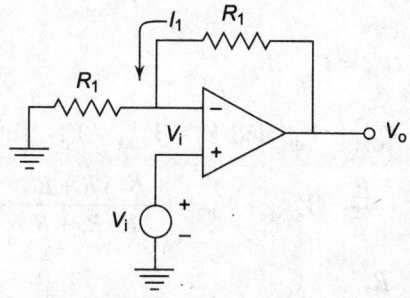

Fig. B2.4 *Circuit Diagram for Ex. 2.6*

$$A_{CL} = 1 + R_f/R_1$$

$$15 = 1 + R_f/10 \text{ K}$$

$$R_f = 140 \text{ k}\Omega$$

(b) Assume $V_i = 1$ V, $R_L = 5$ kΩ;
find V_o, I_L, I_1, and I_o.

(i) $V_o = (1 + R_f/R_1) \, V_i = \left(1 + \dfrac{140}{10}\right) (1) = 15$ V

(ii) $I_L = \dfrac{V_o}{R_L} = \dfrac{15}{5 \text{ k}\Omega} = 3$ mA

(iii) $I_1 = \dfrac{V_i}{R_1} = \dfrac{V_o - V_i}{R_f} = \dfrac{15-1}{10 \text{ K}} = 1.4$ mA

(iv) $I_o = I_L + I_1 = 3 + 1.4 = 4.4$ mA

Ex. 2.7 For a differential amplifier using op-amp, find output voltage V_o.
   (a) Find V_o if V_{CM} is zero. Assume $R'/R = R_2/R_1$.
   (b) Find CMRR of the amplifier.

$$V_o = -\frac{R'}{R} \, V_2 + \left(\frac{R+R'}{R}\right)\left(\frac{R_2}{R_1+R_2} V_1\right) \qquad (1)$$

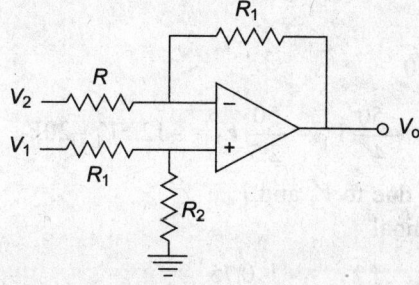

Fig. B2.5 *Circuit Diagram for Ex. 2.7*

(a) $V_{CM} = \dfrac{V_1 + V_2}{2}$, $\qquad V_d = V_1 - V_2$

So, $\qquad\qquad V_1 = V_{CM} + V_{d/2}$ and $V_2 = V_{CM} - V_{d/2}$, substitute V_1 and V_2 in Eq. (1)

$$V_o = -\frac{R'}{R}\ (V_{CM} - V_{d/2}) + \frac{R_2}{R}\frac{R + R'}{R_1 + R_2}\ (V_{CM} + V_{d/2}) \qquad\qquad (2)$$

if $\qquad\qquad \dfrac{R'}{R} = \dfrac{R_2}{R_1}$,

$$\frac{R'}{R} + 1 = \frac{R_2}{R_1} + 1 \quad \text{or} \quad \frac{R' + R}{R} = \frac{R_1 + R_2}{R_1}$$

$$V_o = \left(\frac{R'}{R} + \frac{R_2}{R_1}\right)\frac{V_d}{2} = \left(\frac{R_2}{R_1}\right) V_d$$

(b) $\text{CMRR} = \dfrac{A_{DM}}{A_{CM}} = \dfrac{R'(R_1 + R_2) + R_2(R + R')}{R'(R_1 + R_2) - R_1(R + R')}$

Ex. 2.8 Find V_o for the adder–subtractor shown below:

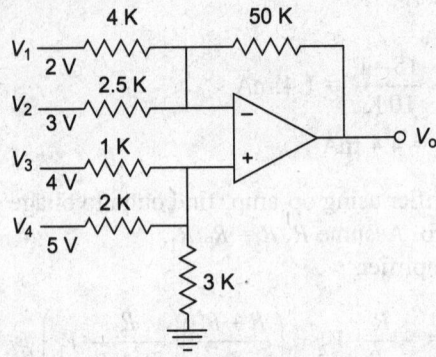

Fig. B2.6 *Circuit Diagram for Ex. 2.8*

Solution:

Adder, keeping $V_3 = V_4 = 0$

$$V'_o = \frac{-50}{4}\ V_1 - \frac{50}{2.5}\ V_2 = -12.5V_1 - 20V_2$$

Now $V_1 = V_2 = 0$, find V_o due to V_3 and V_4.

Let V^+ *I/P* at positive terminal

$$V^+ = \frac{1.2}{1 + 1.2}\ V_3 + \frac{0.75}{2.75}\ V_4$$

$$V^+ = \frac{1.2}{2.2}\ V_3 + \frac{0.75}{2.75}\ V_4$$

Now, the circuit can be redrawn as

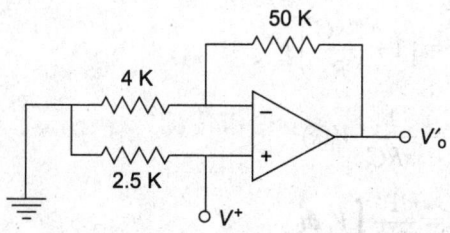

Fig. B2.7 *Circuit Diagram for Ex. 2.8*

$$V'_o = (1 + R_f/R)V^+$$

$$= \left(1 + \frac{50}{4112.5}\right) V^+$$

$$= \left(1 + \frac{50}{1.5}\right) V^+$$

so, $$V'_o = \left(1 + \frac{50+1.5}{1.5}\right)\left(\frac{1.8}{2.2}V_3 + \frac{0.75}{2.75}V_4\right)$$

$$= \frac{51.5}{1.5} \times \frac{1.2}{2.2} V_3 + \frac{51.5}{1.5} \times \frac{0.75}{2.75} V_4$$

$$= 18.72V_3 + 9.36V_4$$

$$V_o = -12.5V_1 - 20V_2 + 18.72V_3 = 9.36V_4$$

$$= -95 + 74.88 + 46.8$$

$$V_o = 26.68 \text{ V}$$

Ex. 2.9 For non-inverting integrator shown below find *O/P* V_o.

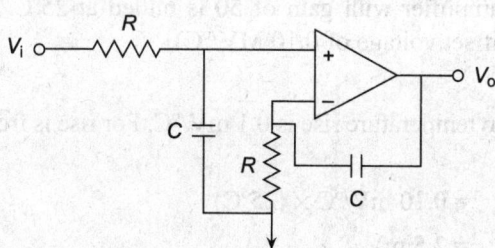

Fig. B2.8 *Circuit Diagram for Ex. 2.9*

Solution:

The voltage at the positive input terminal of op-amp due to potential divider:

$$V^+ = \frac{1/sC}{R + 1/sC} \ V_i(s)$$

The output voltage $V_o(s)$ for the non-inverting amplifier is

$$V_o(s) = \left(1 + \frac{1/sC}{R}\right) V^+$$

$$= \frac{1}{sRC} \, V_i(s)$$

So, $\qquad\qquad V_o = \frac{1}{RC} \int V_i \, dt$

Ex. 2.10   A Schmitt trigger with the upper threshold level $V_{UT} = 0$ V and hysteresis width $V_H = 0.4$ V converts 500 Hz sine wave amplitude 2 V_{PP} into square wave. Calculate the time direction of negative and positive portion of output waveform.

Solution:

$$V_{UT} = 0, \; V_H = V_{UT} - V_{LT}$$

$$= 0.4 \text{ V}, \; V_{LT} = -0.4 \text{ V}$$

$$-0.4 = V_m \sin(\pi + \theta)$$

$$= -1 \sin \theta = -\sin \theta$$

$$\theta = avc \sin 0.1$$

$$T = 1/f = 1/500 = 2 \text{ msec}$$

$$WT_\theta = 2\pi \, (500) \; T_\theta = 0.1$$

$$T_\theta = (0.2/2\pi)^2 \text{ msec} = \left(\frac{0.2}{2}\pi\right) \text{ msec} = (0.1 \, \pi) \text{ msec}$$

$$= 0.314 \text{ msec}$$

$$T_1 = T/2 + T_\theta = 2 + 0.314 = 2.314 \text{ msec}$$

$$T_2 = T/2 - T_\theta = 2 - 0.314 = 1.686 \text{ msec}$$

Ex. 2.11   A non-inverting amplifier with gain of 50 is nulled at 25°C. Find the *O/P* voltage when temperature raises to 50°C (offset voltage of 0.10 MV/°C)

Solution:

I/P offset voltage due to temperature rise is 0.1 mV/°C. For rise is from 25°C to 50°C. (i.e., 25°C), *I/P* offset voltage is

$$= 0.10 \text{ mV/°C} \times (25°C)$$

$$= 2.5 \text{ mV}$$

Give gain $A = 100$
O/P voltage change will be

$$V_o = V_{os} \times A_{CL}$$

$$V_{os} = 2.5 \text{ mV}$$

$$A_{CL} = 100$$

$V_o = 250$ mV (There is a large shift in *O/P* voltage due to temperature increase.)

Ex 2.12 For the circuit shown in figure below, calculate amplitude of triangular wave, square wave, and signal frequency. Given $R = 10$ kΩ, $C = 0.1$ μF, $R_i = 100$ kΩ, $R_f = 800$ kΩ, and $\pm V_{Supply} = 15$ V.

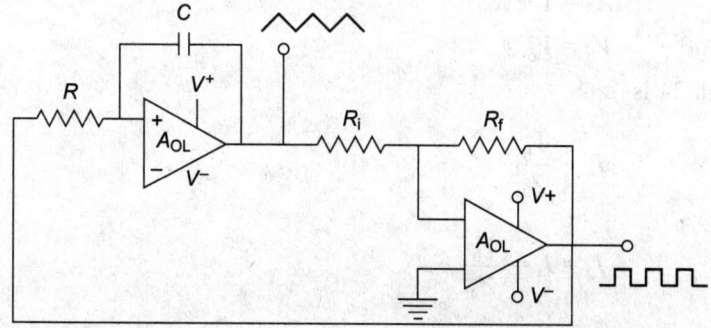

Fig. B2.9 *Circuit Diagram for Ex. 2.12*

(a) Square wave amplifier

$$V_{p-p} = V_{sat} - (-V_{sat}) = 10 \text{ V} - (-10) = 20 \ V_{p-p}$$

(b) Triangular wave amplifier

$$V_{p-p} = \frac{2R_i}{R_f} \ V_{sat}$$

$$= \frac{2(100 \text{ k}\Omega)(10 \text{ V})}{800 \text{ k}\Omega}$$

$$= 2.5 \ V_{p-p}$$

(c) Frequency $f = \dfrac{R_f}{4R_i R_C} = \dfrac{800 \text{ k}\Omega}{4(100 \text{ k}\Omega)(10 \text{ k}\Omega)(0.1 \, \mu\text{F})}$

$$= 2 \text{ kHz}$$

Ex. 2.13 For the second-order active filter model, find transfer function.

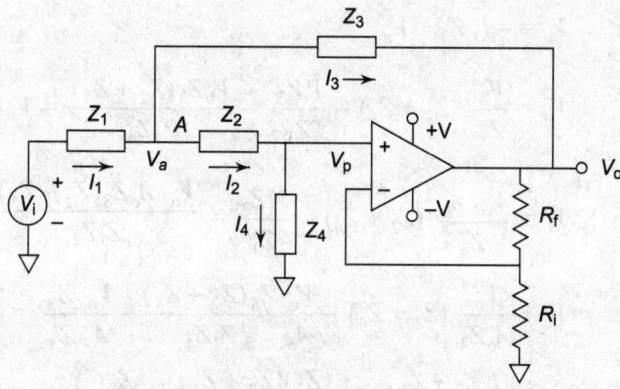

Fig. B2.10 *Circuit Diagram for Ex. 2.13*

The output V_o is given by

$$V_o = A_o V_b \tag{1}$$

$$A_o = 1 + R_f / \hat{R}_i \tag{2}$$

$$V_b = V_o / A_o \tag{3}$$

So, I_4 through 24 is

$$I_4 = \frac{V_b}{Z_4}$$

$$I_2 = I_4 = \frac{V_b}{Z_4} \tag{4}$$

At the node A,

$$V_a = I_4 (Z_2 + Z_4)$$

$$V_a = \frac{V_b}{Z_4} (Z_2 + Z_4) \tag{5}$$

I_1 is the difference in potential across Z_1 divided by Z_1,

$$I_1 = \frac{V_i - V_a}{Z_1} = \frac{V_i}{Z_1} - \frac{V_a}{Z_1}$$

$$= \frac{V_i}{Z_1} - \frac{V_b}{Z_1 Z_4} (Z_2 + Z_4) \tag{6}$$

But write $I_3 = I_1 - I_2$

$$= \frac{V_i}{Z_1} - \frac{V_b (Z_2 + Z_4)}{Z_1 Z_4} - \frac{V_b}{Z_4} \tag{7}$$

Writing equation at node A, O/P write input

$$V_a - I_3 Z_3 - V_o = 0$$

$$V_o = - I_3 Z_3 + V_a$$

$$V_o = \frac{V_b}{Z_4} (Z_2 + Z_4) - \frac{V_i Z_3}{Z_1} - \frac{V_b Z_3 (Z_2 + Z_4)}{Z_1 Z_4} - V_b Z_3 / Z_4$$

$$V_o = \frac{V_o / A_o}{Z_4} (Z_2 + Z_4) - \frac{V_i Z_3}{Z_1} - \frac{V_o / A_o Z_3 (Z_2 + Z_4)}{Z_1 Z_4} \quad V_o / A_o Z_3 / Z_4$$

$$-V_o + \frac{V_o}{A_o Z_4} (Z_2 + Z_4) - \frac{V_o}{A_o} \frac{Z_3 (Z_2 + Z_4)}{Z_1 Z_4} - \frac{V_o}{A_o} \frac{Z_3}{Z_4} = \frac{V_i Z_3}{Z_1}$$

$$V_o \left(\frac{Z_2 + Z_4}{A_o Z_4} - 1 - \frac{(Z_3)(Z_2 + Z_4)}{A_o Z_1 Z_4} - \frac{Z_3}{Z_4 A_o} \right) = V_i \frac{Z_3}{Z_1}$$

$$\frac{V_o}{V_i A_o}\left(\frac{A_o(Z_2+Z_4)}{A_o Z_4}-A_o-\frac{Z_3(Z_2+Z_4)}{Z_1 Z_4}-\frac{A_o Z_3}{A_o Z_4}\right)$$

$$=\frac{Z_3}{Z_1}$$

$$\frac{V_o}{V_i}=\frac{A_o Z_3/Z_1}{\dfrac{Z_2+Z_4}{Z_4}-A_o-\dfrac{Z_3(Z_2+Z_4)}{Z_1 Z_4}-\dfrac{Z_3}{Z_4}}$$

$$=\frac{A_o Z_3 Z_4}{\dfrac{\cancel{Z_1}(Z_1(Z_2+Z_4)-A_o Z_1-Z_3(Z_2+Z_4)-Z_1 Z_3)}{\cancel{Z_1}}}$$

$$\frac{V_o}{V_i}=\frac{A_o Z_3 Z_4}{Z_1 Z_2+Z_1 Z_4-A_o Z_1-Z_2 Z_3-Z_3 Z_3-Z_1 Z_3}$$

Ex.2.14 For the second-order low-pass active filter, find phase, magnitude, and transfer functions.

Solution:
$$A_o=1+R_f/R_1 \tag{1}$$
$$W_o^2=1/R_1 R_2\, C_1 C_2 \tag{2}$$

$$\omega_o=1/\sqrt{R_1 R_2 C_1 C_2}$$

$$\alpha=\text{damping coefficient}=\frac{R_2 C_2+R_1 C_2+R_1 C_1(1-A_o)}{\sqrt{R_1 R_2 C_1 C_2}}$$

System transfer function

$$V_o=\frac{V_i A_o \omega_o^2}{S^2+\alpha\omega_o s+\omega_o^2}$$

$$\frac{V_o}{V_i}=\frac{A_o}{S^2+\alpha s+1},\ \alpha=R_2 C_2+R_1 C_2+R_1 C_1\,(1-A_o),\ \text{subset } S=j\omega$$

$$\frac{V_o}{V_i}=\frac{A_o}{(j\omega)^2+\alpha j\omega+1}=\frac{A_o}{(1-\omega^2)+j\alpha\omega}$$

$$\frac{V_o}{V_i}=\frac{A_o}{(1-\omega^2)+j\alpha\omega}\frac{(1-\omega^2)-j\alpha\omega}{(1-\omega^2)-j\alpha\omega}$$

$$=\frac{A_o(1-\omega^2)}{(1-\omega^2)^2+\alpha^2\omega^2}-j\frac{A_o\alpha\omega}{(1-\omega^2)^2+\alpha^2\omega^2}$$

$$(G)=\sqrt{\text{Red}^2+(\text{imaginery})^2}=\frac{A_o\sqrt{(1-\omega^2)^2+\alpha^2\omega^2}}{(1-\omega^2)^2+\alpha^2\omega^2}$$

$$= \frac{A_o}{\sqrt{(1-\omega^2)^2 + \alpha^2\omega^2}}$$

The phase relationship is
ϕ = arctan imaginary/vcel.

$$= -\arctan \frac{\alpha\omega}{1-\omega^2}$$

Ex. 2.15 Design a circuit that has output frequency of sine wave twice that of input frequency. One wave at f_x, the other sine frequency f_y

$$e_x = E_{DC} + E_x \sin 2\pi f_{xt}$$

$$e_y = E_y \sin 2\pi f_y \, t$$

$$V_{load} = \frac{e_x e_y}{E_{ref}}$$

$$= \frac{1}{E_{ref}} \left[(E_{DC} + E_x \sin 2\pi f_{xt}) (E_y \sin 2\pi f_y t) \right]$$

$$= \frac{1}{E_{ref}} \left[E_{DC} E_y \sin 2\pi f_y t + (E_x \sin 2\pi f_x t) (E_y \sin 2\pi f_y t) \right]$$

$$= \frac{E_{DC} E_y}{E_{ref}} \sin 2\pi f_y t + \frac{E_x E_y}{E_{ref}} (\sin 2\pi f_x t)(\sin 2\pi f_y t)$$

$$V_{load} = \frac{E_{DC} E_y}{E_{ref}} \sin 2\pi f_y t + \frac{E_x E_y}{2 E_{ref}} \cos 2\pi (f_y - t_x)$$

$$- \frac{E_x E_y}{2 E_{ref}} \cos (f_y + f_x)$$

The output at load is the sum and difference of two frequency of input sine wave.

Ex. 2.16 An inverting amplifier with $A = -2$, $V_{ce} = \pm 10$ V, p–p triangular wave. Sketch V_i, V_o, and V_N when $I/P \pm 10$ V and $O/P \pm 20$ V, the line between linear portion and saturation occurs $V_i = \pm 13/2 = \pm 6.5$ V

When between -6.5 V $< V_i < 6.5$ V, op-amp is linear region.

$$V_o = -2V_i, \ V_N \simeq 0 \text{ (virtual ground)}$$

$$V_i > 6.5 \text{ V, op-amp set } V_o = -13 \text{ V}$$

$$V_N = (2/3) \, V_{i+} (1/3) \, (-13), \ V_i = 10 \text{ V}, \ V_N = 2.33 \text{ V}$$

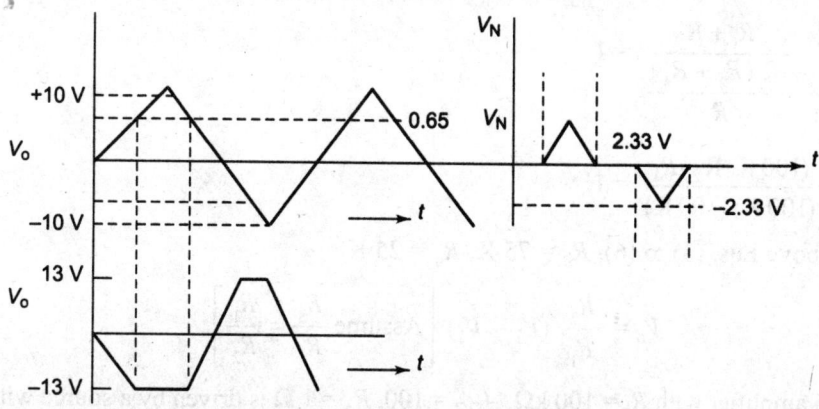

Fig. B2.11 *Waveforms for Ex. 2.16*

Ex. 2.17 Design a circuit such that $V_o = V_2 - 3V_1$,

$$R_{i1} = R_{i2} = 100 \text{ k}\Omega$$

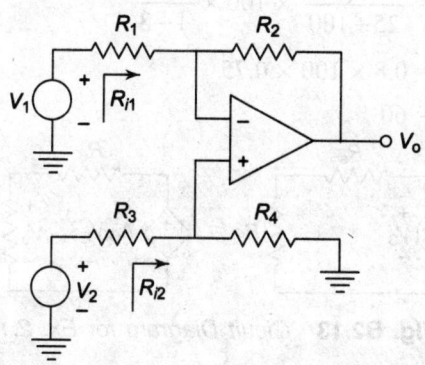

Fig. B2.12 *Circuit Diagram for Ex. 2.17*

$$R_{i1} = R \tag{1}$$
$$R_{i2} = R_3 + R_4 \tag{2}$$
$$R_o = 0 \tag{3}$$
$$V_o = V_2 - 3V_1 \tag{4}$$
$$V_o = \frac{R_2}{R_1}\left(\frac{1+R_1/R_3}{1+R_3/R_4}V_2 - V_1\right) \tag{5}$$

Compare Eqs. (4) and (5)

$$\frac{R_2}{R_1}\left[\left(\frac{1+R_1/R_2}{1+R_3/R_4}\right)\right] = 1$$

$$R_{i1} = R_1 = 100 \text{ k}\Omega$$
$$R_{i2} = R_3 + R_4 = 100 \text{ k}\Omega$$

$$\frac{R_1 + R_2}{\dfrac{R_1(R_3 + R_4)}{R_4}} = 1 \tag{6}$$

$$\frac{(100\,\text{K} + R_2)R_4}{(100\,\text{K})(100\,\text{K})} = 1$$

From above Eqs. (3) to (6), $R_3 = 75$ K, $R_u = 25$ K

$$V_o = \frac{R_2}{R_1}\,(V_2 - V_1)\left[\text{Assume } \frac{R_3}{R_4} = \frac{R_1}{R_2}\right]$$

Ex. 2.18   An amplifier with $R_i = 100\ \text{k}\Omega$, $A_{OC} = 100$, $R_o = 1\ \Omega$ is driven by a source with $R_s = 25\ \text{k}\Omega$, drives a load $R_L = 3\ \Omega$, calculate overall gain.

$$\frac{V_o}{V_s} = \frac{R_i}{R_s + R_i}\,A_{OC}\,\frac{R_L}{R_o + R_L}$$

$$\frac{V_o}{V_s} = \frac{100}{25 + 100} \times 100 \times \frac{3}{1 + 3}$$

$$= 0.8 \times 100 \times 0.75$$

$$A_o = 60$$

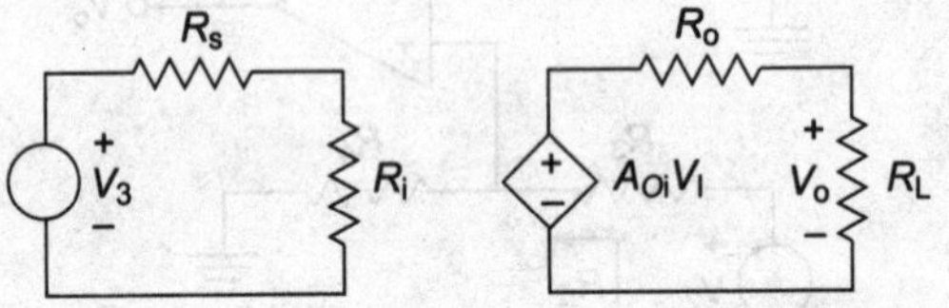

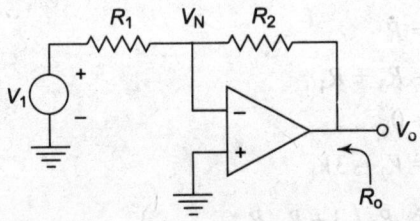

Fig. B2.13   *Cicuit Diagram for Ex. 2.18*

Ex. 2.19   For the circuit shown below,

Fig. B2.14   *Circuit Diagram for Ex. 2.19*

given

$$R_1 = R_2 = 100\ \text{k}\Omega$$

Find A, R_i, and R_o

Solution:

$$A = \frac{-1}{1 + a\beta} = -\frac{R_2}{R_1}\,\frac{1}{1 + 1/\alpha\beta}$$

$$= \frac{-1}{1 + 1/10^5} = -0.99$$

$$R_i = 10^5 + 1 = 100 \text{ k}\Omega$$

$$R_o = \frac{75}{1 + 10^5} = 0.75 \text{ m}\Omega$$

Ex. 2.20 An inverting amplifier with $R_1 = 10$ kΩ, $R_2 = 20$ kΩ, $V_i = 3$ V drives 3 kΩ load. Find the power dissipated inside the op-amp ($I_Q = 0.5$ mA).

$$V_o = -\frac{20}{10} \times 3 = -6 \text{ V}$$

$$I_L = \frac{3 \times 2}{2} = 3 \text{ mA}, I_2 = I_1 = I$$

$$I_1 = I_2 = \frac{3}{10} = 0.3 \text{ mA}$$

$$I_o = I_2 + I_L = 0.3 + 3 = 3.3 \text{ mA}$$

$$I_{EE} = I_{CC} + I_o = 0.5 + 3.3 = 3.8 \text{ mA}$$

$$P = (V_{cc} - V_{EE})I_Q + (V_o - V_{EE})I_o$$
$$= (15 - (-15)) 0.5 + [-6 - (-15)] \times 3.3$$
$$= 44.7 \text{ MW}$$

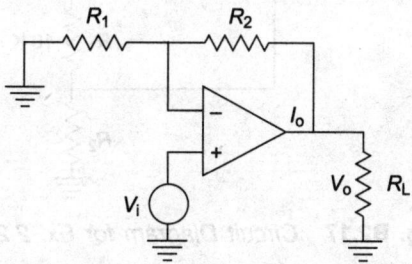

Fig. B2.15 *Circuit Diagram for Ex. 2.20*

Ex. 2.21 Find the transfer function of the given circuit.

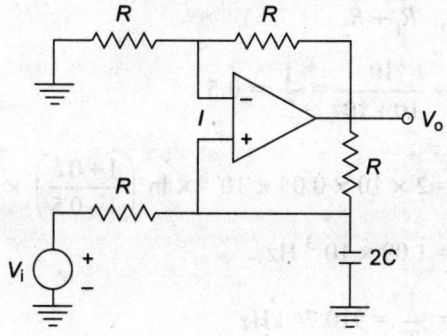

Fig. B2.16 *Circuit Diagram for Ex. 2.21*

$$I = \frac{V_i}{R} \tag{1}$$

$$V_p = \frac{I}{S_2 C} = \frac{V_i}{2sRC} \tag{2}$$

$$V_p = V_i/2sRC, \tag{3}$$

$$V_o = \left(1 + \frac{R}{R}\right) V_p$$

$$= \frac{V_i}{sRC}$$

$$A = \frac{V_o}{V_i} \; 1/sRC, \; H(s) = 1/RCs$$

Ex. 2.22 Design astable multivibrator using IC 741, IC 555.

   (a) IC 741

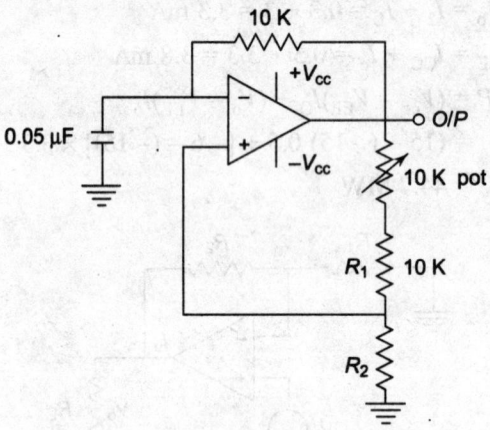

Fig. B2.17 *Circuit Diagram for Ex. 2.22*

$$T = 2 \, RC \ln \left(\frac{1+\beta}{1-\beta}\right)$$

$$\beta = \frac{R_2}{R_1 + R_2}$$

$$\beta = \frac{10}{10 + 10} = \frac{1}{2} = 0.5$$

$$T = 2 \times 10 \times 0.05 \times 10^{-6} \times \ln \left(\frac{1 + 0.5}{1 - 0.5}\right) \times 10^3$$

$$T = 1.09 \times 10^{-3} \; \text{Hz}$$

$$f = \frac{1}{T} = 910.74 \; \text{kHz}$$

$$V_{OT} = \frac{R_1(+V_{sat})}{R_1 + R_2} = 0.5(1\ V) = 0.5\ V$$

$$V_{LT} = \frac{R_1}{R_1 + R_2}(-V_{sat}) = -0.5\ V$$

(b) Using IC 555 timer

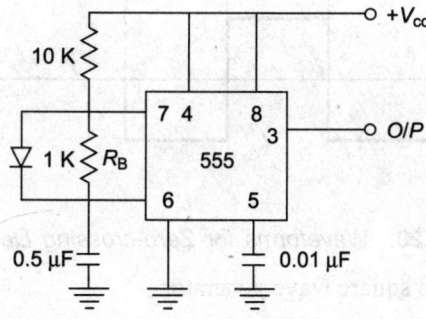

Fig. B2.18 *555 Timer Circuit*

$$f = \frac{1}{0.69(R_A + 2R_B)}$$

$C_1 = 0.5\ \mu F,\ C_2 = 0.01\ \mu F,\ R_A = 10\ K,\ R_B = 1\ K.$

$$f = \frac{1}{0.69(10+2)\times 10^3} = 241.54\ kHz$$

The reference voltage $V_{ref} = 0$, so *I/P* is fine waveform, wherever $V_{in} > V_{ref}$, $V_o = -V_{sat}$ and $V_{in} < V_{ref}$, $V_o = +V_{sat}$.

Ex. 2.23 Design and draw input–output waveforms for positive comparator, negative comparator, and zero-crossing detector.

Solution:

(a) **Zero-crossing detector** $-R_i = 1\ k\Omega$

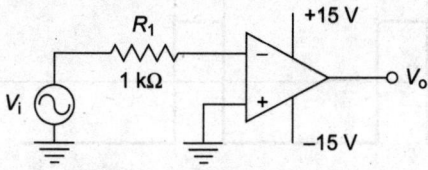

Fig. B2.19 *Zero-crossing Detector*

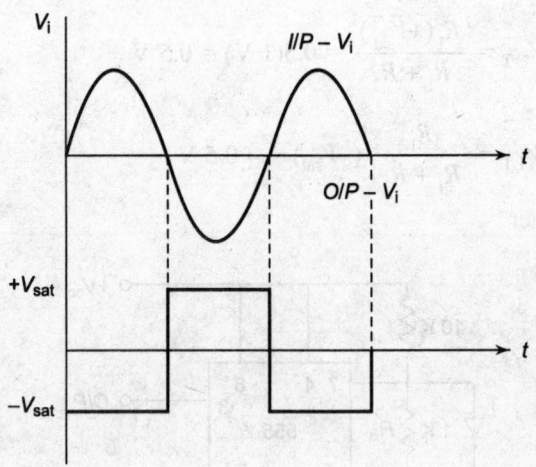

Fig. B2.20   *Waveforms for Zero-crossing Detector*

Zero detector is sine to square wave generator.

(b) **Positive Comparator**

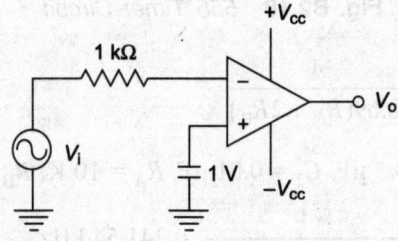

V_{ref} sets to 1 V to non-inverting input.

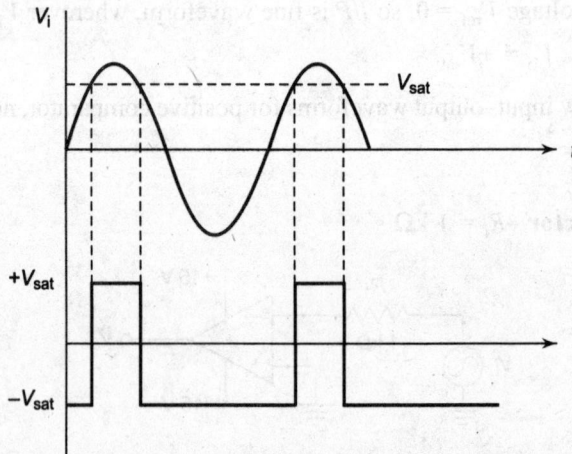

Fig. B2.21   *Waveforms for Positive Comparator*

(c) **Negative Comparator**

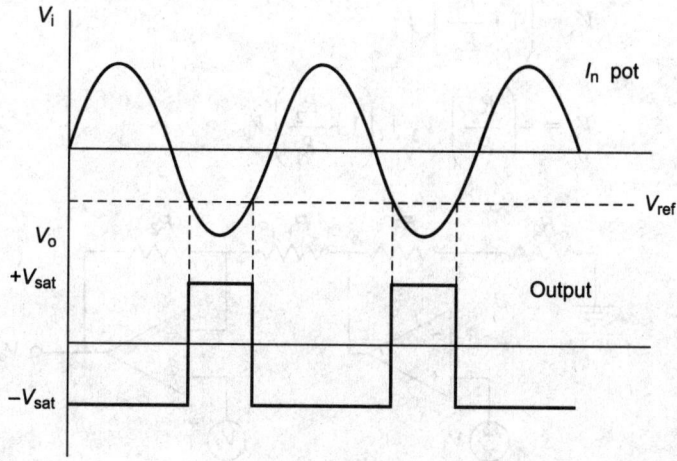

Fig. B2.22 *Waveforms*

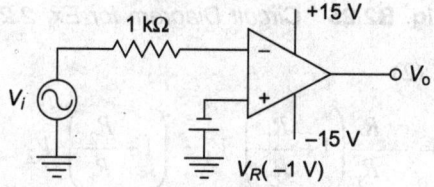

Fig. B2.23 *Circuit Diagram*

Ex. 2.24 Design an instrumentation amplifier whose gain can be varied $1 < A < 10^3$ using 100 kΩ pot.

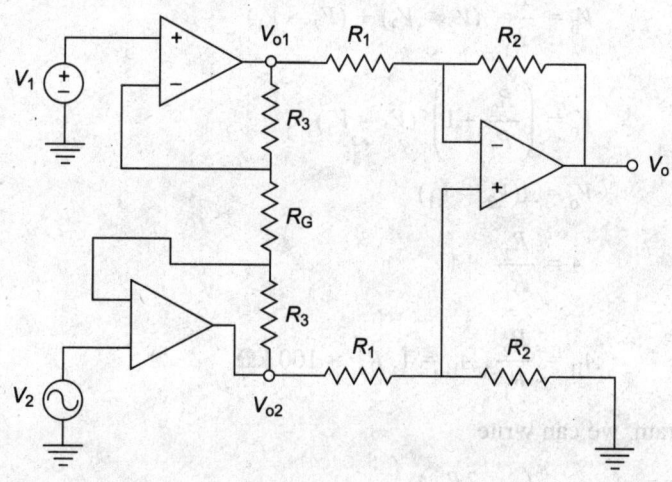

Fig. B2.24 *Circuit Diagram for Instrumentation Amplifier*

Ex. 2.25 Design a dual op-amp instrumentation amplifier and find its output V_o amplifier.

Let

$$V_3 = \left(1 + \frac{R_1}{R_2}\right) V_1 \tag{1}$$

$$V_o = -\left(\frac{R_2}{R_1}\right) V_3 + \left(1 + \frac{R_2}{R_1}\right) V_2 \tag{2}$$

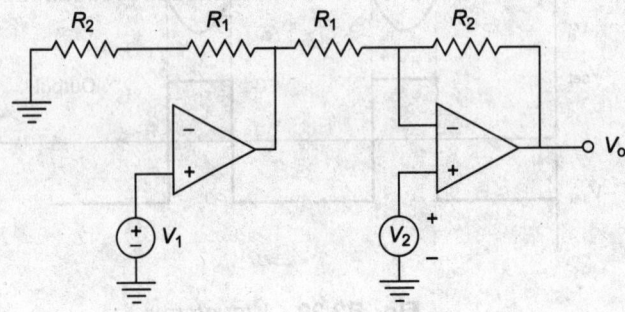

Fig. B2.25 *Circuit Diagram for Ex. 2.25*

Subtituting Eq. (1) in Eq. (2)

$$V_o = -\frac{R_2}{R_1}\left(1 + \frac{R_1}{R_2}\right) V_1 + \left(1 + \frac{R_2}{R_1}\right) V_2$$

$$= -\frac{R_2}{R_1} V_1 - \frac{R_1 R_2}{R_1 R_2} V_1 + V_2 + \frac{R_2 V_2}{R_1}$$

$$V_o = \frac{R_2}{R_1}\ (V_2 - V_1) + (V_2 - V_1)$$

$$V_o = \left(\frac{R_2}{R_1} + 1\right)(V_2 - V_1)$$

$$V_o = A(V_2 - V_1)$$

$$A = \frac{R_2}{R_1} + 1$$

Let

$$A_H = \frac{R_2}{R_1},\ A_H = 1,\ R_{e_1} = 100\ \text{k}\Omega$$

From the diagram, we can write

$$V_{o1} - V_{o2} = \left(1 + \frac{2R_3}{RG}\right)(V_1 - V_2) \tag{1}$$

$$V_o = \frac{R_2}{R_1} (V_{o2} - V_{o1}) \tag{2}$$

From Eqs. (1) and (2)

$$V_o = A(V_2 - V_1)$$

$$A = A_1 \times A_H$$

$$A_H = \frac{R_2}{R_1}$$

$$A_1 = \left(1 + \frac{2R_3}{RG}\right)$$

$$A = \left(1 + \frac{2R_3}{RG}\right) (R_2/R_1)$$

when A varies from 1 to 10^3.

Assume $R_1 = 100$ kΩ, $R_2 = 50$ kΩ, A varies between 2 and 2000

So,

$$2 = 1 + 2 R_3/(RG + 100 \text{ K}) \tag{3}$$

$$2000 = 1 + 2R_3/RG \tag{4}$$

Solving Eqs. (3) and (4)

$$R_3 = 50 \text{ k}\Omega$$

$$RG = 10 \ \Omega \text{ gives perfect design}$$

Ex. 2.26 Explain and design window-detector circuit.

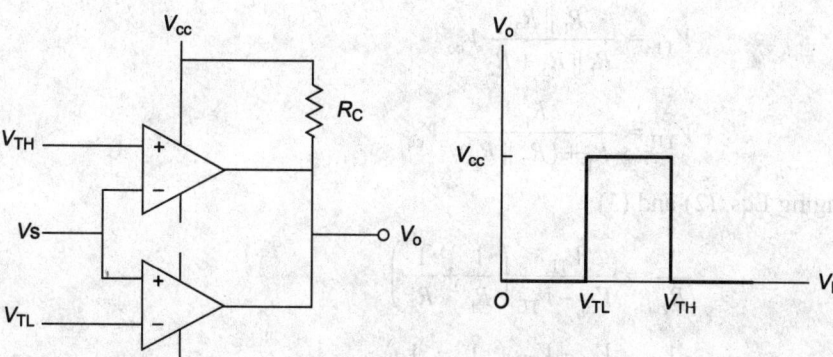

Fig. B2.26

When a given voltage falls within specific bond or window, window detector is used to specify the range of voltage levels.

V_{TH} – Upper threshold level

V_{TL} – Lower threshold level

1. $V_{TL} < V_I < V_{TH}$, $V_o = V_{cc}$, Q_{o1}, Q_{o2} of op-amp arc off condition, R_c pulls V_o to V_{cc}.
2. V_I falls outside the range, Q_{o1} or Q_{o2} in or condition which brings V_o near to zero.

Ex. 2.27 Design single supply-inverting Schmitt trigger circuit.

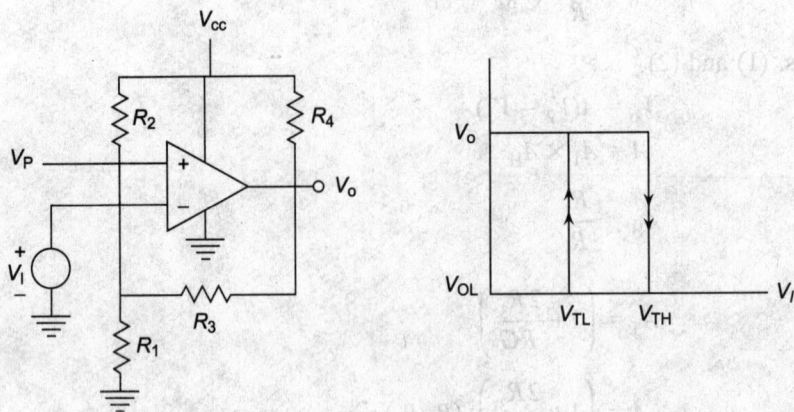

Fig. B2.27

Apply superposition principle. Find V_p.

$$V_p = \frac{R_1 \| R_3}{R_1 \| R_3 + R_2} V_{cc} + \frac{R_1 \| R_2}{R_1 \| R_2 + R_3} V_o \tag{1}$$

Assume $V_{OL} = 0$, $V_{OH} = V_{cc}$,
R_4-high, $R_4 \ll R_3 + R_1 \| R_2$
(1) $V_P = V_{TL}$, $V_o = 0$
(2) $V_p = V_{TH}$, $V_o = V_{OH} = V_{cc}$

So,

$$V_{TL} = \frac{R_1 \| R_3}{R_1 \| R_3 + R_2} V_{cc} \tag{2}$$

$$V_{TH} = \frac{R_1}{R_1 + (R_2 \| R_3)} V_{cc} \tag{3}$$

Rearranging Eqs. (2) and (3)

$$\frac{1}{R_2} = \frac{V_{TL}}{V_{cc} - V_{TL}} \left(\frac{1}{R_1} + \frac{1}{R_3} \right) \tag{4}$$

$$\frac{1}{R_1} = \frac{V_{cc} - V_{TH}}{V_{TH}} \left(\frac{1}{R_2} + \frac{1}{R_3} \right) \tag{5}$$

Ex. 2.28 Design Schmitt trigger with $V_{cc} = 5$ V, $V_{OL} = 0$, $V_{OH} = 5$ V, $V_{TL} = 1.5$ V, and $V_{TH} = 2.5$ V.
Solution:

$$R_3 \gg R_4$$
$$R_3 = 100 \text{ k}\Omega$$

$$\frac{1}{R_1} = \frac{1}{R_2} + \frac{1}{100 \times 10^3} \left(\frac{5 - 2.5}{2.5} \right) \tag{1}$$

$$\frac{1}{R_2} = \left(\frac{1.5}{2.5}\right)\left(\frac{1}{R_1} + \frac{1}{100 \times 10^3}\right) \qquad (2)$$

From Eqs. (1) and (2), solve for R_1 and R_2

$$R_1 = 40 \text{ k}\Omega, \ R_2 = 67 \text{ k}\Omega$$

$$R_3 = 100 \text{ k}\Omega, \ V_{cc} = 5 \text{ V}, \ V_{TH} = 2.5 \text{ V}, \ V_{TL} = 1.5 \text{ V}$$

Ex. 2.29 Design a triangular wave generator of the circuit given below, given $f_0 = 2$ kHz and $V_o = 7$ V.

$$V_{cc} = \pm 15 \text{ V}$$

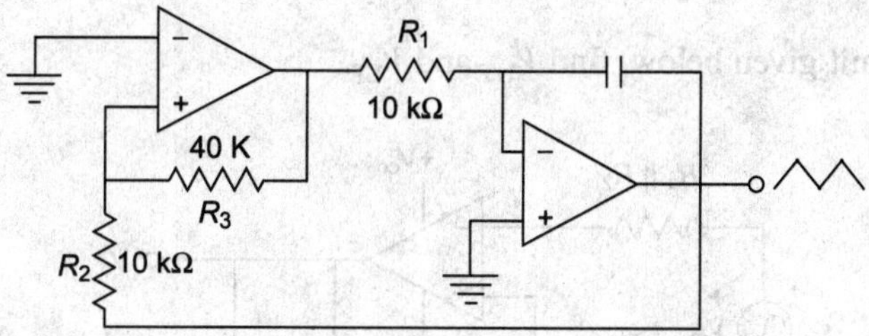

Fig. B2.28 *Circuit Diagram for Ex. 2.29*

$$V_{sat} = 14 \text{ V}$$

$$V_o(PP) = \frac{R_2}{R_3} V_{sat}$$

$$V_{ramp} = -\frac{R_2}{R_3} (-V_{sat})$$

$$\frac{R_2}{R_3} = \frac{7}{2(14)} = \frac{R_3}{4}$$

$$R_2 = 10 \text{ k}\Omega, \ R_3 = 40 \text{ k}\Omega$$

$$f = \frac{40 \text{ k}\Omega}{4RC_1 10 \text{ k}\Omega} \quad \left(f = \frac{R_3}{4R_1C_1R_2}\right)$$

$$C_1 = 0.05 \ \mu f; \ [Y = R_1C_1 = 0.5 \text{ msec}]$$

So, $\qquad f = 2$ kHz.

$$2 \times 10^3 = \frac{40 \text{ K}}{4 \times R_1C_1(10 \text{ k}\Omega)}$$

$$R_1C_1 = 0.5 \text{ msec}$$

$$C_1 = 0.05 \ \mu F$$

$$R_1 = 10 \text{ k}\Omega$$

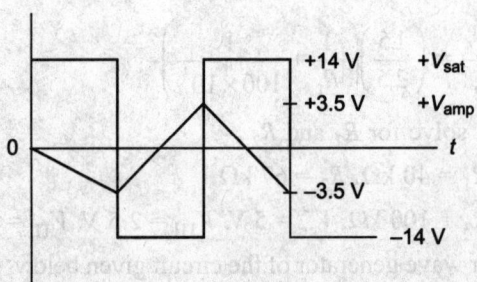

Fig. B2.29   *Waveforms for Ex. 2.29*

Ex. 2.30   For the circuit given below, find V_{UT} and V_{LT}.

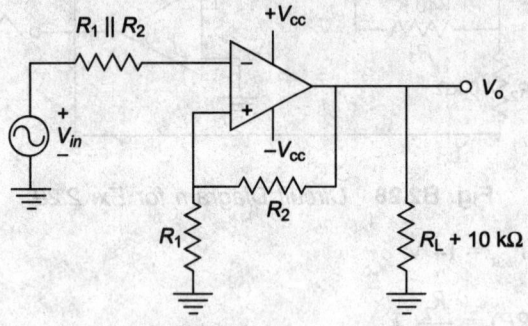

Fig. B2.30   *Circuit Diagram for Ex. 2.30*

$$V_{UT} = \frac{R_1}{R_1 + R_2} \, (+V_{sat})$$

$$V_{LT} = \frac{R_1}{R_1 + R_2} \, (-V_{sat})$$

Let $R_1 = 100 \; \Omega, \, R_2 = 56 \; k\Omega, \, V_i = 1 \; V, \, V_{cc} = \pm 15 \; V$

$$V_{UT} = \frac{100}{56,100} \, (14) = 25 \; mV$$

$$V_{LT} = \frac{100}{56,100} \, (-14) = -25 \; mV$$

Output waveform

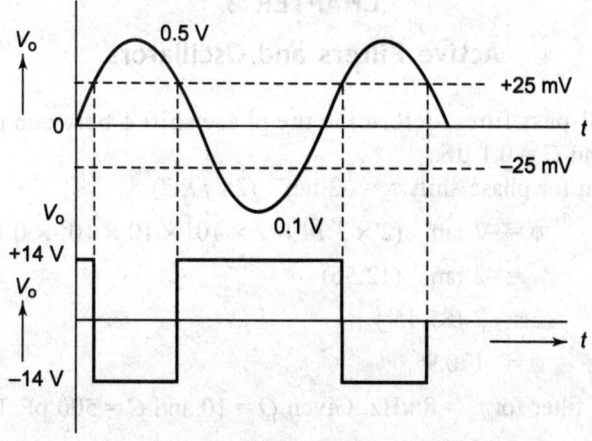

Fig. B2.31 *Input and Output Waveforms for Ex. 2.30*

Ex. 2.31 What is the voltage of points *A* and *B*?

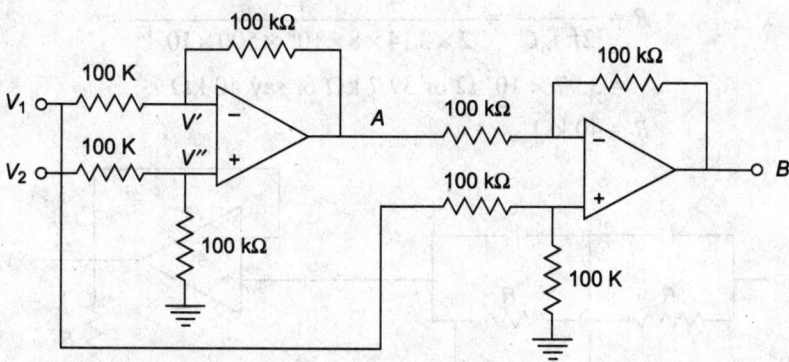

Fig. B2.32 *Circuit Diagram for Ex. 2.31*

At *A*, V_A. Assume $V_1 = 5$ V, $V_2 = 5$ V

$$\frac{V' - V_1}{100} + \frac{V_A - V'}{100} = 0$$

$$\frac{V'' - V_2}{100} + \frac{V''}{100} = 0$$

$$\frac{V_A}{100 \, K} = \frac{V_1 - V_2}{100 \, K}$$

$$V_A = 0.1 \text{ V}$$

$$V_B = -V_A + V_1$$

$$= 5 - 0.1 = 4.9 \text{ V}$$

CHAPTER 3

Active Filters and Oscillators

Ex. 3.1 For a given all-pass filter, determine the phase shift ϕ between the input and output at $f = 2$ kHz. $R = 10$ kΩ and $C = 0.1$ μF.

Solution: Expression for phase shift $\phi = -2 \tan^{-1} (2\pi f RC)$

So,
$$\phi = -2 \tan^{-1} (2 \times 3.14 \times 2 \times 10^3 \times 10 \times 10^3 \times 0.1 \times 10^{-6})$$
$$= -2 \tan^{-1} (12.56)$$
$$= -2 (85.45°)$$
$$\phi = -170.9°$$

Ex. 3.2 Design a notch filter for $f_o = 8$ kHz. Given $Q = 10$ and $C = 500$ pF. Draw the circuit.

Solution: Given $f_o = 8$ kHz; $Q = 10$; $C = 500$ pF.

$$f_o = \frac{1}{2\pi RC};$$

So,
$$R = \frac{1}{2\pi f_o C} = \frac{1}{2 \times 3.14 \times 8 \times 10^3 \times 500 \times 10^{-12}}$$
$$= 3.97 \times 10^4 \ \Omega \text{ or } 39.7 \text{ k}\Omega \text{ or say } 40 \text{ k}\Omega$$
$$R = 40 \text{ k}\Omega$$

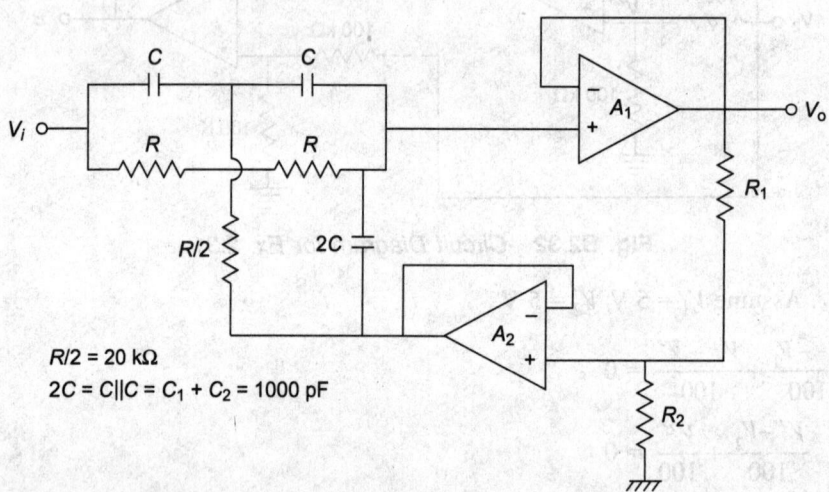

$R/2 = 20$ kΩ
$2C = C\|C = C_1 + C_2 = 1000$ pF

Fig. B3.1 *Circuit Diagram for Notch Filter (Ex. 3.2)*

Ex. 3.3 For a given narrow BPF $f_c = 2$ kHz, $Q = 20$, and $A_F = 10$. If the centre frequency is to be changed to 1 kHz, keeping gain and bandwidth constant, what changes are to be made in the circuit?

Solution: Given $f_c = 2$ kHz; $Q = 20$; $A_F = 10$

Let $C = 0.1$ μF

$$R_1 = \frac{Q}{2\pi f_c C A_F} = \frac{20}{2 \times 3.14 \times 2 \times 10^3 \times 0.1 \times 10^{-6} \times 10}$$

$$= 1.5 \text{ k}\Omega$$

$$R_2 = \frac{Q}{2\pi f_c C(2Q^2 - A_F)} = \frac{20}{2 \times 3.14 \times 2 \times 10^3 \times 0.1 \times 10^{-6}(800 - 10)}$$

$$= 200 \ \Omega$$

$$R_3 = \frac{Q}{\pi f_c C} = \frac{20}{3.14 \times 2 \times 0.1 \times 10^3 \times 10^{-6}} = 29.4 \text{ k}\Omega$$

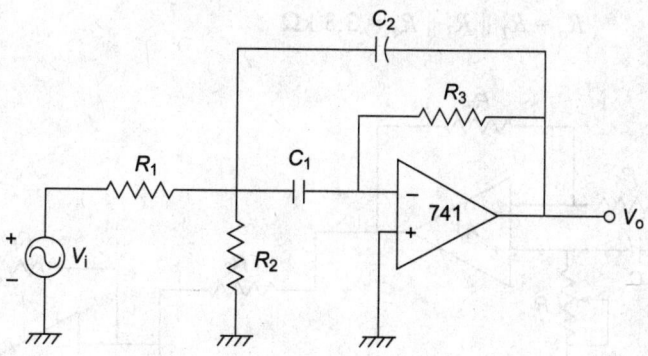

Fig. B3.2 *Circuit Diagram for Ex. 3.3*

To change centre frequency f_c keeping gain and bandwidth *BW* constant, R_2 is to be changed. Let R_2' be the new value of R_2 to change the centre frequency from 2 kHz to 1 kHz.

So,

$$R_2' = R_2 \left(\frac{f_c}{f_c'}\right)^2 = 200 \left(\frac{2}{1}\right)^2 = 400 \ \Omega$$

New value $R_2' = 400 \ \Omega$

Ex. 3.4   Design a wide BRF having $f_H = 200$ Hz and $f_L = 1$ kHz. Draw the circuit and assume necessary data.

Solution: Given: $f_H = 200$ kHz

$f_L = 1$ kHz

Let the gain be 2

So, $\text{Gain} = 1 + \left(\frac{R_F}{R_1}\right) = 2$

Therefore, $\frac{R_F}{R_1} = 1$

Hence, $R_F = R_1 = 10 \text{ k}\Omega$

and $R_F' = R_1' = 10 \text{ k}\Omega$

Let $\qquad C' = 0.05\ \mu F;$ given $f_H = 200$ Hz

So, $\qquad R' = \dfrac{1}{2\pi f_H C} = \dfrac{1}{2 \times 3.14 \times 0.05 \times 10^{-6}} = 15.9\ k\Omega$

Therefore, $\qquad R' = 15.9\ k\Omega$

Let $\qquad C = 0.1\ \mu F;$ given $f_L = 1$ kHz

So, $\qquad R = \dfrac{1}{2\pi f_L C} = \dfrac{1}{2 \times 3.14 \times 1 \times 10^3 \times 0.1 \times 10^{-6}} = 15.9\ k\Omega$

$$R_2 = R_3 = R_F'' = 10\ k\Omega$$

$$R_L = 10\ k\Omega$$

$$R_o = R_2 \parallel R_3 \parallel R_4 = 3.3\ k\Omega$$

Circuit diagram:

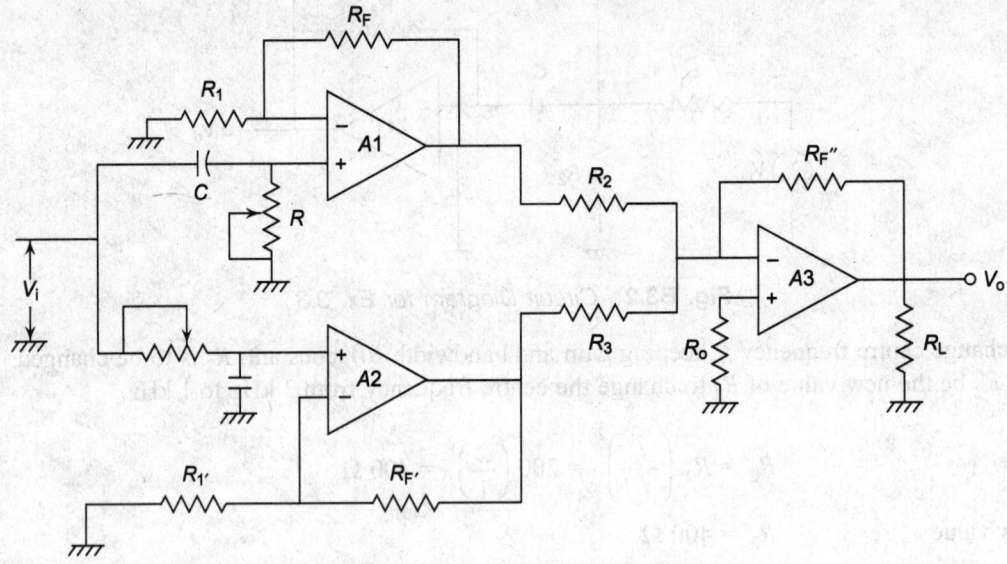

Fig. B3.3 *Circuit Diagram for Ex. 3.4*

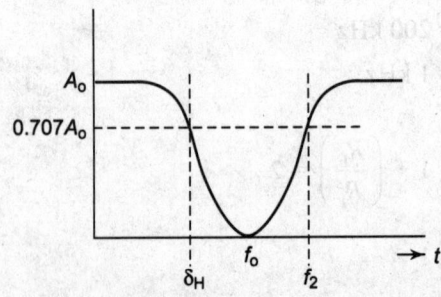

Fig. B3.4 *Frequency Response of Band-Reject Filter (BRF)*

Ex. 3.5 Determine the order of Butterworth LPF so that at $\omega = 1.5\ \omega_{3dB}$ the magnitude response is down by at least 30 dB.

Solution: Given: $\omega = 1.5\ \omega_{3dB}$

Magnitude response: 30 dB

$$\left|\frac{1+(j\omega)}{A_o}\right| = \frac{1}{\sqrt{1+\left(\dfrac{\omega}{\omega_n}\right)^{2n}}}$$

$$20 \log \frac{H(j\omega)}{A_o} = -30\ \text{dB} \;\Rightarrow\; \frac{H+(j\omega)}{A_o} = \left(10^{-\frac{30}{20}}\right)$$

$$= 10^{-3/2} = 0.0316$$

$$\left|\frac{1+(j\omega)}{A_o}\right|^2 = (0.0316)^2 = \frac{1}{1+\left(\dfrac{\omega}{\omega_n}\right)^{2n}}$$

$$= \frac{1}{1+(1.5)^{2n}} = 9.9927 \times 10^{-4}$$

$$1+(1.5)^{2n} = \frac{10^4}{9.9927} \simeq 1000.7 \simeq 1001$$

$$(1.5)^{2n} = 1000 \Rightarrow 2n \log_{10}(1.5) = \log_{10}(1000) = 3$$

So, $2n = 17.03$ or $n = 8.5$ or say 9.

The order of Butterworth filter is 9.

Ex. 3.6 Find 3-dB frequency and phase shift for first-order LPF.

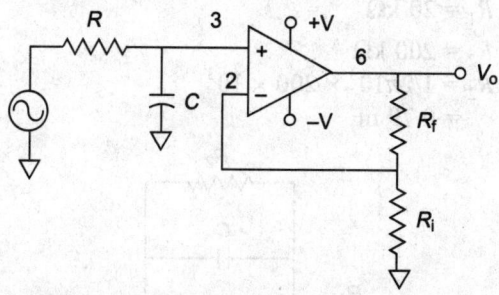

Fig. B3.5 *Circuit Diagram for Ex. 3.6*

Gain $A_o = 1 + R_f/R_i$

$$\frac{V_o}{V_i} = \frac{A_o}{RC_s + 1}$$

$$\omega_o = 1/RC,$$

$$\frac{V_o}{V_i} = \frac{A_o}{S/\omega_o + 1},\quad \frac{V_o}{V_i} = \frac{A_o\omega_o}{S+\omega_o}$$

The magnitude of gain, $|G| = \dfrac{A_o}{\sqrt{1+\omega^2 R^2 C^2}}$

For 3 dB frequency

$$|G| = 0.707\, A_o = \frac{A_o}{\sqrt{1 + \omega_{-3dB}^2 R^2 C^2}}$$

$$(0.7)^2 A_o = \frac{A_o}{(1 + \omega_{3dB}^2 R^2 C^2)^2}$$

$$0.5 = 1/(1 + \omega_{3dB}^2\, R^2 C^2)^2$$

So, $\omega_{-3dB}^2\, R^2 C^2 = 1$

$$\omega_{-3dB}^2 = 1/R^2 C^2, \ 1/RC = \omega_{3dB}^2$$

$$\omega_{-3dB} = 1/RC$$

$$\omega = 2\pi f$$

$$f_{-3dB} = 1/2\pi\, RC$$

At critical frequency
Phase shift is

$$\phi_o = - \arctan \omega_o\, RC$$

$$= - \arctan (1/RC)\, RC$$

$$= - \arctan = 45°$$

Ex. 3.7   As shown in figure below, design active LPF for –3 dB frequency of 1 kHz with DC gain of 20 dB and $R_1 = 20$ kΩ. At what frequency does gain drop to 0 dB?

Solution: 20 dB → gain = 10

$$R_2 = 10\, R_1$$

$$R_1 = 20 \text{ k}\Omega$$

So, $R_2 = 200$ kΩ

$$C = 1/\omega_o R_2 = 1/2\pi 10^3 \times 200 \times 10^3$$

$$= 0.79 \text{ nF}$$

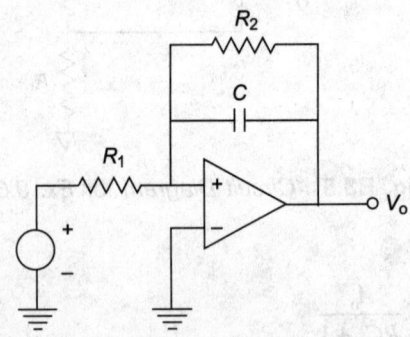

Fig. B3.6   *Circuit Diagram for Ex. 3.7*

Ex. 3.8   Design BPF circuit using op-amp with a gain of 20 dB.

$$R_2 = 10\, R_1, \ A = 20 \text{ dB}$$

$$R_1 = 10 \text{ k}\Omega, \ R_2 = 100 \text{ k}\Omega, \ \omega_L = 2\pi f = 2\pi \times 20 \text{ rad/sec}$$

$$C_1 = 1/2\pi f \times R = 0.795 \ \mu\text{F}.$$

$$W_H = 2\pi \times 20 \text{ rad/sec}, \ C_2 = 1/2\pi\, R_1 f = 100 \text{ pF}$$

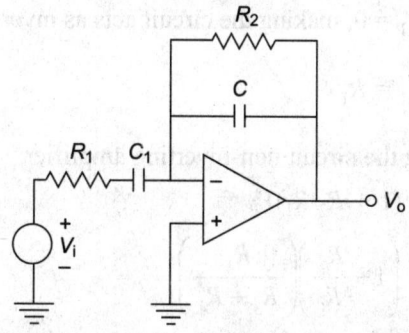

Fig. B3.7 *Circuit Diagram for Ex. 3.8*

Ex. 3.9 Using 741 op-amp, design an audio amplifier with gain of 650 dB and find actual $B\omega$.
Since 60 dB is gain

(i) A for OA_1 OA_2 is 10^3
(ii) $f_B > 20$ kHz (audio amp)

$$A = A_1 \times A_2; A_1 \text{ is gain of } OA_1 \text{ and } A_2 \text{ is gain of } OA_2.$$

$$A = 1000$$

$$= \sqrt{1000} = 31.6 \text{ V}, f_{B1} = f_B = 10^6/31.62 = 31.62 \text{ kHz}$$

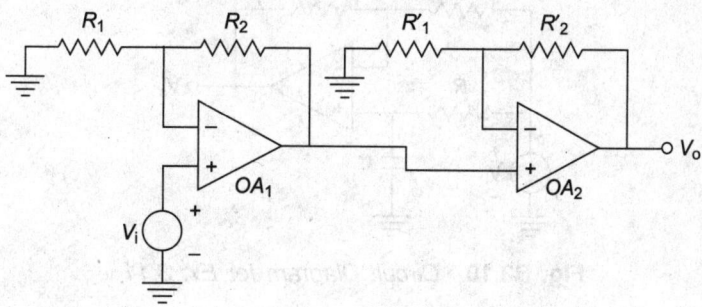

Fig. B3.8 *Circuit Diagram for Ex. 3.9*

Ex. 3.10 Find the output of the given circuit.

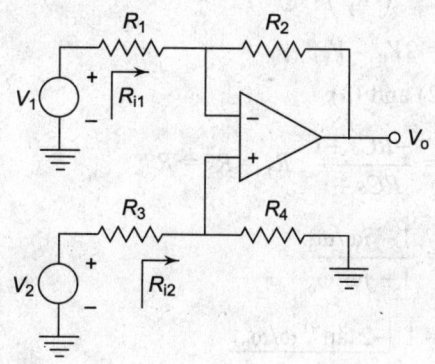

Fig. B3.9 *Circuit Diagram for Ex. 3.10*

(i) Letting $V_2 = 0$ yields $V_P = 0$, making the circuit acts as inverting amplifier with respect to V_1.

So, $V_{o1} = -\dfrac{R_2}{R_1} V_1$, $R_{i1} = R_1$

(ii) Letting $V_1 = 0$, making the circuit non-inverting amplifier

$$V_{o2} = (1 + R_2/R_1)V_P$$

$$= \left(1 + \frac{R_2}{R_1}\right)\left(\frac{R_4}{R_3 + R_4}\right) V_2$$

$$V_o = V_{o1} + V_{o2}$$

$$= \frac{R_2}{R_1}\left(\frac{1 + R_1/R_2}{1 + R_3/R_4}V_2 - V_1\right)$$

Ex. 3.11   For the given phase shifter circuit, find $H(s)$, $H(j\omega)$, and magnitude angle ω.
Let V_P voltage related to V_i by low-pass function,

$$\frac{V_P}{V_i} = \frac{1}{RCS + 1} \tag{1}$$

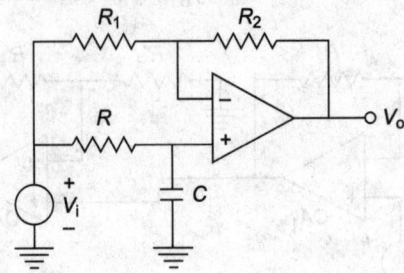

Fig. B3.10   *Circuit Diagram for Ex. 3.11*

V_o of *LPf* using op-amp is

$$V_o = -\left(\frac{R_2}{R_1}\right) V_i + (1 + R_2/R_1)V_P \tag{2}$$

$$V_o = 2V_P - V_i \tag{3}$$

Eliminate V_P from Eqs. (2) and (3)

$$\frac{V_o}{V_i} = \frac{-RCs + 1}{RCs + 1}, \; R_1 = R_2 = R$$

$$H(j\omega) = \frac{1 - j\omega/\omega_o}{1 + j\omega/\omega_o}$$

$$= 1 \; \underline{|-2\tan^{-1}\omega/\omega_o}$$

Ex. 3.12 Design Wien-bridge oscillator circuit for $f_0 = 965$ Hz.
Solution:

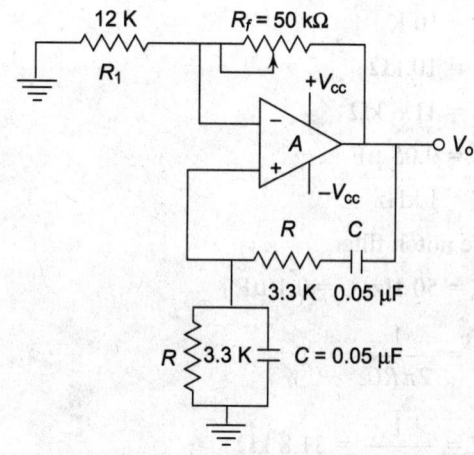

Fig. B3.11 *Circuit Diagram for Ex. 3.12*

Let
$$C = 0.05 \ \mu\text{F}, f_0 = \frac{0.159}{RC}$$

$$R = \frac{0.159}{5 \times 10^{-8} \times 965} = 3.3 \ \text{k}\Omega$$

$R_1 = 12 \ \text{k}\Omega,$

$R_f = 2(12 \ \text{k}\Omega) = 24 \ \text{k}\Omega$

$(R_f = 50 \ \text{k}\Omega \ \text{Pot})$

Ex. 3.13 Design square-wave oscillator if $f_0 = 1$ kHz, $V_{ce} = \pm15$ V

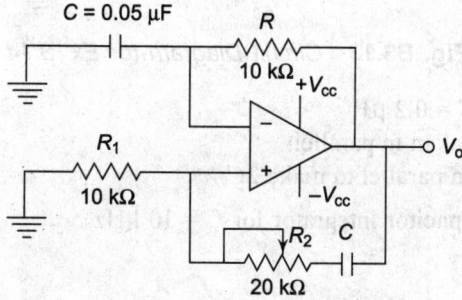

Fig. B3.12 *Circuit Diagram for Ex. 3.13*

Let

$R_2 = 1.16 \ R_1$

$R_1 = 10 \ \text{k}\Omega$

$R_2 = 11.6 \ \text{k}\Omega$

$C = 0.05 \ \mu\text{F}$

$$R = \frac{1}{10 \times 10^{-8} \times 10^3} = 10 \text{ k}\Omega$$

$$R = 10 \text{ K}$$

$$R_1 = 10 \text{ k}\Omega$$

$$R_2 = 11.6 \text{ k}\Omega$$

$$C = 0.05 \text{ }\mu\text{F}$$

$$f = 1 \text{ kHz}$$

Ex. 3.14   Design 50-Hz active notch filter.

Solution: $f_0 = 50$ Hz, $C = 0.1$ μF

$$f_0 = \frac{1}{2\pi RC}$$

$$R = \frac{1}{2\pi fC} = 31.8 \text{ k}\Omega$$

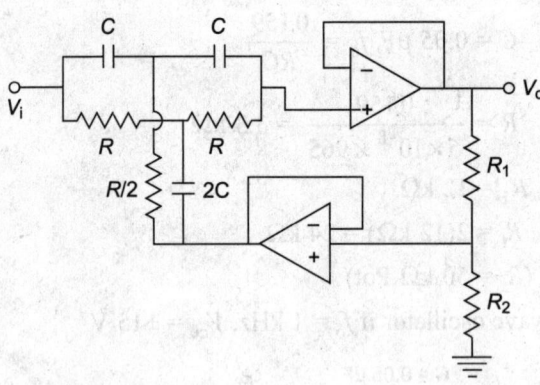

Fig. B3.13   *Circuit Diagram for Ex. 3.14*

Take = $R/2 = 16$ kΩ; $2C = 0.2$ μF
The resistors of 31.8 K taken in parallel.
0.1 μF of the capacitor in parallel to make $2C$.

Ex. 3.15   Design switched capacitor integrator for $f_0 = 10$ kHz.
Solution:

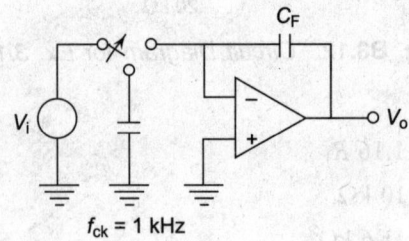

Fig. B3.14   *Circuit Diagram for Ex. 3.15*

$$f_{CK} = 1 \text{ kHz}$$

Let
$$\frac{C_F}{C_1} = \frac{f_{ck}}{2\pi f_o} = \frac{10^3}{2\pi} \times 10 = 15.9$$

Assume
$$C_1 = 1 \text{ pF}, C_F = 15.9 \text{ pF}$$
$$R_1 = 1 \text{ m}\Omega$$

$$C_F = \frac{1}{2\pi R_1^2 f_o} = \frac{1}{2 \times 3.14 \times 1.6 \times 1.6 \times 10} = 10 \ \Omega F$$

So,
$$C_F = 10 \ \Omega F$$
$$C_1 = 1 \text{ pF}$$
$$R_1 = 1 \text{ m}\Omega$$

CHAPTER 4
Time and Phase-Locked Loops (PLLs)

Ex. 4.1 Determine the DC control voltage V_c at lock, if signal frequency $f_s = 10$ kHz, V_{co} free-running frequency is 10.66 kHz, and the voltage-to-frequency transfer coefficient of V_{co} is 6600 Hz/V.

Solution: Given
$$f_s = 10 \text{ kHz}$$
$$f_o = 10.66 \text{ kHz}$$
$$K_v = 6600 \text{ Hz/V}$$

V_c is DC control voltage

$$f_s = f_o + K_V V_c$$

$$V_c = \frac{f_s - f_o}{K_v} = \frac{0.66 \times 10^3}{6600} = 0.11 \text{ V}$$

So,
$$V_c = 0.11 \text{ V}.$$

Ex. 4.2 Design a 555 astable multivibrator to operate at 10 kHz with 40% duty cycle.

Solution: Given
$$f = 10 \text{ kHz}$$

Duty cycle:
$$D = 40\% = 0.4$$

$$f = \frac{1.45}{(R_A + 2R_B)C}$$

$$D = \frac{R_B}{R_A + 2R_B}$$

Let
$$C = 0.01 \ \mu F$$

$$0.4 = \frac{R_B}{R_A + 2R_B}, \ 0.4 \ R_A + 0.8 \ R_B = R_B$$

So,
$$0.4 \ R_A = 0.2 \ R_B$$

Therefore,
$$R_A = 0.5 \ R_B$$

$$f = 10 \text{ kHz} = \frac{1.45}{(0.5R_B + 2R_B)C}$$

$$10 \times 10^3 = \frac{1.45}{2.5R_B \times 0.01 \times 10^{-6}}$$

$$10^4 = \frac{1.45 \times 10^8}{2.5R_B}$$

or

$$R_B = \frac{1.45 \times 10^8}{2.5 \times 10^4}$$

$$R_B = 0.58 \times 10^4 = 5.8 \text{ k}\Omega$$

So,

$$R_A = 0.5 \ R_B = 2.9 \text{ k}\Omega$$

Ex. 4.3 In the monostable multivibrator circuit using 555 timer IC, $C = 0.01 \ \mu\text{F}$ and $R_A = 2.7 \text{ k}\Omega$, connected between pins 4 and 7, calculate the duration of the output pulse width t_p.

Solution:

$$t_p = 1.1 \ R_A C$$
$$= 1.1 \times 2.7 \times 10^3 \times 0.1 \times 10^{-6}$$
$$= 0.297 \ \mu\text{sec}$$

Ex. 4.4 A monostable multivibration circuit is to be used as a divide by three networks. The frequency of the input trigger is 10 kHz. If $C = 0.06 \ \mu\text{F}$, determine the value of R_A.

Solution: Given:

$$f = 10 \text{ kHz}, \ C = 0.06 \ \mu\text{F}, \ R_A = ?$$

$$T = \frac{1}{f} = \frac{1}{10 \times 10^3} = 0.1 \text{ m sec}$$

Pulse width t_p during which output remains high is,

$$t_p = 1.1 \ R_A C$$

For divide by three operation, let $t_p = 2.2 \ T$

$$= 2.2 \times 0.1 \times 10^{-3} = 0.22 \text{ msec}$$

So,

$$R_A = \frac{t_p}{1.1 \ C} = \frac{0.22 \times 10^{-3}}{1.1 \times 0.06 \times 10^{-6}} = 3.33 \text{ k}\Omega$$

$$R_A = 3.33 \text{ k}\Omega$$

Ex. 4.5 For the astable multivibrator circuit using 555 timer IC, $R_A = 6.8 \text{ k}\Omega$, $R_B = 1.2 \text{ k}\Omega$, and $C = 1 \ \mu\text{F}$. Determine the positive pulse width, negative pulse width, and free-running frequency. What is the duty cycle of the output waveform?

Solution: Given $R_A = 6.8 \text{ k}\Omega; \ R_B = 1.2 \text{ k}\Omega$

$$C = 1 \ \mu\text{F}, \ t_c = ? \ t_d = ? \ f_o = ?$$

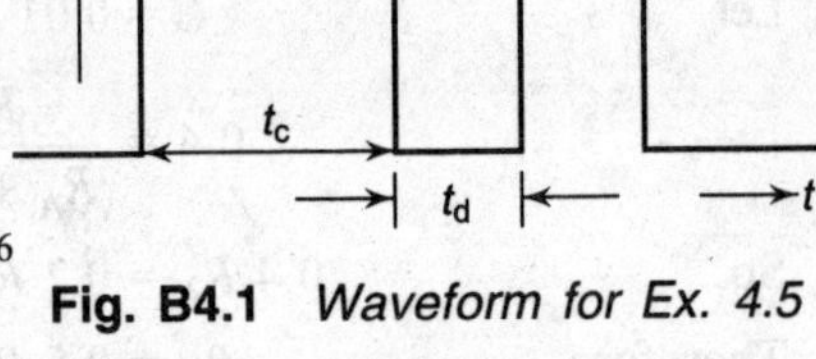

Expression for charging time or positive pulse width

$$= t_c = 0.69 \ (R_A + R_B)C$$
$$= 0.69 \ (6.8 + 1.2) \ 10^3 \times 1 \times 10^{-6}$$
$$= 0.69 \times 8.0 \times 10^{-3}$$

Fig. B4.1 *Waveform for Ex. 4.5*

$$t_c = 5.52 \text{ msec}$$

Negative pulse width $t_d = 0.69 \, R_B C$

$$= 0.69 \times 1.2 \times 10^3 \times 1 \times 10^{-6}$$

$$t_d = 0.83 \text{ msec}$$

$$T = t_c + t_d = 5.52 + 0.83 = 6.35 \text{ msec}$$

$$f_o = \frac{1}{T} = \frac{1}{6.35 \text{ msec}} = 157.5 \text{ Hz}$$

or

$$f_o = \frac{1.45}{(R_A + 2R_B)C} = \frac{1.45}{[6.8 + (2 \times 1.2)]10^3 \times 1 \times 10^{-6}}$$

$$f_o = \frac{1.45}{(6.8 + 2.4) \times 10^{-3}} = \frac{1.45 \times 10^3}{9.2} = 157.5 \text{ Hz}$$

Percentag duty cycle $= \dfrac{t_c}{T} \times 100\%$

$$= \frac{5.52 \times 10^{-3}}{6.35 \times 10^{-3}} \times 100 = 86.92\%$$

$$= 86.92\%$$

Ex. 4.6 For the 565 PLL circuit, determine the range of output frequency if R_1 is varied from 2 kΩ to 20 kΩ. $C_1 = 0.01 \, \mu F$.

Solution: Given, R_1 varies from 2 kΩ to 20 kΩ

$$C_1 = 0.01 \, \mu F$$

Expression for $f_{out} = \dfrac{1.2}{4R_1 C_1}$

for

$$R_1 = 2 \text{ k}\Omega, f_{out} = \frac{1.2}{4 \times 2 \times 0.01} = 15 \text{ kHz}$$

for

$$R_1 = 20 \text{ k}\Omega, f_{out} = \frac{1.2}{4 \times 20 \times 10^3 \times 0.01 \times 10^{-6}} = 1.5 \text{ kHz}$$

f_{out} varies from 1.5 kHz to 15 kHz.

Ex. 4.7 In the case of 565 PLL circuit, if $f_{in} = 500$ Hz, what is the value of R_1? 565 is being used as multiply by 5 circuit $C_1 = 0.01 \, \mu F$.

Solution: If $\qquad f_{in} = 500$ Hz, $\quad f_{out} = 2.5$ kHz.

Expression for $f_{out} = \dfrac{1.2}{4R_1 C_1}$

$$2.5 \times 10^3 = \frac{1.2}{4 \times R_1 \times 0.01 \times 10^{-6}}$$

So,

$$R_1 = \frac{1.2}{4 \times 2.5 \times 10^3 \times 0.01 \times 10^{-6}}$$

$$R_1 = \frac{10^5 \times 1.2}{10} = 12.0 \text{ k}\Omega$$

$$R_1 = 12 \text{ k}\Omega$$

Ex. 4.8 565 IC is being used as ramp generator. If $R = 100$ kΩ, determine the frequency f_0 of the ramp waveform. Given $V_{cc} = +5$ V. $C = 0.05$ µF.

Solution: $\qquad f_0 = \dfrac{3I_C}{V_{cc}.C}$; $I_c = (V_{cc} - V_{BE})/R = \dfrac{(5 - 0.2)}{100 \text{ k}\Omega} = 4.8 \times 10^{-5}$ A

So, $\qquad\qquad f_0 = \dfrac{3 \times 4.8 \times 10^{-5}}{5 \times 0.05 \times 10^{-6}} = 516$ Hz.

Ex. 4.9 Design an astable multivibrator to get output waveform at 10 kHz with a duty cycle of 75% using 555 IC.

Solution: Given $\qquad f = 10$ kHz; Duty cycle = 75%

$$R_A = ? \quad R_B = ?$$

Let $\qquad\qquad\qquad C = 0.1$ µF

$$\text{Duty cycle} = \frac{t_c}{t_c + t_d} \times 100 = \frac{t_c}{T} \times 100$$

$$T = \frac{1}{f} = \frac{1}{10 \times 10^3} = 0.1 \text{ msec}$$

$$\frac{75}{100} = \frac{t_c}{0.1 \text{ msec}} = \frac{t_c}{0.1 \times 10^{-3}}$$

or $\qquad\qquad t_c = 0.75 \times 0.1 \times 10^{-3} = 0.075$ msec

$$T = t_c + t_d$$

$$0.1 \text{ msec} = 0.075 \text{ msec} + t_d$$

So, $\qquad\qquad t_d = (0.1 - 0.075)$ msec

$$t_d = 0.025 \text{ msec}$$

$$t_d = 0.69\, R_B C$$

$$0.025 \times 10^{-3} = 0.69 \times R_B \times 0.1 \times 10^{-6}$$

Therefore, $\qquad R_B = \dfrac{0.025 \times 10^{-3}}{0.69 \times 0.1 \times 10^{-6}} = 36\Omega$ (say 47 Ω)

$$t_c = 0.69\,(R_A + R_B)C$$

$$0.075 \times 10^{-3} = 0.69\,(R_A + 36)\,0.1 \times 10^{-6}$$

Hence, $\qquad R_A = 1.05$ kΩ (say 1 kΩ), 1 kΩ is a standard value of resistor.

Ex. 4.10 Design a ramp generator having an output frequency of 5 kHz.

Solution: Given: ramp output wave-form at 5 kHz

$$R = ?$$

Let $\qquad\qquad V_{cc} = +5$ V, $\quad C = 1$ µF

$$f_0 = \frac{3I_c}{V_{cc}C}$$

$$I_c = \frac{(V_{cc} - V_{BE})}{R}$$

$$5 \text{ kHz} = \frac{3I_c}{5 \text{ V} \times 1 \text{ μF}}$$

So, $$I_c = \frac{5 \times 10^3 \times 5 \times 1 \times 10^{-6}}{3}$$

$$I_c = 8.33 \text{ mA}$$

$$8.33 \times 10^{-3} = \frac{(5 - 0.6)}{R}$$

So, $$R = \frac{4.4}{8.3 \times 10^{-3}} = 516.2 \text{ Ω}$$

or say $$R = 510 \text{ Ω}$$

Also choose, $$C_1 = 0.01 \text{ μF}$$

$$C_3 = 10 \text{ μF}$$

Use 1N 914 diode, 2N 404 transistor and NE555 timer IC.

Circuit diagram:

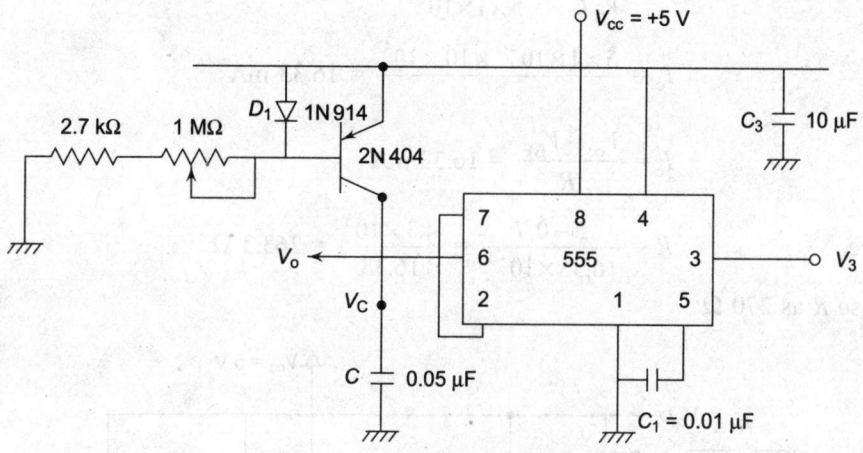

Fig. B4.2 *Circuit Diagram for Ex. 4.10*

Ex. 4.11 Draw a circuit to use 555 timer IC for generating ramp waveforms. Design the circuit to get ramp output at a frequency of approximately 10 kHz. $V_{cc} = 5$ V; $C = 1$ μF.

Solution: The astable multivibrator circuit with 555 timer IC can be used as free-running ramp generator when resistors R_A and R_B are replaced by a current-mirror circuit. The current-mirror circuit starts charging capacitor C towards V_{cc} at a constant rate. When the voltage across C equals $\frac{2}{3} V_{cc}$, comparator 1 turns transistor Q_1 ON. So, the capacitor rapidly discharges through transistor Q_1. When the discharge voltage across C is approximately equal to $\frac{1}{3} V_{cc}$, comparator 2 switches transistor Q_1 OFF, and then the capacitor C starts charging up again. Like this, the charge–discharge cycle keeps repeating and ramp waveform is produced across the capacitor. The discharging time of the capacitor is negligible compared to its charging time. Hence, the time of the ramp waveform is approximately equal to the charging time.

The expression for the frequency of the ramp waveforms is, voltage across the capacitor, as it charges from $\frac{1}{3}V_{cc}$ to $\frac{2}{3}V_{cc}$,

$$= \frac{2}{3}V_{cc} - \frac{1}{3}V_{cc} = \frac{1}{3}V_{cc}$$

$$V_c = \frac{Q}{C}; \quad Q = I_cT; \quad \text{so, } V_c = \frac{I_cT}{C}$$

$$V_c = \frac{1}{g}V_{cc}; \quad T = \frac{1}{f}; \quad \text{so, } T = \frac{1}{f} = \frac{CV_c}{I_c} = \frac{CV_{cc}}{3I_c}$$

or

$$f = \frac{3I_c}{V_{cc}C}$$

$$f_0 = 10 \text{ kHz}; \quad V_{cc} = 5 \text{ V}; \quad C = 1 \text{ μF}$$

$$f_0 = \frac{3I_c}{V_{cc}C} = \frac{3 \times I_c}{5 \times 1 \times 10^{-6}} = 10 \times 10^3$$

So,

$$I_c = \frac{5 \times 1 \times 10^{-6} \times 10 \times 10^3}{3} = 16.33 \text{ mA}$$

$$I_c = \frac{V_{cc} - V_{BE}}{R} = 16.33 \text{ mA}$$

Therefore,

$$R = \frac{5 - 0.7}{16.33 \times 10^{-3}} = \frac{4.3 \times 10^3}{16.33} = 263.3 \text{ Ω}$$

So, choose R as 270 Ω

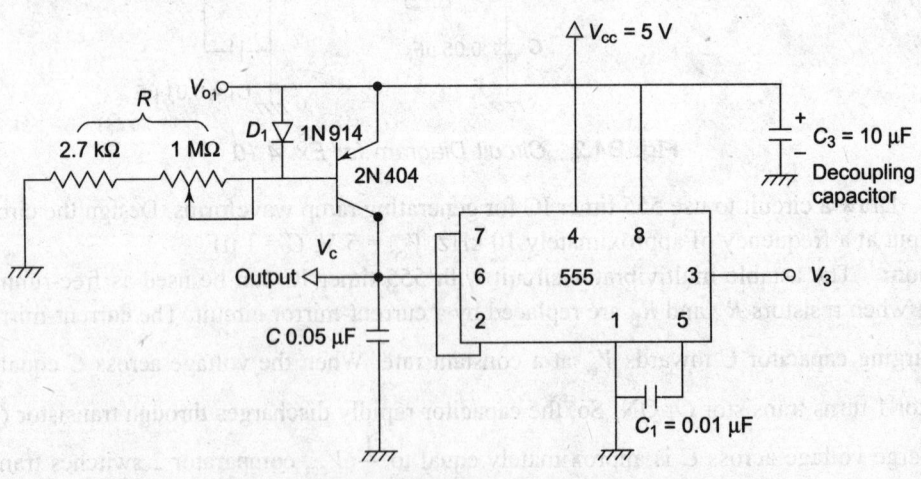

Fig. B4.3 *Circuit Diagram for Ex. 4.11*

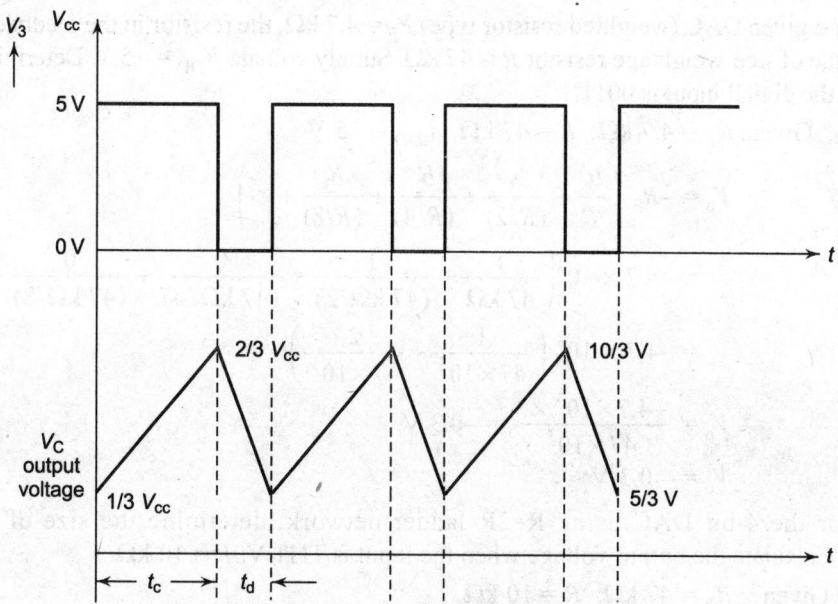

Fig. B4.4 *Waveforms for Ex. 4.11*

CHAPTER 5
ADCs and DACs

Ex. 5.1 IC 9400 is a F/V converter. Determine the output voltage V_0 if $f_{in} = 6$ kHz, $V_{Ref} = -5$V, $C_{ref} = 56$ pF, and $R_{int} = 1$ mΩ.

Solution: Given $f_{in} = 6$ kHz

The expression for V_0 for 9400 IC is,

$$V_0 = [V_{ref} C_{ref} R_{int}] f_{in}$$
$$V_0 = 5 \times 56 \times 10^{-12} \times 1 \times 10^6 \times 6 \times 10^3$$
$$V_0 = 1.68 \text{ V}$$

Ex. 5.2 For a given DAC with binary weighted resistors, determine the size of each step if $R_f = 2.2$ kΩ. Supply voltage is +8 V. Value of full-weightage resistor R is 12 kΩ. What is the output voltage when bit b_3 is at logic 1 (5 V)?

Solution: Given: $R_F = 2.2$ kΩ, $V_{DD} = +8$ V, R = 12 kΩ

Size of each step $= \dfrac{R_F}{R} 5 \text{ V} = \dfrac{2.2 \text{ k}\Omega}{12 \text{ k}\Omega} 5\text{V} = 0.92$ V

Expression for $V_0 = -R_F \left(\dfrac{b_0}{R} + \dfrac{b_1}{(R/2)} + \dfrac{b_2}{(R/4)} + \dfrac{b_3}{(R/8)} + \cdots \right)$

So, $b_3 = +5$ V (logic 1)

Therefore, $V_0 = -\dfrac{R_F}{(R/8)} b_3 = -\dfrac{2.2 \text{ k}\Omega}{\left(\dfrac{12 \text{ k}\Omega}{8}\right)} 5\text{V}$

$V_0 = 7.33$ V

Ex. 5.3   For a given DAC (weighted resistor type) $R_F = 4.7$ kΩ, the resistor in the feedback path of op-amp. The value of free-weightage resistor R is 47 kΩ. Supply voltage $V_{DD} = +5$ V. Determine the output voltage V_o if the digital input is 0011.

Solution: Given: $R_F = 4.7$ kΩ, $R = 47$ kΩ, $V_{DD} = +5$ V

$$V_o = -R_F \left(\frac{b_0}{R} + \frac{b_1}{(R/2)} + \frac{b_2}{(R/4)} + \frac{b_3}{(R/8)} + \cdots \right)$$

$$= -4.7 \times 10^3 \left(\frac{1}{47 \text{ k}\Omega} + \frac{1}{(47 \text{ k}\Omega/2)} + \frac{0}{(47 \text{ k}\Omega/4)} + \frac{0}{(47 \text{ k}\Omega/8)} \cdots \right)$$

$$= -4.7 \times 10^3 \left(\frac{1}{47 \times 10^3} + \frac{2}{47 \times 10^3} \right)$$

$$V_o = \frac{-4.7 \times 10^3 \times 3}{47 \times 10^3} = -0.3 \text{ V}$$

$$V_o = -0.3 \text{ V}$$

Ex. 5.4   For the 4-bit DAC using R–2R ladder network, determine the size of each step if $R_F = 47$ kΩ. Calculate the output voltage when the input is 1111 V; $R = 10$ kΩ.

Solution: Given $R_F = 47$ kΩ; $R = 10$ kΩ

Let $\qquad V_{DD} = +5$ V

$$\text{Size of each step} = \frac{R_F}{16\,R} \times V_{DD} = \frac{47 \times 10^3}{16 \times 10 \times 10^3} \times 5 \text{ V}$$

for 4 bit DAC (R–2R ladder type)

$$\text{Step size} = 1.47 \text{ V}$$

for input 1111, V_o is,

$$V_o = -R_F \left(\frac{b_3}{2R} + \frac{b_2}{4R} + \frac{b_1}{8R} + \frac{b_0}{16R} \right)$$

$$= -47 \times 10^3 \left(\frac{5 \text{ V}}{2 \times 10 \times 10^3} + \frac{5 \text{ V}}{4 \times 10 \times 10^3} + \frac{5 \text{ V}}{8 \times 10 \times 10^3} + \frac{5 \text{ V}}{16 \times 10 \times 10^3} \right)$$

$$= -47 \left(\frac{10}{10 \times 10^3} \right) \left(\frac{1}{2} + \frac{1}{4} + \frac{1}{8} + \frac{1}{16} \right) 5 \text{ V} = -47 \times \frac{1}{10} \times 0.9375$$

$$= 4.371 \times 5 \text{ V} = 21.85 \text{ V}$$

Ex. 5.5   Design a phase shift oscillator to oscillate at 200 Hz. $C = 0.1$ µF.

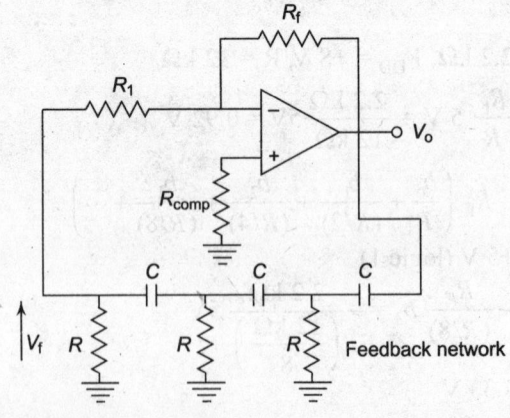

Fig. B5.1   *Circuit Diagram for Ex. 5.5*

$$F_0 = \frac{1}{\sqrt{6}(2\pi RC)}, \quad R = \frac{1}{\sqrt{6}(2\pi\,200 \times 0.1 \times 10^{-6})}$$

$$= 0.32 \text{ k}\Omega$$

$$R_1 \le R.10$$

$$R_1 \le 3.2 \text{ K}$$

$$R_1 = 3.2 \text{ K}$$

$$R_1 = 10\,R = 3.2 \text{ K}$$

$$R_f = 29\,R_1$$

$$R_f = 29.8 \text{ k}\Omega$$

Ex. 5.6 Find the transfer function of *RC* phase-shift oscillator using op-amp and also find loop gain A_v.

$$\beta = \frac{V_f}{V_0}$$

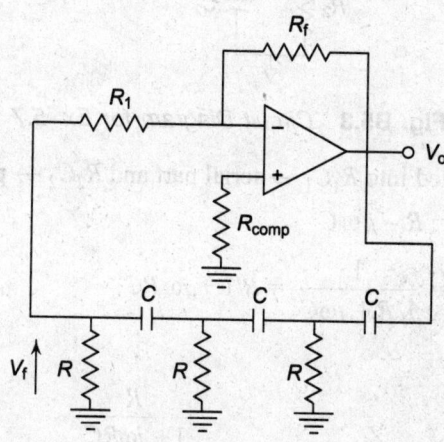

Fig. B5.2 *Circuit Diagram for Ex. 5.6*

$$= \frac{1}{1 + 6/sRC + 5/s^2R^2C^2 + 1/s^3R^3C^3}$$

Put $S = j\omega$

$$\beta = \frac{1}{1 - 5(f_1/f)^2 - j[6(f_1/f) - (f_1/f)^3]}$$

$$f_1 = 1/2\pi RC$$

$A_v\beta = 1$, imaginary part equal to zero

$$6(f_1/f) - (f_1/f)^3 = 0$$

$$f_1/f = \sqrt{6}$$

So,
$$f_o = \frac{1}{\sqrt{6}(2\pi RC)}$$

$$A_v\beta = 1$$

$$A_v \frac{1}{1 - 5(f_1/f_o)^2} = 1$$

$$A_v \geq 29, \text{ so, } R_f = 29R_1$$

Ex. 5.7 For the given Wien-bridge oscillator, find the frequency of oscillation.

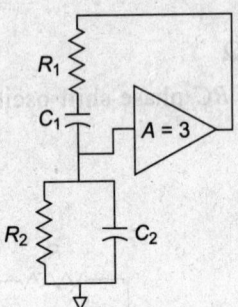

Fig. B5.3 *Circuit Diagram for Ex. 5.7*

The above circuit is divided into $R_1C_1 \rightarrow$ serial part and $R_2C_2 \rightarrow$ parallel part.

Let
$$Z_1 = R - j/\omega C$$

$$Z_2 = \frac{1}{1/R + j\omega C} = R/1 + j\omega RC$$

Apply volt-divider law

$$V_{load} = \frac{Z_2}{Z_1 + Z_2} I_{in} = \frac{\dfrac{R}{1 + j\omega RC}}{R - j/\omega c + \dfrac{R}{1 + j\omega RC}} e_{in}$$

$$\frac{V_{load}}{e_{in}} = \frac{R}{3R + j(\omega R^2 C - 1/\omega C)}, \text{ so, } A\beta = 1, \hat{\beta} = V_{load}/e_{in}$$

The imaginary part in the above equation must be zero

$$\omega R^2 C - 1/\omega C = 0, f_o = 1/2\pi RC, \omega = 1/RC$$

When $R_1 \neq R_2$, $C_1 \neq C_2$, frequency is

$$f_o = 1/2\pi\sqrt{R_1 R_2 C_1 C_2}$$

LM301 OP-AMP SPECIFICATIONS

It is a general-purpose op-amp with improved performance over industry standards like LM 709. The improved features are as follows:

- Offset voltage 3 mV maximum
- Input current 100 mA maximum
- Offset current 20 mA
- Guaranteed drift characteristic
- Slew rate 10 V/msec as a summing amplifier

Absolute Maximum Ratings:

- Supply voltage: ±18 V
- Power dissipation: 500 mW
- Differential input voltage: ±30 V

Table B5.1   Electrical Characteristics

Parameter	Min.	Typical	Max.	Unit
Input offset voltage		2	7.5	mV
Input offset current		3	50	nA
Input bias current		70	250	nA
Input resistance	0.5	2		MΩ
Input bias current			0.3	μA
CMMR	70	90		dB
SVRR	70	96		dB

LM311 VOLTAGE COMPARATOR IC

- Total supply voltage: 36 V
- Output to negative supply voltage: 40 V
- Differential input voltage: ±30 V
- Power dissipation: 500 mW
- Output short-circuit duration: 10 sec

Table B5.2   Electrical Characteristics

Parameter	Condition	Min.	Typical	Max.	Unit
Input offset voltage	$T_A = 25°C$		2.0	7.5	mV
	$R_s \leq 50\ K$				
Input offset current	$T_A = 25°C$		6.0	50	nA
Input bias current	$T_A = 25°C$		100	250	nA
Voltage gain	$T_A = 25°C$	40	200		V/mV
Positive supply current	$T_A = 25°C$		5.1	7.5	mA
Negative supply current	$T_A = 25°C$		4.1	5.0	mA

555 TIMER IC

Absolute Maximum Ratings:

Supply voltage   : +18 V
Power dissipation   : 600 mW

Table B5.3   Electrical Characteristics

Parameter	Test conditions	Min.	Typ.	Max.	Unit
Supply voltage		4.5		16	V
Supply current	V_{cc} = 5V, R_L = ¥	3	6		mA
	V_{cc} = 15 V, R_L = ¥	10	15		mA
Trigger voltage	V_{cc} = 15 V		5		V
Rise time of output				100	nsec
Fall time of output			100		nsec

PROGRAMMABLE TIMER/COUNTER XR2240

It can produce ultra-long delays without sacrificing accuracy. The IC consists of an internal time-base oscillator, a programmable 8-bit counter, and a control flip-flop. The time delay is set by an external *RC* network and can be programmed to any value from 1 *RC* to 255 *RC*.

Absolute maximum ratings:
- Supply voltage: 18 V
- Power dissipation ceramic package   : 750 mW
- Plastic package   : 625 mW

THREE-TERMINAL ADJUSTABLE REGULATOR IC: LM117

LM 117/217/317 are adjustable three-terminal positive voltage regulators capable of supplying in excess of 1.5 A over a 1.2 V to 37 V output range.

Features:
- Adjustable output down to 1.2 V
- Guaranteed 1.5-A output current
- Line regulation typically 0.01% V
- Load regulation typically 0.1%
- Standard three lead package
- 80-dB ripple rejection

Absolute Maximum Ratings:
- Power dissipation: internally limited
- Input–output voltage differential   : 40 V
- Storage temperature   : −65°C to +150°C

Table B5.4   Electrical Characteristics

Parameter	Conditions	Min.	Typ.	Max.	Unit
Line regulation	T_A = 25°C		0.01	0.02	%/V
Temperature stability	$T_{min} \leq T_i \leq T_{max}$		1		%
Maximum load current	$(V_{in} - V_{out})$ = 40 V		3.5	10	mA
Ripple rejection ratio			65		dB
Thermal resistance					
Junction to case	T package		4		°C/W

CHAPTER 6

Voltage Regulators

Ex. 6.1 Design a step-down switching regulator to give $V_o = 12$ V at 1 A maximum current. Ripple voltage should not exceed 1% of V_o. $V_{in} = 18$V DC.

Solution: Given $\qquad V_{in} = 18$-V DC; $\quad V_o = 12$ V; $\quad I_L = 1$ A max.

$$\text{Ripple} = 1\% \text{ of } V_o.$$

Let us choose 78540 IC_x switching regulator. For this IC,

$$V_D = 1.25 \text{ V}; \quad V_{ref} = 1.245 \text{ V}; \quad V_s = 1.1 \text{ V}; I_B = 200 \text{ mA max.}$$

$$I_{PK} = 2I_o = 2 \times 1A = 2A$$

$$R_{SC} = \frac{0.33}{I_{PK}} = \frac{0.33}{2A} = 0.165 \ \Omega$$

$$\frac{t_{ON}}{t_{OFF}} = \frac{V_o + V_D}{V_{in} - V_S - V_o} = \frac{12 + 1.25}{18 - 1.1 - 12} = \frac{13.25}{4.9}$$

$$\frac{t_{ON}}{t_{OFF}} \simeq 2.7; \quad T = t_{ON} + t_{OFF}$$

$$T = (2.7 \ t_{OFF} + t_{OFF})$$

Let $\qquad T = 50$ μsec. So, $t_{OFF} (2.7 + 1) = 50$ μsec.

or $\qquad t_{OFF} = \dfrac{50}{3.7} = 13.51$ μsec.

$$t_{ON} = T - t_{OFF} = 50 - 13.51 = 36.49 \text{ μsec.}$$

$$C_T \text{ in μF} = 45 \times 10^{-5} \ t_{OFF} \text{ (μsec)}$$

$$= 45 \times 10^{-5} \times 13.51$$

$$C_T = 0.0061 \text{ μF}; \quad \text{use } 0.006 \text{ μF}$$

$$L = \frac{V_o + V_D}{I_{PK}} \times t_{OFF}$$

$$L = \left(\frac{12 + 1.25}{2}\right) \times 13.51 \times 10^{-6}$$

$$L = 89.5 \text{ μH}$$

$$C_o = \frac{I_{PK} \times (t_{ON} + t_{OFF})}{8 \ V_{ripple}}; \quad V_{ripple} = \frac{1}{100} \times 12 = 0.12 \text{ V}$$

$$C_o = \frac{2 \times (36.49 + 13.51)}{8 \times \left(\dfrac{1}{100} \times 12\right)} = 104.2 \text{ μF}$$

Use 100 μF capacitor.

Let
$$I_2 = 0.1 \text{ mA}; \quad R_2 = \frac{V_o}{I_2}$$

$$R_2 = 12.45 \text{ k}\Omega, \text{ use } 12 \text{ k}\Omega$$

$$V_{R2} = V_{ref} = \frac{R_2(V_o)}{(R_1 + R_2)}$$

so,
$$R_1 = 103.7 \text{ k}\Omega$$

Circuit diagram for 12-V step-down switching regulator:

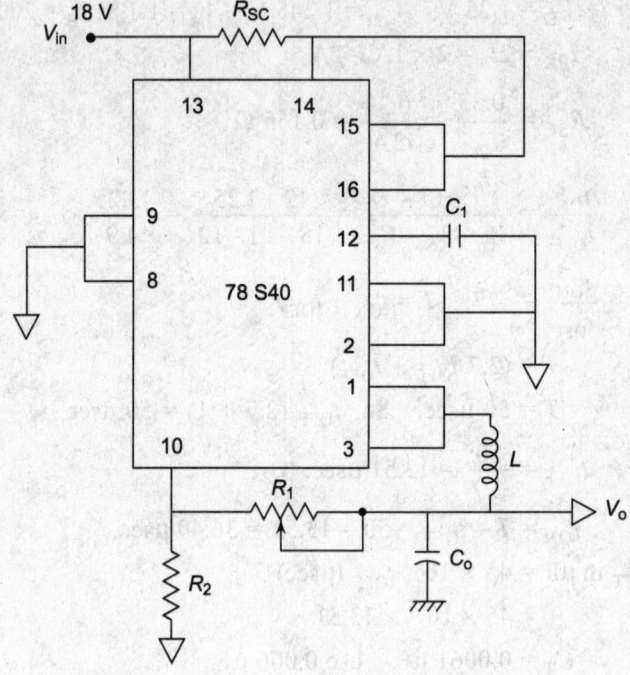

Fig. B6.1 *Step-Down Switching Regulator for Ex. 6.1*

Ex. 6.2 Employing 7805 C voltage regulator, design a current source that will deliver 150-mA current to 10 Ω, 10 W load.

Solution: Given:
$$I_L = 180 \text{ mA}, \quad R_L = 10 \text{ }\Omega, 10 \text{ W}$$

$$\text{IC} = 7805 \text{ C}; \quad \text{Drop-out voltage for } 7805 \text{ C} = 2 \text{ V}$$

For   7805 C IC, $V_R = 5$ V; $I_Q = 4.3$ mA

$$I_L = \frac{V_R}{R} + I_Q$$

$$V_o = V_R + V_L$$

$$V_{in} = V_o + \text{Drop-out voltage (2 V)}$$

$$180 \times 10^{-3} = \frac{5}{R} + 4.3 \times 10^{-3}$$

$$(180 \times 10^{-3} - 4.3 \times 10^{-3}) = \frac{5}{R}$$

$$R = \frac{5}{175.7 \text{ mA}} = 28.45 \text{ }\Omega$$

$$R \simeq 28 \text{ }\Omega$$

$$V_0 = V_R + V_L$$

$$= 5 + R_L I_L$$

$$= 5 + 10 \times 180 \times 10^{-3}$$

$$V_0 = 6.8 \text{ V}$$

$$V_{in} = 6.8 + 2 \text{ V}$$

$$V_{in} = 8.8 \text{ V}$$

Circuit:

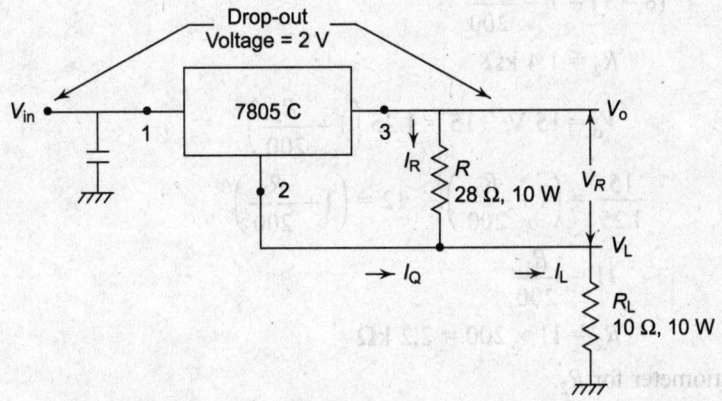

Fig. B6.2 *Circuit Diagram for Ex. 6.2*

Ex. 6.3 Using LM317, design an adjustable voltage regulator to satisfy the following specifications. Output voltage $V_0 = 10$ to 15 V; $I_0 = 0.5$ A.

Solution: Given: $\qquad V_0 = 10$ to 15 V, $\quad I_0 = 0.5$ A, $\quad I_c = $ LM 317

For LM317 voltage regulator IC, the maximum value of adjustment pin current I_{ADJ} is 100 μA.

$$V_0 = R_1 I_1 + R_2 (I_1 + I_{ADJ})$$

where $\qquad I_1 = \dfrac{V_{REF}}{R_1}$

$\qquad R_1 = $ current I_1 set resistor

$\qquad R_2 = $ output V_0 set resistor

$\qquad I_{ADJ} = $ adjustment pin current

$$V_0 = V_{REF} \left(1 + \frac{R_2}{R_1}\right) + I_{ADJ} R_2$$

where $V_{REF} = 1.25$ V = reference voltage between the output and adjustment terminals. But the drop across R_2 due to I_{ADJ} can be neglected as it is very small.

So,
$$V_o = 1.25 \left(1 + \frac{R_2}{R_1}\right)$$

Let
$$R_1 = 200 \ \Omega$$

Therefore, for
$$V_o = 10 \text{ V},$$

$$10 = 1.25 \left(1 + \frac{R_2}{200}\right)$$

$$\frac{10}{1.25} = 1 + \frac{R_2}{200}$$

$$8 = 1 + \left(\frac{R_2}{200}\right)$$

$$(8 - 1) = 7 = \frac{R_2}{200}$$

So,
$$R_2 = 1.4 \text{ k}\Omega$$

For
$$V_o = 15 \text{ V}, \quad 15 = 1.25 \left(1 + \frac{R_2}{200}\right)$$

$$\frac{15}{1.25} = \left(1 + \frac{R_2}{200}\right); \quad 12 = \left(1 + \frac{R_2}{200}\right)$$

$$11 = \frac{R_2}{200}$$

or
$$R_2 = 11 \times 200 = 2.2 \text{ k}\Omega$$

Use 3 kΩ potentiometer for R_2.
The circuit is as shown.

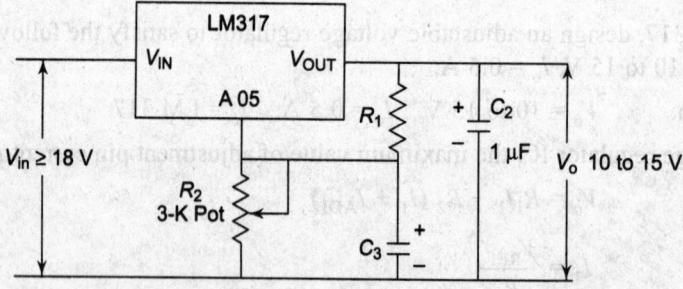

Fig. B6.3 *Current Diagram for Ex.6.3*

Appendix C
Design Problems

1. Calculate (i) ε_{high}, (ii) ε_{low}, and (iii) free-running frequency for the timer circuit shown below:

Solution: (i) $\varepsilon_{high} = 0.695 (R_A + R_B)C$

$$R_A = 6.8 \text{ K}, R_B = 3.3 \text{ K}, C = 0.1 \text{ }\mu\text{F}$$

$$\varepsilon_{high} = 0.7 \text{ msec}$$

(ii) $\varepsilon_{low} = 0.695 (R_B)C$

$$= 0.695 (3.3 \text{ K}) 0.1 \text{ }\mu\text{F}$$

$$= 0.23 \text{ msec}$$

(iii) free-running frequency, $f = \dfrac{1.44}{(R_A + 2R_B)C}$

$$f = \frac{1.44}{\left[6.8 + 2(3.3)0.1 \times 10^{-6}\right]} = 1 \text{ kHz}$$

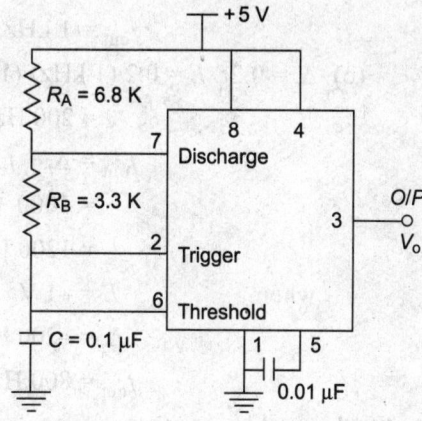

Fig. C.1 *Circuit for Problem 1*

2. Define and calculate duty cycle for the values given below, and also find the minimum value of R_A so that discharge transistor is on and current through it is limited to low value. The ratio of lower time of the wireform ε_{low} to the total time of a one cycle?

Solution:
$$D = \frac{\varepsilon_{low}}{T} = \frac{R_B}{R_A + 2R_B}$$

Given

$$R_B = 3.3 \text{ k}\Omega$$
$$R_A = 6.8 \text{ k}\Omega$$

$$D = \frac{3.3 \text{ K}}{6.8 \text{ K} + 2(3.3 \text{ K})} = 0.25$$

So, timer output is lower for 25% of T. The minimum value of R_A

$$R_A = \frac{V_{cc}}{I_a} \qquad \text{(Given } I_A = 0.2 \text{ A)}$$

$$= \frac{5}{0.2 \text{ A}} = 25 \text{ }\Omega$$

R_A 25 Ω is low value such that V_{CE} and pin 7 shorted together. Therefore, R_A should be in kΩ of minimum 1 kΩ.

IC-555 APPLICATIONS: FREQUENCY SHIFTER

3. In a frequency shifter, calculate (a) charge current I for $E = 0$, (b) center frequency f_c when $E = 0$, and (c) frequency shift when $E = \pm 1$ and f_{out}.

Solution: Given $R_E = 3$ kΩ, $V_{cc} = 5$ V, $C = 1$ μF

Let

(a) $I = \dfrac{V_{cc} - E}{R_E} = \dfrac{5 - 0}{3 \text{ K}} = 1.67$ mA

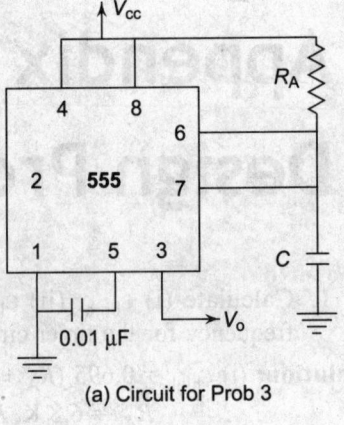

(a) Circuit for Prob 3

(b) $f_c = \dfrac{3}{R_{EC}} = \dfrac{3}{(3K)(1 \times 10^{-6} F)} = \dfrac{3}{3 \text{ msec}} = 1$ kHz

$$f_{out} = 1 \text{ kHz, when } E = 0$$

(c) $\Delta_f = 0.2 f_c \, E = 0.2 \, (1 \text{ kHz}) \, (1)$

$$= 200 \text{ Hz}$$

$$f_{out} = \Delta_f + f_c$$

$$= 1000 + 200$$

$$= 1200 \text{ Hz}$$

when $\qquad E = -1$ V

$$\Delta_f = -200 \text{ Hz}$$

$$f_{out} = 800 \text{ Hz}$$

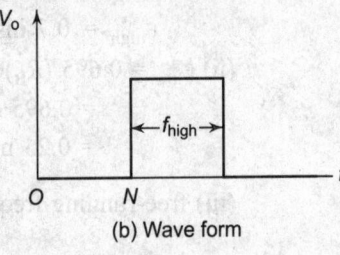

(b) Wave form

Fig. C.2

4. Monostable operation

$$\varepsilon_{high} = 1.1 \, R_A C$$

If $\qquad R_A = 9$ kΩ

find C for given diagram

Solution: $\qquad C = \dfrac{f_{high}}{1.1 R_A} = \dfrac{1 \times 10^{-3}}{1.1 \, (9 \times 10^3)} = 0.1$ μF

5. For IC-555 monostable operation, $R_A = 10$ kΩ, $C = 0.2$ μF. Find ε_{high} and time constant.

Soluion: (i) $\varepsilon_{high} = 1.1 \, R_A C$

$$= 1.1 \, (10 \text{ K}) \, (0.2 \times 10^{-6})$$

$$= 2.2 \text{ msec}$$

(ii) time constant $= R_A C_i \quad (C_i = 0.001 \, \mu F)$

$$`u' = (10 \text{ K}) \, (0.001 \times 10^{-6})$$

$$= 0.01 \text{ msec}$$

6. In the circuit shown below, specify suitable components for $f_o = 50$ kHz, $D = 75\%$.
Assume $C = 1$ nF

Soluion: $R_A + 2R_B = 1.44/f_o \, C$

$$= \dfrac{1.44}{50 \times 10^3 \times 1 \times 10^{-9}}$$

$$= 28.85 \text{ k}\Omega$$

So, $D = \dfrac{R_A + R_B}{R_A + 2R_B}$

$$= 0.75$$

Therefore, $R_A = 2R_B$

Hence, $R_A = 14.4 \text{ k}\Omega$, $R_B = 7.21 \text{ k}\Omega$

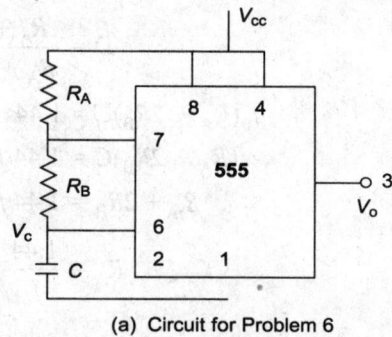

(a) Circuit for Problem 6

Fig. C.3 *Circuit for Problem 6*

O/P Waveform

$$I = T_L + TH = R_B\, C \ln_2 + (R_A + R_B)\, C \ln \dfrac{V_{cc} - V_{TH}/2}{V_{cc} - V_{TH}}$$

$V_{TH} = 2/3\ V_{cc}$

$f = 1/T$

Precentage $D = 100\,TH/T_L + TH$

$$f_0 = \dfrac{1.44}{(R_A + 2R_B)C}$$

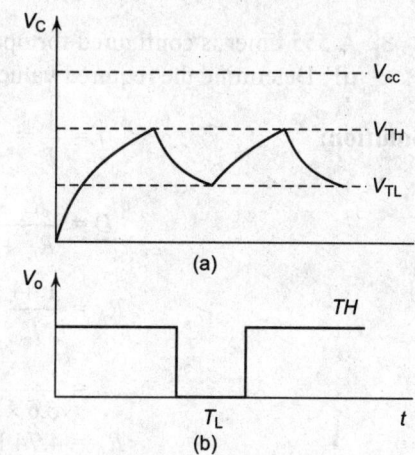

(a)

(b)

Fig. C.4 *Waveforms*

7. The 555 timer in Fig. C.5 is to produce the output waveform with pulse repetitive rate of 5 kHz. Determine the required value of R_A.

Solution:

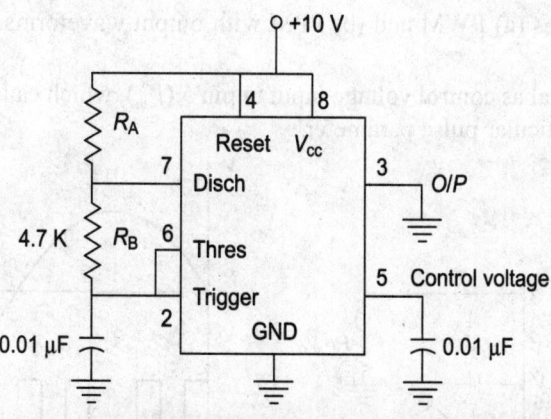

Fig. C.5 *Circuit for Problem 7.*

Pulse repetitive frequency

$$= \dfrac{1}{T} = \dfrac{1.44}{(R_A + 2R_B)C}$$

$$D = \frac{R_A + R_B}{R_A + 2R_B}$$

$$f_\theta(R_A + 2R_B)C = 1.44$$

$$(R_A + 2R_B)C = 1.44/f$$

$$R_A + 2R_B = 1.44/f_c$$

$$R_A = \frac{1.44}{f_c} - 2R_B$$

$$= \frac{1.44}{5 \times 10^3 \times 0.01 \times 10^{-6}} - 2(4.7 \times 10^3)$$

$$R_A = 19.4 \text{ k}\Omega$$

8. A 555 timer is configured for operation as an astable multivibrator with $R_B = 2.2$ kΩ, $C = 0.022$ µF. Determine the required value of R_A such that $P_{RR} = 5.6$ kHz. Also determine the duty cycle.

Solution:

$$f = \frac{1}{T} = \frac{1.44}{C(R_A + 2R_B)}$$

$$D = \frac{R_A + R_B}{R_A + 2R_B}$$

$$R_A = \frac{1.44}{f_c} - 2R_B$$

$$= \frac{1.44}{5.6 \times 10^3 \times 0.022 \times 10^{-6}} - 2 \times 2.2 \times 10^3$$

$$R_A = 4.74 \text{ k}\Omega$$

$$D = \frac{4.74 + 2.2}{4.74 + 2 \times 2.2} = \frac{6.94}{9.14} = 0.86$$

9. Draw IC-555 timer as (a) PWM and (b) PPM with output waveforms.

Solution: (a)   **PWM**

(i) Analog signal as control voltage input to pin 5 (V_m), which causes proportional variation in some particular pulse parameter.

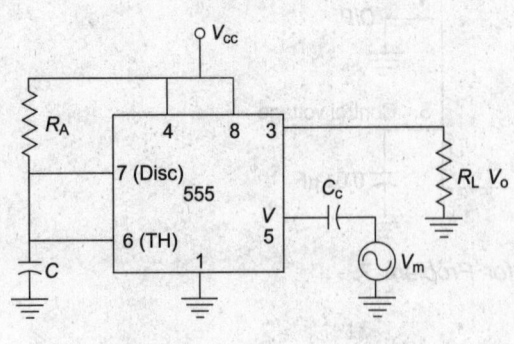

Fig. C.6 *Pulse-width modulator (Prob. 9)*

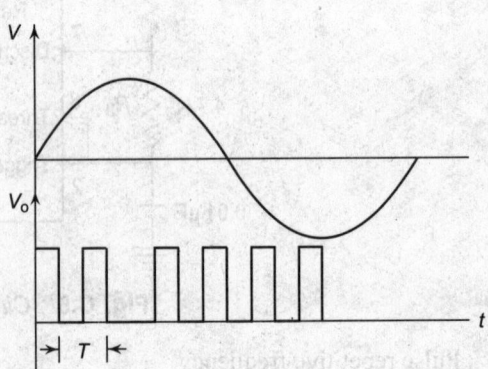

Fig. C.7 *Pulse-width O/P waveform*

(b) **Pulse-position modulator**

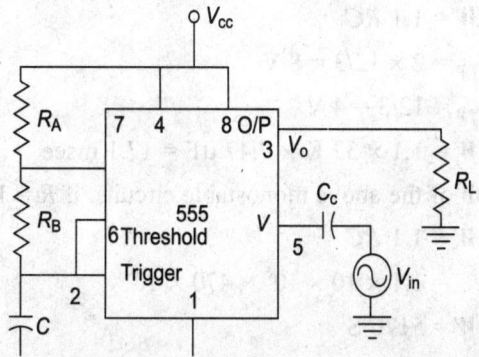

Fig. C.8   *PPM using IC 555*

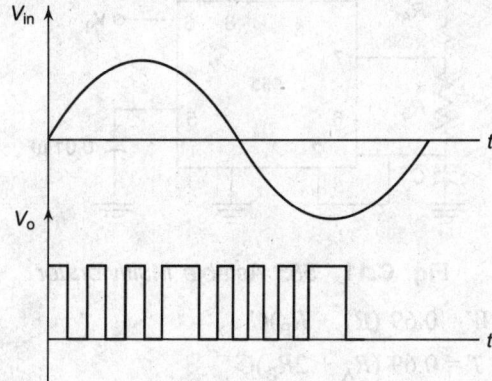

Fig. C.9   *PPM*

10. For a monostable timer circuit as shown in Fig. C.10, $V_{cc} = 12$ V, $R = 33$ kΩ, and $C = 0.47$ μF. What is the minimum trigger voltage that produces an output pulse? What is the maximum capacitor voltage? What is the width of the output pulse?

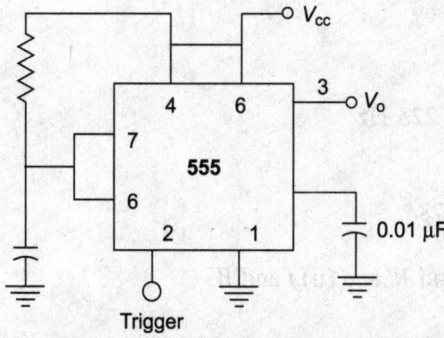

Fig. C.10   *Monostable timer*

$$U_{TP} = \frac{2}{3} V_{cc}$$

$$L_{TP} = V_{cc}/3$$

$$W = 1.1 \, RC$$

So,
$$U_{TP} = 2 \times 12/3 = 8 \text{ V}$$

$$L_{TP} = 12/3 = 4 \text{ V}$$

$$W = 1.1 \times 33 \text{ K} \times 0.47 \text{ μF} = 17.1 \text{ msec}$$

11. What is the pulse width of the above monostable circuits, if $R = 10 \text{ m}\Omega$, $C = 470 \text{ μF}$?

Solution:
$$W = 1.1 \, RC$$

$$= 1.1 \times 10 \times 10^6 \times 470 \times 10^{-6}$$

$$W = 5170 \text{ S}$$

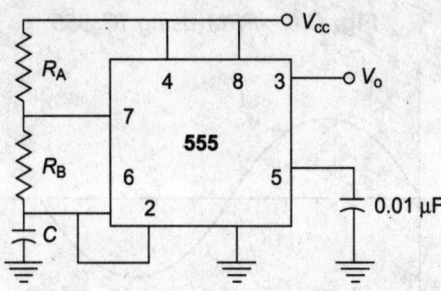

Fig. C.11 *555 Astable Multivibrator*

$$W = 0.69 \, (R_A + R_B)C$$

$$T = 0.69 \, (R_A + 2R_B)C$$

$$f = \frac{1.44}{(R_A + 2R_B)C}, \quad D = \frac{R_A + R_B}{R_A + 2R_B}$$

Given
$$V_{cc} = +12 \text{ V}, \ R_A = 75 \text{ K}, \ R_B = 30 \text{ K}$$

Find U_{TP}, L_{TP}, W, f, and D.

(a) $U_{TP} = 2V_{cc}/3 = 8 \text{ V}$

$L_{TP} = V_{cc}/3 = 4 \text{ V}$

(b) $f = \dfrac{1.44}{(75 + 60)\text{K}(47\Omega)} = 225 \text{ Hz}$

(c) $D = \dfrac{75 \text{ K} + 30 \text{ K}}{75 \text{ K} + 60 \text{ K}} \doteqdot 0.778$

(d) Relation between (i) T and W and (ii) f and W

   (i) $T = W + 0.69 \, R_B C$

   (ii) $f = \dfrac{1}{W + 0.69R_B C}$,

   (iii) $W = -(R_A + R_B)C, \ln \dfrac{V_{cc} - V_{Con}}{V_{cc} - 0.5 \, V_{Con}}$

12. IC-555 VCO (Voltage-Controlled Oscillator)

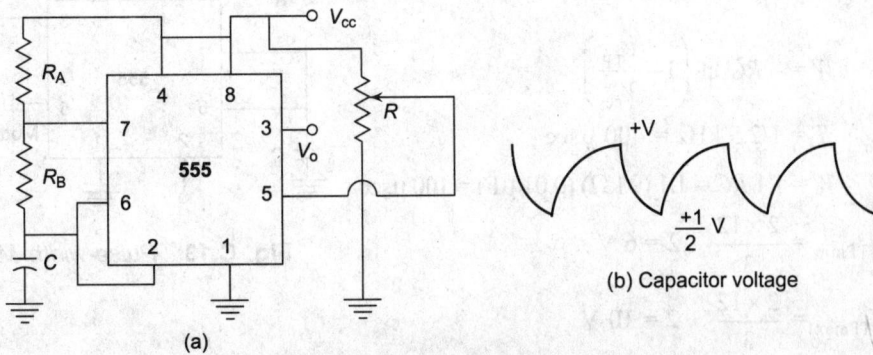

(a)

(b) Capacitor voltage

Fig. C.12 *Voltage-Controlled Oscillator Circuit*

V_{CO} using IC 555

$$W = (R_A + R_B) \, C \ln \frac{V_{cc} - V}{V_{cc} - 0.5 \, V}$$

$$T = W + 0.69 \, R_B C$$

(a) Find the f and D when V is 11 V. Given $R_1 = 75$ kΩ, $R_2 = 30$ K, $C = 47$ nF

(b) Find f and D when V is 1 V?

Solution:

(a) $W = - (75 + 30)\text{K} \; 47 \text{ nF} \ln \dfrac{12 \text{ V} - 11 \text{ V}}{12 - 5.5} = 9.24 \; \mu\text{sec}$

$\qquad T = W + 0.69 \, (30 \text{ K}) \, (47 \text{ nF}) = 10.2 \; \mu\text{sec}$

$$D = \frac{W}{T} = \frac{9.24 \, \mu\text{sec}}{10.2 \, \mu\text{sec}} = 0.906$$

$$f = \frac{1}{T} = 98 \text{ Hz}$$

(b) $W = - (75 + 30)\text{K} \; 47 \text{ nF} \ln \dfrac{12 - 1}{12 - 0.5} = 0.2 \; \mu\text{sec}$

$\qquad T = 0.2 \; \mu\text{sec} + 0.69 \, (30 \text{ K}) \, (47 \text{ nF})$

$\qquad\quad = 1.2 \; \mu\text{sec}$

$$D = \frac{W}{T} = \frac{0.2}{1.2} = 0.183, f = \frac{1}{T} = 833 \text{ Hz}$$

13. A pulse-width modulator shown below has $V_{cc} = 120$ V, $R = 9.1$ kΩ, and $C = 0.01 \; \mu$F. The clock has a frequency of 2.5 kHz. If a modulating signal has a peak value of 2 V, what is the period of output pulse? What is quiescent pulse width? What are the minimum and maximum pulse width? What are the minimum and the maximum duty cycles?

Solution: $\quad T = 1/f_{\text{desk}}$

$$U_{TP} = \frac{2V_{cc}}{3} + V_{\text{nod}}$$

$$D = \frac{W}{T}$$

$$W = -RC \ln\left(1 - \frac{V_{TP}}{V_{cc}}\right)$$

$$T = 1/2.5 \text{ kHz} = 400 \text{ } \mu\text{sec}$$

$$W = 1.1 \, RC = 1.1 \, (9 \text{ k}\Omega) \, (0.01 \text{ } \mu\text{F}) = 100 \text{ } \mu\text{sec}$$

$$U_{TPmin} = \frac{2 \times 12}{3} - 2 = 6 \text{ V}$$

$$U_{TPmax} = \frac{2 \times 12}{3} + 2 = 10 \text{ V}$$

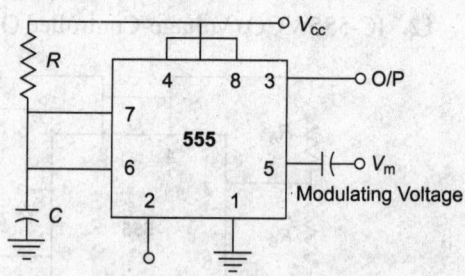

Fig. C.13 *Pulse-width Modulator*

$$\dot{W}_{min} = -(9 \text{ k}\Omega) \, (0.0114 \text{ F}) \ln\left(1 - \frac{6}{12}\right)$$

$$= 63 \text{ } \mu\text{sec}$$

$$W_{max} = -(9 \text{ k}\Omega)(0.0114 \text{ F}) \ln\left(1 - \frac{10}{12}\right)$$

$$= 163 \text{ } \mu\text{sec}$$

$$D_{min} = 63/400 = 0.158$$

$$D_{max} = 163/400 = 0.408.$$

14. A pulse-position modulator is as shown in Fig. C.14. $V_{cc} = 12$ V, $R_1 = 3.9$ kΩ, $R_2 = 3$ kΩ, and $C = 0.01$ μF. What are the quiescent width and period of the output pulses? If a modulating signal has a peak value of 1.5 V, what are the minimum and maximum pulse width? What is the space between pulse?

Solution:

$$U_{TP} = \frac{2}{3} V_{cc} + V_{nod}$$

$$W = -(R_A + R_B) \ln \frac{V_{cc} - V_{TP}}{V_{cc} - 0.5 \, V_{TP}}$$

$$T = W + 0.69 \, R_B C$$

$$\text{Space} = 0.69 \, R_B C$$

(a) $W = 0.69 \, (3.9 \text{ K} + 3 \text{ K}) \, (0.01 \text{ } \mu\text{F})$
    $= 47.8 \text{ } \mu\text{sec}$

(b) $T = 0.69 \, (3.9 \text{ K} + 6 \text{ K}) \, (0.01 \text{ } \mu\text{F})$
    $= 68.6 \text{ } \mu\text{sec}$

(c) $U_{TPmin} - 8 \text{ V} - 1.5 \text{ V} = 6.5 \text{ V}$
    $U_{TPmax} = 8 \text{ V} + 1.5 \text{ V} = 9.5 \text{ V}$

(d) $W_{min} = -(3.9 \text{ K} + 3 \text{ K}) \, (0.01 \text{ } \mu\text{F}) \ln \dfrac{12 - 6.5}{12 - 3.25}$

    $= 32 \text{ } \mu\text{sec}$

$W_{max} = -(3.9 \text{ K} + 3 \text{ K}) \, (0.01 \text{ } \mu\text{F}) \ln \dfrac{12 - 9.5}{12 - 4.75}$

    $= 73.5 \text{ } \mu\text{sec}$

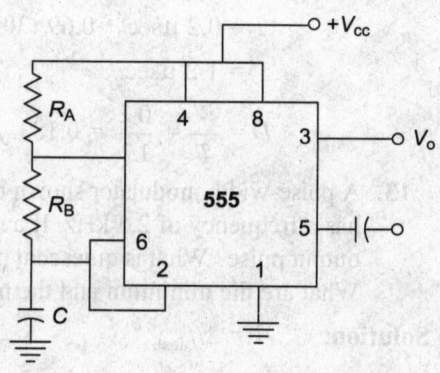

Fig. C.14 *Pulse-Position Modulator*

(e) $T_{min} = 32$ μsec $+ 0.69$ (3 K) (0.01 μF) $= 52.8$ μsec

$T_{max} = 73.5$ μsec $+ 0.9$ (3 K) (0.01 μF) $= 94.3$ μsec

15. The ramp generator of Fig. C.15 has constant collector current of 1 μA. If $V_{cc} = 15$ V and $C = 100$ nF. What is the slope of output ramp? What is its peak value? What is its direction?

Solution:

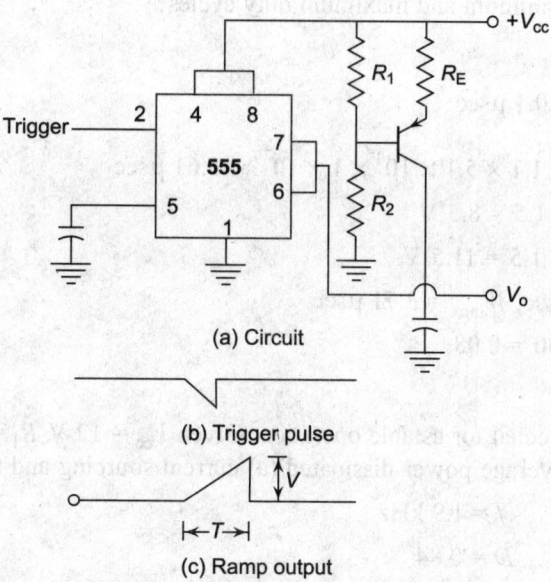

(a) Circuit

(b) Trigger pulse

(c) Ramp output

Fig. C.15 *Ramp-Generator Circuit*

$$I_C = \frac{V_{cc} - V_E}{R_E}$$

$$S = \frac{I_C}{C}$$

$$V = 2/3 \ V_{cc}$$

$$T = \frac{2}{3} \frac{V_{cc}}{S}$$

(a) $S = \dfrac{1\,mA}{100\,nF} = 10$ V/μsec

(b) $V = 2(15\text{ V})/3 = 10$ V

(c) $T = 2/3 \left(\dfrac{15}{10}\right) = 1$ μsec

16. The ramp generator has a constant collector current 0.5 mA. If $V_{cc} = 12$ V and $C = 68$ nF. What is the slope of the output ramp? What is its peak value? What is its direction (see figure C.15)?

(a) $S = \dfrac{0.5 \times 10^{-3}}{68 \times 10^{-9}} = 7.3$ V/μsec   (b) $V = \dfrac{2}{3} (12) = 8$ V

(c) $T = \dfrac{2}{3} \dfrac{(12)}{7.3} = 1.09$ μsec

17. A pulse-width modulator as in Problem 14 has $V_{cc} = 10$ V, $R = 5.1$ kΩ, $C = 1$ nF, $f = 10$ kHz, and modulator signal = 1.5 V.

   (a) What is the period of output pulse?
   (b) What is quiescent pulse width?
   (c) What are the minimum and maximum pulse width?
   (d) What are the minimum and maximum duty cycles?

Solution:

   (a) $T = \dfrac{1}{10 \text{ kHz}} = 0.1$ μsec

   (b) $W = 1.1 \ RC = 1.1 \times 5.1 \times 10^3 \times 1 \times 10^{-9} = 5.61$ μsec

   (c) $U_{TP \ min} = 10 - 1.5 = 8.5$ V

   $U_{TP \ max} = 10 + 1.5 = 11.5$ V

   $W_{min} = 8.66$ μsec, $W_{max} = 3.71$ μsec

   (d) $D_{max} = 8.66/100 = 0.08$

   $D_{min} = 0.03$

18. A 555 timer is corrected for astable operation. Given $V_{cc} = 12$ V, $R_1 = 10$ kΩ, $R_2 = 8.3$ kΩ, and $C = 0.1$ μF. Find average power-dissipated (a) current sourcing and (b) current sinking.

Solution:
$$f = 1.9 \text{ kHz}$$
$$D = 0.84$$

   (a) $P_{avg} = \dfrac{1}{T_1 + T_2} \displaystyle\int_0^{T_1} (V_{cc}/R_L)^2 \ dt$

$$= (V_{cc}/R_L)^2 \times \dfrac{T_1}{T_1 + T_2}$$

$$= (12/10^3)^2 \times 0.84$$

$$= 0.12 \text{ MW} \rightarrow \text{current sourcing}$$

   (b) Current sinking

$$P_{av} = \dfrac{1}{T_1 + T_2} \displaystyle\int_0^{T_2} (V_{cc}/R_L)^2 \ dt$$

$$= \dfrac{T_2}{T_1 + T_2} (V_{cc}/R_L)^2$$

$$= (1 - D) \times (V_{cc}/R_L)^2$$

$$= 0.023 \text{ MW}$$

19. A 555 timer is connected for astable operation. If a 6-Hz triangular waveform weighs 4-V peak to peak is applied to the control pin through 100 μF capacitor. Calculate the output frequency range.

   (a) Short-circuit frequency volt source and capacitor acts as open circuit.

$$V_{A1} = (2/3) \times 15 = 10 \text{ V}$$

(b) Short circuit DC power supply

$$V_{A2} = \frac{(2/3)R}{(2/3)R + 1/j(2\pi \times 6 \times 100\ \mu F)}$$

$$\simeq 4$$

(c) $V_A = V_{A1} + V_{A2} = 14\ V$

$$V_{max} = 14\ V$$

$$V_{min} = 6\ V$$

(d) $T_{max} = (R_1 + R_2)C \ln\left(\dfrac{1 - V_A/2V_{cc}}{1 - V_A/V_{cc}}\right)$

$$= 3.84\ \mu sec$$

$$T_{min} = 0.64\ \mu sec$$

$$T_2 = 0.69\ R_2, C = 0.104\ \mu sec$$

(e) $f_{max} = \dfrac{1}{T_{min} + T_2} = 2\ kHz, f_{min} = \dfrac{1}{T_{max} + T_2} = 6.3\ kHz$

20. An op-amp is configured in multivibrator, R_2 is fixed R_1 is *POT*, $R_2/R_1 \simeq 1.8$ to 9. Calculate output frequency range, $4 = 1\ \mu sec$

$$\frac{R_2}{R_1} = 1.8\ \text{to}\ 9$$

$$\beta = \frac{R_1}{R_1 + R_2} = 0.35 + 0.1$$

$$T_{max} = 4 \ln\left(\frac{1+\beta}{1-\beta}\right) = 2 \times 1 \ln\left(\frac{1+0.35}{1-0.35}\right)$$

$$= 1.49\ \mu sec$$

$$f_{min} = 669\ Hz$$

$$T_{min} = 24 \ln\left(\frac{1+0.1}{1-0.1}\right) = 0.4\ \mu sec$$

$$f_{max} = 2.5\ kHz$$

So, frequency range is 669 Hz to 2.5 kHz.

MULTIPLE-CHOICE QUESTIONS

1. Upper threshold and lower threshold of comparators are

   (a) $V_{UT} = \dfrac{R_2}{R_1 + R_2}\ (+V_{sat})$

   $V_{LT} = \dfrac{R_2}{R_1 + R_2}\ (-V_{sat})$

   (b) $V_{UT} = \dfrac{R_2}{R_1 + R_2}\ (-V_{sat})$

   $V_{LT} = \dfrac{R_2}{R_1 + R_2}\ (V_{sat})$

   (c) $V_{UT} = \dfrac{R_2}{R_1 + R_2}\ V_{sat}$

   (d) $V_{UT} = \dfrac{R_2}{R_1 + R_2}$

$$V_{LT} = \frac{R_2}{R_1 + R_2} \qquad\qquad V_{LT} = \frac{R_2}{R_1 + R_2}(-V_{sat})$$

2. The hysterisis and center voltage V_{Cts} of zero defectors are
   (a) $V_H = V_{UT} + V_{LT}$
   $\quad V_{Ctr} = V_{UT} - V_{LT}$
   (b) $V_H = V_{UT} - V_{LT}$
   $\quad V_{Ctr} = V_{UT} + V_{LT}/2$
   (c) $V_H = V_{UT} + V_{LT}/2$
   $\quad V_{Ctr} = V_{UT} - V_{LT}/2$
   (d) $V_H = V_{UT}/2 + V_{LT}/2$
   $\quad V_{Ctr} = V_{UT}/2 - V_{LT}/2$

3. The function of window detector is
   (a) to monitor input and output voltage
   (b) to find the input and output voltage
   (c) to monitor an input voltage and indicate when this voltage goes either above or below prescribed limits.
   (d) none of the above

4. The propagation delay of op-amp circuits can be tested using
   (a) differentiator
   (b) LPF
   (c) comparator
   (d) difference amplifier

5. The circuit in Fig. C.16 is called
   (a) comparator
   (b) zero-crossing detector
   (c) voltage-level detector
   (d) Schmitt trigger

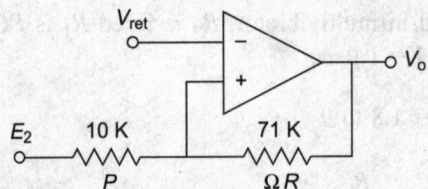

Fig. C.16   *Circuit for Question 5*

6. The circuit in Fig. C.17 represents
   (a) comparator
   (b) zero-crossing detector
   (c) window detector
   (d) waveform generator-square waveform

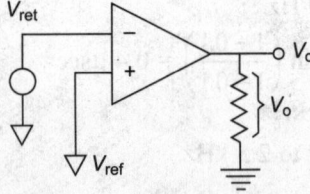

Fig. C.17   *Circuit for Qustion 6*

7. Effect of noise in comparator circuits can be eliminated by _____ feedback.
   (a) negative feedback
   (b) positive feedback
   (c) no feedback
   (d) none of the above

8. The frequency of triangular wave generator oscillation is given by
   (a) $f = 4\,PRC$
   (b) $f = \dfrac{PR}{4\,C}$
   (c) $f = \dfrac{P}{4\,RC}$
   (d) $f = P/RC$

9. Sawtooth waveform can be obtained from _____ circuit.
   (a) ramp generator
   (b) triangle waveform generator
   (c) square waveform generator

10. The function of dead-zone circuit is
    (a) to check if a signal is below or above a particular reference voltage
    (b) to compare the two signals
    (c) to check if a signal is below or above reference voltage
    (d) none of the above

11. The function of amplitude limiter circuit is that it
    (a) clips off all signals above positive reference voltage
    (b) clips off all signals below positive reference voltage
    (c) clips off all signals above negative reference voltage
    (d) clips off all signals below positive reference voltage

12. An inexpressive first cousin to the instrumentation amplifier is the basic
    (a) differential amplifier   (b) instrumentation amplifier
    (c) V to I converter   (d) I to V converter

13. What are floating leads of op-amp?
    (a) One terminal connected to ground other floating.
    (b) Both terminals are not connected to ground.
    (c) Both Q terminals are connected to ground.
    (d) none of the above.

14. What is used to eliminate the effects of offset voltage Q?
    (a) Resistor feedback   (b) Feedback circuit with passive device
    (c) Voltage offset null circuit   (d) none of the above

15. Drift of op-amp is defined as
    (a) change in offset current and voltage due to external noise
    (b) change in offset current and voltage due to temperature
    (c) change in offset current and voltage due to power supply
    (d) none of above

16. The maximum slew rate of IC 741 is
    (a) 0.1 V/μsec   (b) 0.8 V/μsec   (c) 1 V/μsec   (d) 0.5 V/μsec

17. The maximum frequency f_{max} at which we can obtain an undistorted output voltage with peak value of V_{OP} is determined by slew rate in accordance with

    (a) $f_{max} = SR/6.28 \times V_{OP}$ (b) $f_{max} = \dfrac{SR}{V_{OP}}$

    (c) $f_{max} = SR \times 6.28 \times V_{OP}$ (d) $f_{max} = V_{OP}/6.28$

18. The frequency compensation technique of op-amp are
    (a) internal frequency compensation   (b) external frequency compensation
    (c) feed-forward frequency compensation   (d) all of the above

19. External frequency compensation can be done by
    (a) using frequency-compensating capacitor
    (b) using external resistance connected to power supply
    (c) using POT corrected to power supply
    (d) none of the above

20. What is the typical open-loop gain of 741 op-amp at very low frequencies?
    (a) 100,000   (b) 200,000   (c) 20,000   (d) 500,000

21. The transient-response rise time of an op-amp is 0.07 μsec. Find small signal bandwidth.
    (a) 4 mHz   (b) 10 mHz   (c) 5 mHz   (d) 8 mHz

22. Find the maximum frequency for a sine output voltage of 10 V peak with op-amp whose slew rate is 1 V/μsec
    (a) 16 kHz   (b) 15 kHz   (c) 14 kHz   (d) 12 kHz

23. If we increase the compensating capacitor, the unity-gain bandwidth will
    (a) increase   (b) decrease
    (c) remain the same   (d) none of the above

24. What is the noise gain for a five-input-inverting adder?
    (a) 5   (b) 7   (c) 4   (d) 6

25. The important feature of instrumentation amplifier
    (a) high CMRR   (b) high gain accuracy
    (c) high gain stability   (d) all of the above

26. Compensating networks are used in op-amp to
    (a) improve the stability   (b) increase the BW
    (c) decrease the gain   (d) none of the above

27. The gain of op-amp decreases to high frequencies because of
    (a) resistance in the feedback   (b) capacitor within op-amp
    (c) feedback network with active device   (d) none of the above

28. The op-amp integrator and differentiator are useful for
    (a) wave-shaping circuit   (b) clipping circuit
    (c) clamping circuit   (d) none of the above

29. A precision diode can be used as
    (a) half-wave rectifier   (b) full-wave rectificer
    (c) clipper and damper   (d) all of the above

ANSWERS

1. (a)	2. (b)	3. (c)	4. (c)
5. (c)	6. (b)	7. (b)*	8. (c)
9. (a)	10. (c)	11. (a)	12. (a)
13. (c)	14. (c)	15. (b)	16. (d)
17. (a)	18 (d)	19. (a)	20. (b)
21. (c)	22. (a)	23. (b)	24. (b)
25. (d)	26. (a)	27. (b)	28. (a)
29. (d)			

* Positive feedback does not eliminate noise, but less response to it.

OBJECTIVE-TYPE QUESTIONS

IC-555 TIMER

1. The frequency of oscillation of astable multivibrator using IC 555 is
   (a) $1.44/(R_A + 2R_B)C$
   (b) $1.44 (R_A + 2R_B)C$
   (c) $\dfrac{(R_A + 2R_B)C}{1.44}$
   (d) $\dfrac{1.44}{(R_A + 2R_B)^2 C^2}$

2. The T_{high} and T_{low} of timer circuit using IC 555 is (R_A = 6.8 K, R_B = 3.3 K)
   (a) 0.35 μsec, 0.12 μsec
   (b) 0.7 μsec, 0.23 μsec
   (c) 1.4 μsec, 0.46 μsec
   (d) 0.3 μsec, 0.23 μsec

3. The duty cycle of astable operation D is defined as
   (a) $R_A (2R_A + R_B)$
   (b) $R_A/R_A + R_B$
   (c) $\dfrac{R_A}{R_A + 2R_B}$
   (d) $R_A(1 + R_B)$

4. IC 555 is used as one shot monostable operation, the output waveform is high for a time given by
   (a) $t_{high} = 1.1 R_{AC}$
   (b) $t_{high} = R_{AC}$
   (c) $\varepsilon = R_{AC} + 1$
   (d) $\varepsilon = R_{AC} 0.69$

5. If IC 555 can be used as frequency shifter of voltage controlled circuit, IC 555 oscillation at a centre frequency f_c is determined by
   (a) $f_c = \dfrac{RC}{3}$
   (b) $3/RC$
   (c) $3RC$
   (d) $f_c = RC$

6. One shot application of IC 555 is
   (a) touch switch
   (b) frequency divider
   (c) pulse divider
   (d) all of the above

7. The number of pins in IC 555 are
   (a) 14
   (b) 16
   (c) 8
   (d) none of the above

8. Operating modes of 555 timer
   (a) free-running operation
   (b) one-shot operation
   (c) both a and b
   (d) none of the above

9. The valve of output voltage of IC-555 timer in the low state is
   (a) 0 V
   (b) 0.1 V
   (c) 1 V
   (d) 0.7 V

10. The output switches high when IC 555 timer is operated as one shot multivibrator due to
    (a) negative-going trigger pulse to the timer
    (b) positive-going trigger pulse to the timer
    (c) none of the above

11. The number of operating states for IC 555 are
    (a) 2
    (b) 3
    (c) 1
    (d) 0

12. The lower threshold voltage V_{LT} and upper threshold voltage for IC 555 timer is
    (a) $V_{cc}/3, 2V_{cc}/3$
    (b) $2V_{cc}/3, V_{cc}/3$
    (c) $V_{cc}/2, V_{cc}/3$
    (d) $2V_{cc}, V_{cc}/3$

13. The control voltage for IC 555 can be used as
    (a) change threshold and trigger voltage levels
    (b) change threshold only
    (c) change trigger voltage only
    (d) none of the above

14. T_{high} and T_{low} for astable node operation of IC 555 are
    (a) $t_{high} = 0.695 (R_A + R_B)C$ (b) $t_{high} = 0.695 \, RA_C$
         $t_{low} = 0.695 (R_A + 2R_B)C$ $t_{low} = 0.695 \, R_B C$
    (c) $t_{high} = 0.695(R_A + R_B)C$ (d) $t_{low} = 0.695 (R_A + R_B)C$
         $t_{low} = 0.695 \, R_B C$ $t_{high} = 0.695 \, R_B C$

15. When $R_A = R_B$ in IC 555 timer circuit, then total times for output waveform
    $T = 0.695 (R_A + R_B)C$, duty cycle for the circuit is
    (a) 80%   (b) 50%   (c) 25%   (d) 100%

16. If $R_A = 10$ K, $C = 0.1$ µF, calculate the timing interval for IC 555 monostable operation
    (a) 2 µsec   (b) 2.1 µsec   (c) 2.2 µsec   (d) 2.3 µsec

17. Voltage regulators are equipped with _____ circuit-output overload protection.
    (a) current overload   (b) second breakdown
    (c) thermal overload   (d) all of the above

18. The efficiency of the regulator is defined as
    (a) $\eta = R_o/P_I$ (b) $\eta = \dfrac{V_o I_o}{V_I I_o}$

    (c) $\eta(\%) = 100 \dfrac{V_o}{V_I}$ (d) all of the above

19. Output overload protection for voltage regulator can be done by _____ methods.
    (a) short-circuit protection
    (b) current foldback protection
    (c) both a and b

20. The role of op-amp in configuring a regulator as power voltage source is to
    (a) eliminate CMRR errors
    (b) eliminate PSRR errors
    (c) both a and b

21. The duty cycle of a switching regulator is
    (a) $D = \dfrac{t_{on}}{f_s}$ (b) $D = f_s \, t_{on}$

    (c) $\dfrac{t_{on}}{t_{on} + t_{off}}$ (d) all of the above

22. The duty cycle of switching regulators can be adjusted using
    (a) P_{WM}, f_s is kept fixed, t_{on} is adjusted   (b) P_{FM}, t_{on} is fixed, f_s adjusted
    (c) all of the above

23. The monolithic switching regulators are advantageous due to a
    (a) wide range of performance specifications
    (b) switch provided externally for higher switching frequencies
    (c) smaller energy-storage elements
    (d) all the above

24. The function of error amplifier in volt regulator for volt node control is
    (a) high loop gain only
    (b) stability for phase margin
    (c) high gain and stability
    (d) high gain and stability and regulation

ANSWERS

1. (a)	2. (b)	3. (c)	4. (a)
5. (b)	6. (d)	7. (c)	8. (c)
9. (b)	10. (a)	11. (a)	12. (a)
13. (b)	14. (c)	15. (b)	16. (b)
17. (d)	18. (d)	19. (c)	20. (c)
21. (d)	22. (c)	23. (d)	24. (d)

PROBLEMS ON PHASE-LOCKED LOOP (PLL)

1. For a PLL, select the filter components whose lowest frequency from phase detector is 5 Hz. A 0.5% variation in frequency is allowable. How long will it take for output from the filter to stabilize?

   Solution: For a frequency of 5 Hz, $T = 0.2$ sec

   $$Y = \frac{50\% \times 0.2}{0.5\%} = 20$$

   $$Y = RC \times 20 \text{ sec}$$

   if $\qquad C = 100 \ \mu F$

   $R = 200 \ k\Omega$. Therefore, simple, passive LPF from the phase detector has R and C above values. Hardly for RC circuit to settle, it requires 5τ.

   $$5\tau = 5 \times 20 = 100 \text{ sec}$$

   It will take 100 sec before changes in the capacitor voltage. After 3τ, the capacitor falls within 5% of initial value.

2. For the given circuit shown in Fig. C.18, calculate all the indicated quantities. Assume PLC in locked condition. The diagram has CLK, counter, EX–OR gate, Schmitt trigger, motor, and Encoder.

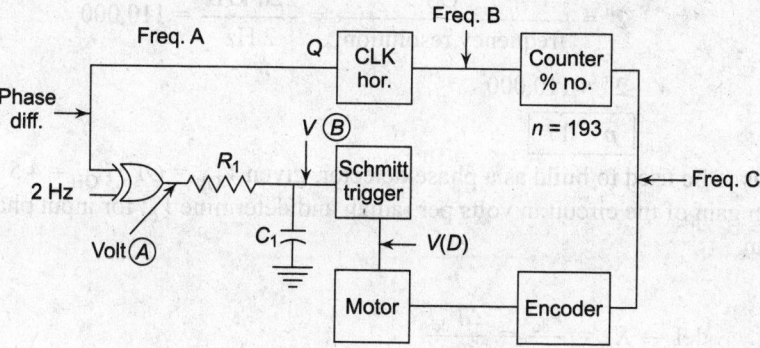

Fig. C.18 *Circuit for Problem 2*

Solution:

Motor frequency = 0.5 Hz/rpm

(a) Assume loop-locked frequency of EX–OR gate is 2 Hz.

Flip-flop produces in the circuit 50% duty cycle.

(i) frequency = 2 frequency = 4 Hz

(ii) multiply by $n = n \times 4 = 193 \times 4 = 772$ Hz

(iii) Speed of the encoder shaft

$$\text{Speed} = \frac{\text{frequency}}{\text{No of rpm}} = \frac{772}{0.5 \text{ Hz/rev}} = 1544 \frac{\text{rev}}{\text{min}}$$

(iv) Assume voltage, 1 V produces 150 rpm

$$V_o = \frac{1}{150 \text{ rpm}} \times 1544 \text{ rpm} = 10.3 \text{ V}$$

(v) Let us assume inside SC, Trigger. V_E and V_D be voltage. $R_2 = 200$ K, $R_3 = 100$ K.

$$V_E = \frac{R_3}{R_2 + R_3}, \quad V_D = \frac{100 \text{ K}}{220 \text{ K} + 100 \text{ K}} \times 10.3 = 3.2 \text{ V}$$

$$V_B = V_E = 3.2 \text{ V}$$

Freq $A \to 2$ Hz, Freq $B \to 4$ Hz, Freq $C \to 772$ Hz

Voltage $\to V_A = V_B = 3.2$ V, $V_D = 10$ V

3. In digital frequency synthesizer, using PLL, determine the frequency of the reference oscillator and the number of bits needed in the phase accumulator for maximum output frequency of 100 kHz, resolution 2 Hz.

Solution:

Max frequency in digital frequency synthesizer

$$f_{\text{max}} = \frac{f_{\text{ref}} OSC}{2.2}$$

$$f_{\text{rot}} OSC = 2.2 \times f_{\text{max}}$$

$$= 2.2 \times 100 \text{ kHz} = 220 \text{ kHz}$$

No. of bits required in lowest synthesizer

$$\text{Frequency resolution} = f_{\text{lowest}} = \frac{f_{\text{ref}} \text{ oscillator}}{2}$$

$$2^n = \frac{f_{\text{ref}} OSC}{\text{frequency resolution}} = \frac{220 \text{ kHz}}{2 \text{ Hz}} = 110,000$$

$$2^n = 110,000$$

$$\boxed{n = 17}$$

4. A 7486 has to be used to build as a phase detector, given $V_{OL} = OV$, $V_{OH} = 4.5$ V. Determine the conversion gain of the circuit in volts per radian and determine V_{cc} for input phase differential of 2.36 radian.

Solution:

$$\text{def} \to K_\varphi = \frac{V_{dc}}{\varphi} = \frac{\Delta V_{dc}}{\Delta Q}$$

$$K\varphi = \frac{\Delta V_{dc}}{\Delta\varphi} = \frac{4.5 \text{ V}}{\pi \text{ vcd}} = 1.43 \text{ V/rad}$$

$$V_{dc} = K\varphi \, \Delta Q = \text{circuit in V/rad}$$

$$= 1.43 \text{ V/rad} \times 2.36 \text{ rad}$$

$$= 3.37 \text{ V}$$

5. Assume the frequency is 250 kHz, it is desired that V_{CO} operate at a frequency of 6.25 kHz. Determine the modulus of counter.

Solution:

Given

$$f_{out} = N f_{in}$$

$$N = \frac{f_{out}}{f_{in}}$$

$$= \text{mod } \frac{6.25 \text{ m}}{250 \text{ K}} = \text{mod } 25$$

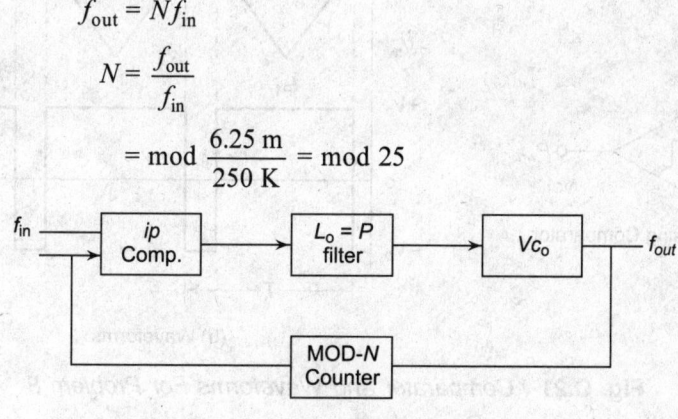

Fig. C.19

$$F_{out} = N f_{in}$$

6. Write down the equations for
   (i) V_{CO}-free-running frequency f_o
   (ii) Loop gain
   (iii) Tracking range
       Design equations for 565 ave.

Solution:

(i) $f_o = \dfrac{1}{3.7 R_o C_o}$, V_{CO} free-running frequencuy.

(ii) Loop gain $= \dfrac{33.6 f_o}{V_{cc} + V_{EE}}$ H_2/V

(iii) Tracking range $= \dfrac{8 f_o}{V_{cc} + |V_{EE}|}$

7. Generate frequency from 1 to 99,999 Hz with 1 Hz increments, using PLL, 50-Hz line frequency as reference.

Solution:
   (a) Use the above circuit, 50-Hz line source as in the place of *OSC*, $n_1 = 50$ (input frequency divider)

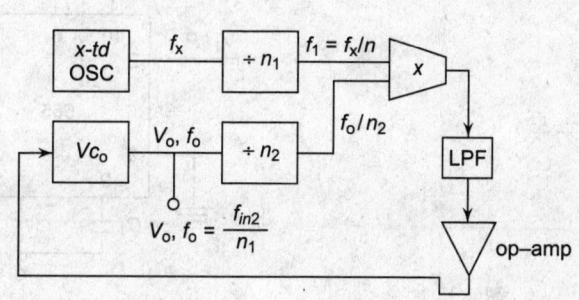

Fig. C.20 *Frequency Synthesizer using PLL*

$$f_a = \frac{50}{50} = 1 \text{ Hz}$$

(b) Replace the output frequency divider n_2 with 5 decades (Programmed with *BCD* switch) for increment from 1 to 99,999 Hz.

8. Use comparator as pulse-width modulator circuit giving triangular wave as input. Find ramp speed, pulse width, and duty cycle.

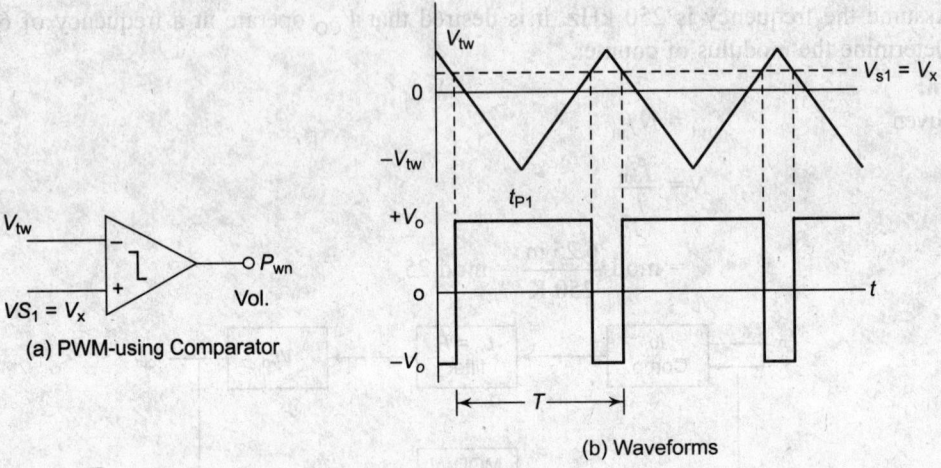

Fig. C.21 *Comparator and Waveforms For Problem 8*

(a) Pulse repetition frequency $F_{fw} \gg f_s$

  (i) Ramp speed $4\,V_{tw}/T$

  (ii) $t_{p1} = (2\,V_{tw} + 2V_x)\,\dfrac{T}{4V_{tw}}$

  (iii) $\delta = \dfrac{t_{P1}}{T} = \dfrac{1}{2}\left(\dfrac{V_x}{V_{tw}} + 1\right)$

9. Design a *PLL* for the given specifications, $f_o = 2.5$ kHz, $f_c = 50$ Hz, $V_{cc} = -V_{EE} = 12$ V.

Solution:

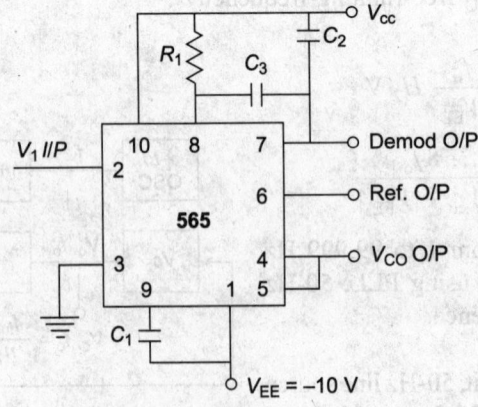

Fig. C.22 *PLL Circuit*

$$R_1 = \frac{1.2}{4C_1 f_o}$$

(a) Choose suitable value C_1, $C_1 = 0.01\ \mu F$

(b) Find the value of R_1

$$R_1 = \frac{1.2}{4C_1 f_o} = \frac{1.2}{4 \times 0.01\ \mu F \times 2.5 \times 10^3} = 12\ k\Omega$$

(c) Find lock range

$$f_L = \frac{8 \times 2.5\ K}{12 - (-12)} = 833\ Hz$$

(d) Find C_2

$$C_2 = \frac{f_L}{2\pi \times 3.6 \times 10^3 f_c^2} = \frac{833}{2\pi \times 3.6 \times 10^3 \times 50^2} = 14.17\ \mu F$$

$$\boxed{C_2 = 14\ \mu F}$$

10. An f_m signal is being modulated over the range of 1 mHz ± 10 kHz with modulating frequency of 1 kHz. Using PLL, design a circuit to demodulate such a signal VCO.

Solution:

(a) $f_o = 1$ mHz, Choose $2f_R$

$$2f_R = 0.5\ mHz$$

(b) $K_o = \dfrac{2\pi 2 f_R}{2.8} = 1.122 \times 10^6$ rad/sec

(c) $R_1 = 95\ k\Omega$, $R_2 = 130\ k\Omega$, $C = 100$ pF

(d) Choose lead leg. parameter, $Y = 0.707, f_{-3dB} > f_m\ f_{-3dB} = 10\ K$

So, $W_p = 553$ rad/sec

$W_z = 22.5$ rad/sec

11. Find out the transfer function $Q_o(s)/Q_i(s)$ for the synthesizer shown in Figure C.23, if the synthesizer uses a PLL to synthesize a 1 mHz signal. From a 25-kHz reference frequency, calculate N. Find the resulting transfer function for this frequency and bandwidth of the synthesizer if $K_d = 2$ V/rad, $K_o = 100$ Hz/V.

Solution:

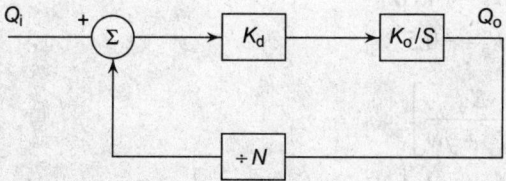

Fig. C.23 *PLL Block Schematic*

Open-loop transfer function, $h(s) = \dfrac{K_o K_d}{S}$

Feedback path transfer function

$$H(s) = \frac{1}{N}$$

Closed loop transfer function.

$$\frac{Q_o(s)}{Q_i(s)} = \frac{h(s)}{1 + h(s)H(s)}$$

$$= \frac{K_o K_d / s}{1 + \left(\dfrac{K_o K_d}{s}\right)\left(\dfrac{1}{N}\right)} = \frac{K_o K_d}{s + K_o K_d / N} = \frac{N K_v}{S + K_v}$$

Let $K_v = \dfrac{K_d K_o}{N}$ is synthesizer bandwidth. In order to realize output frequency of 1 mHz.

$$N = \frac{10^6}{25 \times 10^3} = 40$$

Subset N as above, K_d and K_o given, we get

$$\frac{Q_o}{Q_i} = \frac{(2 \times 100) 2\pi}{S + (2 \times 100 \times 2\pi)/40}$$

$$K_v = \frac{2 \times 100}{40} = 5 \text{ Hz}$$

Therefore,   Synthesizer $BH = 5$ Hz

12. A PLL has K_o of 2π (1 kHz)/V, a K_v of 500/sec, free-running frequency of 500 Hz.

(a) For a constant input signal frequency of 250 Hz and 1 kHz find V_f.

(b) If the input signal is frequency modulated so that $W_i(\varepsilon) = (2\pi)$ 500 Hz $(1 + 0.1 \sin(2\pi \times 10^2)t]$. Find $V_f(t)$.

Solution:

(a) $V_f = \dfrac{W_o(\varepsilon) - W_o}{K_o}$, unlocked condition $W_o(t) = W_i$

$$= \frac{W_i - W_c}{K_o}, \quad W_i = 250 \text{ Hz}$$

$$V_f = \frac{2\pi(250) - 2\pi(500)}{2\pi(1 K + 2)/V} = -0.25 \text{ V}$$

$$W_i = 1 \text{ kHz}$$

$$V_f = +0.5 \text{ V}$$

(b) $\dfrac{V_f(jw)}{W_1(jw)} = \dfrac{1}{K_o}\left[\dfrac{K_v}{K_v + jw}\right]$

$$= \frac{1}{K_o}\left[\frac{K_o}{K_v + j(2\pi \times 100)}\right]$$

$$= \frac{1}{2\pi(1 \text{ kHz})/V}\left[\frac{500}{500 + j628}\right]$$

$$= \frac{1}{2\pi(1 \text{ kHz})/V}\left[0.39 - j0.48\right]$$

$$W_i(jw) = (0.1)\ (500 \text{ Hz})\ (2\pi) = 100\ \pi$$

$$V_f(jw) = 50/1000 \ (0.62 \ \lfloor -51°]$$

$$V_f(t) = 0.031 \sin (2\pi \times 10^2 t - 51°)$$

13. A PLL has free-running frequency $W_c = 500$ kHz. Bandwith of LPF $= 10$ kHz, suppose an input signal of frequency 600 kHz is applied. Will the loop acquire lock? What is V_{CO} output frequency?

Solution: Phase detector output is

$$f_\pi \ tf_c = 600 + 500 = 1100 \text{ kHz}$$

$$f_\pi - f_c = 600 - 500 = 100 \text{ kHz}$$

Both sum and difference frequency comparator are outside the pass band of LPF. Hence, loop will not acquire lock. V_{CO} frequency will be free-running frequency.

14. A synthesizer uses PLL $K_v = 5\pi$ rad/sec, what value of LPF bandwidth should be used so that closed-loop system approximates a second-order Butterworth filter?

Solution:

(a) Damping ratio $\varepsilon = 0.707$

$$2\varepsilon = \sqrt{w_L / K_v} = 2 \times 0.707$$

$$\left(\frac{w_L}{K_v} \right)^{1/2} = 1.414$$

$$W_L = 10 \ \pi \text{ rad/sec.}$$

(b) Therefore $B_w = W_n = (K_v \ W_L)^{1/2} = 7\pi$

(c) System rise time $= \dfrac{2.2}{W_n} = 98.8$ msec

15. A PLL has V_{CO} with $K_o = 25$ kHz/V and $f_c = 50$ kHz. The amplifier gain is $A = 2$ and the phase detector has a maximum output voltage swing of ± 0.7 V. Find lock range of PLL output volt swing.

Solution:

(i) $V_{d(man)} = \pm K_d \ (\pi/2) = \pm \left(\dfrac{\pi}{2} \right) 0.7$

(ii) $V_f \rightarrow$ Control voltage

$$V_{f(max)} = (\pi/2) \ K_d \ A$$

$$= 1.4 \text{ V}$$

(iii) V_{CO} frequency swing

$$(f_o - f_c)_{max} = \pm K_o \ V_{f(max)}$$

$$= \pm 35 \text{ kHz}$$

(iv) Signal frequency over which *PLL* locked is $f_\pi = f_c \pm \Delta f_c$

$$= (50 \pm 35) \text{ kHz}$$

$$= 15 \text{ to } 85 \text{ kHz}$$

(v) Lock range, $2\Delta f_L = 2(35 \text{ kHz}) = 70 \text{ kHz}$

16. Determine free-running frequency, Δf_L, Cap. range Δf_{cap} of PLL [IC 565]

$$\Delta Af_{Cap} = \sqrt{f_1 \Delta f_L}$$
$$2\Delta f_1 = 1.39 \times f_1$$

1. $f_1 = \dfrac{1.2}{4(10)(10^3)(10^{-8})} = 3 \text{ kHz}$

2. $2 \times \Delta f_2 = 1.39 \times f_1 = 4.2 \text{ kHz}$

3. $\Delta f_{cap} = \sqrt{3 \times 4.2 \times 10^6} = 103 \text{ Hz}$

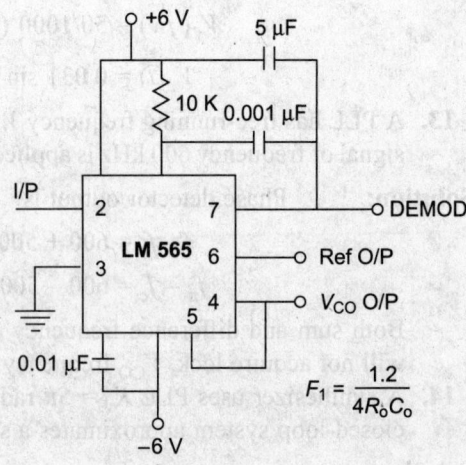

Fig. C.24

17. Design a circuit that accepts 1-kHz reference frequency and synthesize all frequencies between 1 MHz and 2 MHz in 1 kHz steps.

Solution:

A programmable contour

$$N_{min} = 10^6/10^3 = 1000$$

$$N_{max} = 2000$$

For $V_{CO}, f_o = 1.5 \text{ mHz}$

$$2f_R = 3 \text{ mHz}$$

$$K_o = 3.366 \times 10^6 \text{ rad/sec (given)}$$

Therefore, V_{CO} components value, $R_1 = 28 \text{ k}\Omega$, $R_2 = 287 \text{ k}\Omega$, $C = 100 \text{ pF}$ (given).

$$K_d = \frac{5}{4\pi} \text{ V/rad}$$ (Given)

$$N_{mean} = \sqrt{N_{min} N_{max}} = 1414$$

$$K_{V(Mean)} = K_d K_o/N_{mean} = 947/\text{sec}$$

$$Y = 0.707$$

$$W_n = W_I/20 - 2\pi 10^3/20 = \pi \, 100 \text{ rad/sec}$$

$$W_p = 104 \text{ rad/sec}$$

$$W_z = 290 \text{ rad/sec}$$

Therefore, $f_{or} \, N = 1000$, $Y = 0.78$

$$N = 2000, \, Y = 0.65$$

Non-linear AMP, PLL

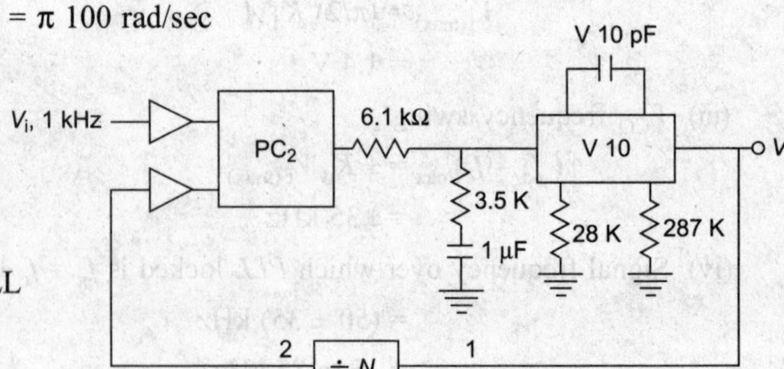

Fig. C.25 *PLL Frequency Synthesizer*

18 . Design type-I phase comparator and draw the waveforms, output as a function of input phase difference, and typical waveforms for the type-I phase comparator in locked condition.

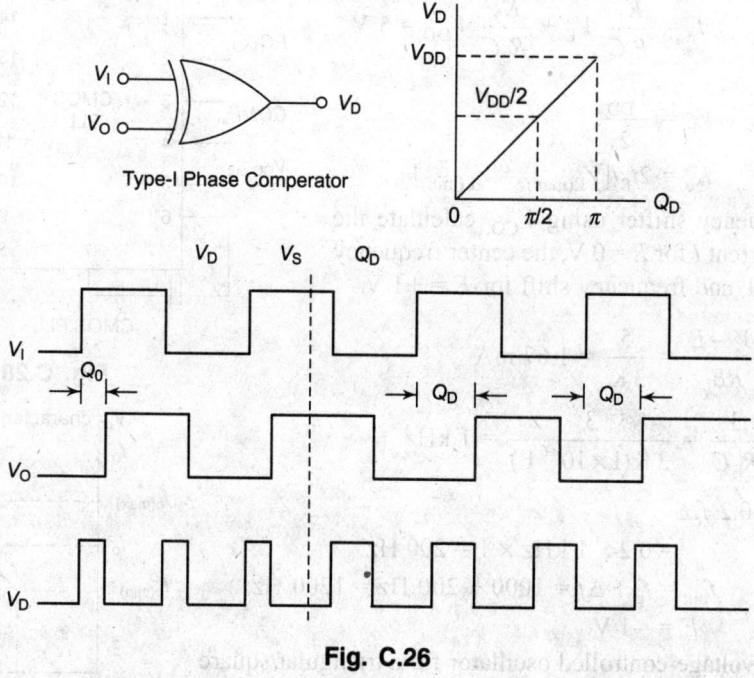

Fig. C.26

19. Design type-III phase comparator and draw output voltage versus phase difference and typical waveforms for type-III phase comparator in the locked condition.

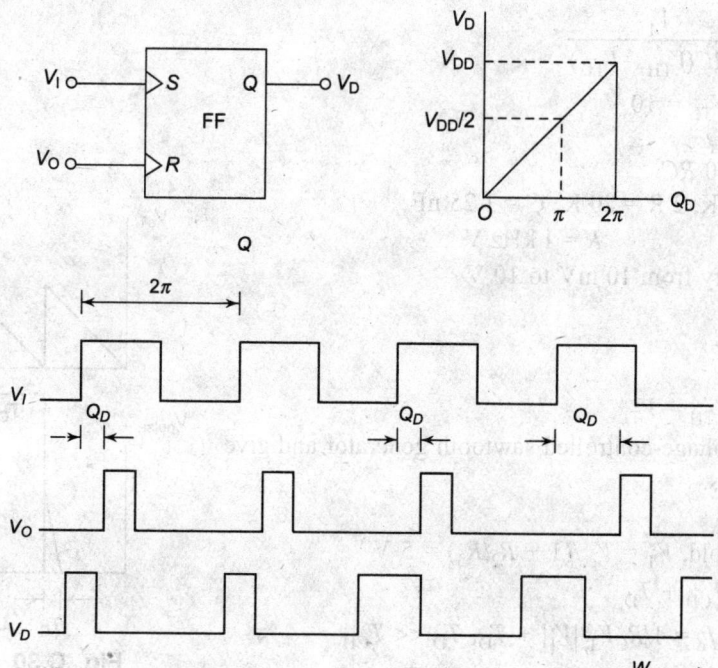

Fig. C.27 *Typical waveforms for Type-III comparator*

Monolithic PLL
Voltage-controlled oscillator

$$f_o = \frac{K_1}{R_1 C} V_E + \frac{K_2}{R_2 C} V_{DD} = 5 \text{ V}$$

$$V_E = \frac{V_{DD}}{2}$$

$$K_o = 2 f_R / [V_{E(max)} - V_{E(min)}]$$

20. In a frequency shifter using V_{CO}, calculate the charge current I for $E = 0$ V, the center frequency f_c at $E = 0$, and frequency shift for $E = \pm 1$ V.

(a) $I = \dfrac{5V - E}{RE} = \dfrac{5}{3 \text{ K}} = 1.67 \text{ mA}$

(b) $f_c = \dfrac{3}{R_E C} = \dfrac{3}{3 \text{ K}(1 \times 10^{-6} \text{ F})} = 1 \text{ kHz}$

(c) $\Delta f = 0.2 \, f_c E$

$$= 0.2 \times 1 \text{ kHz} \times 1 = 200 \text{ Hz}$$

$$f_{out} = f_c + \Delta f = 1000 + 200 \text{ Hz} = 1200 \text{ Hz}.$$

$$E = -1 \text{ V}$$

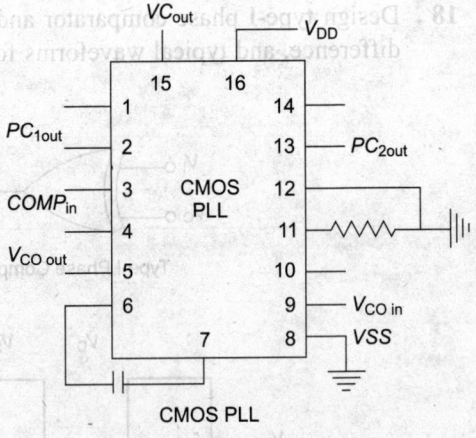

Fig. C.28

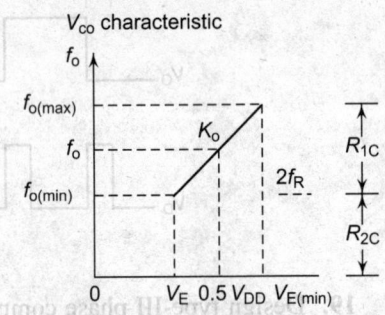

Fig. C.29

21. Design a voltage-controlled oscillator for a triangular/square wave generation.

Solution: Equations

(i) $f_o = \dfrac{V_I}{8RC(V_{TH} - V_{TL})}$

(ii) $V_{TH} - V_{TL} = 10 \text{ V}$

(iii) $f_o = K V_I$

(iv) $K = 1/80 \, RC$

(v) $R = 10 \text{ K}, \ 2 \text{ R} = 20 \text{ K}, \ C = 1.25 \text{ nF}$

$$K = 1 \text{ kHz/V}$$

V_I range vary from 10 mV to 10 V

(vi) $I = \dfrac{V_I}{4 \text{R}}$

(vii) $\Delta V = V_{TH} - V_{TL}$

22. Design a voltage-controlled sawtooth generator and give its equations.

Solution:

(i) Threshold, $V_T = V_{cc}/(1 + R_2/R_3) = 5 \text{ V}$

(ii) $f_o = 1/T_{CH} + T_D$

$$f_o = 1/R_C V_T |V_I| + T_D, \ T_D \ll T_{CH}$$

(iii) $T_D \gg T_{CH}, f_o = V_i/R_C V_T, \ I = V_i/R$

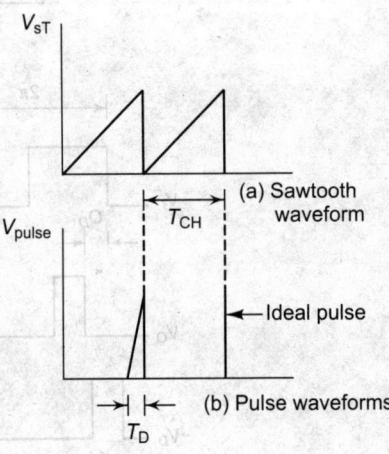

(a) Sawtooth waveform

←— Ideal pulse

(b) Pulse waveforms

Fig. C.30 *Waveforms*

$$f_0 = V_i / R_C V_T$$

$R = 90 \text{ k}\Omega$, $C = 2.2 \text{ nF}$, $f_0 = K |V_i|$, $K = 1 \text{ kHz}$.

$V_i \rightarrow$ up to 10 V.

23. Design voltage-controlled triangular square wave generator for square wave with peak values of ± 10 V and continuously variable from 10 Hz to 10 kHz.

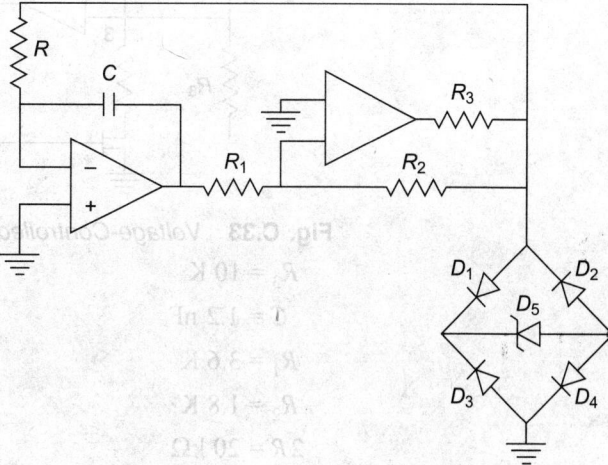

Fig. C.31 *Voltage-Controlled Triangular/Square wave Generator*

Solution:

(i) $V_{25} = 5 - 2 \times 0.7 = 3.6$ V

(ii) $R_2/R_1 = 0.5$

$\qquad R_1 = 20 \text{ K}$, $R_2 = 10 \text{ K}$

(iii) $R = R_{min}$

$R_{min} \simeq R_s = 2.5 \text{ kHz}$

$\qquad C = 0.5/10^4 \times 4 \times 2.5 \times 10^3 = 5 \text{ nF}$

(iv) R_3 to provide current R,

$$I_{R1(max)} = \frac{V_{denp}}{R_{min}} = \frac{5}{2.5} = 2 \text{ mA}$$

$$I_{R2(max)} = \frac{V_{dep}}{R_2} = \frac{5}{R_2} = 0.5 \text{ mA}$$

(v) $R_3 = \frac{13-5}{4.5} = 1.77 \text{ k}\Omega$

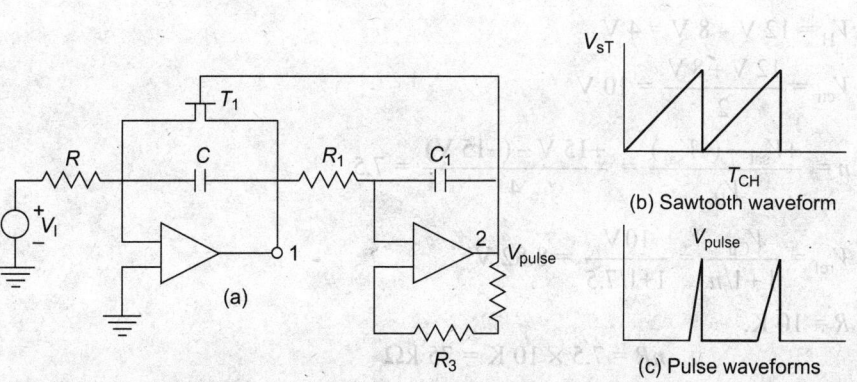

(a)

(b) Sawtooth waveform

(c) Pulse waveforms

Fig. C.32 *Voltage-Controlled Sawtooth Generator*

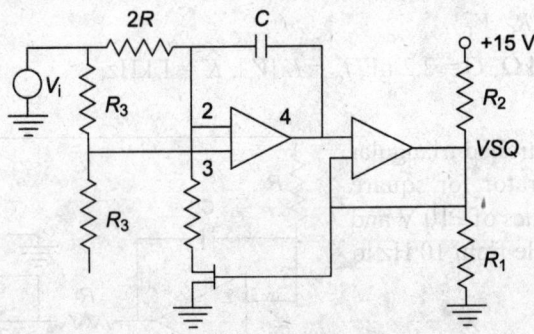

Fig. C.33   *Voltage-Controlled Triangular Generator*

$$R_3 = 10 \text{ K}$$
$$C = 1.2 \text{ nF}$$
$$R_1 = 3.6 \text{ K}$$
$$R_2 = 1.8 \text{ K}$$
$$2R = 20 \text{ k}\Omega$$

COMPARATORS

24. Design the circuit to give $V_{\text{VT}} = 12$ V, $V_{25} = 8$V, $\pm V_{\text{sat}} = \pm 15$ V.

Solution:

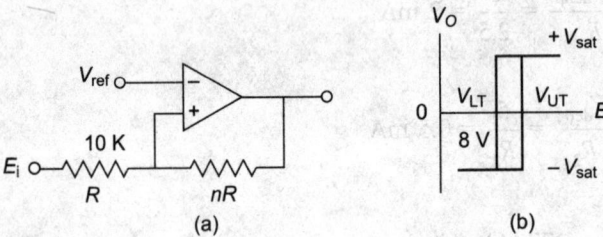

Fig. C.34   *Comparator Circuit and Waveforms*

(i) $V_H = 12 \text{ V} - 8 \text{ V} = 4 \text{ V}$

(ii) $V_{\text{ctr}} = \dfrac{12 \text{ V} + 8 \text{ V}}{2} = 10 \text{ V}$

(iii) $n = \dfrac{+V_{\text{sat}} - (-V_{\text{sat}})}{V_H} = \dfrac{+15 \text{ V} - (-15 \text{ V})}{4} = 7.5$

(iv) $V_{\text{ref}} = \dfrac{V_{\text{Ctr}}}{1 + 1/n} = \dfrac{10\text{V}}{1 + 1/7.5} = 8.82 \text{ V}$

(v) $R = 10 \text{ K,}$

$$nR = 7.5 \times 10 \text{ K} = 75 \text{ k}\Omega$$

25. Design a voltage comparator for $V_{\text{UT}} = 12$, $V_{\text{LT}} = 8$ V, $\pm V_{\text{sat}} = \pm 15$ V, $V_{\text{cft}} = 10$ V, $V_H = 4$ V.

Solution:

(i) $n = \dfrac{+(V_{\text{sat}}) - (-V_{\text{sat}})}{V_H} - 1 = \dfrac{15 - (-15)}{4} - 1 = 6.5$

(ii) $V_{ref} = \dfrac{n+1}{n} V_{ctr} = \dfrac{6.5+1}{6.5} \times 10 = 11.5$ V

Assume $\qquad R = 10$ kΩ,

$\qquad\qquad nR = 6.5 \times 10$ K $= 65$ kΩ

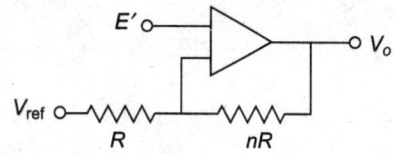

Fig. C.35 *Voltage-Comparator Circuit*

Comparator with Independent Adjustments

26. Design a comparator with the following specifications:
Battery voltage 10.5, $V_{LC} = 10.5$, $V_{UT} = 13.5$, $\pm V_{sat} = \pm 13.0$ V.
Find (a) V_H, (b) V_{ctr}, (c) MR, and (d) RR

(a) $\qquad\qquad V_H = V_{UT} - V_{LT} = 13.5 - 10.5 = 3$ V

(b) $V_{ctr} = \dfrac{V_{VT} + V_{LT}}{2} = 12$ V

$\qquad m = \dfrac{-V_{ref}}{V_{ctr}} = \left(\dfrac{-15\,\text{V}}{12} \right) = 1.25$

(c) $MR = 1.25 \times 10$ K $= 125$ K

(d) RR

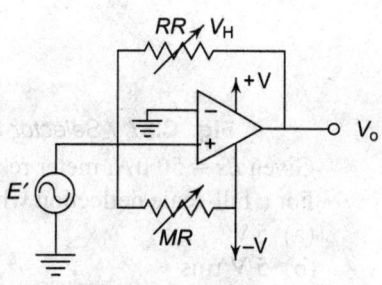

Fig. C.36 *Comparator Circuit with Adjustable Resistor*

$$n = \frac{+V_{sat} - (-V_{sat})}{V_H}$$

$$= \frac{13\,\text{V} - (-13\,\text{V})}{3}$$

$$= 8.66$$

$$RR = 8.66 \times 10\text{ K} = 86.6\text{ k}\Omega$$

Universal High-Resistance Voltmeter

27. Design procedure
(i) *DC* voltmeter

$$R_i = E_{dc}/Ifs$$

(ii) RMS AC voltmeter

$$R_i = 0.9 \frac{\text{full-scale } E_{rms}}{Ifs}$$

(iii) Peak reading voltmeter

$$R_i = 0.636 \frac{\text{full-scale } E_{Peak}}{Ifs}$$

(iv) Peak-to-peak AC voltmeter

$$R_i = 0.318 \frac{\text{full-scale } E_P}{Ifs}$$

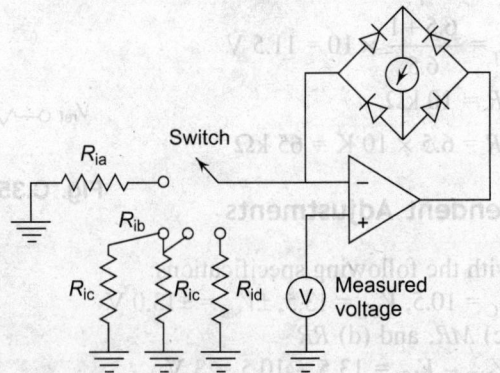

Fig. C.37 *Selector-Switch Circuit to read RMS, Peak-to-Peak Voltages*

Given $Ifs = 50 \mu A$, meter resistance = 5 kΩ

For a full-scale deflection when the voltage to be measured

(a) 5 V

(b) 5 V rms

(c) 5 V peak

(d) 5 V p–p

Solution:

(a) $R_{ia} = \dfrac{\text{full-scale } E_{dc}}{Ifs}$

$$= \dfrac{5\,V}{50\,\mu A} = 100 \text{ k}\Omega$$

(b) $R_{ib} = 0.9 \dfrac{\text{full-scale } E_{rms}}{Ifs}$

$$= 0.9 \times \dfrac{5\,V}{50\,\mu A} = 90 \text{ k}\Omega$$

(c) $R_{ic} = \dfrac{0.63\,\text{full-scale } E_{peak}}{Ifs}$

$$= 0.63 \times \dfrac{5\,V}{50\,\mu A} = 63 \text{ k}\Omega$$

(d) $R_{id} = \dfrac{0.318 \times \text{full-scale } E_P / P}{Ifs}$

$$= 31.8 \text{ k}\Omega$$

28. Design high-input *DC* voltmeter with E_i 0.5 V of micrometer 50 μA. Given *Efs* = 5 V.

Solution:

(i) $I_m = E_i/R_i$, $R_i = Efs/Ifs$

(ii) $R_i = \dfrac{5\,V}{50 \times 10^{-6}} = 100 \text{ k}\Omega$

(iii) $I_n = \dfrac{I_i}{R_i} = \dfrac{0.5}{100\,K} = 5 \mu A$

29. Design *V*-to-*I* converter using zener diode.
Given $V_o = 10.3$ V, $E_p = $ IV, $R_p = 1$ kΩ.
Find zener current and zener voltage.

Solution:

$$I = \frac{E_i}{R_i} = \frac{5V}{1K} = 5 \text{ mA}$$

$$V_o = V_2 + E_i$$

$$V_2 = V_o - E_i$$

$$= 10.3 \text{ V} - 5 \text{ V}$$

$$= \textbf{5.3 V}$$

Therefore, $V_2 = 5.3, I = 5$ mA, $V_o = 10.3$ V

$$I_L = \frac{E_1 - E_2}{R}$$

$$V_L = I_L R_L$$

$$V_o = 2V_L - E_2$$

$$I_m = \frac{E_i}{R_i}, R_{i_i} = \frac{Efs}{Ifs}$$

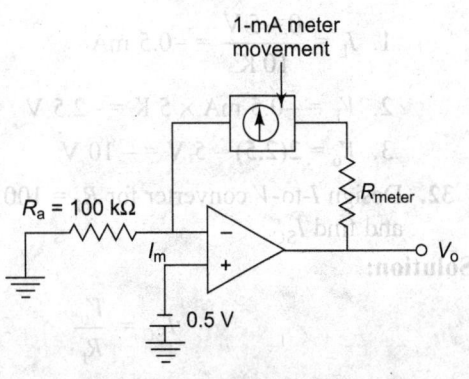

Fig. C.38 *High-Input Resistance DC Voltmeter*

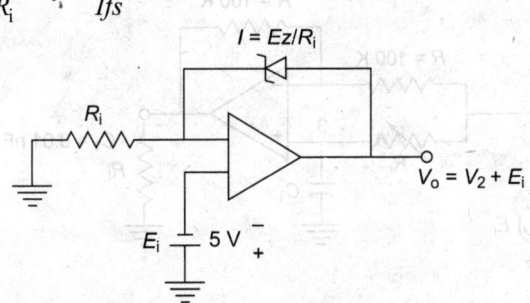

Fig. C.39 *Voltage-to-Current Converter Circuit*

30. Design differential voltage-to-current converter for
$R = 10$ kΩ, $R_L = 5$ kΩ, $E_3 = 5$ V. Find
(a) I_L, (b) V_L and (c) V_o.

Solution:

(a) $I_L = E_1 - E_2/R$

$$I_L = \frac{5 \text{ V} - 0}{10 \text{ K}} = 0.5 \text{ mA}$$

(b) $V_L = 0.5$ mA × 5 K
       $= 2.5$ V

(c) $V_o = 2 \times 2.5$ V = 5 V

So, $V_L = 2.5$ V, $V_o = 5$ V, $I_L = 0.5$ mA

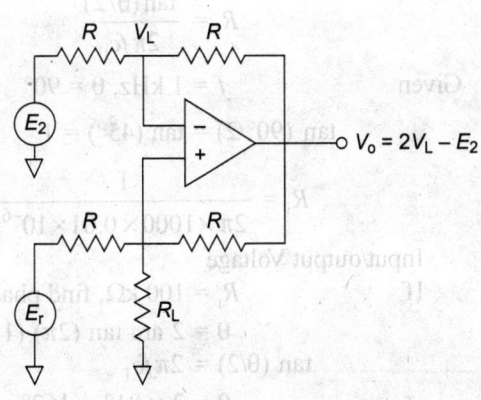

Fig. C.40 *Differential Voltage-to-Current Converter*

31. Design V-to-I converter for $R = 10$ K, $E_2 = 5$ V, $E_1 = 0$, $R_L = 5$ K and find (1) I_L, (2) V_L, (3) V_o.

Solution:

1. $I_L = \dfrac{0 - 5\,V}{10\,K} = -0.5$ mA

2. $V_L = -0.5$ mA $\times 5$ K $= -2.5$ V

3. $V_o = 2(2.5) - 5$ V $= -10$ V

32. Design I-to-V converter for $R_f = 100$ K, $V_o = 5$ V and find I_{SC}

Solution:

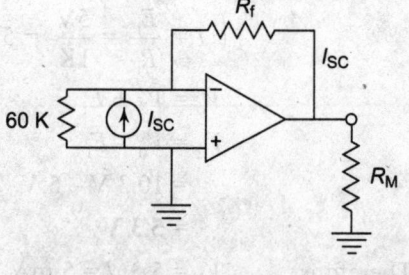

Fig. C.41   *Current-to-Voltage Converter*

$$I_{SC} = \frac{V_o}{R_f}$$

Let $V_o = 5$V, $R_f = 100$ K

$$I_{SC} = \frac{V_o}{R_f} = \frac{5}{100\,K} = 50\text{ mA}$$

33. Find R_i as shown in figure, V_o will leg E_i by 90°.

Solution:

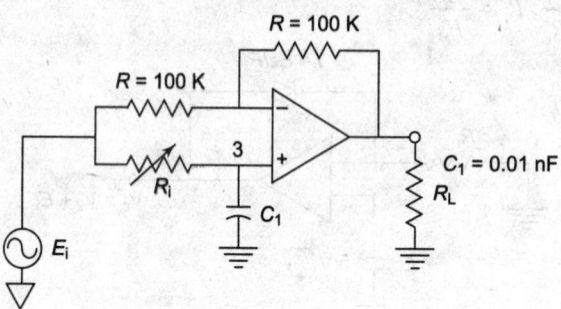

Fig. C.42   *Phase-Shifter Circuit*

Phase shifter

$$R_i = \frac{\tan(\theta/2)}{2\pi f c_1}$$

Given $f = 1$ kHz, $\theta = 90°$

$$\tan(90°/2) = \tan(45°) = 1$$

$$R_i = \frac{1}{2\pi \times 1000 \times 0.01 \times 10^{-6}} = 15.9\text{ k}\Omega$$

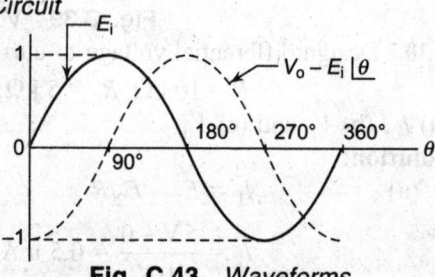

Fig. C.43   *Waveforms*

Input/output Voltage

If $R_i = 100$ kΩ, find phase angle

$$\theta = 2 \text{ arc tan } (2\pi)\,(1 \times 10^3)\,(100 \times 10^3)\,(0.01 \times 10^{-6})$$

$$\tan(\theta/2) = 2\pi f c_1$$

$$\theta = 2 \times 81° = 162°$$

$$V_o = E_i \,\lfloor -162°$$

34. For the circuit shown in Fig. C.44, find V_{UT} and V_{LT} for the given values of R_1 and R_2.
Solution:

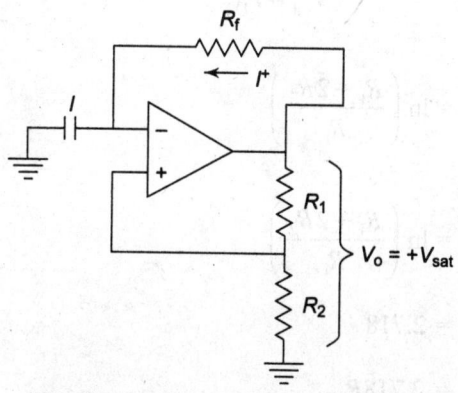

Fig. C.44 *Square Wave Generator*

$$V_{UT} = \frac{R_2}{R_1 + R_2} \; (+V_{sat})$$

$$V_{LT} = \frac{R_2}{R_1 + R_2} \; (-V_{sat})$$

(a) if $R_1 = 100 \text{ k}\Omega$, $R_2 = 85 \text{ k}\Omega$, $+V_{sat} = 15$ V, $-V_{sat} = -15$ V, find (a) V_{UT} and (b) V_{LT}.

$$V_{UT} = \frac{86 \text{ k}\Omega}{100 + 86 \text{ k}\Omega} \times 15 \text{ V} = 7 \text{ V}$$

$$V_{LT} = \frac{86 \text{ k}\Omega}{100 + 86 \text{ k}\Omega} \times -15 \text{ V} = -7 \text{ V}$$

(b) If $R_F = 100 \text{ k}\Omega$, $C = 0.1 \text{ } \mu\text{F}$. Find the period T of the multivibrator

$$T = 2 \, R_F \, C$$

$$R_2 = 0.86 \, R_1$$

$$f = \frac{1}{T} = \frac{1}{2R_FC}$$

T is in seconds, R_F in ohms, C in farad.

$$T = 2R_FC = 2 \times 100 \text{ K} \times 0.1 \times 10^{-6}$$

$$= 0.020 \text{ sec} = 20 \text{ msec}$$

$$f = \frac{1}{2R_FC} = \frac{1}{T} = \frac{1}{20 \times 10^{-3}} = 50 \text{ Hz}$$

(c) Prove that

$$T = 2R_FC \text{ when } R_2 = 0.86 \text{ R}$$

The time required for a capacitor C to change through resistor R_F from start to stop voltage.

$$t = R_FC \ln \left(\frac{V_a - V_{st}}{V_a - V_{stop}} \right)$$

$$= R_F C \ln \left(\frac{V_{sat} - V_{LT}}{V_{sat} - V_{UT}} \right)$$

$$\ln \left(\frac{V_{sat} - \dfrac{R_2}{R_1 + R_2}(-V_{sat})}{V_{sat} - \dfrac{R_2}{R_1 + R_2}(V_{sat})} \right) = \ln \left(\frac{R_1 + 2R_2}{R_1} \right)$$

$$\ln \left(\frac{15 + 7}{15 - 7} \right) = \ln \left(\frac{R_1 + 2R_2}{R_1} \right)$$

$$\frac{R_1 + 2R_2}{R_1} = 2.718$$

$$R_1 + 2R_2 = 2.718R_1$$

$$1.718R_1 = 2R_2$$

$$R_2 = 0.86R_1$$

$$T = t_1 + t_2 = 2R_F C$$

$$T = 2R_F C$$

35. A triangle wave generator oscillates at a frequency of 1000 Hz with peak values of +5 V. Calculate the required values for PR, R_i, and C.

Solution:

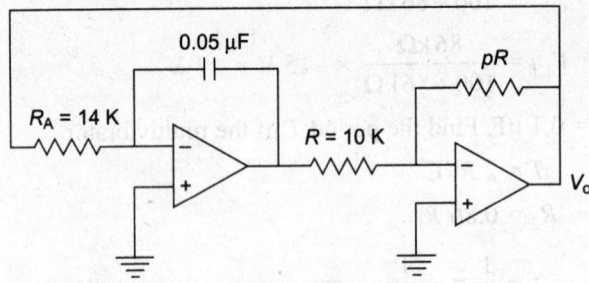

Fig. C.45 *Triangular Waveform Generator*

$$V_{UT} = \frac{+V_{sat}}{P}, \quad +V_{sat} = 14.2 \text{ V}$$

$$V_{LT} = \frac{-V_{sat}}{P}, \quad -V_{sat} = -13.8 \text{ V}$$

$$P = \frac{pR}{R}$$

$$f = \frac{p}{4R_i C}$$

$$P = \frac{p}{4R_i C} = \frac{-13.8 \text{ V}}{5 \text{ V}} = 2.8$$

Choose
$$R = 10 \text{ k}\Omega, \; C = 0.05 \text{ μF}$$

$$pR = 2.8 \times 10 \text{ K} = 28 \text{ K}$$

$$R_i = \frac{P}{4_F C} = \frac{2.8}{4(1000)(0.05 \times 10^{-6})} = 14 \text{ k}\Omega$$

$$P = \frac{pR}{R}, \quad V_{UT} = \frac{-V_{sat} + 0.6 \text{ V}}{P}$$

$$f = \frac{p}{2R_i C}$$

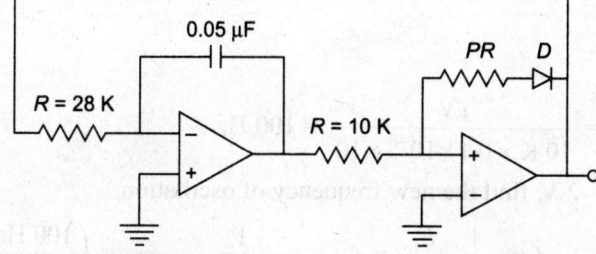

Fig. C.46 *Symmetrical Triangular Waveform Generator*

$$p = \frac{28 \text{ K}}{10 \text{ K}} = 2.8$$

$$V_{UT} = -\left(\frac{-V_{sat} + 0.6}{P}\right) = -\left(\frac{-13.8 \text{ V} + 0.6}{2.8}\right)$$

$$= 4.7 \text{ V}$$

$$f = \frac{P}{2R_i C} = \frac{2.8}{2(2.8 \text{ K})(0.05 \text{ } \mu\text{F})}$$

$$= 1000 \text{ Hz}$$

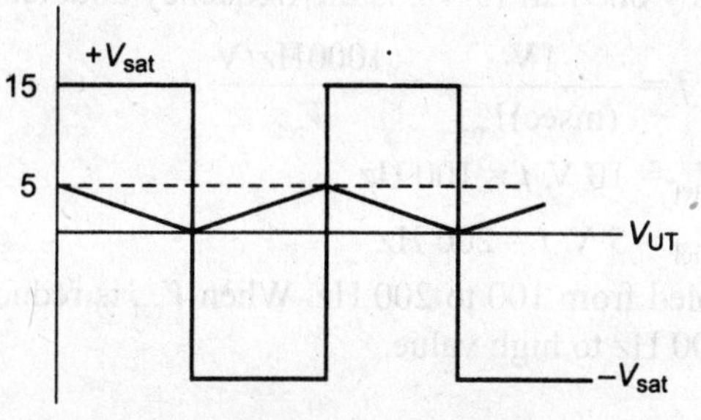

Fig. C.47 *Waveforms*

36. Design a sawtooth wave generator to have 10-V peak output and frequency of 100 Hz.
Solution:

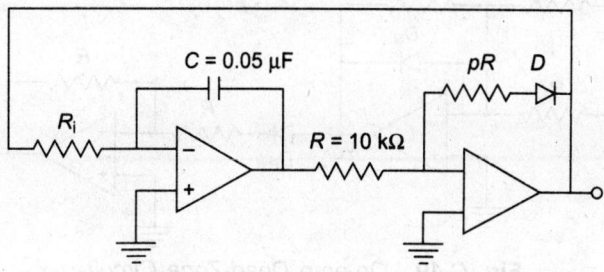

Fig. C.48 *Sawtooth Wave Generator*

(a) Design a voltage divider to give a reference voltage $V_{ref} = +10$ V.

(b) Select a ramp rate rise of 1 V/rms. R_iC combination to give 1 msec. Select $R_i = 10$ kΩ, $C = 0.1$ μF.

(c) E_i from voltage divider and voltage follower to make an ideal voltage source.

(d) Time of rise $= \dfrac{\text{distance}}{\text{speed}}$

(e) V_o ramp $= \dfrac{E_i}{R_iC}$

(f) $T = V_{ref}/E_i/R_iC$

(g) $f = \dfrac{1}{R_iC}\dfrac{E_i}{V_{ref}} = \dfrac{1V}{10 \text{ K} \times 0.1 \times 10^{-6} \times 10} = 100$ Hz

2. If E_i is doubled to -2 V, find the new frequency of oscillation.

$$f = \frac{1}{R_iC}\frac{E_i}{V_{ref}} = \frac{1}{(10 \times 10^3)(0.1 \times 10^{-6})} = \left(\frac{100 \text{ Hz}}{\text{Volt}}\right)E_i$$

$$E_i = 1 \text{ V}$$

$$f = 100 \text{ Hz}, E_i = -2 \text{ V}$$

$$f = (2 \text{ V})(100) = 200 \text{ Hz}$$

$$E_i = 10 \text{ V}$$

$$f = 1 \text{ kHz}$$

E_i changes from 1 to 10 V
Frequency changes from 0 to 1 kHz.

3. If the voltage reduced by one half (5 V), is the frequency doubled or halved?

$$f = \frac{1V}{(\text{msec})V_{ref}} = \frac{1000\text{Hz/V}}{V_{ref}}$$

$$V_{ref} = 10 \text{ V}, f = 100 \text{ Hz}$$

$$V_{ref} = 5 \text{ V}, f = 200 \text{ Hz}$$

Therefore, frequency is doubled from 100 to 200 Hz. When V_{ref} is reduced to 0 V, frequency increased from 100 Hz to high value.

Op-amp with Diodes, Dead-Zone Circuit

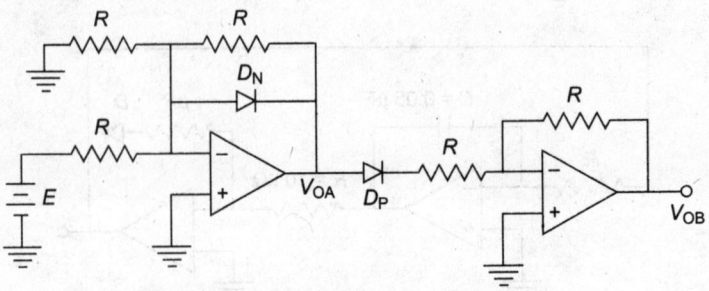

Fig. C.49 *Op-amp Dead-Zone Circuit*

$$V_{ref} = +15 \text{ V}/3$$
$$= 5 \text{ V}$$
$$V_{OA} = -E_1 - V_{ref} = -(-10 \text{ V}) - 5 = +5 \text{ V}$$
$$V_{OB} = -5 \text{ V}$$

All input signal above $-V_{ref}$ fall in a dead zone and are eliminated from the output.

Basic Differential Amplifier Circuit

37. Given:

$$n = \frac{mR}{R} = \frac{100 \text{ k}\Omega}{1 \text{ k}\Omega} = 100$$

Find V_o for $E_1 = 10$ mV, $E_2 = 10$ mV, $E_2 = -20$ mV

Solution:

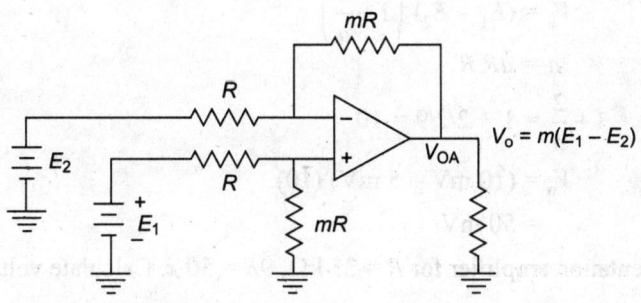

Fig. C.50 *Op-amp Differential Amplifier Circuit*

$$V_o = mE_1 - mE_2$$
$$= m(E_1 - E_2)$$

(a) $E_1 = 10$ mV

$$V_o = 100 (10 - 10) = 0$$

(b) $E_2 = 10$ mV

$$V_o = 100 (10 - 0) \text{ mV}$$
$$= 1 \text{ V}$$

(c) $E_2 = -20$ mV

$$V_o = 100 [10 - (-20)]$$
$$= 100 (30 \text{ mV}) = 3 \text{ V}$$

38. Design a buffered differential amplifier

$$E_1 = 10 \text{ mV}, E_2 = \text{mV. Find } V_o.$$

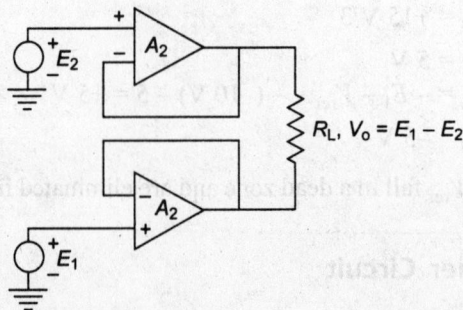

Fig. C.51 *Buffered Differential Amplifier*

Solution:

$$aR = 2 \text{ k}\Omega, R = 9 \text{ k}\Omega$$

$$\frac{aR}{R} = \frac{2\,\text{K}}{9\,\text{K}} = \frac{2}{9} = a$$

$$V_o = (E_1 - E_2)\left(1 + \frac{2}{a}\right)$$

$$a = aR/R$$

$$1 + \frac{2}{a} = 1 + 2/2/9 = 10$$

$$V_o = (10 \text{ mV} - 5 \text{ mV})\,(10)$$

$$= 50 \text{ mV}$$

39. Design instrumentation amplifier for $R = 25 \text{ k}\Omega$, $9R = 50\ x$. Calculate voltage gain.

Solution:

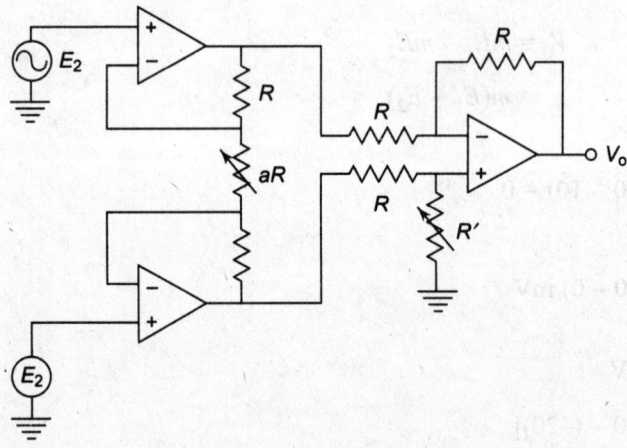

Fig. C.52 *Instrumentation Amplifier Circuit*

$$\frac{V_o}{E_1 - E_2} = 1 + \frac{2}{a}$$

$$\frac{aR}{R} = \frac{50}{25\,K} = \frac{1}{500} = a$$

$$\frac{V_o}{E_1 - E_2} = 1 + \frac{2}{a} = 1 + \frac{2}{1/500}$$

$$= 1 + (2 \times 500)$$

$$= 1001$$

$$V_o = 1001\,(E_1 - E_2)$$

Case 1:

If aR is removed from instrumentation amplifier, what is voltage gain?

$$\frac{V_o}{E_1 - E_2} = 1 + \frac{2}{a}$$

a is removed, $a = \infty$

$$\frac{V_o}{E_1 - E_2} = 1 + \frac{2}{\infty} = \mathbf{1}$$

Case 2:

Assume gain of 1001. Find V_o for $E_1 = 5.001$ V, $E_2 = 5.002$ V.

$$V_o = 1001\,(E_1 - E_2)$$

$$= 1001\,(5.001 - 5.002)\text{ V}$$

$$= 1001\,(-0.001)$$

$$= -1.001\text{ V}$$

Case 3:

If $E_1 = 5.001$ V, $E_2 = 5.000$ V, find V_o

$$V_o = 1001\,(5.001 - 5.000) = 1001(0.001) = 1\text{ V}$$

Case 4:

$$E_1 = -1.001\text{ V}, E_2 = -1.002\text{ V}$$

$$V_o = 1001\,[-1.001 - (-1.002)]\text{ V}$$

$$= 1001\,(0.001)\text{ V}$$

$$= 1.001\text{ V}$$

40. Analyze the circuit.

Assume $E_i = 2$ V, capacitor C charged, so current flows through R_i and voltage drop is zero,

(a) $E_i = 2$ V
(b) $V_F = E_i$
(c) $I = V_F/R_B$

$$V_o = I(R_A + R_B) = \frac{V_F}{R_B}\,(R_A + R_B)$$

$R_A = R_B = 10 \text{ k}\Omega, V_0 = 2V_F$

(d) Op-amp gain of -1 V,

$V_0 = -V_R$

$V_R = -V_0$

(e) Capacitor voltage, $E_i - V_R$

So, $V_0 = 2V_F = 2E_i = -V_R$

$V_{cap} = E_i - V_R = 3E_i$

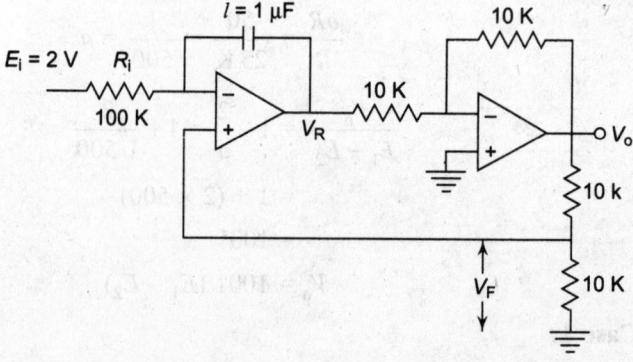

Fig. C.53   *Op-Amp Servo Amplifier Circuit*

(i) Calculate the equilibrium voltages for the servo amplifier.

(a) $E_i = 2$ V,   so, $V_f = 2$ V

(b) V_F forces to V_0

V_0 to $2V_F = 4$ V

(c) V_0 forces V_R to -4 V

(d) V_{cap} stabilizes at $3E_i = 6$ V

(ii) When E_i stopped to 4 V, find the new equilibrium voltages.

(a) $E_i = 4$ V, $V_F = 4$
$V_0 = 2 \ V_F = 2 \times 4 = 8$ V

(b) V_R decreases to -8 V

(c) $V_{cap} = 12$ V

(iii) How long V_0 takes to reach equilibrium in servo amplifier?

$$T = 3R_iC$$

$$R_i = 100 \text{ k}\Omega$$

$$C = 1 \times 10^{-6} \text{ F}$$

$$T = 3 \times 100 \text{ K} \times 1 \times 10^{-6}$$

$$= 0.3 \text{ sec}$$

Equilibrium time $= 5T$

$$= 5 \times 0.35$$

$$= 1.55$$

$$V_0 = \begin{cases} E_1 \ (-R_F/R_1) \text{ Channel 1 gain} \\ E_2 \ (-R_F/R_1) \text{ Channel 2 gain} \\ E_3 \ (-R_F/R_3) \text{ Channel 3 gain} \end{cases}$$

$$V_{RF} = (I_1 + I_2 + I_3)R_F$$

$$I_1 = \frac{E_1}{R_1}, I_2 = \frac{E_2}{R_2}, I_3 = \frac{E_3}{R_3}$$

$$V_0 = -\left(E_1 \frac{R_F}{R_1} + E_2 \frac{R_F}{R_2} + E_3 \frac{R_F}{R_3} \right)$$

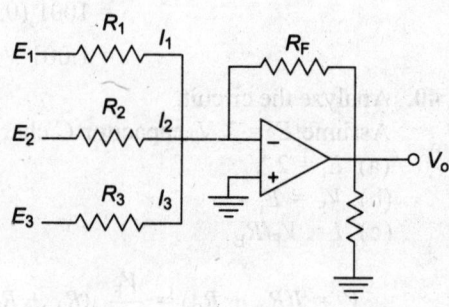

Fig. C.54   *Multichannel Amplifier*

$$A_{CL_1} = -\frac{R_F}{R_1}, A_{CL_2} = -\frac{R_F}{R_2}, A_{CL_3} = -\frac{R_F}{R_3}$$

$$V_o = E_1 A_{CL_1} + E_2 A_{CL_2} + E_3 A_{CL_3}$$

41. Design a three channel-inverting amplifier, the gains for each channel will be

Channel number	Voltage gain
1	−10
2	−5
3	−2

Solution:

(a) Select 10 kΩ resistor for input resistance of the channel with highest gain. Choose $R_1 = 10$ kΩ, A_{CL_1} is largest.

(b) Calculate feedback resistor R_F

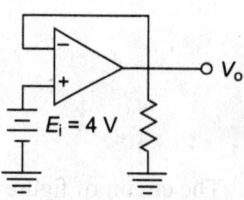

$$A_{CL} = -\frac{R_F}{R_1}, -10 = \frac{R_F}{10\text{ K}}$$

$$R_F = 100\text{ k}\Omega$$

Fig. C.55 *Multi Channel Inverting Amplifier*

(c) $R_2 = 20$ kΩ, $R_3 = 50$ kΩ

(d) If $R_1 = R_2 = R_3 = R = 100$ kΩ

$$R_F = 100\text{ k}\Omega/3 = 33\text{ k}\Omega$$

$$E_1 = +5\text{ V}, E_2 = +5\text{ V}, E_3 = -1\text{ V, find } V_o$$

$$V_o = -\left[\frac{E_1 + E_2 - E_3}{3}\right]$$

$$= -\left[\frac{5+5-1}{3}\right] = -\frac{9}{3} = -3\text{ V}$$

Voltage Follower

1. $V_o = E_i = 4$ V

2. $I_L = \dfrac{V_o}{R_L} = \dfrac{4}{10\text{ K}} = 0.4$ mA

3. $I_o = I + I_L = 0 + 0.4 = 0.4$ mA

If E_i reversed, the role unity of V_o also, direction of currents would also be reversed.

Voltage Regulator

1. Voltage sources

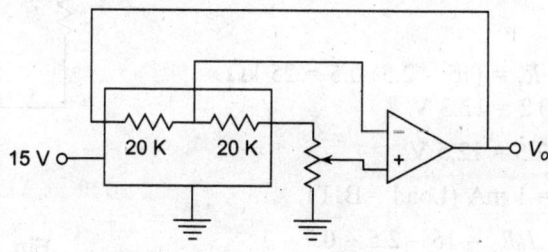

Fig. C.56 *Voltage Regulator Circuit*

Variable voltage reference $-10\ V \le V_o \le 10\ V$

The above diagram shows variable voltage reference which can be termed as current source.

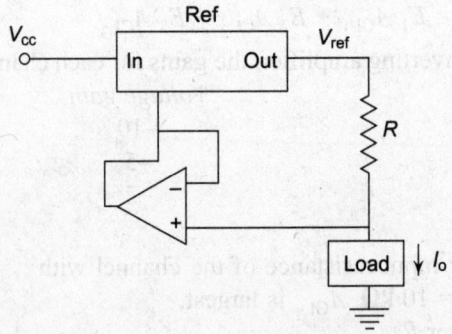

Fig. C.57   *Voltage Regulator Circuit*

42. The circuit of figure uses 5-V reference with $T_c = 20\ MV/°C$ line regulation = 50 MV/V, drop of voltage $V_{DO} = 3\ V$, $CMRR_{dB} = 100_{dB}$, specify R for $I_o = 10\ mA$. Find the worst case values of $TC(I_o)$ and resistance R_o. Assume $\pm 15\ V$. Find the voltage compliance.

Solution:

(a) $R = \dfrac{5\ K}{10} = 500\ \Omega$

(b) 1°C change in T cause worst case change in the voltage across R of $20 + 5 = 25\ \mu V/°C$.

I_o is $25 \times 10^{-6}/500 = 50\ nA/°C$

A 1 V change in V_L causes $50\ \mu V/V$

Change in V_{ref}, $10^{-200/20} = 10\ \mu V/V$

Worst case change in I_o of $\dfrac{(50+10)10^{-6}}{500} = 120\ nA/V.$

$$R_{o(min)} = \frac{1}{120 \times 10^{-9}} = 8.33\ m\Omega$$

(c) $V_L \le V_{cc} - V_{DD} - V_{ref}$

$$= 15 - 3 - 5 = 7\ V$$

43. See the circuit of the Figure C.58. Current source was 741 op-amp with $V_{cc} = 15\ V$, with bias current of 0.5 mA. With BJT of R_2 1 kΩ, find R and R_1 for $I_o = 100\ mA$, also find the voltage compliance of the source.

(i) $R = \dfrac{2.5}{0.1} = 25\ \Omega$

$$R_1 = (15 - 2.5)/0.5 = 25\ k\Omega$$

(ii) $V_L \le 15 - 2.5 - 0.2 = 12.3\ V$

I/P to 741, $15 - 2.5 = 12.5\ V$

Assume $\beta = 100$, $I_B = 1\ mA$ (Load – BJT)

$V_{CC} - V_{ref} - V_{BE(on)} - I_B R_2 = 15 - 2.5 - 0.7 - 1$

$$= 10.8\ V$$

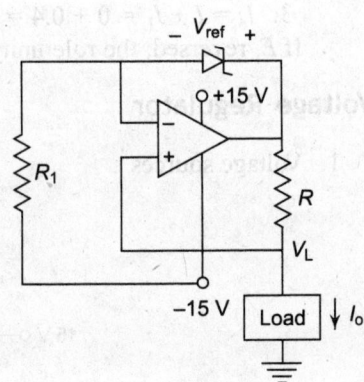

Fig. C.58   *Current Source Circuit*

Sink current is in A.

44. Let R_B = 510 Ω, R_E = 3.3 kΩ. Find (a) R_2/R_1 for V_o = 5.0 V, (b) error amplifier output drive needed to provide I_o = 1A, and (c) dropout voltage V_{DO}.

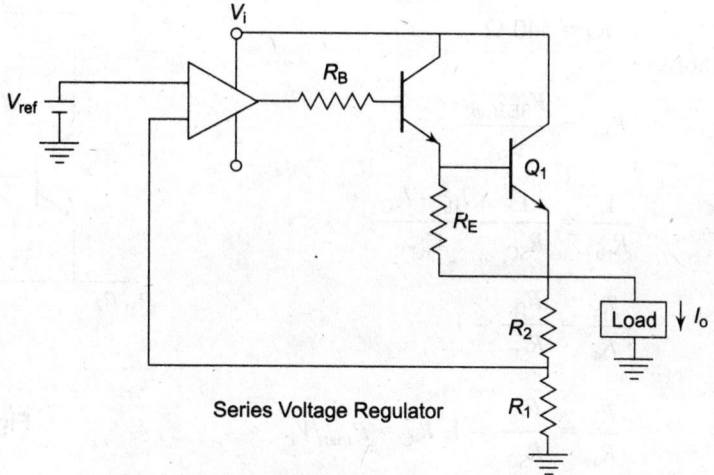

Fig. C.59 *Series Voltage-Regulator Circuit*

Solution:

(a) $\left(\dfrac{1+R_2}{R_1}\right) V_{ref} = V_o$

$$5 = (1 + R_2/R_1)\, 1.28$$
$$R_2/R_1 = 2.9$$

(b) I_o = 1 A, $I_{B1} = I_{E1}/(1 + \beta) = \dfrac{1}{21} \approx 48$ mA

$$I_{E2} = I_{B1} + V_{BE(on)}/R_E = 48 \text{ mA}$$
$$V_{OA} = V_{RB} + V_{BE2(on)} + V_{BE1(on)} + V_o$$
$$= 0.51 \times 0.47 + 0.7 + 1 + 5$$
$$= 7$$

(c) $V_i \geq 7.5$ V

$$V_{DO} = 7.5 - 5 = 2.5 \text{ V}.$$

45. A 5-V regulator with V_i = 8 V uses 12 W series pass BJT. (a) Assume typical BJT parameters, specify suitable components for output short-circuit protection.

Solution:

Short-circuit protection:
(a) $I_{SC} = V_{BE}/R_{SC}$

$$I_{SC} = 12/8 = 1.5 \text{ A}$$
$$R_S = V_{BE}/I_{SC} = 0.7/1.5 = 0.47 \text{ Ω}$$

(b) $I_{fb} = 12/8 - 5 = 4$ A

$$R_{fb} = [0.47 - (4 - 1.5)/5]^{-1} = 0.6 \text{ Ω}$$

$$R_3/R_4 = \frac{0.61}{0.47} - 1 = 0.3$$

$$R_3 = 160 \ \Omega$$

$$R_4 = 540 \ \Omega$$

Design elevations

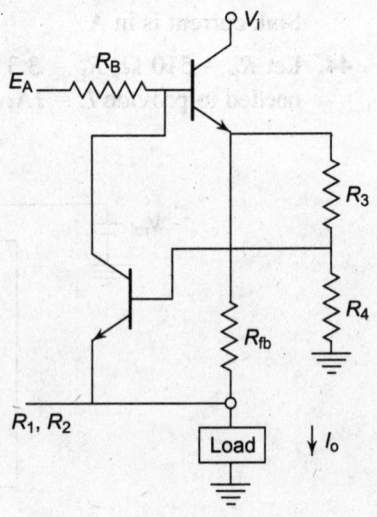

$$R_{SC} = \frac{V_{BE3(on)}}{I_{SC}}$$

$$\frac{1}{R_{fb}} = \frac{1}{R_{SC}} - \frac{I_{fb} - I_{SC}}{V_{ref}}$$

$$\frac{R_3}{R_4} = \frac{R_{fb}}{R_{SC}} - 1$$

$$\frac{R_3}{R_4} = \frac{R_{fb}}{R_{SC}} - 1, \ I_{SC} = P_{man}/V_i.$$

Fig. C.60

Linear Regulator Applications

46. Use 7805 regulator as shown in figure C.61, specify suitable components for $V_o = 15$ V.

Solution:

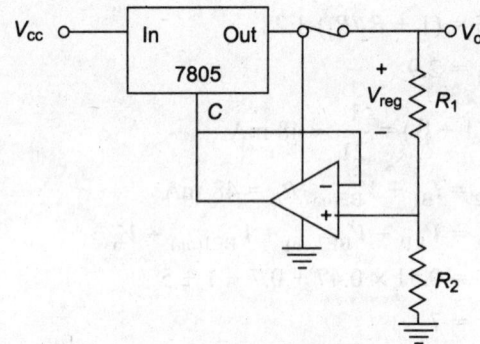

Fig. C.61   *Voltage Source Circuit*

Power voltage source:

Use 741 op-amp, $R_1 = 10 \ k\Omega$, $R_2 = 20 \ k\Omega$

$$V_o = \left(1 + \frac{R_2}{R_1}\right) V_{ref}$$

Adjustable power current source $V_{DO} = 2$ V, line regulation = 0.07%. Assume 10 kΩ Pot. $CMRR \geq 70$ dB, ±15 V supply. Find R for adjustable current 0 to 1 A.

$$I_o = K \ \frac{V_{ref}}{R}$$

Given 2 W current $\rightarrow$ 1 A, $K = 1$

$$R = V_{\text{reg}}, R \simeq 1.2 \text{ V}$$
$$V_L \le V_{\text{cc}} - V_{\text{DO}} - kV_{\text{reg}}$$
$$V_L \le 15 - 2 - 1.25$$
$$V_L = 11.75 \text{ V}$$
$$I_o = 1 \text{ A}$$
$$R_{\text{min}} = 1/0.9 \text{ mA} = \mathbf{1 \text{ k}\Omega}$$

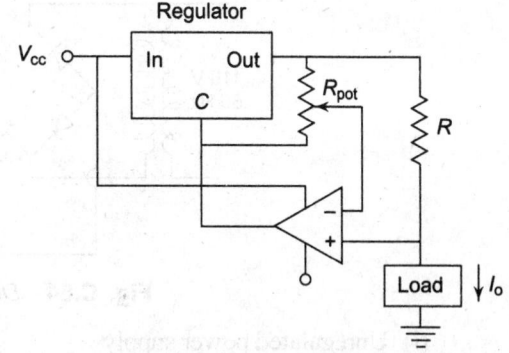

Fig. C.62 *Regulator Circuit*

Thermal Consideration

$$T_j - T_A = Q_{jA} P_D$$

$P_D \rightarrow$ dissipated power

$T_j \rightarrow$ junction temperture

$T_A \rightarrow$ ambient temperture

$\quad T_A = 25°\text{C}, P_D = 2 \text{ W}, T_j = T_A + QP_D, \theta = 50°\text{C/W}.$

$\quad T_j = 25 + 50 \times 2 = 125°\text{C}$

Given $T_{j(\text{max})} = 150°\text{C}, T_{A(\text{max})} = 50°\text{C}.$

47. Find maximum power corresponding to case temperature, also find maximmum current $V_i = 8$ V.

Solution:

$$P_{D(\text{max})} = T_{j\text{max}} - T_{A\text{max}}/\theta_{jA}$$
$$= 150 - 50/60$$
$$= 1.67 \text{ W}$$

$$T_c = T_j - Q_c P_D$$
$$= 150 - 3 \times 1.67 = 145°\text{C}$$

$$P_D = (V_i - V_o) I_o$$
$$I_o \le 1.67/8 - 5 = 0.556 \text{ A}$$

Suitable component values are as follows:

$$W_1 = \frac{1}{R_4 C_2}, W_2 = \frac{1}{R_2 C_3}, W_3 = \frac{1}{R_3 C_3}, W_4 = \frac{1}{R_4 C_1}$$

$V_i = 12 \text{ V}, f_s = 100 \text{ kHz}, V_{sm} = 1 \text{ V}, L = 100 \text{ mH}$

$C_1 = 240 \text{ pF}, C_2 = 1.08 \text{ nF}, C_3 = 17.3 \text{ nF}.$

$R_1 = R_2/V_o$ (Ref $- 1$)

$R_2 = 10 \text{ k}\Omega, R_3 = 850 \Omega, R_4 = 16 \text{ k}\Omega$

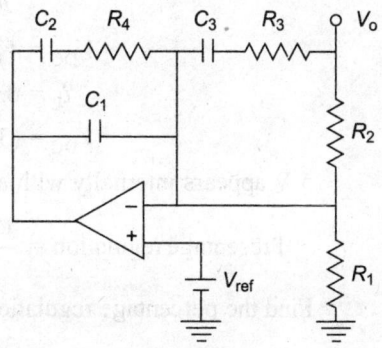

Fig. C.63 *Buck-Regulator Circuit*

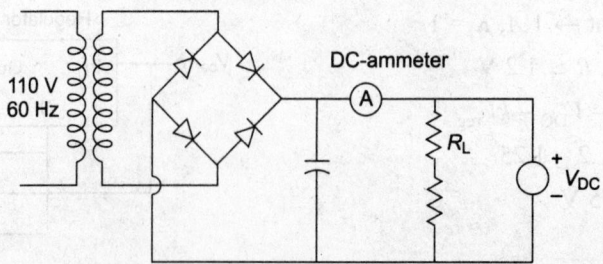

Fig. C.64   *DC Power Supply Circuit*

(i) Unregulated power supply

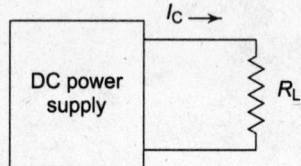

Fig. C.65   *Unregulated Power Supply*

(ii) Full load DC voltage supply

48. Calculate output resistance R_o from no load and full-load measurement of above diagram.
$I_L = 0.5$ A

Solution:

(a) At no load $I_C = 0$, $V_{DCNL} = 34$, full load $I_C = 1$ A and $V_{DCFL} = 24$ V

$$R_o = \frac{V_{DCNL} - V_{DCFL}}{I_{LFL}}$$

$$= \frac{34 - 24}{1} = 10 \ \Omega$$

(b) $V_{DCNL} = V_{DCFL} + I_{LFL}R_o$

$$V_{DCFC} = \frac{R_L}{R_o R_L} \ V_{DCNL}$$

$$V_{DC} = V_{DCNL} - I_L R_o$$

$$I_L = 0.5 \text{ A}, R_o = 10 \ \Omega, V_{DCNL} = 34 \text{ V}$$

$$V_{DC} = (34V - .5 \times 10\Omega) = 29 \text{ V}$$

5 V appears internally with a power supply to course 5 V × 0.5 A = 2.5 W of heat.

$$\text{Precentage regulation} = \frac{V_{DCNL} - V_{DCFL}}{V_{DCFL}} \times 100$$

Find the percentage regulation of DC power supply of above problem.

$$\text{Percentage regulation} = \frac{34 - 24}{24} \times 100 = 41.7\%.$$

49. A full-wave bridge rectifier has (i) full load correct of 1 A, (ii) full load voltage of 24 V, and (iii) filter capacitor of 1000 μF. Calculate (a) peak-to-peak rms value of ripple voltage at full load and (b) minimum output voltage.

Solution:

(a) $\Delta V_o = \dfrac{1\,A}{200(1000 \times 10^{-6}\,F)} = 5$ V

$\Delta V_o = \dfrac{I_L}{200\,C}$

(b) $\Delta V_o = 3.5\,V_{rms}$

(c) $V_L = V_{DCFL} - \dfrac{\Delta V_o}{2}$

(a) $\Delta V_o = 5$ V

(b) $V_{rms} = \dfrac{5}{3.5} = 1.43$ V

(c) $V_L = 24 - 5/2 = 21.5$ V

(c) Given $V_{DCFL} = 24$ V, $V_{rms} = 1.43$, find precentage ripple

$$\text{Precentage ripple} = \dfrac{1.43}{24} \times 100 = 6\%.$$

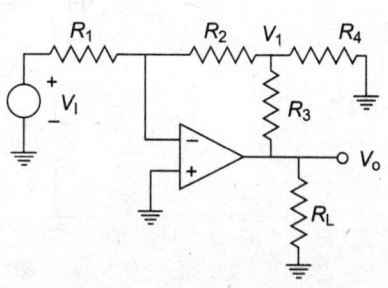

Fig. C.66   *Op-Amp with Feedback*

50. Find A_{ideal}, actual gain if op-amp $V_d = 1$ mΩ.

Solution:

Given $a = 10^5$ V/V, $r_o = 100$ Ω

$R_1 = R_2 = 1$ mΩ, $R_3 = 100$ K, $RU = 1$ kΩ, $R_L = 2$ kΩ, $T = 314$

$$A_{ideal} = -\dfrac{R_2}{R_1}\left(1 + \dfrac{R_3}{R_2} + \dfrac{R_3}{R_4}\right)$$

$$= -101 \text{ V/V}$$

$$R_A = r_d \parallel R_1 = 500 \text{ k}\Omega$$

$$A = \dfrac{-101}{1 + 1/T}$$

$$T = -V_R/V = 314$$

$$A = -100$$

51. An inverting amplifier with $R_1 = 10$ K, $R_2 = 20$ K, $V_i = 3$ V drives 2-kΩ load. (a) Assuming $I_Q = 0.5$ mA, find i_{ce}, i_{EE} and I_o. (b) Find the power dissipated in op-amp.

Solution:

$$V_o = -\dfrac{R_2}{R_1} V_i$$

$$= -\dfrac{20}{10}\,3 = -6 \text{ V}$$

$$i_L = \dfrac{6}{2} = 3 \text{ mA}$$

$$i_2 = i_1 = \dfrac{3}{10} = 0.3 \text{ mA}$$

$$i_o = i_2 + i_1 = 3.3 \text{ mA}$$
$$i_{ce} = i_Q = 0.5 \text{ mA}$$
$$i_{EE} = i_{ce} + i_o = 0.5 + 3.3 = 3.8 \text{ mA}$$

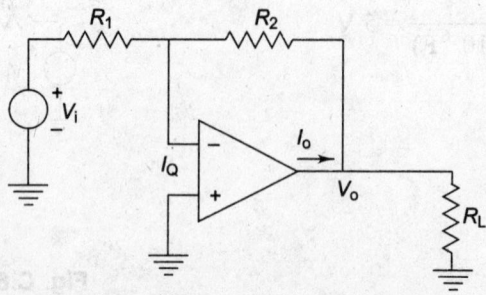

Fig. C.67 *Op-amp Inverting Amplifier*

Power, $\qquad P = V_i$

$$P_{OA} = (V_{cc} - V_{EE})I_Q + (V_o - V_{EE})\, i_o$$
$$= 30 \times 0.5 + [-6 - (-15)] \times 3.3$$
$$P_{OA} = 44.7 \text{ mW}$$

52. When experimenting with op-amps, it is hard to have variable source over the range $-10 \text{ V} \le V_s \le 10 \text{ V}$.

    (a) If V_s is set to 10 V, how much does it change when we connect a 1 kΩ load to the source?
    (b) Design a resistive network to produce an adjustable voltage over the range -10 V to $+10$ V. This -25 kΩ resistors to drop 5 V.

$$V_A = 10 \text{ V}, \ V_B = -10 \text{ V}$$

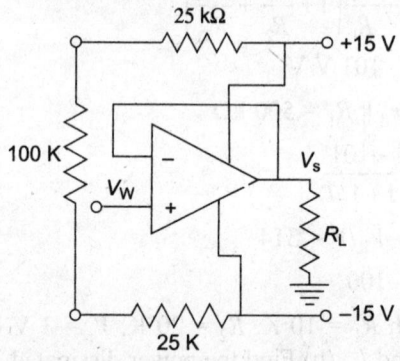

Fig. C.68 *Circuit for Problem 52*

VAR Source

$$-10 \text{ V} \le V_W \le 10 \text{ V}$$

(b) Correcting 1 kΩ load will draw a current $i_L = \dfrac{10}{1} = 10 \text{ mA}$

The output resistance is $R_o = r_o/1 + T = \dfrac{75}{1 + 200000} = 0.375 \text{ m}\Omega$

$$\Delta V_s = R_o \, \Delta i_L = 0.375 \times 10^{-3} \times 10 \times 10^{-3}$$
$$= \textbf{3.75 mV}$$

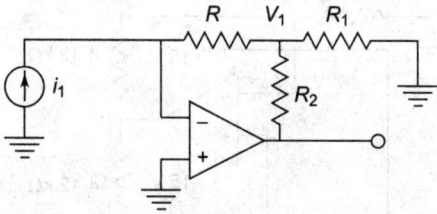

Fig. C.69 *Current-to-Voltage Converter Circuit*

53.   Design the circuit in Fig. C.69 to achieve sensitivity of 0.1 V/nA.

Solution:

$$V_o = -KR_i. \text{ Given } KR = 0.1 \text{ V/nA (sensitivity)}$$

Assume $R = 1 \text{ m}\Omega, R_1 = 1 \text{ k}\Omega$

$$1 + R_2/10^3 + R_2/10^6 = 100$$

$$R_2 = 99 \text{ k}\Omega \rightarrow i_1 \text{ is in mA.}$$

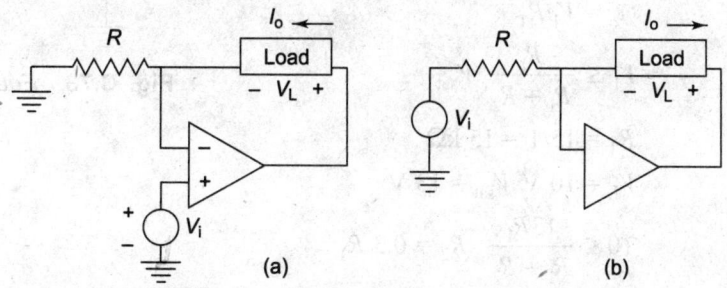

Fig. C.70 *Floating Load-Converter Circuit*

$$V_i = 5 \text{ V}$$
$$R = 10 \text{ k}\Omega, \pm V_{sat} = \pm 13 \text{ V}$$

54. Find i_o, voltage compliance, R_L

Solution:

(a) $i_o = \dfrac{V_i}{R} = \dfrac{5}{10} = 0.5 \text{ mA}$

In Fig. C.70(a) flowing from right to left
In Fig. C.70(b) left to right

(b) $-8 \text{ V} < V_L < 8 \text{ V}, -13 \text{ V}, < V_L < +13 \text{ V}$

(c) $R_L = \dfrac{8}{0.5} = 16 \text{ k}\Omega$ for Fig. C.70(a).

$$R_L = \dfrac{13}{0.5} = 26 \text{ k}\Omega \text{ for Fig. C.70(b).}$$

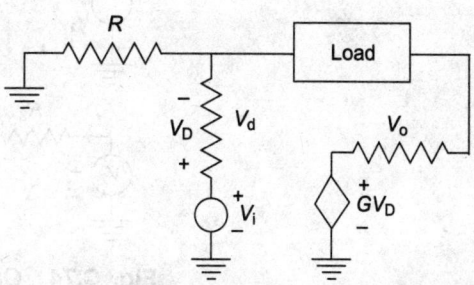

Fig. C.71 *Equivalent Circuit of Op-amp Converter*

55. Using 741 op-amp powered from ±15-V regulated supplies, design 1-mA DC source having 10-V voltage compliance.

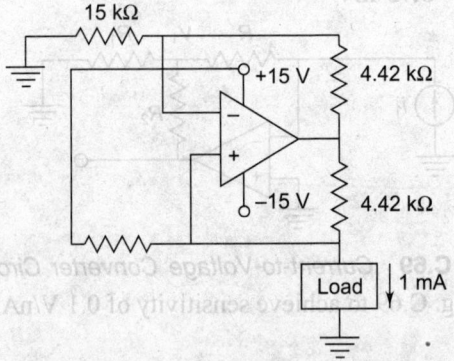

Fig. C.72 *Op-amp Current Source*

Solution:

1-mA source op-amp I_C 741

$$V_i = +15 \text{ V},$$

$$i_o = V_I/R_1$$

$$V_L \leq \frac{R_1}{R_1 + R_2} V_{sat}$$

$$R_1 = 15/1 = 15 \text{ k}\Omega$$

$$V_L = 10 \text{ V}, V_{sat} = 13 \text{ V}$$

$$10 \leq \frac{13R_1}{R_1 + R_2}, R_2 = 0.3 R_1$$

$$R_1 = R_3 = 15 \text{ k}\Omega$$

$$R_2 = R_4 = 0.3 \times 15 = 4.5 \text{ k}\Omega$$

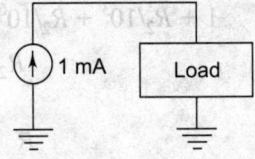

Fig. C.73 *Equivalent Circuit*

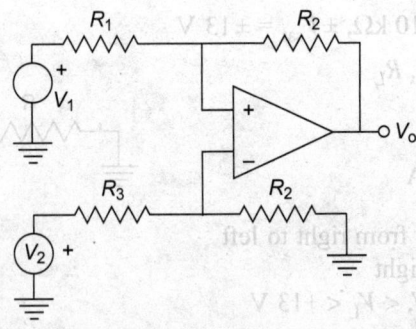

Fig. C.74 *Op-amp Difference Amplifier*

$$R_3 = R_1$$

$$R_2 = R_4$$

56. In the above circuit let $R_1 = R_3 = 10$ kΩ, $R_2 = R_4 = 100$ kΩ. Find V_o for each of the following input voltage pairs.

Solution:

(i) $(V_1, V_2) = (-0.1$ V, $+0.1$ V)

$$V_o = \frac{100}{10}(V_2 - V_1)$$
$$= 10(V_2 - V_1), \ 0.2 \text{ V} = V_2 - V_1$$
$$V_o = 10 \times 0.2 = 2 \text{ V}$$

(ii) $(V_1, V_2) = (4.9$ V, 5.1 V)

$$V_o = 10(0.2) = 2 \text{ V}$$

(iii) $(9.8$ V, 10.1 V)

$$V_o = 10(0.3) = 3 \text{ V}$$

(iv) $R_1 = 10$ K, $R_2 = 98$ K, $R_3 = 9.9$ K, $R_4 = 103$ K

$$V_o = A_2 V_2 - A_1 V_1, \ A_2 = (1 + R_2/R_1)/1 + R_3/R_4 = 9.853.$$
$$A_1 = 9.8$$

So, $\qquad (V_1, V_2) = (-0.1$ V, $+0.1$ V), $V_o = 1.965$ V

Effect of Resistance Mismatches in Difference Amplifier:

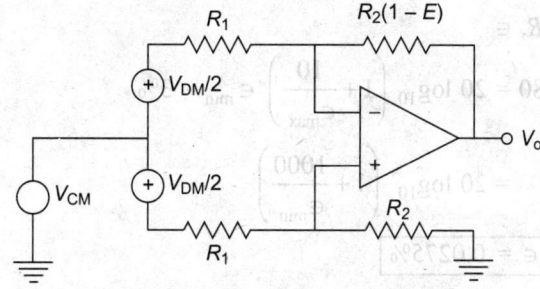

Fig. C.75

Fourth resistor $R_2(1 - \epsilon) \rightarrow$ imbalance. Apply super position principle

$$V_o = -\frac{R_2(1-\epsilon)}{R_1}\left(V_{CM} - \frac{V_{DM}}{2}\right) + \frac{R_1 + R_2(1-\epsilon)}{R_1} \times \frac{R_2}{R_1 + R_2}\left(V_{CM} + \frac{V_{DM}}{2}\right)$$

$$V_o = A_{dm} V_{DM} + A_{cm} V_{CM}$$

$$A_{dm} = \frac{R_2}{R_1}\left(1 - \frac{R_1 + 2R_2\epsilon}{R_1 + R_2 2}\right)$$

$$A_{cm} = \frac{R_2}{R_1 + R_2}\epsilon \quad (\epsilon \text{ near to zero})$$

$$CMRR_{dB} = 20 \log_{10}\left|\frac{A_{dm}}{A_{cm}}\right|$$

$$\frac{A_{dm}}{A_{cm}} \simeq \frac{R_2}{R_1}/R_2 \ \epsilon / R_1 + R_2$$

$$CMRR_{dB} \simeq 20 \log_{10} \left| \frac{1 + R_2/R_2}{\epsilon} \right|$$

$\frac{R_2}{R_1}$ larger $\rightarrow$ Difference gain is high.

57. Design op-amp with $R_1 = R_3 = 10$ kΩ and $R_2 = R_4 = 100$ kΩ (a) Discuss the implications of using 1%. (b) Estimate the resistance tolerance for a guaranteed *CMRR* of 80 dB.

Solution:

(a)   Theoretical $|\epsilon|_{max} \simeq 4P$, P is percentage tolerance.

$$P = 1\% = 0.01, |\epsilon|_{man} = 0.04$$

$$A_{dm} = \frac{100}{10}\left(1 - \frac{210}{110} \times \frac{0.04}{2}\right)$$

$$= 9.62 \text{ V}$$

$$A_{cm} = \frac{100}{110} \times 0.04 = 0.0364$$

$$CMRR = 20 \log_{10} \frac{9.62}{0.0364}$$

$$= 48.4 \text{ dB}$$

(b) To increase *CMRR*, ϵ

$$80 = 20 \log_{10} \left(1 + \frac{10}{\epsilon_{max}}\right) \epsilon_{min} = 3\%$$

$$= 20 \log_{10}^{+} \left(1 + \frac{1000}{\epsilon_{min}}\right)$$

$$\boxed{\epsilon = 0.0275\%}$$

With $V_{DM} = 0$, $V_{CM} = 10$ V

Output error can be large as V_o

$$V_o = A_{CM \, (max)} \times V_{CM}$$

$$= 0.0364 \times 10 = 0.364 \text{ V}$$

Linear Voltage-Buck Regulator

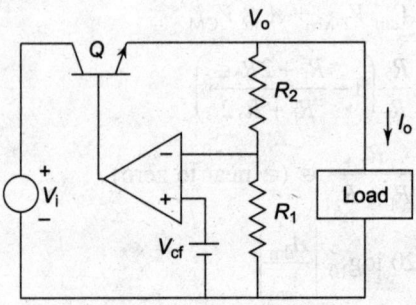

Fig. C.76 *Buck-Regulator Circuit*

Duty cycle to the switch

$$D = \frac{t_{on}}{T_s} = \frac{t_{on}}{T_{on} + T_{off}} \tag{1}$$

$$V_o = \frac{D}{1-D}(V_i - V_{sat}) + V_F \tag{2}$$

$$V_o = DV_i \text{ when } V_{sat} \to 0, V_F \to 0$$

$$V_o = \frac{D}{1-D} V_i$$

58. Given, Buck regulator V_i = 12 V, V_o = 5 V. Find D if (a) switch and diode are ideal, (b) V_{sat} = 0.5 V, V_F = 0.7 V, (c) repeat a and b if 8 V ≤ V_i ≤ 16 V.

Solution:

(a) V_o = 5 V, V_i = 12 V
D = 5/12 = 41.7%

(b) $V_o = D(V_i - V_{sat}) - (1-D)V_F$
$5 = D(5 - 0.5) - (1-D) 0.7$
D = 46.7%

(c) 8 V ≤ V_i ≤ 16 V   (b) 8 V ≤ V_i ≤ 16 V
(a) D_{min} = 31.2% D_{min} = 35.2%
    D_{max} = 62.5% D_{max} = 69.5%

The efficiency of switching regulator

$$\eta(\%) = \frac{P}{P_o + P_{dis}}$$

P_o–Power delivered to load
P_{diss}–Power dissipation
$P_{diss} = P_{SW} + P_D + P_{coil} + P_{CCP} + P_{controller}$

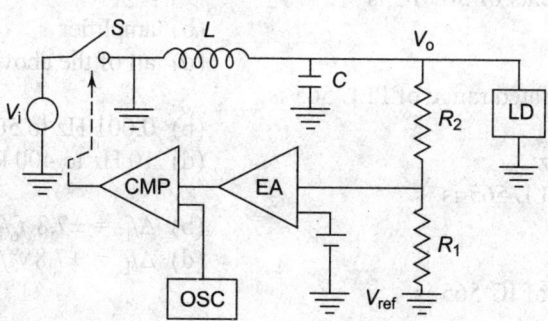

Fig. C.77 *Monolithic Switching Regulator*

HW–Control to output transfer function

$$= \frac{V_o}{V_i} = \frac{V_1}{V_{SM}} \times \frac{1 + jw/w_2}{1 - (w/w_0)^2 + (jw/w_0)/Q}$$

$$W_0 = \frac{1}{\sqrt{LC}}, \quad W_2 = \frac{1}{RC}$$

$$Q = \frac{1}{R_{o\,coil} + RC\sqrt{C/L}}$$

$V_{SM} \rightarrow$ Peak value of output waveform

H_{EA}–Transfer function $= \dfrac{V_C}{V_o}$.

MULTIPLE-CHOICE QUESTIONS

A phase-locked loop (PLL) is a frequency selective circuit designed to synchronize incoming signal and maintain synchronization inspite of noise or variations in the incoming signal frequency.

1. PLL consists of
   - (a) phase detector
   - (b) VCO
   - (c) phase defector
   - (d) all of the above
2. VCO is designed so that at zero voltage it is oscillating at some initial frequency W_0 called
   - (a) cut-off frequency
   - (b) free-cycle frequency
   - (c) free-running frequency
   - (d) none of the above
3. The time-takes for a PLL to capture the incoming signal is called
   - (a) pull-out time
   - (b) capture time
   - (c) lock-out time
   - (d) none of the above
4. The filter bandwidth of PLL reduces one of the following effects.
   - (a) slows down the capture process
   - (b) increases the pull time
   - (c) increases the capture range
   - (d) decreases the interjection-injection capabilities
5. The output frequency of VCO in PLL is given by
   - (a) $f_0 = \dfrac{0.25}{R_T C_T}$
   - (b) $\dfrac{R_T C_T}{4}$
   - (c) $4R_T\,C_T$
   - (d) $\dfrac{R_T C_T}{0.25}$
6. The basic components of 565 IC is
   - (a) phase vector
   - (b) amplifier
   - (c) VCO
   - (d) all of the above
7. The maximum operated range of PLL 565 is
   - (a) 0 to 500 kHz
   - (b) 0.001 Hz to 500 kHz
   - (c) 100 to 400 kHz
   - (d) 10 Hz to 400 kHz
8. Lock-in range of PLL 565 is
   - (a) $\Delta f_L = 7.8\,f_0/V$
   - (b) $\Delta f_L = \pm 7.8\,f_0/V$
   - (c) $\Delta f_L = -7.8\,f_0/V$
   - (d) $\Delta f_L = \pm 7.8 V/f_0$
9. The capture range of IC 565 is
   - (a) $\Delta f_c \simeq \sqrt{f_1 \Delta f_L}$
   - (b) $\Delta f_c = 2\sqrt{f_1 \Delta f_L}$
   - (c) $2\Delta f_c \simeq 2\sqrt{f_1 \Delta f_L}$
   - (d) $\Delta f_c \simeq \sqrt{f_1 \Delta f_L/2}$
10. A frequency multiplier using PLL has VCO output frequency f_o is given by
    - (a) Nf_s
    - (b) f_s/N
    - (c) f_s
    - (d) $1/Nf_s$
11. The frequency shift in frequency translator at the output of the VCO in PLL
    - (a) $f_o = f_s - f_L$
    - (b) $f_o + f_s = f_L$
    - (c) $f_o = f_s + f_1$
    - (d) $f_o = f_s - f_1$

12. The main advantage of PLL as a detector with other amplifier detector is
    (a) high degree of selectivity and noise immunity
    (b) low noise immunity with high degree of selectivity
    (c) high degree of selectivity only
    (d) noise immunity only
13. The important characteristics of PLL are
    (a) lock-in range   (b) capture range
    (c) pull-in time   (d) all of the above
14. _____ filter controls the capture range and lock range of PLL.
    (a) LPF   (b) HPF
    (c) BPF   (d) none of the above
15. The other name for phase detector is
    (a) comparator   (b) multiplier
    (c) adder   (d) none of the above
16. The applications of *PLL* are
    (a) frequency multiplier   (b) AM, FM demodulator
    (c) frequency demodulator   (d) all of the above
17. The operating voltage of range of IC 565 is
    (a) ± 6 V to ± 12 V   (b) ± 10 to ± 12 V
    (c) ± 8 to ± 12 V   (d) ± 12 V
18. An external capacitor connected across 565 will act as
    (a) passive device   (b) low-pass filter
    (c) charging device   (d) discharging device
19. Let f_o free-running frequency, V_C voltage shift from VCO, the new frequency shift from VCO in a PLL is
    (a) $f_o + K_V V_C$ (b) $f_o - K_V V_C$
    (c) KV_C (d) f_o
20. The transfer function of LPF in PLL is

    (a) $\dfrac{1}{1+j(f_1/f)}$ (b) $\dfrac{1}{1+jff_1}$ (c) $\dfrac{1}{1+jf/f_1}$ (d) $1/1+jf_1$

21. The voltage V_C required to derive VCO is
    (a) $V_e T(f)A$ (b) $V_e/T(f)A$
    (c) $AV_e/T(f)$ (d) $T(f)/V_e A$
22. The maximum frequency shift of VCO in PLL is
    (a) $K_V K_\varphi (\pi/2) A(f_1/\Delta f)$ (b) $K_V K_\varphi/A(f_1/\Delta f)$

    (c) $\dfrac{A(f_1/\Delta f)\pi/2}{K_V K_\varphi}$ (d) $\dfrac{A(f_1/\Delta f)K_V}{K_\varphi}$

23. The capture range is _____ located with respect to VCO free-runing frequency f_o in PLL
    (a) asymmetrical   (b) symmetrical
    (c) opposite   (d) none of the above
24. The output waveform of PLL 566 is
    (a) square and triangular   (b) square and sine wave
    (c) triangular and sine wave   (d) none of the above
25. The number of pins in IC 566
    (a) 10   (b) 16   (c) 12   (d) 8

26. The output frequency of VCO can be changed by
    (a) R_T (b) C_T (c) V_C (d) all of the above
27. The voltage-to-frequency converter of VCO is defined as
    (a) $\Delta f_o / \Delta V_C$ (b) $\Delta V_C / \Delta f_o$
    (c) Δf_o (d) ΔV_C

ANSWERS

1. (d)	2. (c)	3. (b)	4. (a, b)
5. (a)	6. (d)	7. (b)	8. (b)
9. (c)	10. (a)	11. (c)	12. (a)
13. (d)	14. (a)	15. (b)	16. (d)
17. (a)	19. (a)	20. (c)	21. (a)
22. (a)	23. (b)	24. (a)	25. (d)
26. (d)	27. (a)		

VOLTAGE REGULATOR

1. The output voltage of linear regular to using op-amp is

    (a) $V_o = \left(1 + \dfrac{R_2}{R_1}\right) V_{ref}$ (b) $V_o = -\left(1 + \dfrac{R_2}{R_1}\right) V_{ref}$

    (c) $V_o = -\dfrac{R_2}{R_1} V_{ref}$ (d) $V_o = \dfrac{R_2}{R_1} V_{ref}$

2. The name linear regulator is due to
    (a) transistor in regulator conducts in cut-off region
    (b) transistor in regulator conducts in linear region
    (c) transistor in saturation region
    (d) none of the above
3. A voltage regulator is a circuit that provides
    (a) stable DC voltage
    (b) stable DC voltage independent of load current
    (c) stable DC voltage
    (d) stable DC voltage with no load current variations.
4. Voltage regulators are classified as _____
    (a) series regulators   (b) parallel regulators
    (c) square regulators   (d) rectangle regulators
5. Series regulators are called
    (a) non-linear regulators   (b) linear regulator
    (c) switched regulator   (d) none of the above
6. Switching regulator has advantage over series regulator because of
    (a) improved efficiency   (b) high switch on/off
    (c) both a and b   (d) none of the above
7. The three-terminal regulator has output following limitations:
    (a) no short-circuit protection (only)   (b) O/P voltage is fixed (only)
    (c) both 1 and 2   (d) none of these

8. A regulated power supply has
   (a) reference voltage circuit
   (b) error amp
   (c) series pass transistor
   (d) all of the above

9. Purpose of combining reference voltage source, error op-amp, and pass transistor in IC regulator is due to
   (a) short-circuit limiting
   (b) overload protection
   (c) thermal overload protection
   (d) all of the above

10. The advantage of switched power supply regulators over other regulators
    (a) size
    (b) cost
    (c) size and cost
    (d) none of the above

11. The pass transistor used in regulators has switching frequency
    (a) below 20 kHz
    (b) 20 kHz
    (c) above 20 kHz
    (d) none of the above

12. Linear voltage regulators have low-line frequency due to
    (a) low values of filter capacitors required
    (b) large values of filter capacitors required
    (c) large values of filter capacitors to reduce the ripple
    (d) none of the above

13. The output voltage V_o of IC-723 regulator is
    (a) $V_o = V_{ref} (1 + R_1/R_2)$
    (b) $V_o = 1 + R_1/R_2$
    (c) $V_o = 8 (1 + R_1/R_2)$
    (d) $V_o = R_1/R_2$

14. The limitation of IC 723 is that it has
    (a) no SC protection
    (b) output voltage
    (c) no in-built thermal protection
    (d) none of the above

15. The constant output voltage and current from IC regulator is due to
    (a) constant current source
    (b) reference amplifier
    (c) constant current source and reference amplifier (d) none of the above

16. The important characteristics of IC regulators are
    (a) line regulation
    (b) load regulation
    (c) ripple rejection
    (d) all of the above

17. The basic important blocks of IC 555 timer
    (a) voltage source
    (b) flip-flop
    (c) resistors
    (d) switch

18. The duty cycle D of astable operation using IC 555

    (a) $R_A + R_B/2R_A + R_B$

    (b) $\dfrac{R_A + R_B}{2(R_A + R_B)}$

    (c) $\dfrac{R_A + R_B}{R_A + R_B/2}$

    (d) $\dfrac{R_A + R_B}{R_A + 2R_B}$

19. IC 555 timer as an astable multivibrator. Given $f_o = 50$ kHz, $D = 75\%$, $C = 1$ nF, find R_A, R_B.
    (a) 14 kΩ, 7 kΩ
    (b) 14.1 kΩ, 7.1 kΩ
    (c) 14.4 kΩ, 7.21 kΩ
    (d) 14.38 kΩ, 7.24 kΩ

20. The pulse width T of 555 as a monostable multivibrator is

    (a) $T = RC \ln \dfrac{V_{cc}}{V_{cc} - V_{TH}}$

    (b) $T = RC \ln \left(\dfrac{V_{cc} - V_{TH}}{V_{cc}} \right)$

    (c) $T = RC \ln V_{cc} - V_{TH}$

    (d) $T = RC \ln \left(\dfrac{V_{TH} - V_{cc}}{V_{TH}} \right)$

21. Types of triggering action in monostable operation using IC 555 are
    (a) retriggerable
    (b) non-retriggerable
    (c) none of the above
    (d) both a and b
22. Advantage of CMOS timer is
    (a) low power consumption
    (b) high input impedance
    (c) rail-to-rail input swing
    (d) all of the above
23. The timing accuracy of 555 stable approaches
    (a) 1.1% with temperature stability
    (b) 10% with temperature stability
    (c) 0.1% with temperature stability
24. The power supply stability of IC 555 is
    (a) 0.1% V
    (b) 0.05% V
    (c) 0.5% V
    (d) 1% V
25. A duty cycle of 50% (half the frequency) of timer will give
    (a) triangular waveform
    (b) square waveform
    (c) pulse-type waveform
    (d) sawtooth waveform

ANSWERS

1. (a)	2. (b)	3. (b)	4. (a)
5. (b)	6. (c)	7. (c)	8. (d)
9. (d)	10. (c)	11. (c)	12. (c)
13. (a) $-V_0 = V_{ref} R_2/R_1 + R_2$		14. (c)	15. (c)
16. (d)	17. (d)	18. (d)	19 (c)
20. (a)	21. (d)	22. (d)	23. (a)
24. (b)	25. (b)		

Appendix D
Objective-Type Questions

1. A transconductance amplifier is one which has
   (a) Input current I and output voltage V
   (b) Input voltage V and output current I
   (c) Input voltage V_I and output voltage V_o
   (d) Input current I_{in} and output current I_o
2. A transresistance amplifier is one which has
   (a) Input current I and output voltage V
   (b) Input voltage V and output current I
   (c) Input voltage V_I and output voltage V_o
   (d) Input current I_{in} and output current I_o
3. IC 741 op-amp has typical gain of
   (a) 110 dB   (b) 100 dB
   (c) 106 dB   (d) 90 dB
4. Ideal terminal conditions of IC op-amp are
   (a) $V_d = \infty$, $V_o = 0$, $i_P = i_N = $ currents drawn by inverting and non-inverting inputs. $i_P = i_N = 0$.
   (b) $V_d = 0$, $V_o = \infty$, $i_P = i_N = \infty$
   (c) $V_d = 0$, $V_o = 0$, $i_P = i_N$
   (d) none of the above
5. The other name of voltage follower is
   (a) differential amplifier   (b) inverting amplifier
   (c) non-inverting amplifier   (d) unity-gain amplifier
6. If the input is a sine wave, the circuit (output) will produce _____ phase shift
   (a) 360°   (b) 0°   (c) 90°   (d) 180°
7. Find V_o
   (a) $-R_f/R_1 V_1$

   (b) $-R_f/R_1 V_1 - R_f/R_2$

   (c) $-\dfrac{R_f}{R_1} V_1 - \dfrac{R_f}{R_2} (-15)$

   (d) $-R_f/R_2 V_1$

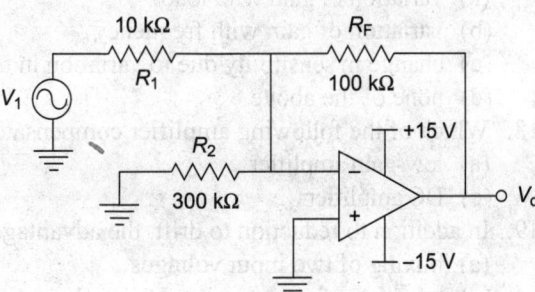

Fig. D.1 *Circuit Diagram for Question 7*

8. An op-amp current-to-voltage converter is also called
   (a) transconductance amplifier
   (b) transresistance amplifier
   (c) transimpedance amplifier
   (d) none of the above

9. A volage-to-current op-amp converter is also called
   (a) transconductance amplifier
   (b) transresistance amplifier
   (c) transimpedance amplifier
   (d) none of the above

10. The differential-mode and common-mode voltage is defined as
    (a) $V_{DM} = V_2 - V_1$, $V_{CM} = V_1 + V_2/2$
    (b) $V_{DM} = V_1 + V_2/2$, $V_{CM} = V_2 - V_1$
    (c) $V_{DM} = V_1/2 = V_2/2$, $V_{CM} = V_2 - V_1/2$
    (d) none of the above

11. A common-mode rejection ratio in dB can be expressed as

    (a) $CMRR_{dB} = 20 \log_{10} \dfrac{A_{cm}}{A_{dm}}$
    (b) $CMRR_{dB} = 20 \log_{10} \dfrac{A_{cm}}{2}$

    (c) $CMRR_{dB} = 10 \log_{10} \dfrac{A_{dm}}{A_{cm}}$
    (d) $CMRR_{dB} = 20 \log_{10} \dfrac{A_{dm}}{2}$

12. For a true difference amplifier, the following holds true:
    (a) $A_{cm} \to 0$, $CMRR_{dB} \to 0$
    (b) $A_{cm} \to \infty$, $CMRR_{dB} \to \infty$
    (c) $A_{cm} \to \infty$, $CMRR_{dB} \to 0$
    (d) $A_{cm} \to 0$, $CMRR_{dB} \to \infty$

13. An ideal op-amp should have
    (a) zero input impedance and output impedance
    (b) infinite input and output impedance
    (c) infinite input impedance and zero output impedance
    (d) zero input impedance and infinite output impedance

14. An ideal op-amp should have
    (a) low gain at low frequencies
    (b) low gain at low frequency and high gain at high frequencies
    (c) high gain at low frequency and low gain at high frequencies
    (d) high gain at all frequencies

15. The actual gain of an op-amp at DC is of the order
    (a) 10 to 100
    (b) 100 to 1000
    (c) 100 to 100,000
    (d) 1000,00,0 to 100,000,000

16. The ideal input impedance range of op-amp is
    (a) 1 kΩ
    (b) 10 kΩ to 10^6 mΩ
    (c) 10^6 mΩ to 10^{12} mΩ

17. Drift of amplifier means
    (a) variation of gain with load
    (b) variation of gain with frequency
    (c) change in sensitivity due to variation in temperature
    (d) none of the above

18. Which of the following amplifier compensates for drift?
    (a) low-gain amplifier
    (b) high-gain amplifier
    (c) DC amplifier
    (d) differential amplifier

19. In addition to reduction to drift, the advantage of a differential amplifier is
    (a) mixing of two input voltages
    (b) an output of either polarity
    (c) rejection of common-mode signal
    (d) all of the above

20. An ideal amplifier should have
    (a) infinite gain at all frequencies   (b) large bandwidth
    (c) zero phase shift   (d) all of the above
21. An amplifier is an unstable condition when
    (a) gain is low   (b) load is variable
    (c) phase shift is 180°   (d) supply is rectified DC
22. Noise in op-amp is due to
    (a) pick up from main supply   (b) sparking in the circuit
    (c) internal generation   (d) all of the above
23. Noise in op-amp can be reduced to
    (a) shielding   (b) use IPF
    (c) proper grounding   (d) all of the above
24. For the op-amp shown in Fig D.2, the output will be
    (a) $-2x$ (b) $-5x$ (c) $-12x$ (d) $-720x$

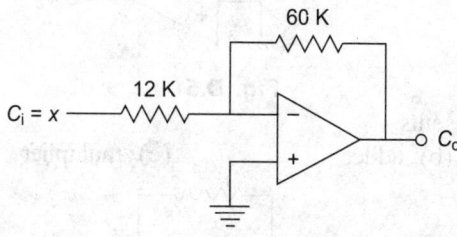

Fig. D.2

25. For which values of R_F and R_i, the output/input ratio of the amplifier shown in Fig. D.3 will be the least?
    (a) $R_F = 100 \text{ k}\Omega$, $R_i = 10 \text{ k}\Omega$ (b) $R_F = 800 \text{ k}\Omega$, $R_i = 200 \text{ k}\Omega$
    (c) $R_F = 1000 \text{ k}\Omega$, $R_i = 10 \text{ k}\Omega$ (d) $R_F = 10 \text{ k}\Omega$, $R_i = 1 \text{ k}\Omega$

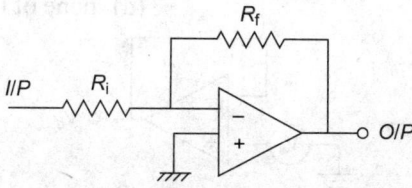

Fig. D.3

26. In the circuit shown below, the feedback factor will be
    (a) 1/3   (b) 2/7   (c) 3/7   (d) 1/7

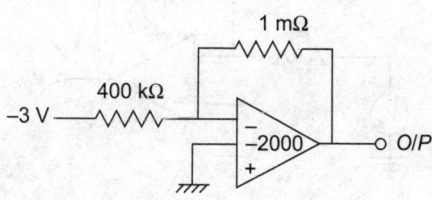

Fig. D.4

27. Loop gain of (Q. 26) is
    (a) 230   (b) 350   (c) 460   (d) 570
28. The output voltage of (Q. 26) will be
    (a) 7.01 V   (b) 6.33 V   (c) 7.487 V   (d) 7.985 V
29. An op-amp is generally not used as
    (a) summer   (b) differentiator   (c) subtractor   (d) integrator
30. Which of the following system is linear?
    (a) log amplifier (b) I-to-V converter (c) V-to-I converter (d) all of the above
31. The circuit shown represents
    (a) adder   (b) integrator   (c) differentiator   (d) subtractor

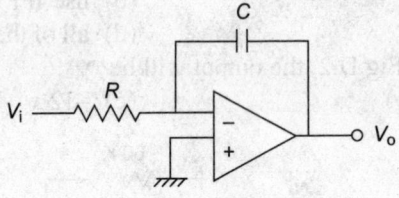

Fig. D.5

32. The circuit shown represents
    (a) differentiator (b) adder   (c) multiplier   (d) sign reversal

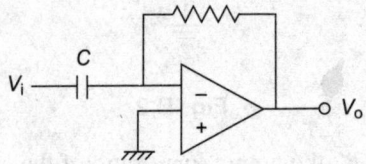

Fig. D.6

33. The circuit shown represents
    (a) DC voltage follower   (b) inverter
    (c) logarithmic amplifier   (d) none of the above

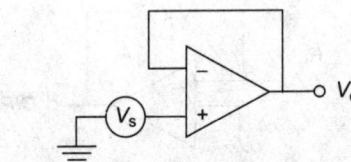

Fig. D.7

34. The circuit shown represents
    (a) I-to-V converter (b) differentiator   (c) log amplifier   (d) analog inverter

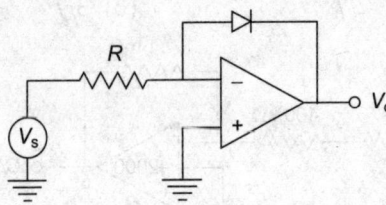

Fig. D.8

35. CMRR for a differential amplifier is the ratio

    (a) $\dfrac{\text{differential gain}}{\text{common-mode gain}}$

    (b) $\dfrac{\text{differential gain}}{\text{integreted gain}}$

    (c) $\dfrac{\text{integrated gain}}{\text{differential gain}}$

    (d) $\dfrac{\text{common-mode gain}}{\text{differential gain}}$

36. A logarithmic amplifier can be used as
    (a) divider
    (b) multiplier
    (c) subtractor
    (d) none of the above

37. An integrator is mostly preferred over a differentiator because of
    (a) more stability
    (b) less voltage drift
    (c) less noise
    (d) all of the above

38. The octave frequency range is specified by

    (a) $\dfrac{\omega_2}{\omega_1} = 2$
    (b) $\dfrac{\omega_2}{\omega_1} = 2$
    (c) $\dfrac{\omega_2}{\omega_1} = 8$
    (d) $\dfrac{\omega_2}{\omega_1} = 10$

39. The decade frequency range is specified by

    (a) $\dfrac{\omega_2}{\omega_1} = 2$
    (b) $\dfrac{\omega_2}{\omega_1} = 8$
    (c) $\dfrac{\omega_2}{\omega_1} = 10$
    (d) $\dfrac{\omega_2}{\omega_1} = 100$

40. A μA 741C op-amp is used in the circuit, the output voltage for the ideal op-amp will be
    (a) –5 V
    (b) –3 V
    (c) –2.5 V
    (d) –2 V

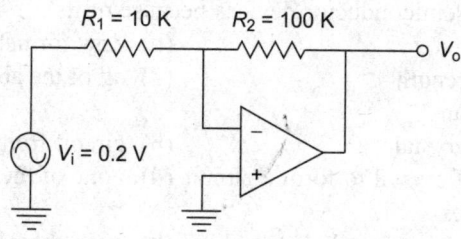

$R_1 = 10\ K$ $R_2 = 100\ K$ V_o

$V_i = 0.2\ V$

Fig. D.9

41. A chip having more than 150 logic gates is known as
    (a) LSI chip
    (b) MSI chip
    (c) SSI chip
    (d) none of the above

42. If a square wave is integrated by integrator using operational amplifier, the output is
    (a) triangular wave
    (b) ramp
    (c) sine wave
    (d) none of the above

43. Monolithic IC waters are typically of
    (a) 1/8-inch diameter
    (b) 1/4-inch diameter
    (c) 1-inch diameter
    (d) 2-inch diameter

44. A good op-amp has
    (a) very high bandwidth
    (b) narrow bandwidth
    (c) high selectivity
    (d) all of the above

45. When an op-amp is operated in common-mode fashion, CMRR should be
    (a) zero   (b) infinitely high   (c) very small   (d) 5 dB
46. Which amplifier will be preferred for highest gain?
    (a) Darlington's pair   (b) operational amplifier
    (c) cascade amplifier   (d) none of the above
47. Decibel is defined in terms of
    (a) voltage ratio   (b) current ratio
    (c) power ratio   (d) none of the above
48. A multivibrator produces
    (a) pure sine waves   (b) square waves
    (c) distortion sine waves   (d) sawtooth voltages
49. A linear system can be described by
    (a) state transition equation   (b) differential equation
    (c) dynamic equation   (d) none of the above
50. At which of the following frequency, the gain of op-amp will be zero?
    (a) β cut-off frequency   (b) α cut-off frequency
    (c) unity-gain cross-over frequency   (d) gain cross-over frequency
51. The loss of precision in a quantity is called
    (a) down time   (b) delay
    (c) unavoidable delay   (d) none of the above
52. A system is critically damped. Now if the gain of the system is increased, the system will behave as
    (a) over-damped   (b) under-damped
    (c) oscillatory   (d) critically defined
53. Silicon is preferred for semiconductor devices because of its
    (a) abundant availability   (b) easy formation of oxide layer
    (c) hood mechanical strength   (d) all of the above
54. A monolithic circuit means
    (a) circuit from single crystal   (b) circuit from more than one crystal
    (c) uses double price of crystal to form a circuit (d) none of the above
55. Photolithography involves
    (a) making photographic mask and photoetching (b) only photoetching
    (c) only masking   (d) none of the above
56. Aluminium is generally used for metallization because it
    (a) forms good mechanical bonds with silicon   (b) is a relatively good conductor
    (c) deposits aluminium films on the surface   (d) all of the above
57. The standard package configurations of manufacturing IC.
    (a) glass metal package   (b) ceramic flat package
    (c) dual-in-line   (d) all of the above
58. Sheet resistance of IC is defined in
    (a) Ohms per square mm   (b) Ohms square mm
    (c) Ohms per mm   (d) Ohms per cubic mm
59. The sheet resistance of emitter diffusion is of the order 200 Ω/sqmm, the emitter resistance is 4 kΩ. Then sheet resistance is
    (a) 20 ohm/sqmm.   (b) 40 ohm/sqmm.   (c) 30 ohm/sqmm.   (d) 15 ohm/sqmm.

60. The preference of polysilicon over aluminium
    (a) lowers threshold voltage
    (b) reduces capacitances
    (c) good bond and mechanical strength, a form of silicon
    (d) all of the above
61. The CMRR of MA 741 is
    (a) 70 dB   (b) 50 dB
    (c) 40 dB   (d) none of the above
62. The common-mode gain and differential-mode gain of op-amp are
    (a) $A_{dm} = A_1 - A_2/2,\ A_{cm} = A_1 + A_2$ (b) $A_{dm} = A_1 - A_2,\ A_{cm} = A_1 + A_2/2$
    (c) $A_{dm} = A_1 + A_2/2,\ A_{cm} = A_1 - A_2$ (d) $A_{dm} = A_1/2 - A_2/2,\ A_{cm} = A_1 - A_2/2$
63. For the below circuit, find CMRR
    (a) $\dfrac{R'(R_1 + R_2) + R_2(R + R')}{R'(R_1 + R_2) + R_1(R + R')}$ (b) $\dfrac{R'(R_1 + R_2) + R_2(R + R')}{R + R'}$

    (c) $\dfrac{R'(R_1 + R_2) + R_2(R + R')}{R'(R_1 + R_2) - R_1(R + R')}$ (d) $\dfrac{R'(R_1 + R_2)}{R'(R_1 + R_2) + R_1(R_1 + R')}$

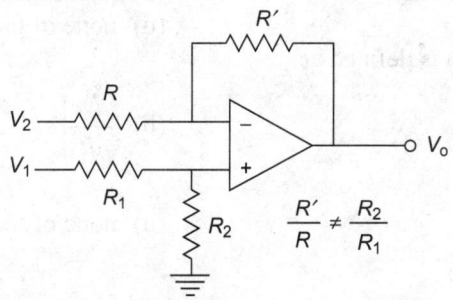

Fig. D.10

64. The basic elements of op-amp is
    (a) differential amplifier   (b) buffer, level translator
    (c) output driver   (d) all of the above
65. For large CMRR, A_{cm} should be
    (a) as large as possible   (b) as small as possible
    (c) zero   (d) level to A_{dm}
66. MOSFET op-amp is preferred over BJT and FET because of
    (a) high input resistance $10^{12}\ \Omega$ (b) low input current, 1 PA
    (c) high slew rate, 10 V/μsec   (d) all the above
67. A current mirror can be used as an active load because it has
    (a) low resistance   (b) high AC resistance
    (c) low AC resistance   (d) high DC resistance
68. DC characters of op-amp are
    (a) input bias and offset current   (b) input offset voltage
    (c) thermal drift   (d) all of the above

69. Thermal drift of op-amp parameters
    (a) force air cooling only
    (b) carefully print circuit board layout
    (c) both a and b
    (d) keep op-amp away from source of heat

70. AC characteristics of op-amp includes
    (a) frequency response
    (b) slew rate
    (c) both a and b
    (d) none of the above

71. An ideal op-amp should have open-loop gain
    (a) around 90 dB constant over audio and R_F frequency
    (b) less than 90 dB, varies on audio and R_F frequency
    (c) 20 dB, constant audio and R_F frequency
    (d) none of the above

72. The upper 3 dB frequency or lower frequency of perodical op-amp is
    (a) $f_1 = 1/2\pi\sqrt{R_o C}$
    (b) $f_1 = 1/2\pi\, R_o C$
    (c) $f_1 = \dfrac{1}{\sqrt{2\pi R_o C}}$
    (d) $f_1 = 1/2\pi RC$

73. The frequency response of op-amp includes
    (a) magnitude character only
    (b) phase character only
    (c) both (a) and (b)
    (d) none of the above

74. The slew rate of op-amp is defined as
    (a) dvc/dt
    (b) $\left.\dfrac{dvc}{dt}\right|_{max}$
    (c) $\dfrac{I_{max}}{C}$
    (d) none of the above

75. The slew rate for IC 741 is
    (a) 0.5 V/μsec   (b) 0.9 V/μsec   (c) 0.8 V/μsec   (d) 1 V/μsec

76. The relation between slew rates, frequency is given by
    (a) $f_{max} = \dfrac{SR}{6.2 \times V_m}$
    (b) $f_{max} = \dfrac{V_n}{S.R \times 6.2}$
    (c) $f_{max} = \dfrac{V_n \times 6.2}{SR}$
    (d) $f_{max} = \dfrac{SR}{V_n}$, SR-slew rate

77. Op-amp 741C cannot be used for high-frequency applications because of
    (a) low slew rate
    (b) high slew rate
    (c) O.T. V/QS
    (d) none of the above

78. Number of diodes required in half-wave rectifier circuit using op-amp is
    (a) 1   (b) 2   (c) 3   (d) 4

79. Number of diodes required in full-wave rectifier using op-amp is
    (a) 2   (b) 3   (c) 1   (d) 4

80. The function of peak detector using op-amp and MOSFET is
    (a) to store the highest value on a capacitor
    (b) highest peak value is stored until the capacitor is discharged

(c) if higher pack signal value comes, the new value is stored in capacity
(d) none of the above

81. The function of clipper is to
    (a) clip off certain portion of input signal to obtain desired output waveform
    (b) shift the DC level of the waveform
    (c) add the required portion to the output waveform
    (d) none of the above

82. The positive clipper can be easily converted into negative clipper by
    (a) reversing the diode and changing the polarity of reference voltage
    (b) reversing the diode only
    (c) changing the polarity
    (d) none of the above

83. The clamper is also known as
    (a) DC restorer   (b) DC inserter
    (c) DC level shifter   (d) all the above

84. The input and output waveform shown below are the input and output of
    (a) positive clamper   (b) positive clipper
    (c) negative clipper   (d) negative clamper

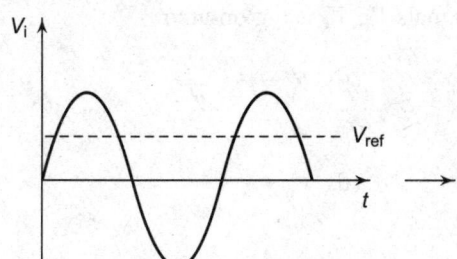 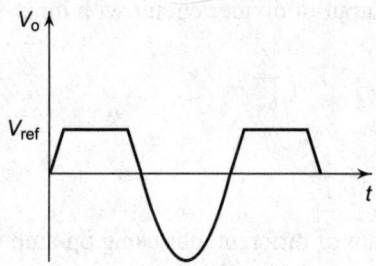

Fig. D.11

85. The input and output waveforms shown below are the input and output of
    (a) negative clipper   (b) positive clamper
    (c) positive clipper   (d) negative clamper

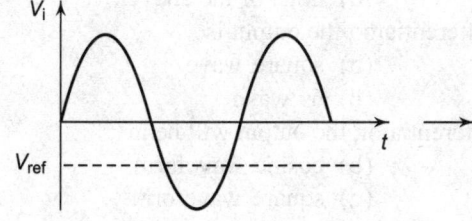 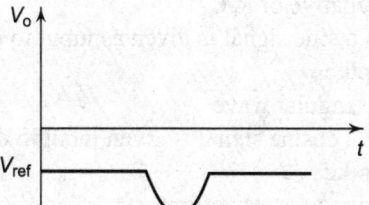

Fig. D.12

86. Sample-and-hold circuit is used in
    (a) digital interfacing   (b) A/D conversion
    (c) pulse-code modulation system   (d) all of the above

87. The output of multiplier circuit is

    (a) $V_o = \dfrac{V_x V_f}{10}$ (b) $V_o = \dfrac{V_f}{V_x} 10$

(c) $\dfrac{V_x}{V_f 10}$ (d) V_f/V_x

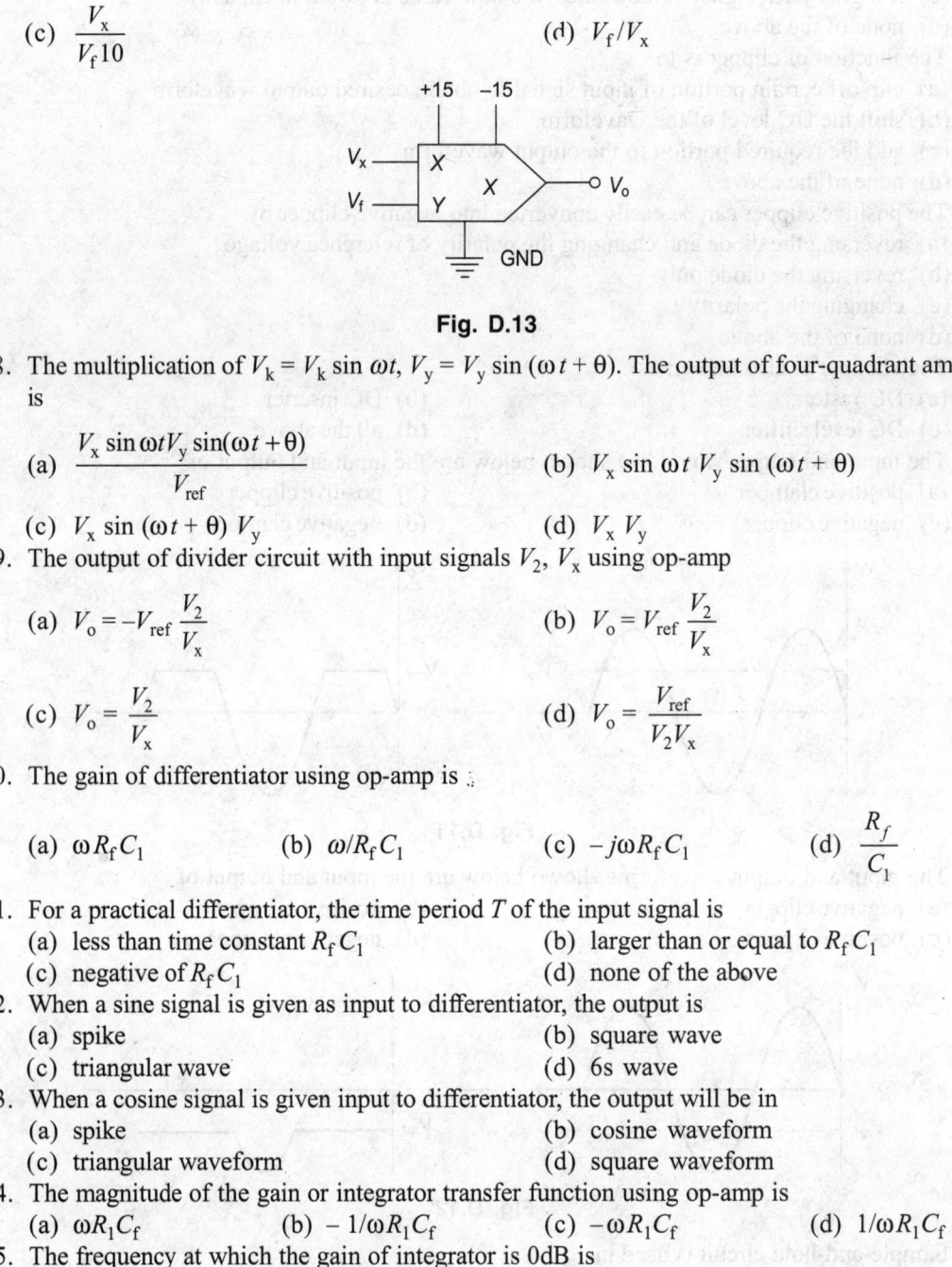

Fig. D.13

88. The multiplication of $V_k = V_k \sin \omega t$, $V_y = V_y \sin(\omega t + \theta)$. The output of four-quadrant amplifier is

  (a) $\dfrac{V_x \sin \omega t\, V_y \sin(\omega t + \theta)}{V_{ref}}$ (b) $V_x \sin \omega t\, V_y \sin(\omega t + \theta)$

  (c) $V_x \sin(\omega t + \theta)\, V_y$ (d) $V_x\, V_y$

89. The output of divider circuit with input signals V_2, V_x using op-amp

  (a) $V_0 = -V_{ref} \dfrac{V_2}{V_x}$ (b) $V_0 = V_{ref} \dfrac{V_2}{V_x}$

  (c) $V_0 = \dfrac{V_2}{V_x}$ (d) $V_0 = \dfrac{V_{ref}}{V_2 V_x}$

90. The gain of differentiator using op-amp is :

  (a) $\omega R_f C_1$ (b) $\omega/R_f C_1$ (c) $-j\omega R_f C_1$ (d) $\dfrac{R_f}{C_1}$

91. For a practical differentiator, the time period T of the input signal is
  (a) less than time constant $R_f C_1$ (b) larger than or equal to $R_f C_1$
  (c) negative of $R_f C_1$ (d) none of the above

92. When a sine signal is given as input to differentiator, the output is
  (a) spike   (b) square wave
  (c) triangular wave   (d) 6s wave

93. When a cosine signal is given input to differentiator, the output will be in
  (a) spike   (b) cosine waveform
  (c) triangular waveform   (d) square waveform

94. The magnitude of the gain or integrator transfer function using op-amp is
  (a) $\omega R_1 C_f$ (b) $-1/\omega R_1 C_f$ (c) $-\omega R_1 C_f$ (d) $1/\omega R_1 C_f$

95. The frequency at which the gain of integrator is 0dB is

  (a) $1/2\pi R_1 C_f$ (b) $1/2\pi R_1$ (c) $\dfrac{R_1}{2\pi C_f}$ (d) $2\pi R_1 C_f$

96. The comparator using op-amp with input sine waveform gives
    (a) cos waveform
    (b) square form
    (c) sine waveform
    (d) triangular waveform
97. The function of window detector is
    (a) to find unknown input between two threshold levels
    (b) to find unknown input at one threshold voltage
    (c) without threshold voltages input is detected
    (d) none of the above
98. The function of phase detector is to measure
    (a) phase angle between two signals
    (b) amplitude between two signals
    (c) only one angle for given signals
    (d) none of the above
99. The output of Schmitt trigger is
    (a) square waveform
    (b) triangular waveform
    (c) sine waveform
    (d) cos waveform
100. The other name of Schmitt trigger is
    (a) regenerative comparator
    (b) square wave generator
    (c) backlash circuit
    (d) all of the above
101. In Schmitt trigger, $R_2 = 100\ \Omega$, $R_1 = 10\ k\Omega$, $V_{ref} = 0$ V, $V_i = 1\ V_{PP}$, $V_{ref} = \pm 14$ V. Find V_{UT} and V_{LT}.
    (a) 28 mV, –28 mV
    (b) 14 mV, –14 mV
    (c) 28 mV, –14 mV
    (d) –28 mV, 14 mV
102. The time T of square wave of astable multivibrator using op-amp is

    (a) $R_C \ln (1 + \beta)$
    (b) $R_C \ln \dfrac{1-\beta}{1+\beta}$
    (c) $2R_C \ln \dfrac{1+\beta}{1-\beta}$
    (d) R_C

103. If $R_1 = R_2$, then $\beta = 0.5$, the total time T of square wave in astable multivibrator using op-amp is
    (a) $T = RC$
    (b) $T = 2\ RC \ln 3$
    (c) $T = 2\ RC \ln 2$
    (d) $T = 2RC \ln (1 + 0.5)$
104. The single output pulse of adjustable time direction in response to triggering signal is from _____ circuit.
    (a) astable multivibrator
    (b) monostable multivibrator
    (c) bistable multivibrator
    (d) none of the above
105. The total time of the pulse from monostable multivibrator is
    (a) $T = 2\ RC$
    (b) $T = 0.3\ RC$
    (c) $T = 0.69\ RC$
    (d) $T = RC \ln (1 + V_D/V_{sat})/1 - \beta$
106. If $R_1 = R_2$, $\beta = 0.5$, $V_{sat} \gg V_D$, then time period of pulse width of monostable multivibrator is
    (a) $T = 2\ RC$
    (b) $T = 0.69\ RC$
    (c) $T = 0.3\ RC$
    (d) $T = RC$
107. The other name of astable multivibrator is
    (a) Schmitt trigger
    (b) free-running oscillator
    (c) regenerative comparator
    (d) none of the above
108. The frequency of oscillation of triangular waveform from generator using op-amp is

    (a) $R_3/4R_1C_1R_2$
    (b) $\dfrac{R_1C_1}{4R_3R_2}$
    (c) $\dfrac{R_1R_2R_3}{4C_1}$
    (d) $C_1/R_1R_2R_4$

109. The frequency of oscillation to of phase shift oscillator using op-amp is given by

    (a) $f_o = 1/2\pi RC$
    (b) $f_o = \dfrac{1}{2\pi\sqrt{RC}}$

(c) $f_o = \dfrac{1}{2\pi RC}$ (d) $f_o = \dfrac{1}{\sqrt{6}\,(2\pi RC)}$

110. The gain of inverting op-amp in a phase shift oscillator (A_v) is
     (a) $A_v \geq -30$ (b) $A_v \geq -29$ (c) $A_v \geq -31$ (d) $A_v \geq 1$
111. A comparator is _____ and gives _____ .
     (a) open-loop op-amp, analog output   (b) open-loop op-amp, no output
     (c) open-loop op-amp, digital output   (d) closed-loop op-amp, digital output
112. Schmitt trigger is comparator _____ feedback.
     (a) no   (b) positive
     (c) negative   (d) none of the above
113. A triangular wave can be generated by integrating
     (a) cosine waveform   (b) sine waveform
     (c) ramp waveform   (d) square waveform
114. For self-sustain oscillations, the conditions to be satisfied for op-amp oscillators are
     (a) $A_v\beta = 1$, phase shift $0°$ (b) $A_v\beta < 1$, phase shift $90°$
     (c) $A_v\beta > 1$, phase shift $0°$ (d) $A_v\beta = 1$ only
115. A regulated power supply has
     (a) error amplifier   (b) series-pass transistor
     (c) feedback network   (d) all of the above
116. The following circuit provides DC voltage independent of load current, temperature, and AC line voltage variations
     (a) regulated power supply   (b) clipper
     (c) clamper   (d) switched regulators
117. _____ regulator operates the power transistors as high frequency on/off switch.
     (a) series   (b) short
     (c) switching   (d) none of the above
118. The percentage change in output voltage for a change in input voltage is called
     (a) load regulation   (b) line regulation
     (c) ripple rejection   (d) none of the above
119. The change in output voltage for a change in load current is called
     (a) line regulation   (b) ripple rejection
     (c) load regulation   (d) none of the above
120. The high-frequency response in RC active filters using op-amp is due to
     (a) BW   (b) gain
     (c) gain, BW product   (d) high frequency
121. The transfer function of active low-pass filter is

     (a) $H(j\omega) = \dfrac{A_o}{1 + jf_h/f}$ (b) $H(j\omega) = \dfrac{A_o}{1 + jf_h/f}$
     (c) $H(j\omega) = f_h/f$ (d) $H(j\omega) = 1/1 + f/f_h$
122. The transfer function of active high pass filter is

     (a) $H(j\omega) = A_o/1 + j\,f/f_h$ (b) $H(j\omega) = \dfrac{A_o}{1 + jf_h/f}$
     (c) $H(j\omega) = f_h/f$ (d) $H(j\omega) = 1/1 + f/f_h$

123. A general second-order filter has roll-off rate
  (a) –20 dB/decade
  (b) –10 dB/decade
  (c) –40 dB/decade
  (d) –30 dB/decade

124. The transfer function of low-pass second-order system is

  (a) $H(s) = \dfrac{A_o \omega_n^2}{S^2 + \alpha \omega_n s + \omega_n^2}$

  (b) $H(s) = \dfrac{A_o \omega_n^2}{S^2 + \omega_n^2}$

  (c) $H(s) = \dfrac{A_o}{S^2 + \omega_n^2}$

  (d) $\dfrac{A_o \omega_n^2}{S^2 + \omega_n s + \omega_n^2}$

125. By cascading LPF and HPF active filter, the resulting circuit is
  (a) BPF
  (b) narrow BPF
  (c) wide BPF
  (d) all of above

126. The transfer function of first order low pass filter
  (a) $A_o/s + \omega_h$
  (b) $A_o \omega_h / S + \omega_h$
  (c) $A_o \omega_h / S + \omega_h^2$
  (d) $A_o \omega_h$

128. The gain of filter is expressed in
  (a) Neper
  (b) degree
  (c) dB
  (d) Bel

129. The following filter gives maximum flat-pass band
  (a) Butterworth filter
  (b) Chebycheve-II
  (c) Chebysheve
  (d) None of the above

130. IC-555 timer
  (a) provides time delay
  (b) provides oscillations
  (c) acts as counter
  (d) all of the above

131. IC-555 can be used as
  (a) monostable multivibrator
  (b) pulse detector
  (c) ramp generator
  (d) all of the above

132. The capacitor waveform of astable circuit using ZC 555 will be in
  (a) square wave
  (b) ramp
  (c) sine wave
  (d) pulse

133. The total time of square wave of astable circuit using IC 555 is
  (a) $T = RC \times 0.69$
  (b) $T = 0.69 (R_A + R_B)C$
  (c) $T = 0.69 (R_A + 2R_B)C$
  (d) none of the above

134. The duty cycle of astable circuit is defined as

  (a) $\dfrac{t_{ON}}{t_{ON} + t_{OFF}}$

  (b) $\dfrac{t_{OFF}}{t_{ON} + t_{OFF}}$

  (c) $\dfrac{t_{ON}}{t_{OFF}}$

  (d) $\dfrac{1}{t_{ON} + t_{OFF}}$

135. The frequency of output waveform of FSK generator using IC 555 is

  (a) $\dfrac{1.45}{R_A + R_B}$

  (b) $\dfrac{1.45}{(R_A + R_B)C}$

  (c) $\dfrac{1.45}{(R_A + 2R_B)C}$

  (d) $\dfrac{R_A}{R_B C}$

136. The VCO is otherwise called as
  (a) free-running multivibrator
  (b) monostable multivibrator
  (c) bistable
  (d) none of the above

137. The frequency deviation of VCO is directly proportional to
  (a) DC control voltage
  (b) applied power supply
  (c) ground
  (d) frequency of the signal

139. The range of frequency over which PLL can maintain lock with incoming signal is called
  (a) capture range
  (b) lock range
  (c) pull-in time
  (d) none of the above

139. The range of frequencies over which PLL can acquire lock with incoming signal is called
     (a) capture range
     (b) lock range
     (c) pull-in time
     (d) none of the above

140. The total time taken by PLL to establish lock is called
     (a) pull-out time
     (b) pull-in time
     (c) rise time
     (d) hold time

141. The filter used in phase-locked loop is
     (a) high-pass filter
     (b) band-pass filter
     (c) low-pass filter
     (d) band-reject filter

141. The following logic gate can be used as phase detector
     (a) EX-OR   (b) EX-NOR   (c) NAND   (d) OR gate

142. IC 566 functions as
     (a) phase detector
     (b) low-pass filter
     (c) VCO
     (d) error amplifier

143. The output frequency of VCO is

     (a) $\dfrac{2(V_{CC} - Q_C)}{C_T R_T V_{CC}}$ (b) $M\dfrac{2V_{CC}}{C_T R_T V_{CC}}$ (c) $\dfrac{2Q_C}{C_T R_T Q_{CC}}$ (d) $\dfrac{2(Q_C - V_{CC})}{C_T R_T V_{CC}}$

144. The voltage to frequency conversion factor in VCO is defined as

     (a) $K_v = \dfrac{\Delta V_C}{\Delta f_o}$ (b) $K_v = \dfrac{\Delta f_o}{\Delta V_C}$
     (c) $K_v = \Delta f_o\, \Delta V_C$ (d) $V_C\, \Delta f_o$

145. IC PLL 565 is called
     (a) PLL
     (b) monolithic PLL
     (c) VCO
     (d) phase detector

146. The lock in range of PLL is defined as
     (a) $\Delta f_L = \pm K_V K_\phi A\, (\pi/2)$ (b) $\Delta f_L = \pm K_V K_\phi A$
     (c) $\Delta f_L = \dfrac{\pm K_V K_\phi}{A(\pi/2)}$ (d) $\Delta f_L = \pm K_V K_\phi$

147. The capture range of PLL is defined as

     (a) $\Delta f_c = \pm\left[\dfrac{\Delta f_L}{2\pi(3.6 \times 10^3)C}\right]^{1/2}$ (b) $\Delta f_c = \pm\left[\dfrac{\Delta f_L}{2\pi C}\right]^{1/2}$

     (c) $\Delta f_c = \pm\left[\dfrac{\Delta f_L}{2\pi(3.6 \times 10^3)C}\right]$ (d) $\Delta f_c = \pm\left[\dfrac{\Delta f_L}{2\pi(3.6 \times 10^3)C}\right]^{1/2}$

148. Select correct statement of PLL.
     (a) capture range smaller than lock range
     (b) lock range smaller than capture range
     (c) capture range is equal to lack range
     (d) none of the above

149. The other name of phase detector is
     (a) adder   (b) subtractor   (c) multiplier   (d) divider

150. ADC, DAC are called
     (a) analog-to-digital converter, digital-to-analog converter,
     (b) digital-to-digital analog-to-analog converters
     (c) analog to amplifiers digital amplifier

151. ADC, DAC otherwise called as DATA
    (a) amplifiers
    (b) rectifiers
    (c) DATA converters
    (d) none of the above
152. Digital-to-analog conversion can be done by
    (a) weighted resistor method
    (b) R–2R ladder
    (c) inverted R–2R–ladder method
    (d) all of the above
153. The main disadvantage of binary weighted resistor method is
    (a) wider range of resistor values
    (b) temperature variations
    (c) less resolutions
    (d) wood length of binary word is small
154. The only two values of resistor required for DAC is
    (a) weighted resistor method
    (b) R–2R ladder
    (c) binary weighted method
    (d) none of the above
155. The R–2R ladder DAC has drawback of
    (a) higher values of resistance required
    (b) less number of word length
    (c) non-linearity due to power dissipation
    (d) none of the above
156. Non-linearity due to power dissipation in DAC can be avoided by
    (a) weighted resistor method
    (b) R–2R method
    (c) inverted R–2R method
    (d) none of the above
157. ADC can be classified as
    (a) direct type, indirect type
    (b) direct type, comparator type
    (c) comparator type, integrating type
    (d) direct type, integrating type
158. The resolution of DAC for 8-bit length is
    (a) 8-bit resolution
    (b) resolution of 0.392 of full scale
    (c) resolution of 1 part in 255
    (d) all of the above
159. The linearity error of DAC should be
    (a) ±1/2 LSB   (b) ±1/2 MSB   (c) ± LSB   (d) ± MSB
160. The accuracy of inverter can be specified in
    (a) percentage of full-scale voltage
    (b) LSB level
    (c) MSB level
    (d) voltage increments
161. DAC essentially requires
    (a) resistors   (b) op-amp   (c) electronic switches (d) all of the above
162. The property of DAC in which analog output increases with digital input is called
    (a) linearity   (b) stability   (c) monotonicity   (d) accuracy
163. ADC 0800/0801/0802 are
    (a) 8-bit ADC   (b) 6-bit ADC   (c) 4-bit ADC   (d) none of the above
164. DAC 0800/0801/0802 are
    (a) 8-bit ADC   (b) 2-bit ADC   (c) 1-bit ADC   (d) none of the above
165. The fastest ADC technique is
    (a) flash type
    (b) parallel comparator
    (c) dual-slope ADC
    (d) charge balancing type
166. A low-speed ADC is
    (a) successive-approximation technique
    (b) parallel comparator
    (c) dual-slope converter
    (d) charge-balancing type

167. For D/A converter, analog output voltage for MSB is

    (a) $V_o = \dfrac{(2^{N-1})}{(2^N - 1)} V_R$ (b) $V_o = \dfrac{2^N}{2^N - 1} V_R$

    (c) $V_o = \dfrac{2^{N-1}}{2^N} V_R$ (d) $V_o = (2^{N-1})V_R$

168. For D/A converter, analog output voltage for LSB is

    (a) $V_o = \dfrac{2^N - 1}{2^N} V_R$ (b) $\dfrac{1}{2^N - 1} V_R$

    (c) $(2^N - 1)V_R$ (d) $V_o = 2^N V_R$

169. The main drawback of dual-slope ADC converters is
    (a) long conversion   (b) high cost
    (c) comparator and DAC are needed   (d) none of the above

170. The advantage of ADC of dual-slope type is
    (a) excellent noise rejection   (b) long conversion time
    (c) fastest in operation   (d) slow varying in nature

171. A V/F convertor circuit is used in
    (a) charge-balancing ADC   (b) dual-slope ADC
    (c) parallel comparators   (d) successive-approximation ADC

172. The disadvantage of parallel comparators (A/D) converters is
    (a) number of comparators require doubles as bit increases
    (b) number of bits increases, complexity increases
    (c) use of bigger resistive divider network
    (d) all of the above

173. The S/H circuit is not needed for _____ ADC.
    (a) direct type   (b) counter type
    (c) integrating type   (d) none of the above

174. D/A converter requires _____ of A/D conversion.
    (a) direct type   (b) counter type
    (c) integrating type   (d) none of the above

175. _____ of ADC gives error when analog signal changes rapidly.
    (a) Direct type   (b) Counter type
    (c) Integrating type   (d) Tracking type

176. How many levels are possible 2-bit DAC?
    (a) 2   (b) 4   (c) 8   (d) 16

177. What is the resolution of output range 0 to 3 V of DAC?
    (a) 6.66% of full scale   (b) 33.33% of full scale
    (c) 99.99% of full scale   (d) none of the above

178. The conversion time of 10-bit successive-approximation A/D converter if input clock is 5 MHz.
    (a) 1 μsec   (b) 2 μsec   (c) 3 μsec   (d) 4 μsec

179. Calculate trigger points if supply voltage $V = \pm 12$ V. Plot the output voltage V_o versus t if V_i is 100 Hz triangular
    (a) 3.686 V, –3.686 V   (b) 3.686 V, 1.843 V
    (c) –1.843 V, 1.843 V   (d) 7.24 V, –7.24 V

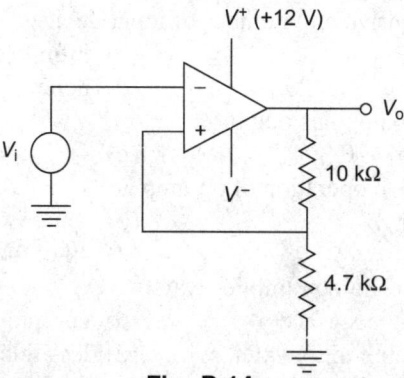

Fig. D.14

180. Find the transfer function of the given circuit.
    (a) $RCS + 1$ (b) $1/RCS + 1$ (c) $RCS - 1$ (d) $1/RCS - 1$

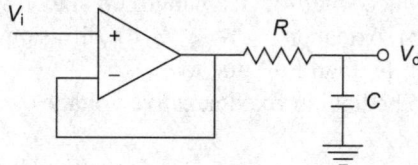

Fig. D.15

181. The frequency of symmetrical waveform generator and unsymmetrical square waveform generator is
    (a) $f_s = 1.45/(R_A + 2R_B)C, f_{us} = 1.45/(R_A + R_B)C$
    (b) $f_s = 1.45/(R_A + 2R_B)C, f_{us} = 1.45/R_AC$
    (c) $f_s = 1.45/(R_A + R_B)C, f_{us} = 1.45/(R_A + 2R_B)C$
    (d) none of the above

182. The gain of an op-amp decreases at high frequency due to
    (a) capacitance   (b) resistors
    (c) gain   (d) none of the above

183. A square wave of peak-to-peak amplitude 500 mV has to be amplified to a peak-to-peak amplitude of 3 V with rise time 4 μsec. Find slew rate.
    (a) 0.5 V/μsec   (b) 0.6 V/μsec   (c) 0.7 V/μsec   (d) 0.8 V/μsec

184. A non-inverting amplifier with a gain of 100 is nulled at 25°C. What will happen to the output voltage if temperature rises to 50°C for an offset voltage drift of 0.15 mV/°C?
    (a) 500 mV   (b) 300 mV   (c) 375 mV   (d) 400 mV

185. The application of op-amp in non-linear region is
    (a) comparators   (b) detectors   (c) amplifiers   (d) all of the above

186. Monolithic op-amp voltage comparators are advantageous than other comparators due to
    (a) high response time   (b) compatible with logic families
    (c) lesser in size   (d) none of the above

187. The response time of comparator increased by _____ op-amp.
    (a) compensated   (b) uncompensated
    (c) monolithic   (d) none of the above

188. Military grade op-amp be operated in the temperature range of
    (a) –55 to 121°C   (b) –50 to 150°C
    (c) 0 to 100°C   (d) at 30°C

189. The reason for op-amp to drive any number of input devices corrected to its output is
     (a) zero output resistance   (b) infinite output resistance
     (c) high output resistance   (d) none of the above

190. Open loop operation of op-amp has output
     (a) $+V_{sat}$ (b) $-V_{sat}$ (c) $-V_{sat}, -V_{sat}$ (d) $V_{sat} = 0$

191. The application of open-loop operation of op-amp is
     (a) zero-crossing detector   (b) square wave generator
     (c) comparator   (d) all of the above

192. The meaning of planner technology implies wafers
     (a) devices fabricated on the surface   (b) cut into piece for different devices
     (c) device fabricated in depth in to wafer   (d) less thickness wafers being used

193. When silicon water is oxidized, the product is
     (a) oxygen, silicon dioxide   (b) hydrogen, silicon dioxide
     (c) silica only   (d) SiO_2 only

194. The technique used to produce small-device pattern on silicon wafer is
     (a) photolithography   (b) oxidation   (c) diffusion   (d) epitaxy

195. nPn transistors preferred in IC than PnP due to
     (a) collector of PnP has to be held in fixed negative voltage
     (b) low current gain
     (c) less process control   (d) all of the above

196. Switching speed is fast in
     (a) PnP Transistor   (b) nPn TV
     (c) Schottky transistor   (d) none of the above

197. Isolation technique is used in IC to
     (a) isolate device and to prevent short circuit   (b) provide high gain
     (c) increase the area   (d) none of the above

ANSWERS

1. (b)	2. (a)	3. (c)	4. (a)	5. (d)	6. (d)	7. (c)
8. (b) or (c)	9. (a)	10. (a)	11. (c)	12. (d)	13. (b)	14. (d)
15. (d)	16. (b)	17. (c)	18. (d)	19. (d)	20. (d)	21. (c)
22. (d)	23. (d)	24. (b)	25. (b)	26. (c)	27. (a)	28. (c)
29. (d)	30. (c)	31. (b)	32. (a)	33. (a)	34. (c)	35. (a)
36. (a) or (b)	37. (d)	38. (c)	39. (c)	40. (d)	41. (a)	42. (a)
43. (c)	44. (a)	45. (a)	46. (c)	47. (c)	48. (c)	49. (d)
50. (c)	51. (d)	52. (b)	53. (d)	54. (a)	55. (a)	56. (d)
57. (d)	58. (a)	59. (a)	60. (d)	61. (a)	62. (a)	63. (c)
64. (d)	65. (b)	66. (d)	67. (b)	68. (d)	69. (c)	70. (c)
71. (a)	72. (b)	73. (c)	74. (c)	75. (a)	76. (a)	77. (c)
78. (a)	79. (a)	80. (b)	81. (a)	82. (a)	83. (d)	84. (c)
85. (a)	86. (d)	87. (a)	88. (a)	89. (a)	90. (c)	91. (b)
92. (c)	93. (a)	94. (d)	95. (a)	96. (b)	97. (a)	98. (a)
99. (a)	100. (d)	101. (a)	102. (c)	103. (b)	104. (b)	105. (d)
106. (b)	107. (b)	108. (a)	109. (d)	110. (b)	111. (c)	112. (c)

113. (d)	114. (a)	115. (d)	116. (a)	117. (c)	118. (b)	119. (c)
120. (c)	121. (a)	122. (b)	123. (c)	124. (a)	125. (c)	126. (b)
127. (c)	128. (a)	129. (d)	130. (d)	131. (b)	132. (c)	133. (a)
134. (c)	135. (a)	136. (a)	137. (b)	138. (a)	139. (b)	140. (c)
141. (a)	142. (c)	143. (a)	144. (b)	145. (b)	146. (a)	147. (a)
148. (a)	149. (c)	150. (a)	151. (c)	152. (d)	153. (a)	154. (b)
155. (c)	156. (c)	157. (d)	158. (d)	159. (a)	160. (a)	161. (d)
162. (c)	163. (a)	164. (a)	165. (b)	166. (a)	167. (a)	168. (b)
169. (a)	170. (a)	171. (a)	172. (d)	173. (c)	174. (a)	175. (d)
176. (b)	177. (b)	178. (b)	179. (a)	180. (b)	181. (a)	182. (a)
183. (b)	184. (c)	185. (d)	186. (b)	187. (b)	188. (a)	189. (a)
190. (c)	191. (d)	192. (a)	193. (b)	194. (a)	195. (d)	196. (c)
197. (a)						

Bibliography

1. Coughlin, R.F. and Frederick F. Driscoll (2001). *Operational Amplifiers and Linear Integrated Circuits, Sixth Edition*. Upper Saddle River, NJ: Prentice Hall.
2. Dailey, Denton J. (1989) *Operational Amplifiers and Linear Integrated Circuits: Theory and Applications*. New York: The McGraw-Hill Companies
3. Franco, Sergio (2002). *Design with Operational Amplifiers and Analog Integrated Circuits*. New York: The McGraw-Hill Companies
4. Gayakwad, Ramakant A. (1999). *Op-Amps and Linear Integrated Circuits. Fourth edition*. Upper Saddle River, NJ: Prentice Hall Inc.
5. Jacob, J. Michael (1996) *Applications and Design with Analog Integrated Circuits*. Upper Saddle River, NJ: Prentice Hall Inc.

INDEX